S0-BLI-364

POLYPHOSPHAZENES
FOR BIOMEDICAL
APPLICATIONS

POLYPHOSPHAZENES FOR BIOMEDICAL APPLICATIONS

Edited by

Alexander K. Andrianov

WILEY

A JOHN WILEY & SONS, INC., PUBLICATION

Chemistry Library

Copyright © 2009 by John Wiley & Sons, Inc. All rights reserved.

Published by John Wiley & Sons, Inc., Hoboken, New Jersey.
Published simultaneously in Canada.

No part of this publication may be reproduced, stored in a retrieval system, or transmitted in any
form or by any means, electronic, mechanical, photocopying, recording, scanning, or otherwise,
except as permitted under Section 107 or 108 of the 1976 United States Copyright Act, without
either the prior written permission of the Publisher, or authorization through payment of the
appropriate per-copy fee to the Copyright Clearance Center, Inc., 222 Rosewood Drive, Danvers,
MA 01923, (978) 750-8400, fax (978) 750-4470, or on the web at www.copyright.com. Requests to
the Publisher for permission should be addressed to the Permissions Department, John Wiley &
Sons, Inc., 111 River Street, Hoboken, NJ 07030, (201) 748-6011, fax (201) 748-6008, or online at
http://www.wiley.com/go/permission.

Limit of Liability/Disclaimer of Warranty: While the publisher and author have used their best
efforts in preparing this book, they make no representations or warranties with respect to the
accuracy or completeness of the contents of this book and specifically disclaim any implied
warranties of merchantability or fitness for a particular purpose. No warranty may be created or
extended by sales representatives or written sales materials. The advice and strategies contained
herein may not be suitable for your situation. You should consult with a professional where
appropriate. Neither the publisher nor author shall be liable for any loss of profit or any other
commercial damages, including but not limited to special, incidental, consequential, or other
damages.

For general information on our other products and services or for technical support, please contact
our Customer Care Department within the United States at (800) 762-2974, outside the United
outside the United States at (317) 572-3993 or fax (317) 572-4002.

Wiley also publishes its books in a variety of electronic formats. Some content that appears in print
may not be available in electronic formats. For more information about Wiley products, visit our
web site at www.wiley.com.

Library of Congress Cataloging-in-Publication Data:

Andrianov, Alexander K.
  Polyphosphazenes for biomedical applications / Alexander K Andrianov.
       p. cm.
  Includes index.
  ISBN 978-0-470-19343-3 (cloth)
1. Polyphosphazenes. 2. Coordination polymers–Synthesis. I. Title.
  QD383.A95A58   2009
  547'.043–dc22

                                                          2008053452

Printed in the United States of America

10   9   8   7   6   5   4   3   2   1

CONTENTS

QD
393
A95
P65
2009
CHEM

CONTRIBUTORS

HARRY R. ALLCOCK, Department of Chemistry, The Pennsylvania State University, University Park, Pennsylvania

ALEXANDER K. ANDRIANOV, Apogee Technology, Inc., Norwood, Massachusetts

LORNE A. BABIUK, Vaccine and Infectious Disease Organization/International Vaccine Center, University of Saskatchewan, Saskatoon, Saskatchewan, Canada; University of Alberta, Edmonton, Alberta, Canada

PONN BENJAMIN, Vaccine and Infectious Disease Organization/International Vaccine Center, University of Saskatchewan, Saskatoon, Saskatchewan, Canada

ROBERTA BERTANI, Dipartimento di Processi Chimici dell'Ingegneria, Università di Padova, Padova, Italy

ANGELO BOSCOLO BOSCOLETTO, Polimeri Europa, Tecnologia Chimica di Base, Venezia, Italy

PAOLO CARAMPIN, Institute of Organic Synthesis and Photoreactivity, C.N.R., Bologna, Italy; School of Pharmacy and Pharmaceutical Sciences, University of Manchester, Manchester, United Kingdom

GABINO A. CARRIEDO, Departamento de Química Orgánica e Inorgánica, Universidad de Oviedo, Oviedo, Spain

MARIA TERESA CONCONI, Dipartimento di Scienze Farmaceutiche, Università di Padova, Padova, Italy

DANIEL P. DECOLLIBUS, Apogee Technology, Inc., Norwood, Massachusetts

ROGER DE JAEGER, Laboratoire de Spectrochimie Infrarouge et Raman, UMR-CNRS 8516, Université des Sciences et Tehnologies de Lille, Villeneuve d'Ascq, France

MENG DENG, Department of Chemical Engineering, University of Virginia, Charlottesville, Virginia; Department of Orthopaedic Surgery, University of Connecticut, Farmington, Connecticut

C. DÍAZ, Departamento de Química, Facultad de Ciencias, Universidad de Chile, Santiago, Chile

JOEL R. FRIED, Department of Chemical and Materials Engineering, University of Cincinnati, Cincinnati, Ohio

DAVID GHATTAS, Faculty of Pharmacy, University of Montreal, Montreal, Quebec, Canada

HELICE A. GILLIS, Apogee Technology, Inc., Norwood, Massachusetts

Mario Gleria,   Dipartimento di Processi Chimici dell'Ingegneria, Università di Padova, Padova, Italy

Claudio Grandi,   Dipartimento di Scienze Farmaceutiche, Università di Padova, Padova, Italy

Jorma Hinkula,   Department of Virology, Swedish Institute for Infectious Disease Control, Solna, Sweden; Department of Microbiology, Tumor and Cell Biology, Karolinska Institute, Stockholm, Sweden; Division of Molecular Virology, University of Linköping, Linköping, Sweden

Claudia Istrate,   Department of Virology, Swedish Institute for Infectious Disease Control, Solna, Sweden; Division of Molecular Virology, University of Linköping, Linköping, Sweden

Kari Johansen,   Department of Virology, Swedish Institute for Infectious Disease Control, Solna, Sweden; Department of Microbiology, Tumor and Cell Biology, Karolinska Institute, Stockholm, Sweden

Elin Johansson,   Department of Virology, Swedish Institute for Infectious Disease Control, Solna, Sweden

Yong Joo Jun,   Center for Intelligent Nano-Biomaterials, Ewha Womans University, Seoul, South Korea

Henry H. Kha,   Apogee Technology, Inc., Norwood, Massachusetts

Byeongyeol Kim,   Department of Chemical and Biological Sciences, Polytechnic Institute of NYU, Brooklyn, New York

Nicholas R. Krogman,   Department of Chemistry, The Pennsylvania State University, University Park, Pennsylvania

Sangamesh G. Kumbar,   Department of Orthopaedic Surgery, University of Connecticut, Farmington, Connecticut; Department of Chemical, Materials and Biomolecular Engineering, University of Connecticut, Storrs, Connecticut

Robert Langer,   Department of Chemical Engineering, Massachusetts Institute of Technology, Cambridge, Massachusetts

Cato T. Laurencin,   Department of Orthopaedic Surgery, University of Connecticut, Farmington, Connecticut; Department of Chemical, Materials and Biomolecular Engineering, University of Connecticut, Storrs, Connecticut

Jean-Christophe Leroux,   Faculty of Pharmacy, University of Montreal, Montreal, Quebec, Canada; Institute of Pharmaceutical Sciences, ETH Zürich, Zürich, Switzerland

Kalle Levon,   Department of Chemical and Biological Sciences, Polytechnic Institute of NYU, Brooklyn, New York

Silvano Lora,   Institute of Organic Synthesis and Photoreactivity, C.N.R., Bologna, Italy

Alexander Marin,   Apogee Technology, Inc., Norwood, Massachusetts

ROBERTO MILANI,   Dipartimento di Scienze Chimiche, Università di Padova, Padova, Italy

GEORGE MUTWIRI, Vaccine and Infectious Disease Organization/ International Vaccine Center, University of Saskatchewan, Saskatoon, Saskatchewan, Canada

LAKSHMI S. NAIR, Department of Orthopaedic Surgery, University of Connecticut, Farmington, Connecticut; Department of Chemical, Materials and Biomolecular Engineering, University of Connecticut, Storrs, Connecticut

SYAM P. NUKAVARAPU, Department of Orthopaedic Surgery, University of Connecticut, Farmington, Connecticut; Department of Chemical, Materials and Biomolecular Engineering, University of Connecticut, Storrs, Connecticut

PIER PAOLO PARNIGOTTO, Dipartimento di Scienze Farmaceutiche, Università di Padova, Padova, Italy

DIDIER PONCET, Virologie Moléculaire et Structurale, CNRS-UMR 2472, INRA-UMR 1157, IFR 115, Gif-sur-Yvette, France

ALOK PRABHU, Department of Chemical and Biological Sciences, Polytechnic Institute of NYU, Brooklyn, New York

LIYAN QIU, College of Pharmaceutical Sciences, Zhejiang University, Hangzhou, China

VLADIMIR SERGEYEV, Department of Chemistry, Moscow State University, Moscow, Russia

YOUN SOO SOHN, Center for Intelligent Nano-Biomaterials, Ewha Womans University, Seoul, South Korea

ALEXANDER STEINER, Department of Chemistry, University of Liverpool, Liverpool, United Kingdom

LENNART SVENSSON, Division of Molecular Virology, University of Linköping, Linköping, Sweden

M. L. VALENZUELA, Departamento de Química, Facultad de Ciencias, Universidad de Chile, Santiago, Chile

PATTY WISIAN-NEILSON, Department of Chemistry, Southern Methodist University, Dallas, Texas

CHENG ZHENG, College of Pharmaceutical Sciences and Institute of Polymer Science, Zhejiang University, Hangzhou, China

PREFACE

Polymers play a key role in the development of drug delivery systems, medical devices, and biosensors. More than ever they face challenging requirements, as clinical science dictates increasingly sophisticated sets of properties and design parameters. Interactions with specific biological targets, biocompatibility, environmental responsiveness, modulated degradation, and formation of supramolecular assemblies are among some of the desired features that have to be integrated in biomedical polymers of the next generation. Yet most synthetic macromolecules used in the biomedical area were not designed originally for these applications and lack the desired chemical flexibility. It has become increasingly evident that the creation of novel macromolecules for life sciences applications can only be realized through the successful merger of a biological rationale with a highly versatile synthetic platform.

Polyphosphazenes, macromolecules with a phosphorus and nitrogen backbone, provide an ideal background for the realization of this objective. A unique synthetic approach, the key feature of polyphosphazene chemistry, allows easy introduction of multiple functionalities and biological modules in a polymer while supporting it with high-throughput discovery methods, which are still largely uncommon in macromolecular chemistry. Unlike many other classes of synthetic polymers, polyphosphazenes offer a hydrolytically degradable backbone with side groups providing reliable "dial-in" controls for rate modulation. Flexibility of the backbone and two side groups at every monomeric unit, which potentially render high functional density, create further opportunities for fine tuning of biologically relevant properties.

These advantages were sufficiently important to trigger a genuine interest in polyphosphazenes as a unique template for constructing biomedical polymers. Still, the attention shows signs of caution as the field advances to provide more robust synthetic approaches, allowing adequate control of macromolecular parameters. The concern is especially valid to industrial scientists, who have to look for efficient and practical solutions to their immediate challenges. The genuine excitement about the opportunities is frequently mixed with reservations about the novel and somewhat unknown behavior of these polymers in living systems. In this book we organize the most recent developments in various areas of biomedical polyphosphazenes to give the reader the opportunity to review the current status of knowledge on the interface of biological sciences and polyphosphazene chemistry.

The book begins, with an introductory section in which general aspects, the most critical advances, and future directions of the technology are discussed. It includes a brief overview of the main synthetic approaches, rational design in polyphosphazene chemistry as it relates to biological applications, and the main representatives of biomedical polyphosphazenes. Further sections are organized based on specific areas of potential clinical applications of polyphosphazenes. A substantial part of the book is dedicated to the most advanced class of biomedical polyphosphazenes, polyelectrolytes, which have been studied extensively in both preclinical and clinical research as vaccine adjuvants. Part II contains an overview of polyphosphazene adjuvants, an extensive collection of in vivo data in various animal models using a variety of antigens, analyzes the importance of the delivery routes, and provides condensed information on their production, control, and potential mechanism of action. It also showcases a critical role of unique structural features possessed by a polyphosphazene family in their interactions with biological targets, including proteins and cells.

Chapters in Part III are focused on the advantages that polyphosphazenes can provide as potential biomaterials. Discussed are applications of polyphosphazenes as scaffolds for tissue engineering, use for surface modification, and in composite and nanofabricated materials. Polymers prepared by condensation polymerization and their evaluation in cytotoxicity studies are also reviewed here. Various aspects of the technology, such as biocompatibility, biodegradability, surface properties of polyphosphazenes, and their buffering capacity in blends are covered in this section.

Drug delivery remains one of the most important and promising areas of biomedical polyphosphazenes. Chapters in Part IV review a broad range of topics, starting with supramolecular polyphosphazene assemblies such as environmentally responsive liposomes and micelles for intracellular targeting and the use of polyphosphazenes for the production of nanostructured materials, and continuing to prodrugs and potential cancer therapies. Although to date computational chemistry and molecular simulations have rarely been applied to biomedical polyphosphazenes, the existing knowledge base on phosphazene membranes and polyelectrolytes is expected to be of interest to the reader. In fact, it can easily be extended to such biologically relevant systems as ionically cross-linked polyphosphazene microspheres and coatings. Thus a review of the relationship between polyphosphazene structure and molecular transport is also included. Detection and monitoring in biological systems and the role that polyphosphazenes can play in such devices are discussed in Part V.

Part VI of the book deals with one of the most fundamental subjects of biomedical polymers: well-defined macromolecular structures and synthetic approaches to their synthesis. A review of chemical regularity in polyphosphazene copolymers can provide an important starting point for scientists interested in the development of polyphosphazenes with multiple functionalities. Finally, a chapter on cyclic phosphazenes is also included, since an indispensible database of biologically relevant properties established for these compounds can open new opportunities for the rational synthesis of new polyphosphazenes for biomedical applications.

The book is intended to reach a broad audience interested in pharmaceutical sciences and biomaterials and to assist researchers and clinicians in enhancing their understanding of polyphosphazene technology. It can also be useful for both graduate and undergraduate students, as it can extend their knowledge base to an important but somewhat less publicized class of biomedical polymers. I hope that the book will help motivate the readers to take a closer look at the fascinating class of polyphosphazene compounds and their potential role in the development of future biomedical polymers.

ALEXANDER K. ANDRIANOV

PART I
Introduction

1 Polyphosphazenes for Biology and Medicine: Current Status and Future Prospects

ALEXANDER K. ANDRIANOV

Apogee Technology, Inc., Norwood, Massachusetts

ROBERT LANGER

Department of Chemical Engineering, Massachusetts Institute of Technology, Cambridge, Massachusetts

Synthetic polymers have played a fascinating role in the successful development of biomedical devices and drug delivery systems. However, until recently, polymers for health care applications were commonly adopted from other industries without their substantial redesign for medical use. Although this strategy helped to resolve many pressing needs and even resulted in successful medical treatments, it is no longer acceptable for modern-day systems, many of which demand combinations of unique biological characteristics. In many life sciences applications, researchers are facing major challenges in creating materials with specific patterns of degradation profiles, biological interactions, release characteristics, and physicochemical and mechanical properties [1]. Today's medical treatments demand macromolecular systems with the ability to participate in cellular signaling processes, modulated interactivity with biomacromolecules, varied environmental responsiveness, and the ability to self-assemble into supramolecular structures. The choice of polymers for life sciences applications, especially those that combine modulated biodegradability and the ease of chemical derivatization, remains scarce. In this regard, synthetic polymers with phosphazene backbones offer unique opportunities in the life sciences arena.

Polyphosphazenes, macromolecules with a phosphorus-nitrogen backbone and organic side groups (Fig. 1), possess a number of properties that make

Polyphosphazenes for Biomedical Applications, Edited by Alexander K. Andrianov
Copyright © 2009 John Wiley & Sons, Inc.

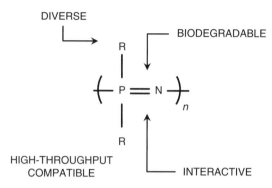

Figure 1 Polyphosphazenes.

them highly attractive for life sciences applications. The following features distinguish them from other classes of biomedical polymers.

1. The inorganic backbone is capable of hydrolytical degradation, which can be modulated through selection of the appropriate side group.
2. The unique synthetic pathway to these polymers, *macromolecular substitution*, allows a huge selection of substituents to be introduced by common organic chemistry methods, free of many ambiguities and restrictions associated with polymerization processes.
3. Such methods lend themselves to high-throughput synthesis, which accelerates the discovery process.
4. The unique flexibility of the backbone and its ability to participate in noncovalent bonding and formation of supramolecular assemblies create new opportunities for an interface with biological systems.

Polyphosphazene chemistry, largely due to pioneering contributions of Professor Harry Allcock, has become an important area of polymer research. Although the features of polyphosphazenes should at least induce curiosity in these compounds and stimulate their testing in challenging applications, these polymers are still infrequent guests on the benches of biomaterials and drug delivery scientists. Limited commercial availability, inadequate information on material quality, and a scarce database on the structure–property relationship make them inaccessible to most application researchers. Despite this, a steady flow of publications, patents, and even clinical trial reports indicate advancement in the field. Multifunctionality and unique biological features are among the key factors in the selection of these macromolecules for biological applications.

Recent discovery of extraordinarily high immunostimulating activity of ionic polyphosphazenes has inspired extensive research in the area and aroused interest in their commercial development. The fact that polyphosphazene polyelectrolytes were overwhelmingly more active than their conventional counterparts, as well as the ability of polyphosphazene backbones to undergo degradation, triggered extensive preclinical and clinical research in the area [2–11]. Recent developments in the field with regard to in vivo activity and chemical control of such systems with emphasis on the unique behavior of these compounds are summarized in several subsequent chapters.

Interestingly, intense development of water-soluble polyphosphazene immunostimulants revealed other important advantages of this class of polymers. Initially, a lead compound, poly[di(carboxylatophenoxy)phosphazene] (PCPP), was introduced in the vaccine industry as a potent immunostimulating excipient for water-soluble formulations (Fig. 2). It was not long before simple, "protein-friendly" aqueous coacervation methods were developed to prepare slow-release microspheres in which a potent immunostimulant, PCPP, also serves as a wall-forming material [12,13]. Slow-release microspheric systems are considered to be important in achieving persistent memory immune responses or as carriers for mucosal immunization. The use of polyphosphazene microspheres not only eliminates the need for an additional microencapsulating agent but allows sustained release of the immunostimulant along with the antigen, a new concept for vaccine delivery which is difficult to achieve with other immunostimulants, such as emulsions or alum.

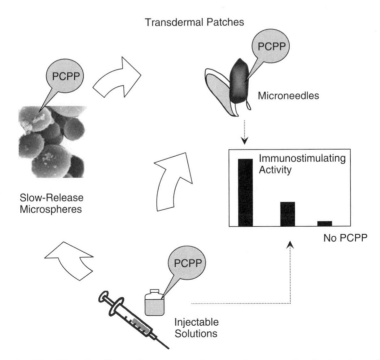

Figure 2 Multifunctionality of polyphosphazene immunostimulants: transdermal, slow release, and solution formulations of PCPP.

Recently, polyphosphazenes have also emerged as a potent class of immunostimulants for intradermal vaccines. Transdermal immunization is one of the promising areas of vaccine development, due to the strength of the skin's defense mechanisms against many infections. Intradermal vaccines have the potentiall to provide improved immune response and antigen sparing. However, topical administration requires the use of special devices, such as arrays of microneedles, due to poor penetration of antigens through the upper layer of the skin: the stratum corneum. Microneedle technology is based on the use of submillimeter structures and often relies on coating such needles with solid-state vaccine formulations. Such systems are designed to be painless, can be self-administered, and in many cases also improve the shelf life of a vaccine over that of its solution formulation. One serious obstacle for the successful development of intradermal vaccines is difficulty in combining the technology with immunostimulants, due either to reactogenicity concerns (alum) or to phase compatibility issues (emulsions). PCPP was shown to be an excellent agent for binding the vaccine formulations to the surface of microneedles (Fig. 2), eliminating the need to use other polymeric excipients, such as carboxymethyl cellulose (CMC). Recent in vivo experiments showed synergy between microneedle and polyphosphazene technologies, with such formulations unexpectedly and dramatically outperforming both CMC microneedle and parenteral polyphosphazene vaccine formulations. These results are discussed thoroughly in Chapter 7. A combination of microspheric and intradermal delivery approach can establish yet another modality of the technology (Fig. 2).

A diverse set of features contributed to the advancement of polyphosphazene polyelectrolytes as immunostimulants. Physicochemical properties linked to immunostimulating activity, such as excellent complex-forming ability, flexibility of the backbone, and high ionic density, were combined successfully with hydrolytically degradable backbone, ionic sensitivity, hydrogel-forming properties, and traditional polymer characteristics such as film-forming properties. This example emphasizes both the unique and multifunctional behavior demonstrated by ionic polyphosphazenes, which led to the introduction of an unparalleled delivery system in this area of the life sciences.

It can be anticipated that polyphosphazenes will also be capable of expanding horizons in other areas of biomedical applications. These expectations are especially high in applications where multicriteria requirements have created considerable obstacles for the successful development of established materials. For example, due to their tailorable biodegradability, polyphosphazenes can provide an important resource for regenerative medicine, a thriving area of research aimed at the treatment of deceased and damaged tissues. The multifaceted strategy employed in the field is directed toward the development of new materials that can interface with tissues structurally, mechanically, and biofunctionally. New materials must provide space, support, and instructive cues while avoiding hostile immune response, and must also degrade slowly to allow for tissue in-growth and removal of the synthetic scaffold.

One of the challenges of the field is that many existing biodegradable synthetic polymers provide minimal biological cues to guide tissue regeneration and have limited bioactivity [14,15]. Structural diversity and synthetic flexibility of biodegradable polyphosphazenes seem to put them in a favorable position since they can provide straightforward routes for the incorporation of peptide sequences with highly specific biofunctionality. Alternatively, sites for protein adsorption can be introduced easily to transform an implant surface into a biological landscape that supports cellular interaction. Recent advances achieved in polyphosphazene use in biomaterials are described in several chapters.

Although most of the research on biodegradable materials is devoted to hydrophobic polyphosphazenes, hydrogels can be as important in such applications. Although their initial mechanical properties cannot compete with those of hydrophobic polymers, hydrogels can better facilitate accelerated tissue formation, due to their aqueous environment. This leads to the rapid development of a natural matrix, which can soon supplement and assume much of the scaffold's mechanical responsibility. A number of preliminary reports indicate that polyphosphazene hydrogels are efficient in cell encapsulation and can provide simple ways for in situ hydrogel formation [16]. Due to the excellent ion-complexing properties of some polyphosphazenes, they can also play important roles in biomineralization processes for bone and cartilage regeneration [17], since they could be a source for the slow supply of calcium and phosphate ions. More research on the use of polyphosphazenes in this field can be expected in upcoming years.

Challenges in drug delivery are another important source of inspiration in polyphosphazene synthesis. Research has been focused primarily on microparticulate and prodrug systems, and some of the insights in the field are reviewed in the present book. Although these developments are vital, new structures and architectures that encompass features currently unachievable in the framework of conventional chemistries pave the road to the recognition of polyphosphazenes in the field. Polyphosphazene alternatives to well-known pharmaceutical carriers currently utilized in drug delivery can potentially introduce superior biological characteristics. For example, originally, PCPP was synthesized to mimic the ionic sensitivity and hydrogel-forming properties of alginic acid. It was introduced successfully as an alternative to alginates in ionic complexation technologies, also bringing the benefits of well-defined structure and high synthetic reproducibility [16,18]. Another important example is a polyphosphazene mimic of poly(vinylpyrrolidone). The latter is used widely in research as a drug carrier. However, its practical use for parenteral administration is severely restricted by its inability to degrade. Recently, polyphosphazene containing pyrrolidone side groups was synthesized [19]. This water-soluble macromolecular system combines functional properties introduced by pyrrolidone groups with hydrolytical degradability and ease of chemical modification.

Delivery of protein drugs is one of the areas where such systems can be of significant interest. The development of many promising protein therapeutics is obstructed by problems associated with their rapid clearance from the body or undesirable side effects, such as immunogenic reactions. Their chemical modification with a synthetic polymer, poly(ethylene glycol) (PEG), proved to be a powerful approach in improving protein's pharmacokinetics, formulation stability, and safety [20–23]. A number of proteins modified with PEG have been approved by regulatory authorities. Advancement of biodegradable polyphosphazene carriers as an alternative to nonbiodegradable PEGs can potentially bring new architectures and functions to these systems. This can also reduce production costs by eliminating the need for careful fractionation, which is imperative for nonbiodegradable PEG. Such approaches may become increasingly important, as new, high-dose treatments such as antibody therapies are becoming more widespread.

Conceptually, biodegradable polyphosphazene with the appropriate protein-complexing properties can even eliminate the need for the sophisticated site-specific attachment of polymer to the protein. Currently, covalent attachment of PEG molecules to proteins is one of the most challenging stages in commercial manufacturing. Synthesis of monofunctional PEG with controlled molecular weight, chemical activation of PEG, a reaction of covalent conjugation, and purification of the synthesized product requires sophisticated technologies and equipment in multiple-step processes and so dictates high development and manufacturing costs. Covalent attachment methods can also result in a loss of biological activity, due to the nonspecific and random linkage of multiple PEG molecules.

Complexation of proteins to a biologically inert polymer through noncovalent links, provided that the resulting assembly does not show high immunogenicity, can potentially reduce sophisticated chemical manufacturing to a routine formulation step. Efforts to develop of polyphosphazene systems that combine PEG-like features with an ability to form complexes with proteins have recently been initiated [24].

Another interesting area where polyphosphazenes can potentially take advantage of their multifunctionality is in the field of shape-memory polymers, where a combination of modulated flexibility with functional diversity can be important. If biodegradability can be built into such materials, they also have potential for use in minimally invasive surgery. Two examples are biodegradable shape-memory polymer as an intelligent suture for wound closure, and insertion of bulky implants into the human body in compressed form through a small incision, with the implants turning into their application-relevant shape within the body [25,26]. Other areas are biodegradable slow-release coating for stents, where biological parameters have to be blended with intrinsic polymer characteristics [27–29], or biocompatible fluorinated coatings, which can be based on multilayer fluorinated systems [30].

There is little doubt that polyphosphazenes represent an important class of biologically relevant polymers which can provide solutions for various challenging life sciences applications. The realization of this potential is profoundly dependent on the synthetic methods of polyphosphazene chemistry, especially on their ability to facilitate rapid synthesis of new compounds and to provide adequate control of molecular characteristics.

The polyphosphazene technology platform seems to be ideally positioned for the development of high-throughput synthesis and combinatorial approaches, due to the structural diversity of the class and the dominance of organic chemistry methods. As soon as the starting point of polyphosphazene synthesis—the macromolecular precursor, poly(dichlorophosphazene) (PDCP)—is synthesized by polymerization, it can be reacted with a multitude of nucleophiles to yield a variety of organic polyphosphazene derivatives [31]. The main challenge to development of the parallel synthesis approach is the inherent hydrolytic instability of PDCP, which dictates the need for frequent time-consuming and labor-intensive polymerization runs. Recent breakthroughs in stabilization chemistry for this hydrolytically sensitive compound made possible new opportunities in this area [32]. In fact, semiautomated synthesis of new polyphosphazene derivatives was conducted successfully for the assembly of an immunoadjuvant library [33,34], a library of protein modification polymers [33,35], and process discovery for the synthesis of sulfonated polymers [36]. Hundreds of new polymers and copolymers were synthesized, and thousands of synthetic runs were conducted using the same lot of PDCP. The downstream processes were sufficiently simplified to complete the synthesis and purification, generally in one or two days. A general representation of biomedical polyphosphazene library construction, which in some cases can be extended to the assembly of the corresponding microspheres, is shown in Figure 3.

The perspectives of high-throughput methods in polyphosphazene chemistry may not be limited to discovery of new molecules and materials. Unprecedented diversity of polyphosphazenes combined with a potential for rapid synthesis can facilitate the construction of extraordinarily large databases, which may be difficult to build for other macromolecular classes. This obviously can be an enormous resource in predicting the properties and functional behavior of other polymers, boosting the role of polyphosphazenes as a unique scientific "toolbox."

For years the development of polyphosphazenes for life sciences applications has been somewhat impeded by challenges in regulating macromolecular characteristics and production consistency. Reproducibility of functional properties, control of structural irregularities, and synthetic by-products are the most critical issues that need to be addressed for the materials to be advanced further into the biomedical arena. There are clear indications that the lack of control in the substitution process can lead to residual moieties on polyphosphazenes, thus significantly affecting degradation profile, shelf life, and eventually, biological characteristics [5]. Although it can still be challenging to achieve adequate control of biologically relevant properties of polyphosphazenes in the research environment, much progress have been made in this area. In this regard, the establishment of the first GMP (good manufacturing practices) process for the manufacture of biomedical polyphosphazenes [5,32] and the following clinical trials [10,11] manifest a critical stage in their commercial development.

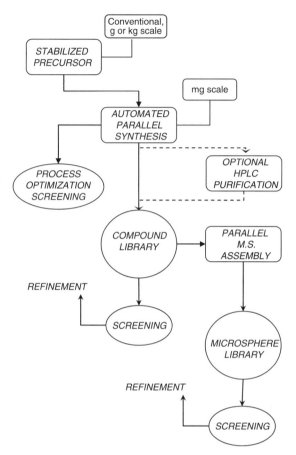

Figure 3 General scheme of high-throughput synthesis and polyphosphazene library construction.

In designing new biomedical polyphosphazenes, it may be important to critically review some of the most traditional approaches presently utilized in the field. The starting point in the blueprinting of any new polyphosphazene is almost always a review of the physicochemical properties of the potential side group. This is obviously important since the material properties of the future polymer will be defined primarily by the characteristics of the substituent. However, the effect of the future side group on the backbone and capability of the letter to interact with other molecules is still almost always neglected. From the chemistry of cyclic compounds it has been known for years that the phosphazene backbone is highly capable of noncovalent bonding and is actively engaged in supramolecular interactions. Obviously, this can be an indispensable resource for macromolecular chemistry as well, since such interactions are proven to be critical for immunostimulating and other biomedical applications of polyphosphazenes.

We look forward to further advancement of biomedical polyphosphazenes and anticipate that the collection of up-to-date articles on their life sciences applications and synthesis in this book will facilitate the process.

REFERENCES

1. Langer, R., Tirrell, D.A. Designing materials for biology and medicine. *Nature*, 2004, 428(6982):487–492.

2. Andrianov, A.K. Polyphosphazenes as vaccine adjuvants. In M. Singh, Ed. *Vaccine Adjuvants and Delivery Systems*. Wiley, Hoboken, NJ, 2007, pp. 355–378.

3. Andrianov, A.K., Marin, A., Chen, J. Synthesis, properties, and biological activity of poly[di(sodium carboxylatoethylphenoxy)phosphazene]. *Biomacromolecules*, 2006, 7(1):394–399.

4. Andrianov, A.K., Marin, A., Roberts, B.E. Polyphosphazene polyelectrolytes: a link between the formation of noncovalent complexes with antigenic proteins and immunostimulating activity. *Biomacromolecules*, 2005, 6(3):1375–1379.

5. Andrianov, A.K., Svirkin, Y.Y., LeGolvan, M.P. Synthesis and biologically relevant properties of polyphosphazene polyacids. *Biomacromolecules*, 2004, 5(5):1999–2006.

6. Mutwiri, G., Benjamin, P., Soita, H., Townsend, H., Yost, R., Roberts, B., Andrianov, A.K., Babiuk, L.A. Poly[di(sodium carboxylatoethylphenoxy)phosphazene] (PCEP) is a potent enhancer of mixed Th1/Th2 immune responses in mice immunized with influenza virus antigens. *Vaccine*, 2007, 25(7):1204–1213.

7. Payne, L.G., Jenkins, S.A., Andrianov, A., Langer, R., Roberts, B.E. Xenobiotic polymers as vaccine vehicles. *Adv. Exp. Medi. Biol.*, 1995, 371(B):1475–1480.

8. Payne, L.G., Jenkins, S.A., Andrianov, A., Roberts, B.E. Water-soluble phosphazene polymers for parenteral and mucosal vaccine delivery. *Pharm. Biotechnol.*, 1995, 6:473–493.

9. Payne, L.G., Jenkins, S.A., Woods, A.L., Grund, E.M., Geribo, W.E., Loebelenz, J.R., Andrianov, A.K., Roberts, B.E. Poly[di(carboxylatophenoxy)phosphazene] (PCPP) is a potent immunoadjuvant for an influenza vaccine. *Vaccine*, 1998, 16(1):92–98.

10. Bouveret Le Cam, N.N., Ronco, J., Francon, A., Blondeau, C., Fanget, B. Adjuvants for influenza vaccine. *Res. Immunol.*, 1998, 149(1):19–23.

11. Kim, J.H., Kirsch, E.A., Gilliam, B., Michael, N.L., VanCott, T.C., Ratto-Kim, S., Cox, J., Nielsen, R., Robb, M.L., Caudrelier, P., El Habib, R., McNeil, J. A phase I, open label, dose ranging trial of the Pasteur Merieux Connaught (PMC) oligomeric HIV-1 Gp160mn/LAI-2 vaccine in HIV seronegative adults. *Abstracts of the 37th Annual Meeting of the Infectious Diseases Society of America*, Philadelphia, 1999.

12. Andrianov, A.K., Chen, J. Polyphosphazene microspheres: preparation by ionic complexation of phosphazene polyacids with spermine. *J. Appl. Polym. Sci.*, 2006, 101(1):414–419.

13. Andrianov, A.K., Chen, J., Payne, L.G. Preparation of hydrogel microspheres by coacervation of aqueous polyphosphazene solutions. *Biomaterials*, 1998, 19(1–3):109–115.

14. Goldberg, M., Langer, R., Jia, X. Nanostructured materials for applications in drug delivery and tissue engineering. *J. Biomater. Sci. Polym. Ed.*, 2007, 18(3): 241–268.

15. Lanza, R.P., Langer, R.S., Vacanti, J. *Principles of Tissue Engineering.* Academic Press, San Diego, CA, 2007.

16. Bano, M.C., Cohen, S., Visscher, K.B., Allcock, H.R., Langer, R. A novel synthetic method for hybridoma cell encapsulation. *Nat. Biotechnol.*, 1991, 9(5): 468–471.

17. Bonzani, I.C., George, J.H., Stevens, M.M. Novel materials for bone and cartilage regeneration. *Curr. Opini. Chemi. Biol.*, 2006, 10(6):568–575.

18. Cohen, S., Bano, M.C., Visscher, K.B., Chow, M., Allcock, H.R., Langer, R. Ionically crosslinkable polyphosphazene: a novel polymer for microencapsulation. *J. Am. Chem. Soc.*, 1990, 112(21):7832–7833.

19. Andrianov, A.K., Marin, A., Peterson, P. Water-soluble biodegradable polyphosphazenes containing N-ethylpyrrolidone groups. *Macromolecules*, 2005, 38(19): 7972–7976.

20. Fee, C.J., Van Alstine, J.M. PEG-proteins: reaction engineering and separation issues. *Chemi. Engi. Sci.*, 2006, 61(3):924–939.

21. Morar, A.S., Schrimsher, J.L., Chavez, M.D. PEGylation of proteins: a structural approach. *Biopharm. Int.*, 2006, 19(4).

22. Veronese, F.M., Harris, J.M. Introduction and overview of peptide and protein pegylation. *Adv. Drug Deliv. Rev.*, 2002, 54(4):453–456.

23. Veronese, F.M., Pasut, G. PEGylation, successful approach to drug delivery. *Drug Discov. Today*, 2005, 10(21):1451–1458.

24. Andrianov, A.K. Functionalized water-soluble polyphosphazene and uses thereof as modifiers of biological agents, WO2005099724, 2005.

25. Behl, M., Lendlein, A. Shape-memory polymers. *Materi. Today*, 2007, 10(4): 20–28.

26. Lendlein, A., Langer, R. Biodegradable, elastic shape-memory polymers for potential biomedical applications. *Science* 296(5573):1673.

27. Sousa, J.E., Serruys, P.W., Costa, M.A. New frontiers in cardiology: drug-eluting stents: I. *Circulation*, 2003, 107(17):2274.

28. Kamath, K.R., Barry, J.J., Miller, K.M. The Taxus™ drug-eluting stent: a new paradigm in controlled drug delivery. *Adv. Drug Deliv. Rev.*, 2006, 58(3): 412–436.

29. Zilberman, M., Eberhart, R.C. Drug-eluting bioresorbable stents for various applications. *Annu. Rev. Biomed. Eng.*, 2006, 8:153–180.

30. Andrianov, A.K., Marin, A., Peterson, P., Chen, J. Fluorinated polyphosphazene polyelectrolytes. *J. Appl. Polym. Sci.*, 2007, 103(1):53–58.

31. Allcock, H.R. *Chemistry and Applications of Polyphosphazenes.* Wiley, Hoboken, NJ, 2002: p. 725.

32. Andrianov, A.K., Chen, J., LeGolvan, M.P. Poly(dichlorophosphazene) as a precursor for biologically active polyphosphazenes: synthesis, characterization, and stabilization. *Macromolecules*, 2004, 37(2):414–420.

33. Andrianov, A.K. Water-soluble biodegradable polyphosphazenes:—emerging systems for biomedical applications. *Polym. Prepri.*, 2005, 46(2):715.

34. Andrianov, A.K. Design and synthesis of functionalized polyphosphazenes with immune modulating activity. *PMSE Prepr.*, 2003, 88.

35. Phase III Medical announces the successful creation of a core polymer library. *Business Wire*, May 13, 2004.

36. Andrianov, A.K., Marin, A., Chen, J., Sargent, J., Corbett, N. Novel route to sulfonated polyphosphazenes: single-step synthesis using "noncovalent protection" of sulfonic acid functionality. *Macromolecules*, 2004, 37(11):4075–4080.

2 Expanding Options in Polyphosphazene Biomedical Research

HARRY R. ALLCOCK

Department of Chemistry, The Pennsylvania State University, University Park, Pennsylvania

INTRODUCTION

The development of the first synthesis route to stable polyphosphazenes in the mid-1960s [1–3] opened the door to a wide range of useful materials for advanced technology in the aerospace, automotive, energy, and biomedical sectors [4–6]. The role of our research group during the past 40 years has been to pioneer the design and synthesis of different phosphazene polymers, to identify unique properties, to understand the structure–property relationships, and to facilitate the development of uses, often through collaboration with other groups. The outcome has been particularly productive in the field of biomedical materials, which now includes the following topics:

- Bioinert elastomers for dental and cardiovascular applications
- Bioerodible polymers for drug delivery and tissue engineering
- Water-soluble polymers, hydrogels, and responsive membranes
- Tailored surfaces for biomedical uses
- Polymeric drugs and heme–polymer models
- Responsive microspheres for oral drug and vaccine delivery
- Micelles for drug delivery

Each of these subjects is discussed briefly after an introduction to the various synthesis options. Some of the subjects introduced in this chapter at the perspective level have been reviewed in detail in a recent book (reference 5, especially Chapter 15). References to our research in this chapter are from the

Polyphosphazenes for Biomedical Applications, Edited by Alexander K. Andrianov
Copyright © 2009 John Wiley & Sons, Inc.

500-plus publications on phosphazenes that have originated from our laboratory. A more comprehensive list is available at our Web site http://www.chem.psu.edu/faculty/hra.

BASIC POLYMER SYNTHESIS PROCESSES

Five fundamental synthetic routes have now been developed for the molecular assembly of polyphosphazenes:

1. Ring-opening polymerization of hexachlorocyclotriphosphazene to poly(dichlorophosphazene) followed by replacement of the chlorine atoms in this polymer by organic groups [1–11].
2. Ring-opening polymerization of cyclic phosphazenes that already bear the organic side groups that are intended for incorporation into the high polymer [12–20].
3. Polymerization of noncyclic phosphazene monomers that already bear the organic side groups destined for the polymer [21–27].
4. Synthesis of poly(dichlorophosphazene) from linear chlorophosphazene monomers, followed by chlorine replacement at the macromolecular level [28–36].
5. Secondary substitution reactions carried out on the organic side groups introduced by one of the preceding methods [37–42].

A combination of the first and fifth approaches accounts for by far the largest number of known polyphosphazenes and for nearly all of the biomedically useful derivatives. The fourth method, with a variant that involves a living cationic polymerization, has the advantage of providing access to block and graft copolymers between polyphosphazenes and organic macromolecules, and this is proving to be an expanding area with strong possibilities for biomedical developments.

1. *Ring-opening polymerization/macromolecular substitution route.* Reaction sequence (1) provides an outline of the earliest and most widely used route to poly(organophosphazene)s [1–11]. In this process, the commercially available cyclic trimer hexachlorocyclotriphosphazene (**1**), which is produced from phosphorus pentachloride and ammonium chloride, is heated at 220 to 250°C with or without an initiator to give the macromolecular intermediate poly(dichlorophosphazene) (**2**). This remarkable polymer is so reactive to nucleophilic reagents in solution that every chlorine atom along each chain (roughly 30,000 chlorine atoms) can be replaced by organic side groups with the use of amino, alkoxide, or aryloxide reagents to give species of types **3** and **4**. Moreover, two or more different side groups can be introduced by the use of different nucleophiles to give mixed substituent polymers, such as the example

Reaction sequence (1)

shown as **5**. This fact alone explains much of the value of this system, since more than 250 different reagents have been shown to react with polymer **2** to give more than 700 different polymers, all with different properties. Thus, this synthetic approach is ideal for tuning the properties of a polymer to vary solubility, solid-state properties, and especially, biomedical characteristics. A major advantage of this route is the widening commercial availability of the cyclic trimer **1**, which has dramatically increased the opportunities for technological developments in this field.

2. *Ring-opening polymerization of cyclic phosphazenes that already bear organic side groups.* A fairly obvious alternative to the first route is to link the organic side groups to the phosphazene ring before polymerization and bypass the macromolecular substitution process [12–20]. Unfortunately, this approach works effectively for only a few organic side groups that can withstand the high-temperature conditions needed for polymerization. The problem is that most organic groups (other than methyl units) are larger than chlorine atoms. Because ring-opening polymerization brings the side groups closer together in linear macromolecules, there is an enthalpic barrier to polymerization. The compromise solution is to link fewer than six side groups to the phosphazene ring, carry out polymerization at a relatively low temperature, and replace the remaining chlorine atoms at the polymer level. So far, few biomedically

interesting polymers have been produced by this route, although it remains a possibility for future work.

3. *Polymerization of noncyclic phosphazene monomers that already bear organic side groups destined for the polymer.* This approach was pioneered by Wisian-Neilson and Nielson [21–24], Flindt and Rose [25] and by Matyjaszewski et al. [26,27]. It uses organophosphoranimines as monomers and includes the following process:

$$
(CH_3)_3Si\text{—}N\text{=}PR_2(OCH_2CF_3) \xrightarrow[- ClSi(CH_3)_3]{200°C} \left[\begin{array}{c} R \\ | \\ N\text{=}P \\ | \\ R \end{array} \right]_n
$$

$$R = CH_3, C_2H_5, C_6H_6, \text{etc.}$$

Reaction sequence (2)

A valuable feature of this route is that it allows access to some polymers that are difficult to produce by the other methods: for example, species with methyl, ethyl, or phenyl side groups linked directly to the skeleton through phosphorus–carbon bonds. There are limits to the types of side groups that permit this polymerization. However, methyl side groups in the polymer can be lithiated and these sites used to link organic units to the polymer chain. The formation of biomedically useful polymers utilizes this method of post-polymerization organometallic substitution. This approach is reviewed in Chapter 10.

4. *Synthesis of poly(dichlorophosphazene) from linear chlorophosphazene monomers, followed by chlorine replacement at the macromolecular level.* There are two variants of this method: a high-temperature condensation process and a room-temperature living cationic condensation polymerization.

    a. *High-temperature condensation reaction.* This approach was invented by De Jeager et al. in France [28,29] and was later refined by Peterson et al. [30]. The method overcomes some of the drawbacks of the first three approaches but introduces some problems of its own. A detailed review of the method has been published [29]. The overall chemistry is summarized as follows:

$$
Cl_3P\text{=}NP(O)Cl_2 \xrightarrow[- P(O)Cl_3]{240–290°C} \left[\begin{array}{c} Cl \\ | \\ N\text{=}P \\ | \\ Cl \end{array} \right]_n
$$

Reaction sequence (3)

The advantage of this process is the use of a relatively inexpensive monomer, coupled with the attributes of subsequent macromolecular substitution. However, this polymerization gives broad molecular-weight distributions and shorter chain lengths than route 1, and is a highly corrosive process.

b. *Room-temperature living cationic polymerization route.* This process developed in our laboratory at Penn State in collaboration with the group of I. Manners, then at the University of Toronto, is the most recent [31–36]. Reaction sequence (4) illustrates the chemistry. This Lewis acid-catalyzed, living cationic polymerization can give narrow molecular-weight distributions, with chain lengths that are controlled by the ratio of monomer to initiator. Moreover, the process can readily be used to make phosphazene–phosphazene or phosphazene–organic block copolymers by methods such as the one shown in reaction sequence (4). The mild reaction conditions means that there are no corrosion problems and few, if any, safety problems. Poly(organophosphazene)s can also be produced directly by the living cationic process when organophosphoranimines are employed as monomers. Block and graft copolymers made by the living cationic route are currently being utilized in our program for tissue engineering applications and for the production of micelles that are of biomedical interest (see later). The main limitations of this method at

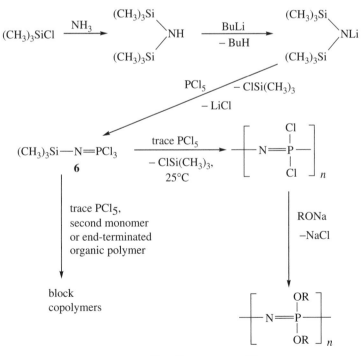

Reaction sequence (4)

present are the need for development work to scale-up the chlorophosphoranimine monomer (**6**) synthesis beyond the scale 10 to 100 g, and the fact that the chain lengths may not be as high as those produced via reaction sequence (1). Once these problems have been solved, this route may be commercially attractive and may be a workable alternative to route 1.

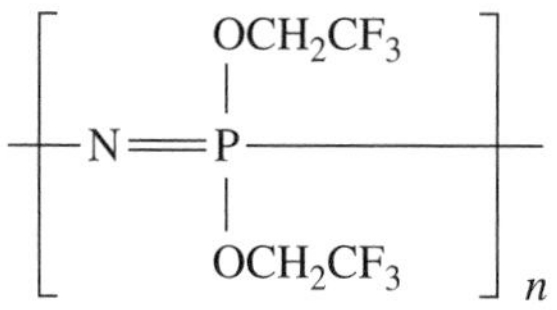

7

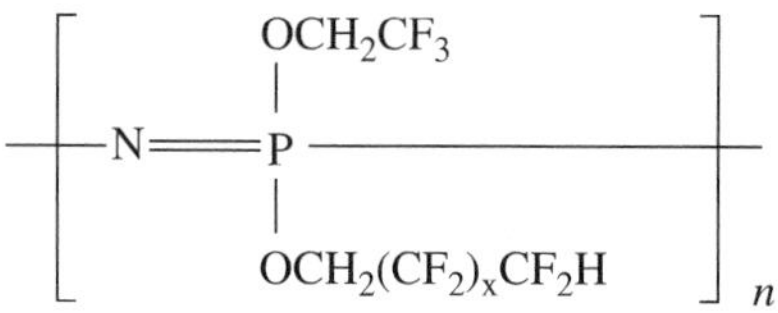

8

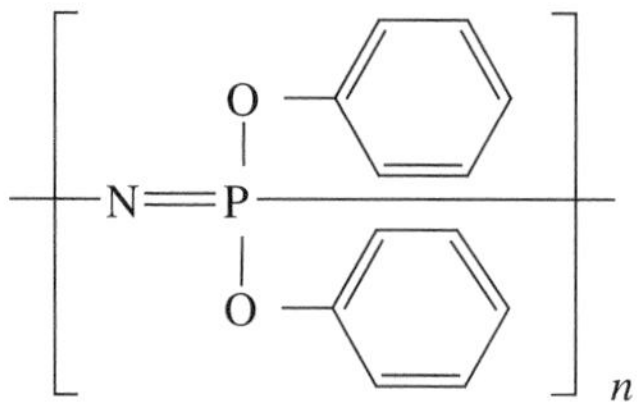

9

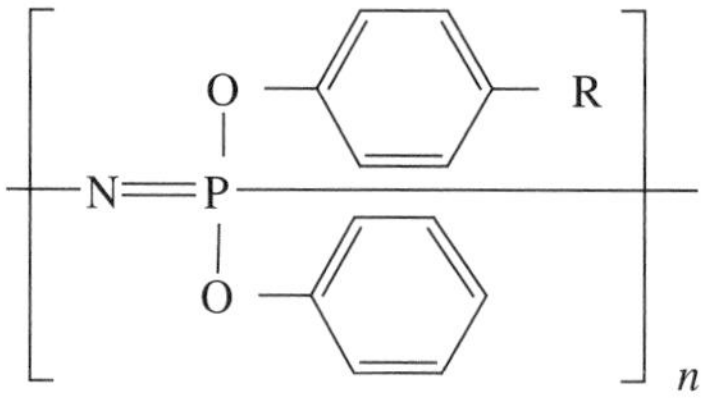

10

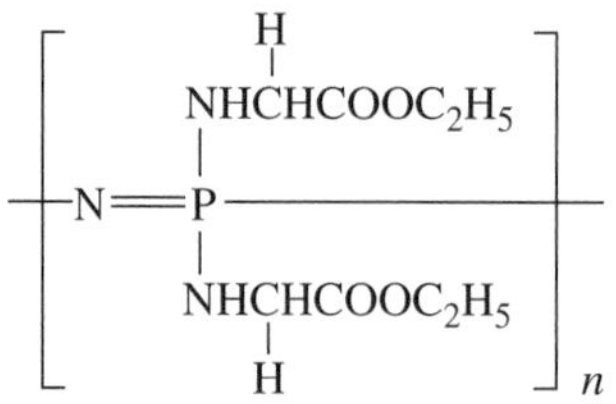

11

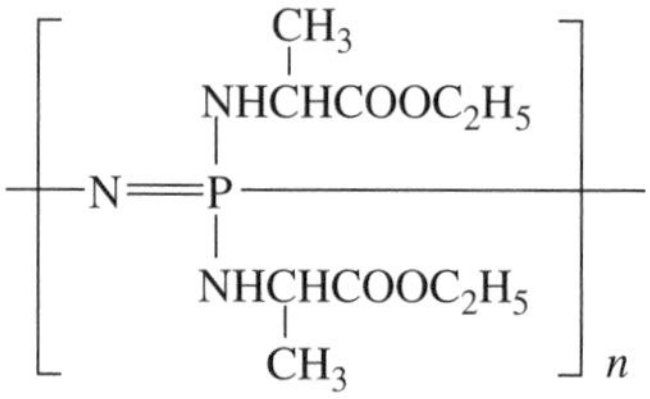

12

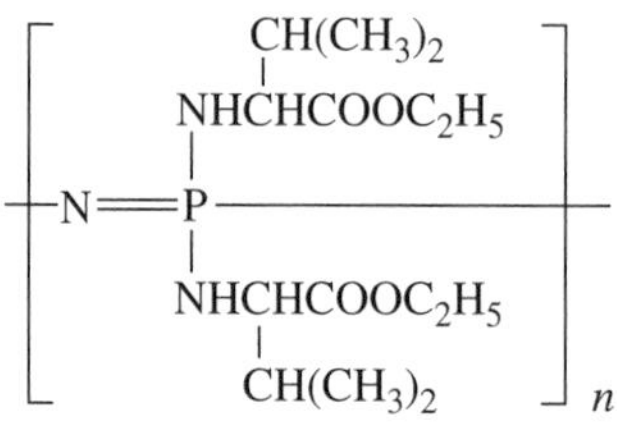

13

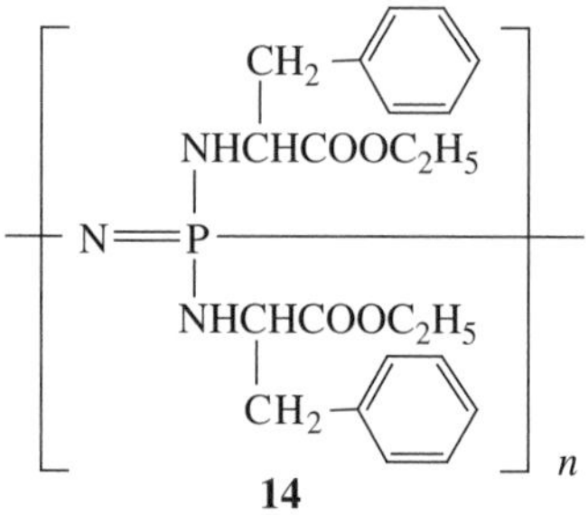

14

CHART 1

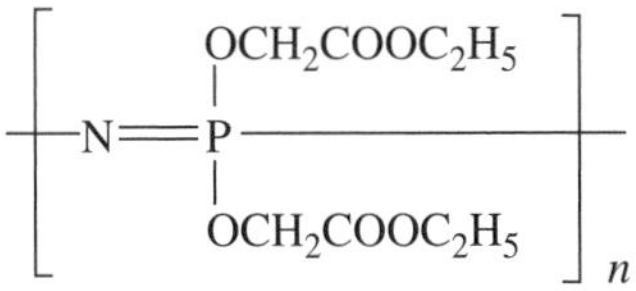

$$\left[\!\!\begin{array}{c} OCH_2COOC_2H_5 \\ | \\ -N\!=\!P \\ | \\ OCH_2COOC_2H_5 \end{array}\!\!\right]_n \qquad \left[\!\!\begin{array}{c} OCH(CH_3)COOC_2H_5 \\ | \\ -N\!=\!P \\ | \\ OCH(CH_3)COOC_2H_5 \end{array}\!\!\right]_n$$

15 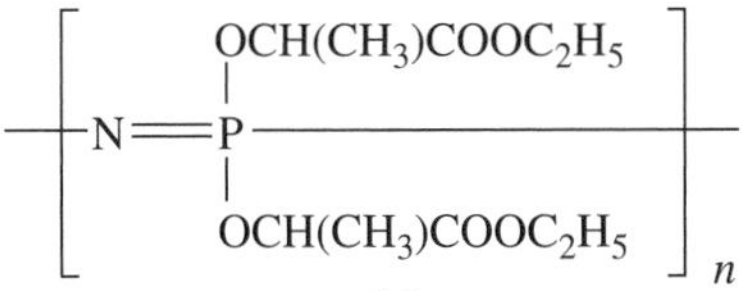   **16**

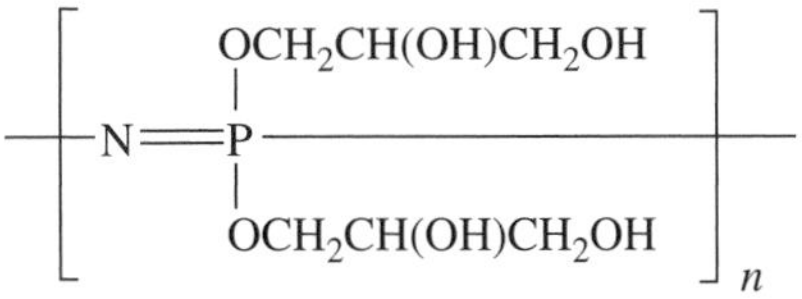

$$\left[\!\!\begin{array}{c} OCH_2CH(OH)CH_2OH \\ | \\ -N\!=\!P \\ | \\ OCH_2CH(OH)CH_2OH \end{array}\!\!\right]_n$$

17

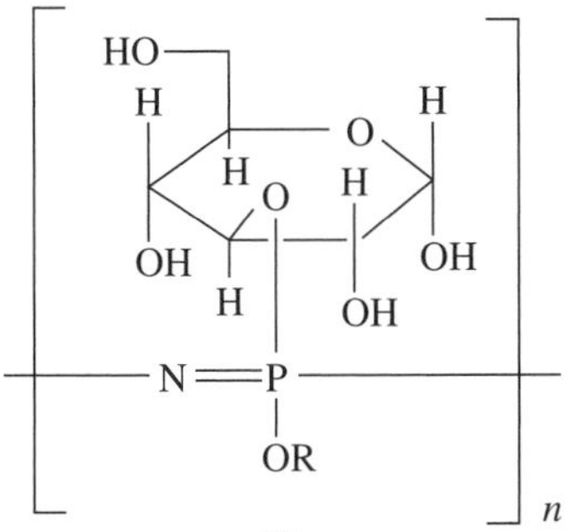

18 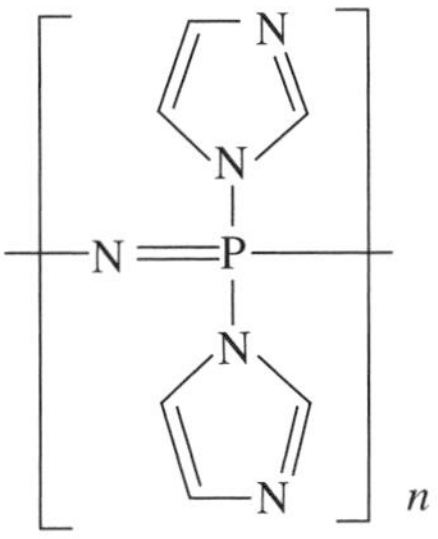   **19**

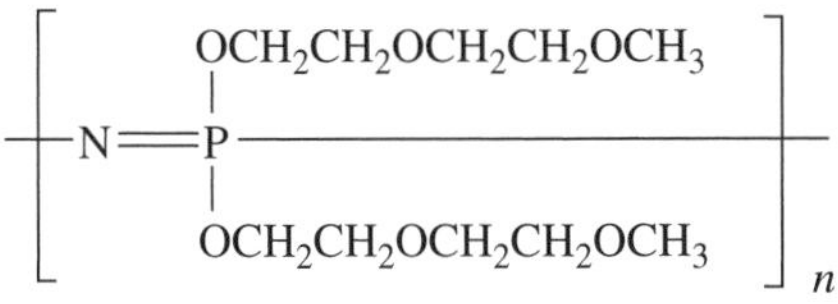

$$\left[\!\!\begin{array}{c} OCH_2CH_2OCH_2CH_2OCH_3 \\ | \\ -N\!=\!P \\ | \\ OCH_2CH_2OCH_2CH_2OCH_3 \end{array}\!\!\right]_n$$

20

$$\left[\!\!\begin{array}{c} CH_2OCH_2CH_2OCH_3 \\ | \\ OCH_2CHOCH_2CH_2OCH_3 \\ | \\ -N\!=\!P \\ | \\ OCH_2CHOCH_2CH_2OCH_3 \\ | \\ CH_2OCH_2CH_2OCH_3 \end{array}\!\!\right]_n$$

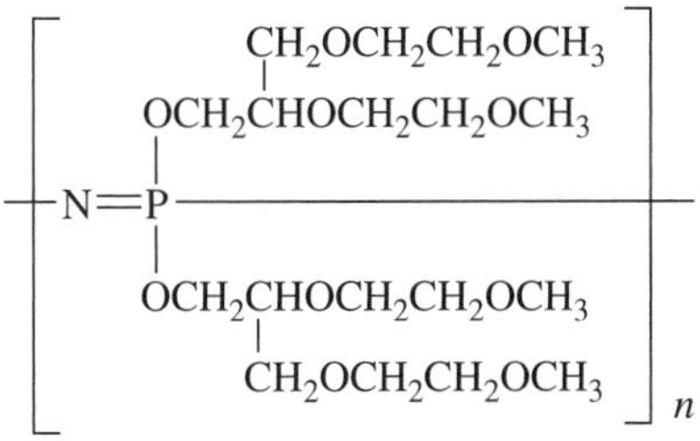

21

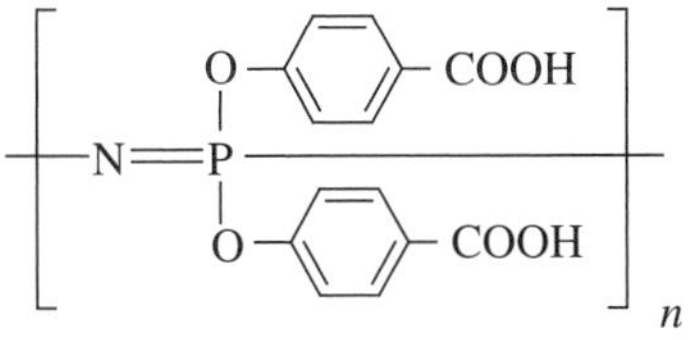

22 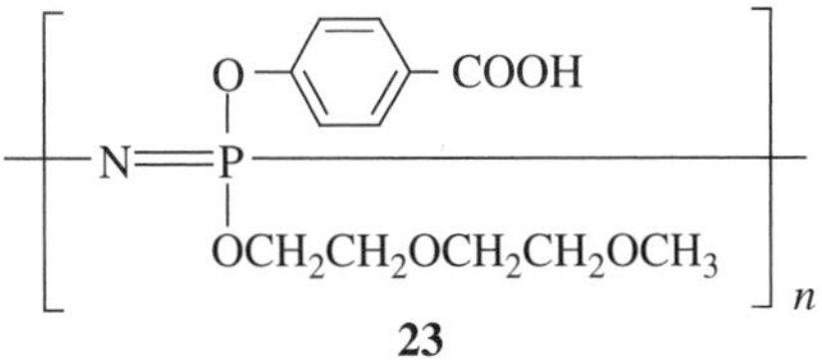   **23**

CHART 1   (*Continued*)

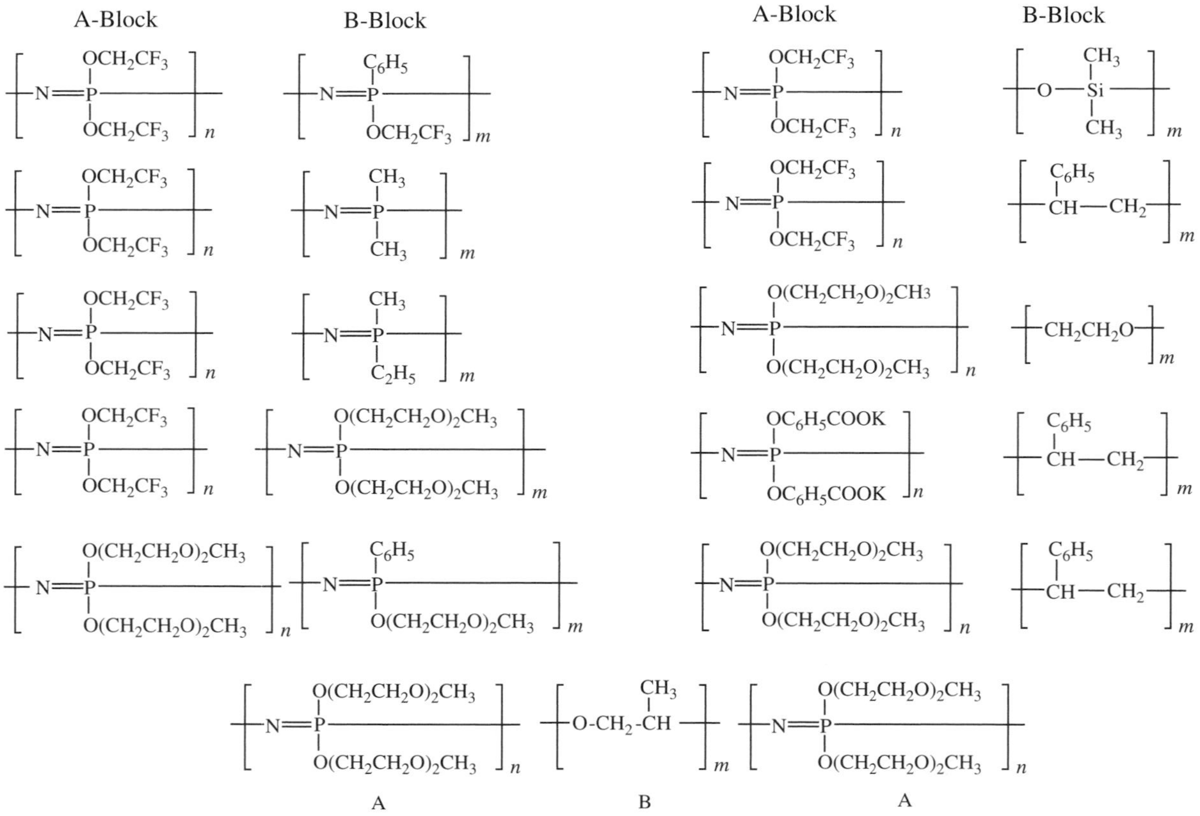

B-Block

A-Block

B-Block

A-Block

$$
\left[\begin{array}{c} \text{CH}_3 \\ | \\ \text{O}-\text{Si}-\text{CH}_3 \\ | \\ \text{CH}_3 \end{array} \right]_m
$$

$$
\left[\begin{array}{c} \text{OCH}_2\text{CF}_3 \\ | \\ \text{N}=\text{P}-\text{OCH}_2\text{CF}_3 \end{array} \right]_n
$$

$$
\left[\begin{array}{c} \text{C}_6\text{H}_5 \\ | \\ \text{N}=\text{P}-\text{OCH}_2\text{CF}_3 \end{array} \right]_m
$$

$$
\left[\begin{array}{c} \text{OCH}_2\text{CF}_3 \\ | \\ \text{N}=\text{P}-\text{OCH}_2\text{CF}_3 \end{array} \right]_n
$$

$$
\left[\begin{array}{c} \text{C}_6\text{H}_5 \\ | \\ \text{CH}-\text{CH}_2 \end{array} \right]_m
$$

$$
\left[\begin{array}{c} \text{OCH}_2\text{CF}_3 \\ | \\ \text{N}=\text{P}-\text{OCH}_2\text{CF}_3 \end{array} \right]_n
$$

$$
\left[\begin{array}{c} \text{CH}_3 \\ | \\ \text{N}=\text{P}-\text{CH}_3 \end{array} \right]_m
$$

$$
\left[\begin{array}{c} \text{OCH}_2\text{CF}_3 \\ | \\ \text{N}=\text{P}-\text{OCH}_2\text{CF}_3 \end{array} \right]_n
$$

$$
\left[\begin{array}{c} \text{CH}_2\text{CH}_2\text{O} \end{array} \right]_m
$$

$$
\left[\begin{array}{c} \text{O}(\text{CH}_2\text{CH}_2\text{O})_2\text{CH}_3 \\ | \\ \text{N}=\text{P}-\text{O}(\text{CH}_2\text{CH}_2\text{O})_2\text{CH}_3 \end{array} \right]_n
$$

$$
\left[\begin{array}{c} \text{CH}_3 \\ | \\ \text{N}=\text{P}-\text{C}_2\text{H}_5 \end{array} \right]_m
$$

$$
\left[\begin{array}{c} \text{OCH}_2\text{CF}_3 \\ | \\ \text{N}=\text{P}-\text{OCH}_2\text{CF}_3 \end{array} \right]_n
$$

$$
\left[\begin{array}{c} \text{C}_6\text{H}_5 \\ | \\ \text{CH}-\text{CH}_2 \end{array} \right]_m
$$

$$
\left[\begin{array}{c} \text{OC}_6\text{H}_5\text{COOK} \\ | \\ \text{N}=\text{P}-\text{OC}_6\text{H}_5\text{COOK} \end{array} \right]_n
$$

$$
\left[\begin{array}{c} \text{O}(\text{CH}_2\text{CH}_2\text{O})_2\text{CH}_3 \\ | \\ \text{N}=\text{P}-\text{O}(\text{CH}_2\text{CH}_2\text{O})_2\text{CH}_3 \end{array} \right]_m
$$

$$
\left[\begin{array}{c} \text{O}(\text{CH}_2\text{CH}_2\text{O})_2\text{CH}_3 \\ | \\ \text{N}=\text{P}-\text{O}(\text{CH}_2\text{CH}_2\text{O})_2\text{CH}_3 \end{array} \right]_n
$$

$$
\left[\begin{array}{c} \text{C}_6\text{H}_5 \\ | \\ \text{N}=\text{P}-\text{O}(\text{CH}_2\text{CH}_2\text{O})_2\text{CH}_3 \end{array} \right]_m
$$

$$
\left[\begin{array}{c} \text{O}(\text{CH}_2\text{CH}_2\text{O})_2\text{CH}_3 \\ | \\ \text{N}=\text{P}-\text{O}(\text{CH}_2\text{CH}_2\text{O})_2\text{CH}_3 \end{array} \right]_n
$$

$$
\left[\begin{array}{c} \text{O}(\text{CH}_2\text{CH}_2\text{O})_2\text{CH}_3 \\ | \\ \text{N}=\text{P}-\text{O}(\text{CH}_2\text{CH}_2\text{O})_2\text{CH}_3 \end{array} \right]_n
\quad\text{A}
$$

$$
\text{A} \quad\left[\begin{array}{c} \text{O}(\text{CH}_2\text{CH}_2\text{O})_2\text{CH}_3 \\ | \\ \text{N}=\text{P}-\text{O}(\text{CH}_2\text{CH}_2\text{O})_2\text{CH}_3 \end{array} \right]_n
\left[\begin{array}{c} \text{CH}_3 \\ | \\ \text{O}-\text{CH}_2-\text{CH} \end{array} \right]_m
\quad\text{B}
$$

CHART 2

5. *Secondary substitution reactions carried out on the organic side groups introduced by one of the preceding methods.* Organic side groups incorporated into a polyphosphazene by one of the preceding methods can themselves be modified by exposure to reagents that introduce additional functionality [37–42]. For example, aryloxy side groups can be sulfonated or nitrated, ester groups can be hydrolyzed to carboxylic acid units, aryloxy groups with chloro or bromo substituents can be lithiated and phosphonated, transition metals can be coordinated with the existing side chains, alkyl groups can be used for cross-linking, and so on. The only limitation on these secondary substitution processes is the need to avoid reactions at the phosphazene backbone, and this has proved to be a restriction for only a small number of these secondary transformations. Many biomedically useful polyphosphazenes have been produced by a combination of route 1 followed by secondary side-group substitution reactions.

As a guide to the discussion in the remainder of this chapter, Chart 1 lists specific polyphosphazenes that are mentioned here. Nearly all of these were synthesized by the route shown in reaction sequence (1), with the block copolymers produced via the living cationic process summarized in reaction sequence (4). Some typical block copolymers produced via the living cationic process are shown in Chart 2.

RANGE OF ORGANIC SIDE GROUPS

One of the most important attributes of polyphosphazenes is the ease with which different side groups can be linked to the phosphorus–nitrogen skeleton, especially by macromolecular substitution and secondary substitutions. A partial list of the various classes of side groups includes hydrophobic fluorinated or organosilicon units, metal-containing groups, groups that lead to polymer, hydrolytic sensitivity, bioactive side groups, groups that generate solubility in water, side groups that are responsible for ion transport and ion-based shape changes, and combinations of these in the same polymer, including block copolymers. It is the ability of the investigator to fine-tune properties through different side groups or combinations of groups that has led to the development of numerous biomedical initiatives. In the following sections we illustrate some of the developments that originated in our program.

BIOINERT POLYPHOSPHAZENES FOR CARDIOVASCULAR, DENTAL, AND RELATED APPLICATIONS

One of the most important polyphosphazenes produced during the initial development of this field was polymer **7**, poly[bis(trifluoroethoxy)phosphazene], prepared by the reaction of sodium trifluoroethoxide with poly(dichlorophosphazene) (**2**) [1,2]. Polymer **7** is highly hydrophobic and, in this respect,

Electrospun [NP(OCH₂CF₃)]ₙ Fibers

Spun Cast Flat Film : WCA 104°
(a)

Electrospun Film : WCA 155°
(b)

FIGURE 1 Nanofibers of electrospun poly[bis(trifluoroethoxy)phosphazene] (**7**) together with the profile of a droplet of water on a spun-cast film and a nanofiber mat. The water contact angle of 155° for the nanofiber mat falls into the category of superhydrophobic materials.

resembles classical fluorocarbon polymers such as Teflon. Yet, unlike Teflon, it is soluble in some organic solvents and can be solution-cast into films or spun into fibers. As shown in Figure 1, this polymer has recently been electrospun into micro- and nanofibers with superhydrophobic surface properties (water contact angle near 160°) [43]. Films and fibers of this polymer are ideal for surface modification using wet chemistry or environmental plasma techniques (see later). The polymer is also completely stable to hydrolysis in neutral biochemical media and has a high degree of stability to ultraviolet and gamma rays.

Soon after the initial description of this polymer was published, other investigators found that the introduction of two different fluoroalkoxy side units (Chart 1, polymer **8**) instead of one type of group changed the properties from those of a microcrystalline thermoplastic to those of a rubbery elastomer [44,45], and this became the basis of a manufacturing process. This elastomer, known variously as PN-F, or Eypel-F, has been developed as a U.S. Food and Drug Administration (FDA)-approved dental liner material [46] because of its antimicrobial characteristics and its unusual impact-absorbing qualities. The same polymer was investigated as a material for artificial heart valves. Interest in this mixed-substituent polymer as a bioinert elastomer stems from its high hydrophobicity, its resistance to hydrolytic breakdown, and its resistance to absorbing lipids from the blood, which apparently is a defect of silicone elastomers. Interruptions in the commercial supply of this elastomer limited its widespread use for a number of years, although it is once again becoming available.

A useful feature of both the single-substituent polymer **7** and the mixed-substituent polymer **8** is their solubility in liquid carbon dioxide, which opens possibilities for environmentally acceptable processing into films and expanded foams [47].

BIOERODIBLE POLYMERS FOR DRUG DELIVERY, TISSUE ENGINEERING, AND SHAPE-MEMORY DEVICES

Amino Acid Ester Derivatives

Quite early in the development of polyphosphazenes (1977), we attempted the synthesis of polymers with amino acid ester side groups [48]. The idea was to link the amino acids to the polyphosphazene chain via the NH_2 terminus, which required that the carboxyl terminus be protected by esterification to prevent this site from reacting with poly(dichlorophosphazene). Initially, the finding that these polymers hydrolyzed slowly in aqueous media was a disappointment until it was realized that this was, in fact, a considerable advantage. The hydrolysis products are phosphate, ammonia, an amino acid, and an alcohol from the ester function. The ammonia is a pH-stabilizing species, and the other products are metabolites. Thus, the use of these polymers in sutures, surgical clips, drug delivery matrices, and as tissue engineering platforms became a possibility. In recent years we have expanded the number of amino acid esters that can be utilized as side groups. These now include the ethyl esters of glycine and alanine (the original examples) as well as tyrosine, serine, threonine, phenylalanine, and several others. Together with variants produced in other research programs [49,50] there are now more than 40 different polyphosphazenes known that have different amino acid ester side groups or combinations of two or more. An important factor is that the rate of hydrolysis can be controlled through the type of amino acid ester, with bulky groups at the α carbon position serving to retard attack by water molecules on the bond between phosphorus and the side-group nitrogen atom [51–53]. The ester function also has an influence on the hydrolytic behavior, with longer-chain aliphatic species also retarding the hydrolytic breakdown.

Access to these polymers has been a cornerstone of our long-term collaboration with C. Laurencin's group at the University of Virginia and University of Connecticut for the development of orthopedic tissue engineering materials, a subject covered in Chapters 8 and 9. Specifically, porous constructs of these polymers, including nanofiber mats, have been used as platforms for the cultivation of osteoblasts for bone regeneration [54–56]. In addition, composites of the amino acid ester polymers and calcium hydroxyapatite provide not only a platform for cell adhesion and growth but also a source of calcium salts for bone regeneration. Other composites have also been examined, including blends of the polyphosphazenes with poly(lactic acid–glycolic acid) [57]. Additional macromolecules that are currently under investigation are block copolymers of amino acid ester polyphosphazenes with poly(lactic acid–glycolic acid), which provide yet another opportunity for fine-tuning the hydrolysis rates and control over the pH of the hydrolysis products [58]. Another aspect of this program is the use of composites of calcium hydroxyapatite with amino acid ester polymers or the polymer poly[di(carboxylatophenoxy)phosphazene] (PCPP; **22**), which forms calcium-derived ionic cross-links that stabilize the skeletal platform in the form of either porous constructs or nanofiber mats [59,60].

Shape-Memory Applications

Shape-memory polymers have a number of potentially important uses in biomedicine. The phenomenon occurs when a polymer is initially fabricated in one shape but reverts to another shape when heated. We have recently shown that specific amino acid ester–substituted polyphosphazenes have the attribute of shape memory. This could be useful for cardiovascular or bile duct stents or other applications where the insertion of a device into the body needs to be followed by a change in shape. It is particularly useful if the device is designed to bioerode and disappear within a specified time frame. Specifically, polymers **12** to **14** are bioerodible and have shape-memory characteristics [61]. The shape-memory behavior may result from the side-group steric hindrance, crystallinity, and/or hydrogen bonding inherent in these polymers. A device such as a polymer helix is fabricated at an elevated temperature and is then

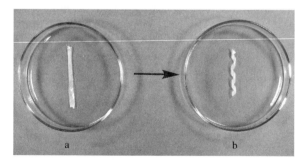

FIGURE 2 Shape-memory response of polymer **14**. The polymer, initially cast as a thin strip (a), is coiled around a dowel and heated above the glass transition temperature. After cooling to room temperature and flattening back to strip form (a), the material will re-form in the spiral configuration (b) when reheated. (Photograph by N. Krogman, The Pennsylvania State University.)

cooled and stretched to an unwound shape at a lower temperature. This unwound configuration is retained until an increase in temperature causes the device to revert to its original coiled shape (Fig. 2).

Other Bioerodible Polyphosphazenes

A few other organic side groups linked to a polyphosphazene chain sensitize the polymer to hydrolysis. These include ethyl glycolate (**15**) [62] ethyl lactate (**16**) [62], glyceryl (**17**) [64], glucosyl (**18**) [41], and imidazolyl (**19**) groups [63]. Imidazolyl side groups have been utilized for controlled drug delivery experiments [64], but the others are also feasible candidates for future work in both tissue engineering and drug delivery.

WATER-SOLUBLE POLYMERS, HYDROGELS, AND RESPONSIVE MEMBRANES

Water-soluble polymers play an important role in biomedicine, from viscosity improvers to the starting materials for hydrogel formation. Hydrogels are water-soluble polymers that have been lightly cross-linked and allowed to absorb water. They may be 90% water, but they have volume and shape and in many ways behave like mammalian soft tissues. Hydrogels are used in many different medical applications, from drug delivery vehicles to soft tissue prostheses. They are also prospective materials for membranes that control

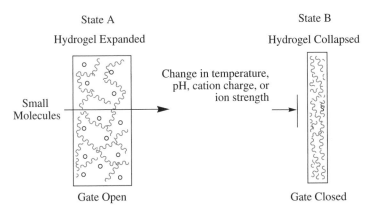

FIGURE 3 Hydrogels derived from polymers **20** to **23** can be fabricated into responsive membranes that are either permeable or impermeable, depending on the temperature, pH, ion strength, or cation charge. This provides a mechanism for the controlled release of drugs, or the timed exposure of enzymes or mammalian cells to small molecules.

the flow of drug molecules or proteins into the body, or which trap mammalian cells within the gel matrix and control their metabolism. These possibilities arise because hydrogels can be designed to be responsive to temperature, cations, and pH. For example, for the control of cell function a hydrogel can be designed to allow access of nutrients to the cells reversibly or to shut down the biological processes by collapsing to an impervious barrier (Fig. 3).

The first water-stable, water-soluble polyphosphazene to be prepared was poly[bis(methoxyethoxyethoxy)phosphazene] (MEEP), shown as structure **20** in Chart 1 [10,65]. MEEP can be cross-linked for gel formation in two ways. First, in the water-free state it is cross-linked by exposure to gamma rays or to strong ultraviolet radiation [66–68]. The number of cross-links, and therefore the degree of water swelling, is controlled by the length of radiation exposure time. A second method requires the introduction of a carboxylic acid substituent function (structure **22**) or the two substituents shown in structure **23** [69–79]. The presence of the carboxylic acidic function in **22** or **23** means that hydrogels expand in basic media (e.g., dilute sodium hydroxide) but contract in acidic media, due to the charge repulsion between the protons in solution and the protonated carboxylic acid groups. Although polymers **22** and **23** are soluble in water in the presence of sodium ions, they are cross-linked by divalent calcium ions. This is illustrated by structures **24** and **25** in reaction (5):

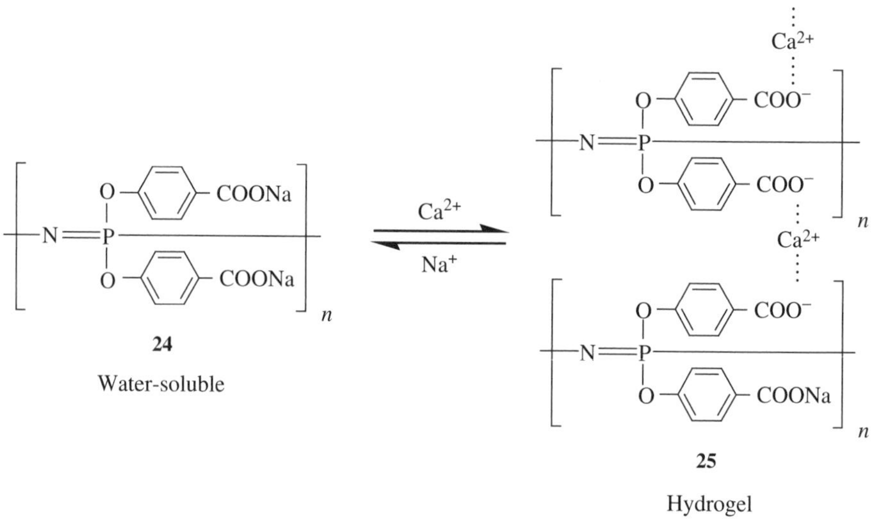

Reaction sequence (5)

Ionic cross-linking within a hydrogel causes collapse of the gel with concurrent extrusion of water, and this step converts a permeable material to an impermeable form. The introduction of sodium ions reverses the process.

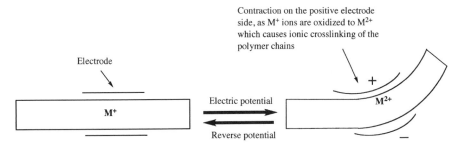

FIGURE 4 Hydrogels derived from polymer **23** in the presence of copper or iron cations are swollen in the presence of Cu^+ or Fe^{2+} but contract when the cations are oxidized to Cu^{2+} or Fe^{3+}, due to the increased density of ionic cross-linking. Electrochemical oxidation of the cations causes bending of the gel toward the oxidation electrode as current is applied, and a reversal of the bending when the current direction is reversed.

Hydrogels based on MEEP have another characteristic. They have a lower critical solution temperature (LCST), which means that below a certain temperature, they are in a water-swollen expanded state, but above the LCST they contract and extrude the water. This means that three mechanisms exist for expanding and contracting a phosphazene gel: temperature changes, pH changes, and changes in the charge on cations in solution. Moreover, we have shown recently [79] that in the presence of a cation that can be reversibly converted electrochemically from a $+1$ to a $+2$ or $+3$ charge state, the gel can be expanded or contracted by passage of an electric current (Fig. 4). Not only does this open possibilities for the controlled release of drugs by electrical stimulation, but also provides a mechanism for gel contractions of the type that occur when muscles are actuated.

Many analogs of MEEP, such as **21**, have been synthesized in our program with different lengths of the ethyleneoxy side chains and the introduction of branched ethyleneoxy units and crown ethers. These changes alter the LCST and the amount of cross-linking needed to generate specific hydrogel properties. Another use of the PCPP system is for the controlled release of vaccines, a topic that is described briefly below and more fully in Chapters 3 to 7.

TAILORED SURFACES FOR BIOMEDICAL USES

The surface presented to a living system by a biomaterial is at least as important as the properties of the material beneath the surface. Properties such as resistance to microbial colonization, protein deposition, and cell adhesion are all surface dependent. Thus, the development of polyphosphazenes as biomedical materials has included the study of ways to control surface behavior.

A variety of surface properties play a role in the response of biosystems to a material:

1. Apart from the question of the overall shape and surface roughness or smoothness, one of the most obvious characteristics is the hydrophobicity, hydrophilicity, or amphilicity of the surface.
2. The chemical stability of the surface is a crucial factor because the interface will change continuously if the material undergoes hydrolytic erosion.
3. There is the question of a direct chemical influence between the functional groups at the surface and mammalian cells or microorganisms in contact with that interface.
4. It is known that many polymers are susceptible to the phenomenon of surface turnover, in which groups that originally were exposed at the surface become buried over time and are replaced from the interior by groups with different properties.

Two approaches to controlling the surface behavior of polyphosphazenes have been explored in our program: the use of wet chemistry to bring about replacement of the side groups that populate the surface of the solid polymer by groups that have different hydrophilicity or hydrophobicity or have specific biological activity; and alteration of polymer surfaces by the use of environmental plasma techniques.

Wet Chemistry to Modify Polyphosphazene Surfaces

Most polymers can be subjected to limited surface modification chemistry, but polyphosphazenes are particularly suited for this process because of the stability of the phosphorus–nitrogen backbone and the ease with which some side groups can be replaced by others. The key requirement for this type of chemistry is that the chemical reactions must be controlled so that the changes are restricted to the surface regions and do not penetrate the core of the material. It also helps if the chemical reactions affect only the polymer side groups and do not disrupt the macromolecular chain. Thus, the objective is to produce a material that has one set of properties within the bulk polymer and a different set at the interface. We have focused on the following reactions.

1. *Replacement of trifluoroethoxy surface groups by other units.* Trifluoroethoxy side groups at the surface of a polymer such as **7** or **8** can be exchanged for other side groups through the use of mild reaction conditions [80–82]. For example, when treated with solutions of $NaOCH_2CF_2CF_2CF_3$, $NaO(CH_2-CH_2O)_xH$, $NaO(CH_2CH_2)_xCN$, $NaO(CH_2CH_2)_xNH_2$, or $NaOH$, polymer **7** yields materials with a surface that contains both OCH_2CF_3 and the new side groups, with a corresponding change in hydrophobicity or hydrophilicity or functional surface character for the linkage of other groups.

2. *Reactions of surface functional groups linked to aryloxy side units.* The linkage of substituted aryloxy groups to a polyphosphazene chain allows many possibilities for surface modification [83–85]. For example, methyl groups connected to the aryl rings are oxidized to –COOH groups, and ester units are hydrolyzed to the same groups. Chlorine or bromine atoms can be lithiated and linked to organic molecules. Moreover, phenoxy groups are readily surface-sulfonated by H_2SO_4 or SO_3 or nitrated with HNO_3 and reduced to NH_2 groups. These surface modifications provide a facile mechanism for the linkage of biologically important groups to a polymer surface.

3. *Linkage of bioactive molecules to polyphosphazene surfaces* [86–89]. Proteins have been immobilized on polyphosphazene surfaces via reactions with –CH_2Br units on phenoxy side groups. Another approach is to link glycidyl methacrylate functionality to aryloxy side groups and then couple a protein via the epoxy unit. Amino groups attached to phenoxy side units also provide sites for surface protein immobilization. Various reagents, such as glutaric dialdehyde, cyanogen bromide, or a diazonium salt, were then used to couple the surface to proteins such as glucose-6-phosphate dehydrogenase or trypsin while maintaining the activity of the protein. Catecholamines such as dopamine have also been linked covalently to an aryloxyphosphazene surface through amino functional groups on the phosphazene. These surfaces are able to inhibit release of the hormone prolactin from pituitary cells in culture, thus demonstrating that penetration of the cell membrane by free dopamine is not essential for this inhibition [88].

4. *Linkage of cyclodextrins to polyphosphazene surfaces* [90]. β-Cyclodextrin and its sulfate have been linked covalently to the surface of a polyphosphazene that bears aryloxycarboxylic acid functional groups. Although the biological properties of these surfaces have yet to be studied, it is anticipated that they may play a role in the inhibition of angiogenisis by reducing excessive capillary growth.

5. *Grafting hydrogels to polymer surfaces* [91]. The formation of a hydrogel on the surface of a hydrophobic biomaterial has many potential uses in biomedicine, ranging from surfaces for cell culture to cardiovascular materials. Such surfaces can be constructed through substitution reactions, as just described. However, an alternative is to graft a water-soluble polymer onto an organic polymer or a polyphosphazene surface. The polyphosphazene MEEP (compound **20**) is especially appropriate for this technique because it can be grafted to a wide variety of polymer surfaces by gamma irradiation or exposure to ultraviolet radiation. The resulting grafted interface is itself cross-linked so that all the attributes of MEEP hydrogels, such as LCST behavior, are incorporated into the system. This is extremely useful for the gel immobilization of enzymes and cells for biochemical reactors [91–94].

Environmental Plasma Processes

An environmental plasma is formed when a radio-frequency field interacts with gas molecules at atmospheric pressure. Under appropriate conditions the gas molecules are converted to ions or free radicals, which attack the surface of a polymer film or fiber. Thus, in a collaboration with S. Kim at Penn State, polymers **7** and **8** and several of the amino acid–substituted derivatives have been subjected to plasmas derived from oxygen nitrogen, methane, or a mixture of tetrafluoromethane and hydrogen [95].

The results from these processes indicate that this is an excellent method for the high-throughput modification of polyphosphazene surfaces. For example, the H_2/CF_4 plasma generates a highly hydrophobic perfluorocarbon surface, whereas the oxygen and nitrogen plasmas lower the surface water contact angles and thus increase hydrophilicity via the formation of hydroxyl, carboxylic, or amino groups. From a biomedical viewpoint this offers the possibility of taking almost any phosphazene polymer film, surface coating, or fiber and introducing fluorocarbon, hydroxyl, carboxylic acid, or amino groups into the interfacial layer, and thereby changing the interaction to water and introducing surface groups for reactions with biological reagents.

POLYMERIC DRUGS AND MYOGLOBIN MODELS

Early work in our program explored the possibility that water-soluble polyphosphazenes might be used as carrier molecules for covalently linked bioactive molecules. The idea was that linkage of the drug to the polymer would retard its excretion from the body and might allow the delivery of drugs to targeted sites in the patient. The ability to design bioerosion into the polymer was an added advantage. Bioactive molecules that were used in this way included steroids [96] cis-platinum drugs [97], and several antibiotics [98,99]. In addition, the possibility was explored that a water-soluble polyphosphazene could be used as a carrier molecule for heme-type structures, with possible use as an emergency blood substitute [100,101].

Research on this topic was deemphasized in our program when it became apparent that each polymer-bound drug was viewed by the authorities as an entirely new drug and was therefore subject to a complete reevaluation even though the drug would be released in the body in its approved molecular form. However, this area of research could become important in a different regulatory environment.

RESPONSIVE MICROSPHERES FOR ORAL DRUG AND VACCINE DELIVERY

A *microsphere* is a micrometer-level particle, usually derived from a polymer, and capable of trapping living cells, drugs, vaccines, or even gas bubbles. Microspheres that enclose mammalian cells can protect those cells from antibodies while allowing nutrients and metabolic products to diffuse into and out of the polymer network. This principle was the basis of the first attempt to utilize the chemistry shown in reaction (5) to produce microspheres for the

FIGURE 5 Micrograph of hybridoma liver cells incorporated into a calcium cross-linked gel of the polymer PCPP (**22**). (From ref. 101.)

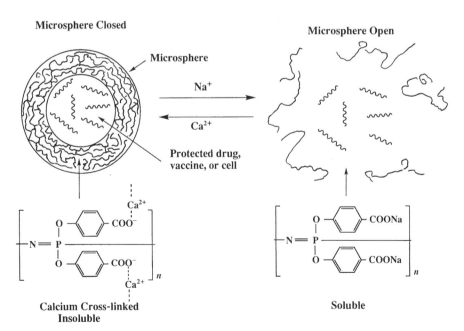

FIGURE 6 Use of microspheres of PCPP (**22**) for the release of trapped biological agents such as drugs or vaccines in the presence of monovalent cations.

encapsulation of hybridoma liver cells for possible applications in artificial liver research [reaction (5) and Figs. 5 and 6] [9,102–104]. This project, carried out as a collaborative effort between our group and R. Langer's group at the Massachusetts Institute of Technology, demonstrated the feasibility of this approach and illustrated the value of this synthetic polymer system compared to the use of alginates for microencapsulation. The same polymer system was later explored for gas bubble encapsulation for use in ultrasound cardiovascular imaging, and more especially for the delivery of vaccine molecules for oral vaccination [105]. In the latter application, the microspheres protect the vaccine molecules from decomposition by stomach acid and release the vaccine in the small intestine. Because polymers **22** and **24** are polyelectrolytes, they have the ability to enhance the activity of the vaccine (they are immunoadjuvants). This research was the basis for the establishment of a company to exploit this chemistry and for the subsequent detailed development work. This initiative is described by A. Andrianov in Chapter 3.

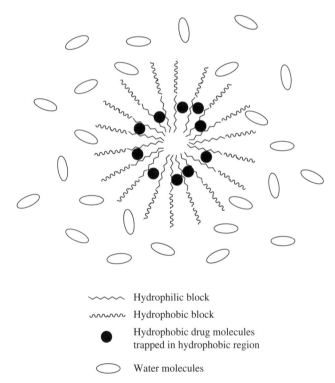

Hydrophilic block

Hydrophobic block

Hydrophobic drug molecules trapped in hydrophobic region

Water molecules

FIGURE 7 Micelle formed from an amphiphilic diblock copolymer in which the hydrophobic core can accommodate hydrophobic small molecules such as drugs, while the hydrophilic corona interacts with water molecules in the surrounding medium. Such nanometer-size particles can pass through the smallest mammalian capillaries to deliver bioactive molecules to sites that are distant from the point of injection. Bioerodible polyphosphazenes provide an almost unique opportunity for developing this technique.

MICELLES FOR DRUG DELIVERY

A *micelle* is a nanometer-size particle usually formed from amphiphilic diblock copolymers. An amphiphilic block copolymer contains a hydrophilic block connected covalently to a hydrophobic block. Micelles are produced when a polymer of this type is sonicated in water. The particles so formed consist of spheres or "worms", with the hydrophilic component on the outside (the corona) and the hydrophobic blocks in the core (Fig. 7). Hydrophobic drug molecules can be trapped in the core during the sonication step, a process that essentially solubilizes molecules that would otherwise be impossible to inject into the body and deliver to points distant from the injection site.

Use of the living cationic polymerization process has allowed us to synthesize a wide range of amphiphilic di- and triblock copolymers of the types shown in Chart 2 [106–111]. Many of these have been converted to spherical micelles, and several are candidates for use as drug delivery vehicles. The most important are those that have at least one bioerodible block, so that the components of each micelle can be excreted or metabolized. Similar principles govern the behavior of dendrimeric phosphazenes [112].

THE VALUE OF MODEL COMPOUND STUDIES

The polymer substitution chemistry described in this chapter was in most cases initially developed by us not with phosphazene high polymers, but with small-molecule cyclic analogs such as the phosphazene cyclic trimer $(NPCl_2)_3$ (compound **1**). The value of this approach is that small-molecule reaction chemistry is much easier to carry out than is the corresponding chemistry at the high-polymer level. Moreover, detailed characterization of small molecules is considerably more certain than is the structural characterization of polymers. This applies particularly to the use of x-ray diffraction and nuclear magnetic resonance techniques. Our experience is that only after the nuances of the small-molecule chemistry have been understood is it possible to carry out pioneering chemistry reliably at the macromolecular level. Attempts to short-circuit this process and proceed directly to high-polymer substitutions frequently lead to the formation of products that have structures different from those envisaged and to materials with suboptimal properties or unexplained biological behavior.

FINAL COMMENTS

The future of biomedicine is critically dependent on the development of new polymer systems to supplement and eventually replace those traditional materials that have well-known drawbacks. A number of new classical organic polymer systems are under development with this objective in mind. Yet few of these systems have the synthetic versatility of the polyphosphazene platform, with its facile property-tuning capabilities; the ability to yield elastomers, fibers, or films with tailored surface character; bioerodibility; responsive membrane behavior; and controlled drug delivery capabilities—all through minor changes in the synthetic protocol. This is a field in which the fundamental scientific work has far outpaced the biomedical developments, mainly because of the essential slowness and cost of biological testing, which itself depends on the availability of development-scale quantities of individual polymers. The scale-up problems are in the process of being solved, and it is anticipated that this will stimulate a rapid expansion of research and development in several areas of phosphazene biomedicine. In this book we point the way for this to happen.

REFERENCES

1. Allcock, H.R., Kugel, R.L. Synthesis of high polymeric alkoxy- and aryloxyphosphonitriles. *J. Am. Chem. Soc.*, 1965, 87:4216–4217.

2. Allcock, H.R., Kugel, R.L., Valan, K.J. High molecular weight poly(alkoxy- and aryloxyphosphazenes). *Inorg. Chem.*, 1966, 5:1709–1715.

3. Allcock, H.R., Kugel, R.L. High molecular weight poly(diaminophosphazenes). *Inorg. Chem.*, 1966, 5:1716–1718.

4. Allcock, H.R., Cook, W.J., Mack, D.P. High molecular weight poly[bis(amino)-phosphazenes] and mixed substituent poly(aminophosphazenes). *Inorg. Chem.*, 1972, 11:2584–2590.

5. Allcock, H.R. *Chemistry and Applications of Polyphosphazenes.* Wiley-Interscience, Hoboken, NJ, 2003.

6. Allcock, H.R. A Perspective of polyphosphazene research. *J. Inorg. Organomet. Polym. Mater.*, 2006, 16(4), 277–294 and Appendix, 437–459.

7. Allcock, H.R., Kim, Y.B. Synthesis, characterization, and modification of poly(organophosphazenes) that bear both 2,2,2-trifluoroethoxy and phenoxy groups. *Macromolecules*, 1994, 6:516–524.

8. Allcock, H.R., Ngo, D.C. Synthesis of poly(bis-phosphazo)phosphazenes bearing aryloxy and alkoxy side groups. *Macromolecules*, 1992, 25:2802–2810.

9. Singler, R.E., Hagnauer, G., Schneider, N.S., LaLiberte, B.R., Sacher, R.E., Matton, R.W. Synthesis and characterization of polyaryloxyphosphazenes. *J. Polym. Sci.*, 1974, 12:433–444.

10. Allcock, H.R., Austin, P.E., Neenan, T.X., Sisko, J.T., Blonsky, P.M., Shriver, D.F. Polyphosphazenes with etheric side groups: prospective biomedical and solid electrolyte polymers. *Macromolecules*, 1986, 19:1508–1512.

11. Maher, A.E., Allcock, H.R. Influence of *n*-hexoxy group on the properties of fluoroalkoxyphosphazene polymers. *Macromolecules*, 2005, 38:641–642.

12. Allcock, H.R., Moore, G.Y. Polymerization and copolymerization of phenyl-halogenocyclotriphosphazenes. *Macromolecules*, 1975, 8:377–382.

13. Allcock, H.R., Patterson, D.B. Ring-ring and ring-chain equilibration of dimethyl-phosphazenes: relationship to phosphazene polymerization. *Inorg. Chem.*, 1977, 16:197–200.

14. Allcock, H.R., Schmutz, J.L., Kosydar, K.M. A new route for poly(organopho-sphazene) synthesis: polymerization, copolymerizaton, and ring-ring equilibration of trifluoroethoxy- and chloro-substituted cyclotriphosphazenes. *Macromolecules*, 1978, 11:179–186.

15. Allcock, H.R., Patterson, D.B., Evans, T.L. Synthesis of alkyl and aryl phospha-zene high polymers. *J. Am. Chem. Soc.*, 1977, 99:6095–6096.

16. Allcock, H.R., Ritchie, R.J., Harris, P.J. Synthesis of alkylphosphazene high polymers via the polymerization of monoalkylpentachlorocyclotriphosphazenes, N3P3Cl5R. *Macromolecules*, 1980, 13:1332–1338.

17. Allcock, H.R., Connolly, M.S. Polymerization and halogen scrambling behavior of phenyl-substituted cyclotriphosphazenes. *Macromolecules*, 1985, 18:1330–1340.

18. Allcock, H.R., Riding, G.H., Lavin, K.D. Polymerization of new metallocenylpho-sphazenes. *Macromolecules*, 1987, 20:6–10.

19. Allcock, H.R., Brennan, D.J., Graaskamp, J.M. Ring-opening polymerization of methylsilane- and methylsiloxane-substituted cyclotriphosphazenes. *Macromole-cules*, 1988, 21:1–10.

20. Allcock, H.R., Dodge, J.A., Manners, I., Riding, G.H. Strain-induced ring-opening polymerization of ferrocenylorganocyclotriphosphazenes: a new synthetic route to poly(organophosphazenes). *J. Am. Chem. Soc.*, 1991, 113:9596–9603.

21. Wisian-Neilson, P., Neilson, R.H. Poly(dimethylphosphazene), (Me$_2$PN)$_n$. *J. Am. Chem. Soc.*, 1980, 102:2848–2849.

22. Neilson, R.H., Wisian-Neilson, P. Poly(alkyl/arylphosphazenes) and their precur-sors. *Chem. Rev.*, 1988, 88:541.

23. Wisian-Neilson, P., Xu, G.E., Wang, T. Fluorinated polyphosphazenes derived from poly(methylphenylphosphazene) and fluorinated aldehydes and ketones. *Macromolecules*, 1995, 28:8657–8661.

24. Wisian-Neilson, P. Poly(alkyl/arylphosphazenes) and their derivatives. In De Jaeger, R., Gleria, M., eds., *Phosphazenes: A Worldwide Insight*, Nova Science, Hauppauge, NY, 2002, Chap. 5.

25. Flindt, E.P., Rose, H. Trivalente–pentavalente Phosphorverbindungen/Phospha-zene: I. Eine neue Methode zur Darstellung von Poly[bis(trifluoräthoxy)pho-sphazenen]. *Z. Anorg. Allgem. Chem.*, 1977, 428:204–208.

26. Montague, R.A., Matyjaszewski, K. Synthesis of poly[bis(trifluoroethoxy)pho-sphazene] under mild conditions using a fluoride initiator. *J. Am. Chem. Soc.*, 1990, 112:6721.

27. Matyjaszewski, K., Moore, M.M., White, M.L. Synthesis of polyphosphazene block copolymers bearing alkoxyethoxy and trifluoroethoxy groups. *Macromole-cules*, 1993, 26:6741.

28. De Jaeger, R., Helioui, M., Puscaric, E. Novel polychlorophosphazenes and process for their preparation, U.S. patent 4,377,558, 1983.

29. De Jaeger, R., Potin, P. In Gleria, M., De Jaeger, R., eds., *Synthesis and Characterization of Poly(organophosphazenes)*. Nova Science, Hauppauge, NY, 2004, pp. 25–48.

30. Peterson, E.S., Luther, T.A., Harrup, M.K., et al. The one-pot synthesis of linear polyphosphazenes. *J. Inorg. Organomet. Polym. Mater.*, 2007, 17:361–366.

31. Honeyman, C.H., Manners, I., Morrissey, C.T., Allcock, H.R. Ambient temperature synthesis of poly(dichlorophosphazene) with molecular weight control. *J. Am. Chem. Soc.*, 1995, 117:7035–7036.

32. Allcock, H.R., Crane, C.A., Morrissey, C.T., Nelson, J.M., Reeves, S.D., Honeyman, C.H., Manners, I. "Living" cationic polymerization of phosphoranimines as an ambient temperature route to polyphosphazenes with controlled molecular weights. *Macromolecules*, 1996, 29:7740–7747.

33. Allcock, H.R., Nelson, J.M., Reeves, S.D., Honeyman, C.H., Manners, I. Ambient temperature direct synthesis of poly(organophosphazenes) via the "living" cationic polymerization of organo-substituted phosphoranimines. *Macromolecules*, 1997, 30:50–56.

34. Allcock, H.R., Reeves, S.D., Nelson, J.M., Crane, C.A., Manners, I. Polyphosphazene block copolymers via the "living" cationic, ambient temperature polymerization of phosphoranimines. *Macromolecules*, 1997, 30:2213–2215.

35. Allcock, H.R., Nelson, J.M., Prange, R., Crane, C.A., de Denus, C.R. Synthesis of telechelic polyphosphazenes via the ambient temperature living cationic polymerization of amino phosphoranimines. *Macromolecules*, 1999, 32:5736–5743.

36. Prange, R., Allcock, H.R. Telechelic syntheses of the first phosphazene siloxane block copolymers. *Macromolecules*, 1999, 32:6390–6392.

37. Allcock, H.R., Fitzpatrick, R.J. Sulfonation of aryloxy- and arylamino-phosphazenes: small-molecule compounds, polymers, and surfaces. *Chemi. Mater.*, 1991, 3:1120–1132.

38. Allcock, H.R., Smith, D.E., Kim, Y.B., Fitzgerald, J.J. Poly(organophosphazenes) containing allyl side groups: cross-linking by hydrosilylation. *Macromolecules*, 1994, 27:5206–5215.

39. Allcock, H.R., Hofmann, M.A., Ambler, C.M., Morford, R.V. Phenylphosphonic acid functionalized poly[aryloxyphosphazenes]. *Macromolecules*, 2002, 35:3483–3489.

40. Allcock, H.R., Austin, P.E. Schiff's-base coupling of cyclic and high polymeric phosphazenes to aldehydes and amines: chemotherapeutic models. *Macromolecules*, 1981, 14:1616–1622.

41. Allcock, H.R., Scopelianos, A.G. Synthesis of sugar-substituted cyclic and polymeric phosphazenes, and their oxidation, reduction, and acetylation reactions. *Macromolecules*, 1983, 16:715–719.

42. Allcock, H.R., Hymer, W.C., Austin, P.E. Diazo coupling of catecholamines with poly(organophosphazenes). *Macromolecules*, 1983, 16:1401–1406.

43. Singh, A., Steely, L., Allcock, H.R. Electrospinning of poly[bis(2,2,2-trifluoroethoxy)phosphazene] superhydrophobic nanofibers. *Langmuir*, 2005, 21:11604–11607.

44. Rose, S.H. Synthesis of phosphonitrilic fluoroelastomers. *J. Polym. Sci. B*, 1968, 6(12), 837–839.

45. Kolich, C.H., Klobucar, W.D., Books, J.T. Process for surface treating phosphonitrilic fluoroelastomers. U.S. patent 4,945,139, 1990.

46. Gettleman, L., Farris, C.L., Rawls, R.H., Lebouef, R.J. Soft and firm denture liner for a composite denture and method for fabricating. U.S. patent 45,43,379, 1985.

47. Steely, L.B., Li, Q., Badding, J.V., Allcock, H.R. Foam formation from fluorinated polyphosphazenes by liquid CO_2 processing. *Polym. Sci. Eng.*, 2008, 48:683–686.

48. Allcock, H.R., Fuller, T.J., Mack, D.P., Matsumura, K., Smeltz, K.M. Synthesis of Poly[(amino acid alkyl ester)phosphazenes]. *Macromolecules*, 1977, 10:824–830.

49. Lemmouchi, Y., Schacht, E., Dejardin, S. Biodegradable poly[(amino acid ester) phosphazenes] for biomedical applications. *J. Bioact. Compat. Polym.*, 1998, 13:4–18.

50. Veronese, F.M., Marsilio, F., Lora, S., Caliceti, P., Passi, P., Orsolini, P. Polyphosphazene membranes and microspheres in periodontal diseases and implant surgery. *Biomaterials*, 1999, 20(1), 91–98.

51. Allcock, H.R., Singh, A., Ambrosio, A.M.A., Laredo, W.R. Tyrosine-bearing polyphosphazenes. *Biomacromolecules*, 2003, 4:1646–1653.

52. Allcock, H.R., Pucher, S.R., Scopelianos, A.G. Poly[(amino acid ester)phosphazenes]: synthesis, crystallinity, and hydrolytic sensitivity in solution and the solid state. *Macromolecules*, 1994, 27:1071–1075.

53. Allcock, H.R., Pucher, S.R., Scopelianos, A.G. Poly[amino acid ester)phosphazenes] as substrates for the controlled release of small molecules. *Biomaterials*, 1994, 15:563–569.

54. Lakshmi, S., Lee, D., Bender, J.D., Barrett, E.W., Greish, Y.E., Brown, P.W., Allcock, H.R., Laurencin, C.T. Synthesis, characterization and in vitro osteocompatibility evaluation of novel alanine-based polyphosphazenes. *J. Biomater. Res.*, 2006, 76a:206–213.

55. Deng, M., Nair, L.S., Nukavarapu, S.P., Kumbar, S.G., Jiang, T., Krogman, N.R., Singh, A., Allcock, H.R., Laurencin, C.T. Miscibility and in vitro osteocompatibility of biodegradable blends of poly[(ethyl alanato)(*p*-phenyl phenoxy)-phosphazene] and poly(lactic acid–glycolic acid). *Biomaterials*, 2007, 29(3), 337–349.

56. El-Amin, S.F., Kwon, M.S., Starnes, T., Allcock, H.R., Laurencin, C.T. The biocompatibility of biodegradable glycine containing polyphosphazenes: a comparative study in bone. *J. Inorg. Organomet. Polym. Mater.*, 2006, 16(4), 387–396.

57. Greish, Y.E., Brown, P.W., Bender, J.D., Allcock, H.R., Lakshmi, S., Laurencin, C.T. Hydroxyapatite–polyphosphazane composites prepared at low temperatures. *J. Am. Ceram. Soc.*, 2007, 90(9), 2728–2734.

58. Krogman, N.R., Steely, L., Hindenlang, M.D., Nair, L.S., Laurencin, C.T., Allcock, H.R. Synthesis and characterization of polyphosphazene-block-polyester and polyphosphazene-block-polycarbonates. *Macromolecules*, 2008, 41:1126–1130.

59. Bhattacharyya, S., Nair, L.S., Singh, A., Krogman, N.R., Greish, Y.E., Brown, P.W., Allcock, H.R., Laurencin, C.T. Electrospinning of poly[bis(ethyl alanato) phosphazene] nanofibers. *J. Biomed. Nanotechnol.*, 2006, 2(1), 36–45.

60. Bhattacharyya, S., Nair, L.S., Singh, A., Krogman, N.R., Bender, J., Greish, Y.E., Brown, P.W., Allcock, H.R., Laurencin, C.T. Development of biodegradable polyphosphazene–nanohydroxyapatite composite nanofibers via electrospinning.

Mater. Res. Soci. Symp. Proc., 2005, 845(Nanoscale Materials Science in Biology and Medicine): 91–96.

61. Krogman, N.R., Stone, D., Allcock, H.R. Bioerodible shape memory polymers derived from poly(amino acid ester polyphosphazenes). Unpublished work.

62. Allcock, H.R., Pucher, S.R., Scopelianos, A.G. Synthesis of poly(organophosphazenes) with glycolic acid ester and lactic acid ester side groups: prototypes for new bioerodible polymers. *Macromolecules*, 1994, 1:1–4.

63. Allcock, H.R., Fuller, T.J. The synthesis and hydrolysis of hexa(imidazolyl)cyclotriphosphazenes. *J. Am. Chem. Soc.*, 1981, 103:2250–2256.

64. Laurencin, C., Koh, H.J., Neenan, T.X., Allcock, H.R., Langer, R.S. Controlled release using a new bioerodible polyphosphazene matrix system. *J. Biomed. Mater. Res.*, 1987, 21:1231–1246.

65. Blonsky, P.M., Shriver, D.F., Austin, P.E., Allcock, H.R. Polyphosphazene Solid electrolytes. *J. Am. Chem. Soc.*, 1984, 106:6854–6855.

66. Allcock, H.R., Gebura, M., Kwon, S., Neenan, T.X. Amphiphilic polyphosphazenes as membrane materials: influence of side group on radiation crosslinking, semipermeability, and surface morphology. *Biomaterials*, 1988, 19:500–508.

67. Bennett, J.L., Dembek, A.A., Allcock, H.R., Heyen, B.J., Shriver, D.F. Radiation crosslinking of poly[bis(2-(2-methoxyethoxy)ethoxy)]phosphazene: effect on solid state ionic conductivity. *Chem. Mater.*, 1989, 1:14–16.

68. Nelson, C.J., Coggio, W.D., Allcock, H.R. Ultraviolet radiation–induced crosslinking of poly[bis(2-(2-methoxyethoxy)ethoxy)phosphazene]. *Chem. Mater.*, 1991, 3:786–787.

69. Allcock, H.R., Kwon, S. An ionically-crosslinkable polyphosphazene: poly[di(carboxylatophenoxy-phosphazene] and its hydrogels and membranes. *Macromolecules*, 1989, 22:75–79.

70. Bano, M.C., Cohen, S., Visscher, K.B., Allcock, H.R., Langer, R. A novel synthetic method for hybridoma cell encapsulation. *Biotechnology*, 1991, 9:468–471.

71. Cohen, S., Bano, M.C., Cima, L.G., Allcock, H.R., Vacanti, J.P., Vacanti, C.A., Langer, R. Design of synthetic polymeric structures for cell transplantation and tissue engineering. *Clin. Mater.*, 1993, 13:3–10.

72. Andrianov, A., Cohen, S., Langer, R., Visscher, K.B., Allcock, H.R. Controlled release using ionotropic polyphosphazene hydrogels. *J. Control. Release*, 1993, 27:69–77.

73. Allcock, H.R. Water-soluble polymers and their hydrogels. In *Hydrophilic Polymers*, Glass, J.E., ed., *ACS Symp. Ser.*, 1996, 248:3–29.

74. Andrianov, A.K., Payne, L.G., Visscher, K.B., Allcock, H.R., Langer, R. Hydrolytic degradation of ionically cross-linked polyphosphazene microspheres. *J. Appl. Polym. Sci.*, 1994, 53:1573–1578.

75. Ten Huisen, K.S., Brown, P.W., Reed, C.S., Allcock, H.R. Low temperature synthesis of a self-assembling composite: hydroxyapatite poly[bis(sodium carboxylatophenoxy)phosphazene]. *J. Mater. Sci. Mater. Med.*, 1996, 7:673–682.

76. Allcock, H.R., Ambrosio, A.M.A. Synthesis and characterization of pH-senstitive poly(organophosphazene) hydrogels. *Biomaterials*, 1996, 17:2295–2302.

77. Allcock, H.R., Ambrosio, A.M.A. Synthesis and characterization of pH responsive poly(organophosphazene) hydrogels. In *Polymer Gels*, Bohidar, H.B., Dubin, P.L., eds. *ACS Symp. Ser.*, 2002, 833:82–101.

78. Barrett, E.W., Phelps, M.V.B., Silva, R.J., Gaumond, R.P., Allcock, H.R. Patterning poly(organophosphazenes) for selective cell adhesion applications. *Biomacromolecules*, 2005, 6:1689–1698.

79. Fei, S.-T., Phelps, M.V.B., Wang, Y., Barrett, E., Gandhi, F., Allcock, H.R. A redox responsive gel based on ionic crosslinking. *Soft Matter*, 2006, 2:397–401.

80. Allcock, H.R., Moore, G.Y. Synthesis of poly(organophosphazene) copolymers and crosslinked polymers by ligand exchange. *Macromolecules*, 1972, 5:231–232.

81. Allcock, H.R., Rutt, J.S., Fitzpatrick, R.J. A surface reaction of poly[bis(trifluoroethoxy)phosphazene] films by basic hydrolysis. *Chem. Mater.*, 1991, 3:442–449.

82. Allcock, H.R., Fitzpatrick, R.J. Functionalization of the surface of poly[bis(trifluoroethoxy)phosphazene] by reactions with alkoxide nucleophiles. *Chem. Mater.*, 1991, 3:450–454.

83. Allcock, H.R., Fitzpatrick, R.J. Sulfonation of aryloxy- and arylaminophosphazenes: small-molecule compounds, polymers, and surfaces. *Chem. Mater.*, 1991, 3:1120–1132.

84. Allcock, H.R., Fitzpatrick, R.J., Salvati, L. Oxidation of poly[di(4-methylphenoxy)phosphazene] surfaces, and chemistry of the surface carboxylic acid groups. *Chem. Mater.*, 1992, 4:769–775.

85. Allcock, H.R., Smith, D.E. Surface studies of poly(organophosphazenes) containing dimethylsiloxane grafts. *Chem. Mater.*, 1995, 7:1469–1474.

86. Allcock, H.R., Morrissey, C.T., Way, W.K., Winograd, N. Controlled formation of carboxylic acid groups at polyphosphazene surfaces: oxidative and hydrolytic routes. *Chem. Mater.*, 1996, 8:2730–2738.

87. Neenan, T.X., Allcock, H.R. Synthesis of a heparinized poly(organophosphazene). *Biomaterials*, 1982, 3(2), 78–80.

88. Allcock, H.R., Hymer, W.C., Austin, P.E. Diazo coupling of catecholamines with poly(organophosphazenes). *Macromolecules*, 1983, 16:1401–1406.

89. Allcock, H.R., Kwon, S. Covalent linkage of proteins to surface-modified poly (organophosphazenes): immobilization of glucose-6-phosphate dehydrogenase and trypsin. *Macromolecules*, 1986, 19:1502–1508.

90. Allcock, H.R., Weiss, P., Draughn, R.L. Immobilization of β-cyclodextrin onto polymer surfaces. M.S. thesis, The Pennsylvania State University, 1998.

91. Allcock, H.R., Fitzpatrick, R.J., Visscher, K. Thin layer grafts of poly[bis(methoxyethoxyethoxy)phosphazene] on organic polymer surfaces. *Chem. Mater.*, 1992, 4:775–780.

92. Allcock, H.R., Pucher, S.R., Visscher, K.B. The activity of urea amidohydrolase immobilized within poly[di(methoxyethoxyethoxy)phosphazene] hydrogels. *Biomaterials*, 1994, 15:502–506.

93. Barrett, E.W., Phelps, M.V.B., Silva, R.J., Gaumond, R.P., Allcock, H.R. Patterning poly(organophosphazenes) for selective cell adhesion applications. *Biomacromolecules*, 2005, 6:1689–1698.

94. Allcock, H.R., Phelps, M.V.B., Barrett, E.W., Pishko, M.V., Koh, W.-G. Ultraviolet photolithographic development for polyphosphazene hydrogel microstructures for potential use in microarray biosensors. *Chem. Mater.*, 2006, 18:609–613.

95. Allcock, H.R., Steely, L.B., Kim, S.H., Kim, J.H., Kang, B.-K. Plasma surface functionalization of poly[bis(trifluoroethoxy)phosphazene films and nanofibers. *Langmuir*, 2007, 23(15), 8103–8107.

96. Allcock, H.R., Fuller, T.J. Phosphazene high polymers with steroidal side groups. *Macromolecules*, 1980, 13:1338–1345.

97. Allcock, H.R., Allen, R.W., O'Brien, J.P. Synthesis of platinum derivatives of polymeric and cyclic phosphazenes. *J. Am. Chem. Soc.*, 1977, 99:3984–3987.

98. Allcock, H.R., Austin, P.E. Schiff's-base coupling of cyclic and high polymeric phosphazenes to aldehydes and amines: chemotherapeutic models. *Macromolecules*, 1981, 14:1616–1622.

99. Allcock, H.R., Austin, P.E., Neenan, T.X. Phosphazene high polymers with bioactive substituent groups: prospective anesthetic aminophosphazenes. *Macromolecules*, 1982, 15:689–693.

100. Allcock, H.R., Greigger, P.P., Gardner, J.E., Schmutz, J.L. Water-soluble polyphosphazenes as carrier molecules for iron(III) and iron(II) porphyrins. *J. Am. Chem. Soc.*, 1979, 101:606–611.

101. Allcock, H.R., Neenan, T.X., Boso, B. Synthesis, oxygen-binding behavior, and Mossbauer spectroscopy of covalently bound phosphazene–heme complexes. *Inorg. Chem.*, 1985, 24:2656–2662.

102. Cohen, S., Bano, M.C., Visscher, K.B., Chow, M., Allcock, H.R., Langer, R. An ionically crosslinkable polyphosphazene: a novel polymer for microencapsulation. *J. Am. Chem. Soc.*, 1990, 112:7832–7833.

103. Bano, M.C., Cohen, S., Visscher, K.B., Allcock, H.R., Langer, R. A novel synthetic method for hybridoma cell encapsulation. *Biotechnology*, 1991, 9:468–471.

104. Cohen, S., Bano, M.C., Cima, L.G., Allcock, H.R., Vacanti, J.P., Vacanti, C.A., Langer, R. Design of synthetic polymeric structures for cell transplantation and tissue engineering. *Clin. Mater.*, 1993, 13:3–10.

105. Andrianov, A.K., Payne, L.G., Visscher, K.B., Allcock, H.R., Langer, R. Hydrolytic degradation of ionically cross-linked polyphosphazene microspheres. *J. Appl. Polym. Sci.*, 1994, 53:1573–1578.

106. Kim, C., Chang, Y., Lee, S.C., Allcock, H.R., Reeves, S.D. An amphiphilic diblock copolyphosphazene: synthesis and micellar characteristics in the aqueous phase. *Polym. Prepr. ACS Div. Polym. Chem.*, 2000, 41:609–610.

107. Chang, Y., Bender, J.D., Phelps, M.V.B., Allcock, H.R. Synthesis and self-association behavior of biodegradable amphiphilic poly[bis(ethyl glycinat-*N*-yl)-phosphazene]–PEO block copolymers. *Biomacromolecules*, 2002, 3:1364–1369.

108. Chang, Y., Powell, E.S., Allcock, H.R., Park, S.M., Kim, C. Thermosensitive behavior of poly[bis(methoxyethoxyethoxy)phosphazene]–poly(ethylene oxide) block copolymers. *Macromolecules*, 2003, 36:2568–2570.

109. Chang, Y., Prange, R., Allcock, H.R., Lee, S.C., Kim, C. Amphiphilic poly-[bis(trifluoroethoxy)phosphazene]-poly(ethylene oxide) block copolymers: synthesis and micellar characteristics. *Macromolecules*, 2002, 35:8556–8559.

110. Allcock, H.R., Powell, E.S., Chang, Y., Kim, C. Synthesis and micellar behavior of amphiphilic polystyrene–poly[bis(methoxyethoxyethoxy)phosphazene] block copolymers. *Macromolecules*, 2004, 37:7163–7167.

111. Allcock, H.R., Cho, S.Y., Steely, L.B. New amphiphilic Poly[bis(2,2,2-trifluoroethoxy)phosphazene]/poly(propylene-glycol) triblock copolymers: synthesis and micellar characteristics. *Macromolecules*, 2006, 39:8334–8338.

112. Cho, S.-Y., Allcock, H.R. Dendrimers derived from polyphosphazene–poly(propyleneimine) systems: encapsulation and triggered release of hydrophobic guest molecules. *Macromolecules*, 2007, 40:3115–3121.

PART II
Vaccine Delivery and Immunomodulation

3 Polyphosphazene Vaccine Delivery Vehicles: State of Development and Perspectives

ALEXANDER K. ANDRIANOV

Apogee Technology, Inc., Norwood, Massachusetts

INTRODUCTION

The search for potent, well-characterized, and safe vaccine adjuvants and delivery vehicles has been widely recognized as a key strategic factor in the development of new and improved vaccines [1]. In this regard, an emerging class of well-defined macromolecules, based on a polyphosphazene backbone (Scheme 1) offers a number of important advantages from both the immunostimulation and delivery standpoints. Impressive immunopotentiation activity and dose-sparing effects reported for these water-soluble molecules [2–11] are augmented with the ease of their assembly into supramolecular microparticulate structures to achieve optimal delivery performance [12–17]. The synthetic origin of polyphosphazene adjuvants and their well-characterized molecular structures assure a high level of reproducibility and ease of quality control [18–20]. Adequate stability, the "mix and fix" aqueous formulation approach, which does not involve covalent conjugation with antigen [3], long-lasting immune responses, and a good safety profile, which includes the results of clinical trials in humans, are among other advantages of this adjuvant system. The polyphosphazene backbone of these molecules allows their biodegradation, which can be modulated through the choice of the side group and results in the release of physiologically benign compounds [4,21–23]. The commercial development of the lead compound is sustained by the existence of a robust GMP (good manufacturing practice) process and the availability of a drug master file to support regulatory applications.

Polyphosphazenes for Biomedical Applications, Edited by Alexander K. Andrianov
Copyright © 2009 John Wiley & Sons, Inc.

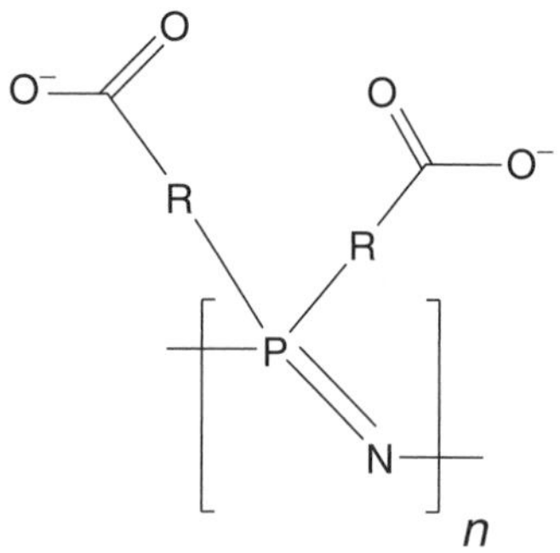

SCHEME 1   A representative structure of a polyphosphazene adjuvant.

The class of polyphosphazene immunoadjuvants appears to be broad, with molecular structures including phosphorus–nitrogen main chains, organic side groups (R), and ionic moieties such as carboxyl groups (Scheme 1). A highly flexible backbone, hydrophobic spacers, high molecular weight, and partially dissociated ionic groups of such structures bring about molecules capable of forming water-soluble complexes with many biological targets, including proteins, which are essential for their immunostimulating activity.

A substantial research effort has already been invested in the field, including work in multiple animal models and with various antigens: both synthetic and mechanistic studies [2–10]. It becomes evident that the polyphosphazene adjuvant technology evolves through the discovery of new, more potent derivatives [24–26], development of microparticulate delivery systems, and the investigation of alternative delivery routes, such as mucosal and intradermal. A number of chapters describing the most recent studies in the field are included in this book. This review is intended to introduce the reader to the most significant accomplishments in the field and to briefly summarize the existing knowledge base for this fascinating system.

IN VIVO ACTIVITY

Immunopotentiating and dose-sparing effects of polyphosphazenes have been documented in a number of publications. In a recent study conducted in mice with formulations containing various doses of influenza X:31 antigen or bovine serum albumin, polyphosphazene derivatives were shown to enhance antibody responses to levels that exceeded those induced by alum-adjuvanted formulations by up to 1000-fold [5]. It has also been reported that a 25-fold reduction in the dose of X:31 antigen had no quantitative or qualitative effect on the antibody responses for polyphosphazene-adjuvanted formulations, indicating a powerful dose-sparing effect [5].

Earlier, the first polyphosphazene adjuvant, poly[di(carboxylatophenoxy)-phosphazene] (PCPP), was shown to be effective with commercial influenza

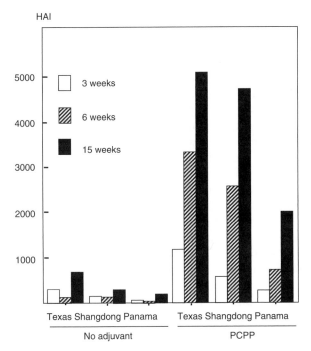

FIGURE 1 HAI immune response kinetics after subcutaneous immunization of BALB/c mice with multivalent influenza vaccine formulated with and without PCPP [5 mice per group; trivalent influenza vaccine: A/Texas/36/91 (H1N1), A Shangdong/9/93 (H3N2), B/Panama/45/90—15 μg HA for each strain; 500 μL injection volume; 100 μg of PCPP; significance testing at week 15: Texas $P = 0.0001$, Shangdong $P = 0.0000001$, Panama $P = 0.00012$]. (From ref. 7, with permission.)

vaccine, enhancing the immune response to all three influenza HA strains in mice [7], including the induction of functional antibodies as assayed by hemagglutination inhibition (HAI), which were approximately 10-fold higher than the levels detected for the vaccine alone (Fig. 1).

A number of researchers emphasized the fast onset of the immune response and long-lasting adjuvant effect of polyphosphazenes. The 10-fold increase in antibody titers for PCPP-formulated X:31 influenza antigen was observed early on, at week 3 following immunization. Similarly, the fast kinetic profile was noted for a weakly immunogenic synthetic peptide (toxin co-regulated pilin TcpA4) formulated with polyphosphazene. In fact, PCPP-formulated peptide was the only formulation capable of inducing significantly higher titers compared to peptide alone, after just one immunization [27]. Sustained levels of antibody titers were also observed with a number of antigens. For example, no decrease in antibody titers for the length of the experiment was reported for X:31 influenza antigen (21 weeks) [7] or for hepatitis B virus surface antigen (HbsAg) (41 weeks) [9].

A potential synergy between PCPP and other adjuvants has also been investigated. In a recent study, a combination of polyphosphazene and CpG oligodeoxynucleotide was studied for its ability to adjuvant intranasally delivered, formalin-inactivated bovine RSV vaccine [28]. The formulation, containing two adjuvants, induced significantly higher levels of BRSV-specific IgG titers in the serum than did vaccine formulations containing individual adjuvants. It was also the only formulation that resulted in a significant reduction in viral replication upon a BRSV challenge. Intranasal immunization of mice with BRSV formulated with CpG oligodeoxynucleotide and polyphosphazene resulted in both humoral and cell-mediated immunity and the development of mucosal immune responses. Indeed, the increases in serum and mucosal IgG, in particular mucosal IgA and virus-neutralizing antibodies, were the most critical differences observed between antigen formulated with both CpG and polyphosphazene compared to individual formulations. This synergistic effect can be an important factor in expanding polyphosphazene utility in vaccine applications.

There have been a number of other studies in addition to those already discussed in which polyphosphazenes were compared to other adjuvants. PCPP has been able to increase antibody titers approximately 20-fold for plasma-derived HBsAg compared to the same antigen formulated on aluminum phosphate, and seven fold for a recombinant HBsAg compared to the aluminum hydroxide–formulated antigen [9]. A comparison of the adjuvant activity of PCPP and aluminum phosphate was also carried out using capsular polysaccharide and polyribosylribitolphosphate (PRP) from *Haemophilus influenzae* type B (Hib) conjugated to tetanus toxoid (Hib-T). Peak titers for PCPP formulations were achieved at week 7 when they were tenfold higher than those achieved with aluminum phosphate [9].

PCPP was compared to QS-21, QS-7, Quil A, and RAS in intramuscular immunization of mice with inactivated rotavirus particles: purified, ultraviolet/psoralen-inactivated murine rotavirus (EDIM) [29]. PCPP stimulated significantly higher titers than other adjuvants for all rotavirus antibodies measured except stool IgA. Twenty-eight days after immunization, BALB/c mice were orally challenged with live EDIM, and virus shedding was measured. The order of rotavirus IgG responses was the following: no adjuvant < RAS < QS-7 < Quil A < QS-21 < PCPP, and it was the same as the order of protection, with the exception of QS-21 and PCPP, which were reversed [29].

PCPP was also compared with another polymeric adjuvant, CRL-1005, a block copolymer of ethylene oxide–propylene oxide, for its ability to adjuvant synthetic peptides: toxin co-regulated pilin TcpA4 and TcpA6, which can present interest for the development of a subunit cholera vaccine [27]. Interestingly, for peptide–PCPP formulations, the protection levels were at 100% against 1 LD_{50} and 75% against 10 LD_{50}, whereas for CRL-1005 the values were 75% and 33%, respectively [27]. Also worth mentioning is a small study of PCPP in primates in which vaccination of two rhesus monkeys with whole inactivated HIV-1 adjuvanted with PCPP protected the animals from becoming infected during a SHIV (simian/human immunodeficiency virus, chimeric virus) challenge [8].

Most studies of polyphosphazene adjuvants were conducted with subcutaneous or intramuscular immunizations; however, some studies indicate that they are also effective when delivered intranasally, with the typical dose range in these experiments usually varying between 25 and 100 μg per mouse [3,5,7,28]. Overall, it has been estimated that polyphosphazenes were tested with 10 bacterial and 13 viral antigens in 11 animal models [11]. PCPP proved to be a potent adjuvant for multiple vaccines and antigens, such as trivalent influenza virus vaccine, hepatitis B surface antigen, herpes simplex virus glycoprotein gD2, tetanus toxoid, polyribosylribitolphosphate from *H. influenzae* type B, toxin co-regulated pilin TcpA4 and TcpA6 synthetic peptides, inactivated rotavirus particles, and formalin-inactivated HIV-1 LAI virus [3,17–25].

MECHANISM OF ACTION AND FORMULATION DEVELOPMENT

The mechanism of action of polyphosphazene immunoadjuvants is still largely under discussion; however, a considerable knowledge base has already been built, which offers critical guidelines for the development of formulations with predictable immunological behavior. This includes both immunological and physicochemical aspects.

In the early stages of polyphosphazene development, it became apparent that polyphosphazenes did not act as a depot, as the excision of the injection site had no detectable effect on the kinetics of antibody induction [7]. Soon thereafter, formation of a water-soluble complex between polyphosphazene and the antigen through noncovalent interactions was discovered (Fig. 2), and the correlation between some complexes' physicochemical characteristics and immunological behavior in vivo was established [3]. In fact, a clear relationship was found between the immunopotentiating activity and the content of "interactive" carboxylic acid groups in the polymer [20].

Regardless of what the detailed immunological mechanism is, it is now clear, based on empirical findings, that the molecular size of polyphosphazene (linked to complex stability) [7], the degree of complex compaction (linked to antigen presentation) [3], and "antigen loading" can have substantial effects on the induction of antibody titers. Further, based on existing knowledge of water-soluble polyelectrolyte complexes [30], it has been proposed that antigen–polyphosphazene complexes can adsorb on cell surfaces, resulting in clustering of membrane proteins, stimulation of intracellular ionic fluxes, and thus eventually, enhancement of the immune response [11,30]. Recent reviews on polymer genomics [31,32], uncovering the role of polymers in the induction of specific genetically controlled responses to antigens and other agents, stimulate further investigations in this direction, with a focus on cooperative interactions of polymers with plasma cell membranes and trafficking of polymers to

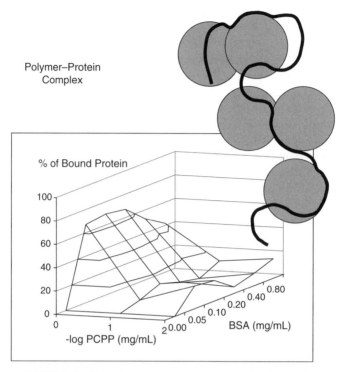

FIGURE 2   Polyphosphazene–antigen complex formation.

intracellular organelles [31,32]. Finally, since polyphosphazene adjuvants are capable of binding multiple antigen molecules (up to several hundreds in certain cases) [3], multimeric antigen presentation [33] can also be an important part of the mechanism, especially for high-antigen-loading formulations.

In summarizing the physicochemical aspects of formulation development, it is important to emphasize that other "nonpolyphosphazene" polyelectrolytes, such as polymethylacrylic and polyacrylic acids, are also known to display immuno-potentiating activity [30]. However, it appears that their activity achieves desirable levels only if the polymer is covalently linked to the antigen [30]. It is clear that contrary to such systems, no covalent attachment of polyphosphazenes to the antigen is required to display an immune response of practical significance [3,9]. This is clearly due to the ability of such polyphosphazenes to form noncovalent complexes with antigens, and thus characterization and control of such complexes remains the main focus of formulation development.

Another critical aspect of the immunopotentiating activity of the adjuvant is its ability to modulate the quality of the immune response, since the immunity to different infectious agents may require distinct types of immune responses. In general, intracellular pathogens tend to involve cellular (Th1) responses, whereas resistance to extracellular pathogens is often associated with humoral

(Th2) responses. From very early studies, the adjuvant effect of a lead polyphosphazene compound, PCPP, has typically been linked predominantly with a Th2 type of response [7]. However, studies on various polyphosphazene derivatives involving copolymers of PCPP [4] and a new homopolymer [2,5] demonstrated that potentially, certain polyphosphazenes can promote a mixed Th1/Th2 response, as suggested based on the ratio of IgG1 and IgG2a isotypes [4,5] and supported further by the detection of high-frequency X:31-specific IFN-γ-secreting cells in immunized mice [5]. Thus, it appears that the quality of the immune response can vary depending on the structure of the polyphosphazene derivative, suggesting the existence of a structure–activity relationship (SAR) for this class of compounds. There have been further discussions on the similarities in the immune responses induced by formulations containing recently synthesized polyphosphazene–poly[di(sodium carboxylatoethylphenoxy)phosphazene] (PCEP) and CpG oligonucleotide adjuvant, which typically achieves its adjuvant effects by activating innate immunity [5]. Authors speculate that innate immunity may be one of the mechanisms by which PCEP mediates its potent adjuvant activity [5,34].

The existence of SAR is potentially a very important finding, which is expected to promote the synthesis of new polyphosphazene derivatives and their immunological studies, establish a quantitative relationship, and potentially, generate new superactive adjuvants with the desired immunological profile. At the same time, it is worth mentioning that some of the structural variations between these derivatives are relatively minor, and it is remarkable that they can produce such significant mechanistic effects. This may be yet another argument in favor of more sophisticated control of supramolecular assemblies in the formulation, which can be at least partially responsible for the effects, since many of the important adjuvant systems are based on emulsions and polymeric microspheres [1]. This also opens new opportunities in terms of designing new vaccine delivery systems, since many of the polyphosphazenes noted in this chapter can easily be formulated into highly controlled microspheres and nanospheres.

PCPP IN CLINICAL TRIALS

Publications on the results of clinical trials involving PCPP remain scarce. A phase I clinical study using PCPP was carried out with influenza vaccine on both young and elderly adults (a total of 96 subjects) [35]. Three doses of PCPP were tested (100, 200, and 500 µg) and a nonadjuvated vaccine was used as a control. No serious adverse events related to the vaccine were reported. The best results were obtained for the A/Johannesburg/33/94 (H3N2) strain, with a 500-µg dose of PCPP found to be most efficient. For this dose, PCPP-adjuvanted vaccine produced a 14.7-fold increase in antibody titers (day 21 versus day 0 post-immunization) compared to a 3.1-fold increase for nonadjuvanted vaccine. The seroconversion rate was found to be 80% for the PCPP formulation as opposed to 51% for the nonadjuvanted vaccine.

PCPP was also examined in clinical trials with oligomeric HIV-1 Gp160mn/LAI-2 vaccine in HIV-seronegative U.S. volunteers [36,37]. Administration of 100 mg of Gp160 vaccine formulated in PCPP resulted in five out of five adults developing Western blot (WB) reactivity to HIV by day 193. In contrast, only three adults out of five who received the same dose of vaccine but formulated in alum showed WB positivity. Similar results were seen in binding antibody studies: Endpoint-dilution GMT (geometric mean titer) for PCPP-formulated vaccine was 25,874:1, whereas for alum-formulated vaccine it was 3986:1. The vaccine was well tolerated, with no serious adverse vaccine-related events. Investigators concluded that vaccine formulated in PCPP was safe and immunogenic [37]. A recent publication also cites the use of PCPP for boost immunization in HIV vaccine clinical trials [38].

HYDROLYTIC DEGRADABILITY OF PCPP

One of the most important advantages of polyphosphazenes over other synthetic polymers is the biodegradability of their backbone. In fact, polyphosphazenes typically degrade hydrolytically, releasing side groups and small amounts of ammonium and phosphate ions (products of a backbone breakdown). Thus, it is important to design polyphosphazene adjuvants having side groups with well characterized biological behavior. PCPP, the lead polyphosphazene adjuvant, is a good example of this approach. Propyl paraben, which is used as a main "building block" to construct a side group for PCPP, is a well-known food preservative and a GRAS (Generally Recognized As Safe) compound for parenteral administration, which has a long history of safe use, whereas hydroxybenzoic acid, which PCPP releases upon degradation, is a metabolite of this GRAS compound [39–41].

It has been established that PCPP undergoes slow hydrolytical degradation in aqueous solutions with a well-pronounced initial rapid degradation phase [20,22,23,42]. The degradation rate can be affected significantly by the presence of residual chlorine atoms and hydroxyl groups: irregularities produced if polyphosphazenes are substituted incompletely. Careful control of the residual groups—"structural defects" or "weak links" in polyphosphazene structures—can open a pathway to the tailoring of polymer degradation characteristics. Degradation profiles of PCPP can also be modulated effectively through the introduction of hydrolytically labile side groups in the polymer structure. It has been demonstrated that mixed-substituent copolymers of PCPP containing N-ethylpyrrolidone [22] or ethyl glycine side groups [23] degrade faster than PCPP, and the rate of hydrolysis is a function of the copolymer composition.

Extensive toxicology studies on PCPP, which were conducted and submitted to the U.S. Food and During Administration (FDA) as part of the Drug

Master File, and which constituted the basis for the transition of PCPP into clinical trials, have not yet been made available to the public.

ADVANCES IN SYNTHETIC CHEMISTRY OF POLYPHOSPHAZENE POLYELECTROLYTES

All polyphosphazenes that are currently known to display immunoadjuvant properties are water-soluble macromolecules containing carboxylic acid groups. Synthesis of such polymers is typically a multistep process which requires sophisticated methods and controls to allow for reproducible biologically important molecular characteristics.

In contrast to traditional polymer chemistry methods, preparation of polyphosphazenes does not depend on polymerization of structurally diverse monomers, but relies on chemical derivatization of the macromolecular precursor poly(dichlorophosphazene) (PDCP) [43]. This highly reactive inorganic intermediate (Scheme 2) is substantially a "naked" phosphorus–nitrogen backbone trimmed with chlorine atoms. Once the intermediate is prepared, the diverse arsenal of organic chemistry can be mobilized to replace chlorine atoms with desired organic side groups (Scheme 2). High reactivity of P–Cl bonds of PDCP results in an unprecedented structural diversity of polyphosphazenes, making it one of the largest classes of synthetic polymers, with almost unlimited opportunities in the synthesis of new molecules.

The advancement of polyphosphazene immunoadjuvants became possible due largely to significant developments that occurred in the area of their synthesis and characterization [2,18,19,44]. It has to be noted here that the very advantage of polyphosphazene chemistry, the reactivity of PDCP, also constitutes its biggest challenge. Hydrolytic reactions occurring with chlorine atoms of the polymer during the synthetic course result in undesirable processes of polymer degradation and cross-linking in the presence of even trace amounts of water [18]. Both of these processes can cause problems with reproducibility of the polymer's molecular weight and lead to a formation of unwanted by-products. Thus, command of hydrolytic reactions of chlorine atoms of PDCP is critical to the successful development of polymers for biological applications.

Development of stabilizing diglyme-containing solvent systems for PDCP had dramatic effects on both achieving consistent molecular characteristics and streamlining the process chemistry [18]. This addressed many of the issues listed above, eliminated the need for frequent PDCP synthesis, allowed production and storage of PDCP in large quantities in a "ready-to-use" form, and led to the development of high-throughput discovery methods [18]. The stabilization technique also resulted in a highly reliable direct analysis of PDCP and development of in-process controls using light-scattering and high-performance liquid chromatography (HPLC) methods [18,19]. Although the exact mechanism of the stabilizing effect is not clear, it was suggested that it is due to the coordination and "inhibition" of water and intermediates with diglyme in the

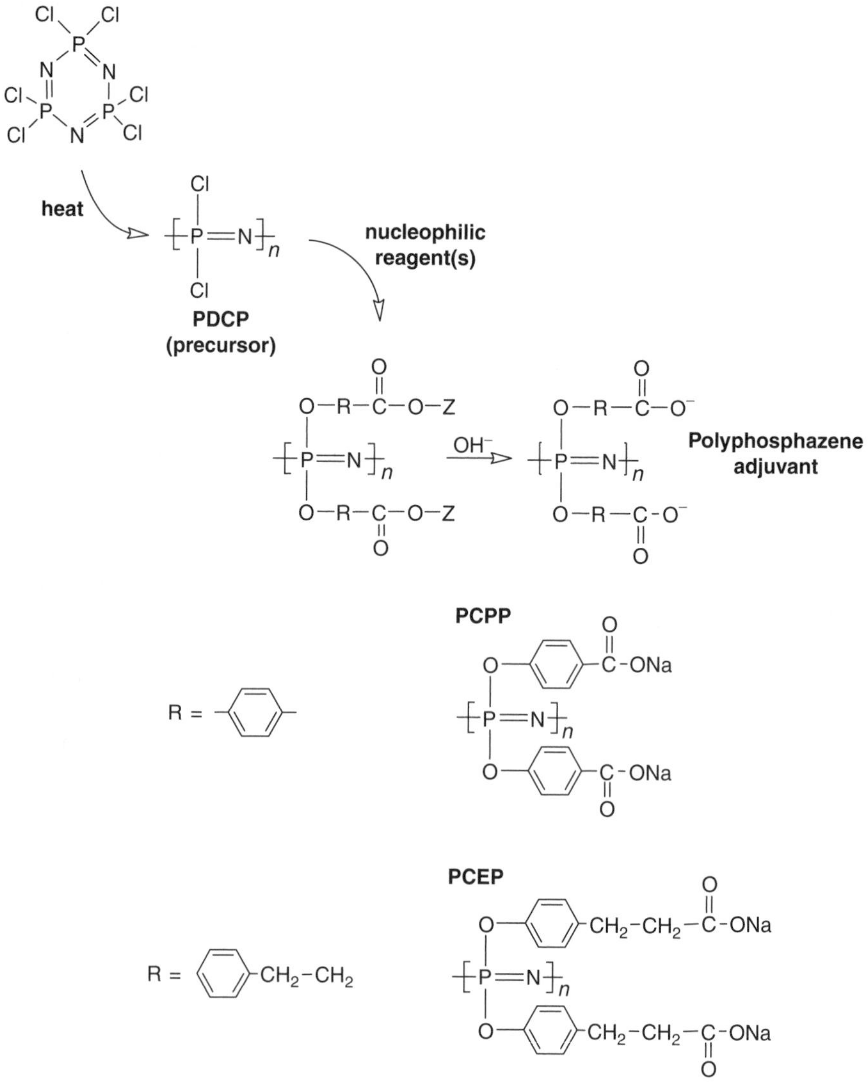

SCHEME 2   Synthesis of polyphosphazene adjuvants.

hydrolysis process [18]. This hypothesis can be supported by the well-known ability of oligo(ethylene oxides) to form complexes with water molecules.

New approaches, based on the use of noncovalent methods of protection, have also emerged in the synthesis of polyphosphazene polyacids [44]. Synthesis of polyphosphazenes containing sulfonic acids can now be performed in a single-step reaction via direct replacement of chlorine atoms in PDCP with a functionalized nucleophile. This method makes use of "noncovalent" protection of sulfonic acid functionality with the hydrophobic dimethyldipalmitylammonium ion, which can then be removed easily after completion of the reaction.

Finally, two new methods of modulating the molecular weight of poly-phosphazene polyelectrolytes have been suggested. The first involves "con-trolled degradation" of the polymer in the substitution reaction through altering the nucleophile/PDCP ratio [18]. Molecular-weight control can also be introduced in the polymerization reaction by making use of the relation-ship between the degree of polymerization and conversion [18].

PRODUCTION OF PCPP AND PROCESS CONTROLS

There are two main stages in the production of polyphosphazene polyelec-trolytes: polymerization and a downstream process of macromolecular sub-stitution, which includes the actual substitution reaction and a step for the removal of protective group to yield carboxylic acid functionality (Scheme 1) [18,20]. A highly reproducible process has been developed for the production of PCPP on a kilo scale, which was later adapted for cGMP (current good manufacturing practice) manufacture of this material for clinical studies.

The polymerization procedure developed for this process included a high-temperature melt polymerization of hexachlorocyclotriphosphazene in a titanium reactor under a nitrogen blanket with constant stirring [18]. The in-process control was established to monitor changes in the electrical current drawn by the stirring motor, which at a constant stirring rate correlated to the viscosity of the reaction mixture. Thus, a kinetic curve was produced and the reaction terminated at a desired viscosity level to achieve consistency in the degree of conversion and molecular weight. Polymerization is a common step for all polyphosphazenes that are of interest as vaccine adjuvants, thus, if a new polyphosphazene derivative goes into the development and production phase, there is no need to make changes in this step.

Synthesis of PCPP and other polyphosphazene polyacids involves the transition from a highly hydrolytically sensitive polymer to a polymer designed for an aqueous environment. In the case of PCPP, the process involves the substitution of chlorine atoms with propyl paraben and the hydrolysis of ester function on the polymer under alkaline conditions (deprotection reaction).

These two reactions were combined in one manufacturing step utilizing a single pot–single solvent approach [20]. The method makes use of diglyme as a single solvent, forced conditions of the substitution reaction, an aqueous-based deprotection reaction, short reaction times, and purification based on salt precipitation of PCPP [20], and eliminates the use of sodium or sodium hydride, commonly used in laboratory methods [45].

The need for synthetic controls in the preparation of polyphosphazene adjuvants should not be underestimated. Both molecular weight and potential irregularities in polymer structures can have profound effects on the biological performance of polyphosphazenes [7,20]. For example, it has been established that the potency of PCPP can be maximized through modulation of the molecular-weight characteristics, with the maximum activity achieved at approximately 1,000,000 g/mol.

When the process is poorly controlled, PCPP macromolecules containing typical "structural defects," propyl ester functionalities, and hydroxyl groups can be anticipated, due to variations in the reaction conditions [20]. Propyl ester–containing by-products, which can result from incomplete deprotection, are easy to detect even at low levels of content using structural analysis methods such as ^{1}H nuclear magnetic resonance (NMR) and fourier transform infrared (FT-IR). Although the reduction in the immune response resulting from the decrease in the content of acid groups in the polymer is not dramatic [20], caution should be exercised, since hydrophobically modified polymers in general can potentially display a trend toward aggregation. The presence of hydroxyl groups in the polymer, the result of an incomplete substitution reaction, can result in an accelerated degradation profile and potentially affect the shelf life of the product [20].

Although one cannot overemphasize the need for synthetic controls, the robustness of the PCPP production process is extremely reassuring. Reproducibility of PCPP synthesis was confirmed for more than 100 samples at a production scale ranging from 200 mg to 2 kg [11]. For PCPP prepared using the synthetic method described above, the weight-average molecular-weight fluctuations were not in excess of 10%. PCPP structure and purity was evaluated using ^{1}H, ^{31}P, ^{13}C NMR, FT-IR, elemental analysis, and multiangle laser light scattering coupled with size-exclusion chromatography, viscometry, Karl-Fisher titration, atomic absorption spectrometry, and inductively coupled plasma/mass spectrometry.

DISCOVERY OF NEW POLYPHOSPHAZENE ADJUVANTS

Polyphosphazenes are one of the most structurally versatile classes of synthetic polymers; new macromolecules can easily be synthesized to identify more potent adjuvants and to establish structure–activity relationships. The diversity of new molecules can be expanded further through the synthesis of mixed-substituent copolymers, containing two or more side groups.

In early studies, mixed-substituent copolymers of PCPP were synthesized containing various amounts of hydrophilic methoxyethoxyethoxy side groups [4]. In vivo studies of these polymers in mice using X:31 influenza antigen showed that some serum IgG titers were almost 10 times, and HAI titers approximately five times, higher than those elicited by PCPP formulation. The activity of newly synthesized copolymers was dependent on polymer composition.

Advancements in high-throughput methods in polyphosphazene synthesis provided a powerful impulse to the discovery of new polyphosphazene adjuvants, and recently, a library of approximately 40 members was built [25].

IgG Response (GMT) X10^{-6}

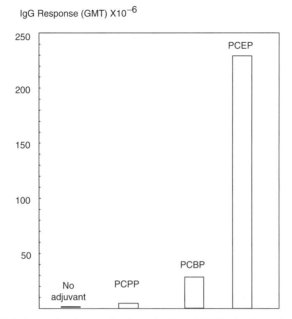

FIGURE 3   IgG titers after immunization of mice with HBsAg formulated with PCPP, HBsAg formulated with new polymer derivatives, and HBsAg alone as a control (five BALB/c mice per group; HBsAg: 1 μg/mouse; polymers: 50 μg/mouse; single-dose intramuscular injection; 16-week data).

Screening some of these compounds in mice using HBsAg demonstrated that at least two of them dramatically surpass the activity of PCPP with this particular antigen (Fig. 3). One of these polymers is a previously mentioned PCEP (Scheme 1), and the other, PCBP, a compound structurally similar to PCPP but with an oxymethylene spacer between the phenoxy group and the backbone. Even more important, as already discussed, some of these polymers, such as PCEP, also seem to be capable of changing the isotope profile and modulating the immune response. Thus, building a SAR database using existing and newly synthesized compounds can potentially result in more powerful adjuvants, provide means for modulating the immune response, and assist in elucidating the detailed mechanisms of action.

CONCLUSIONS

Polyphosphazene polyelectrolytes are uniquely positioned for the development of new vaccine adjuvants. They combine biologically interacting functional groups, precise control of molecular structure, unprecedented structural diversity, high-throughput discovery methods, new derivatives, and a biodegradable backbone.

It is clear that one of the biggest advantages of polyphosphazene polyelectrolytes over other synthetic polyelectrolytes is that they do not require covalent conjugation with the antigen. This is believed to be due to the unique flexibility of the phosphorus–nitrogen backbone and the high density of the functional groups, which allows them to form noncovalent, water-soluble complexes with antigens. In such complexes polyphosphazenes play antigen-transporting and antigen-binding roles and can carry, stabilize, and protect a significant number of ligands. This generates an ideal "interactive" polymer system, which allows tailoring of macromolecular complexes with the desired characteristics, such as protein loading and stability.

In contrast with many other adjuvant and vaccine delivery systems, such as alum, emulsions, or microspheres, and regardless of their mechanism of action, polyphosphazenes are representatives of molecular adjuvants, which inherently means a greater level of control and reproducibility. In fact, development of precise control over polymer characteristics and the production process was critical in establishing the technology. Although the fundamental molecular characteristics of polyphosphazenes can affect their biological performance, they are now sufficiently controlled to allow for consistent and reproducible behavior of polymer–antigen formulations.

The potential of polyphosphazenes is not limited to water-soluble formulations. They can be assembled in supramolecular and microspheric formulations to combine the modalities of the delivery system and biologically interacting molecules. Since polyphosphazenes are synthetic polymers and have excellent film-forming properties, they are also inherently suitable for the formation of solid vaccine–containing coatings for microneedles. This allows new opportunities for intradermal immunization, since many other adjuvant systems, such as emulsions, cannot be used in solid-state formulations. Finally, new derivatives can be expected not only to outperform PCPP in immunopotentiation activity, but also to open new possibilities in modulating the quality of the immune response. These aspects are covered in more detail in subsequent chapters.

REFERENCES

1. Singh, M., ed. *Vaccine Adjuvants and Delivery Systems.* Wiley-Interscience, Hoboken, NJ, 2006, p. 449.

2. Andrianov, A.K., Marin, A., Chen, J. Synthesis, properties, and biological activity of poly[di(sodium carboxylatoethylphenoxy)phosphazene]. *Biomacromolecules,* 2006, 7(1):394–399.

3. Andrianov, A.K., Marin, A., Roberts, B.E. Polyphosphazene polyelectrolytes: a link between the formation of noncovalent complexes with antigenic proteins and immunostimulating activity. *Biomacromolecules,* 2005, 6(3):1375–1379.

4. Andrianov, A.K., Sargent, J.R., Sule, S.S., Le Golvan, M.P., Woods, A.L., Jenkins, S.A., Payne, L.G. Synthesis, physico-chemical properties and immunoadjuvant

activity of water-soluble phosphazene polyacids. *J Bioact Compat. Polym*, 1998, 13(4):243–256.

5. Mutwiri, G., Benjamin, P., Soita, H., Townsend, H., Yost, R., Roberts, B., Andrianov, A.K., Babiuk, L.A. Poly[di(sodium carboxylatoethylphenoxy)phosphazene] (PCEP) is a potent enhancer of mixed Th1/Th2 immune responses in mice immunized with influenza virus antigens. *Vaccine*, 2007, 25(7):1204–1213.

6. Payne, L.G., Jenkins, S.A., Andrianov, A., Roberts, B.E. Water-soluble phosphazene polymers for parenteral and mucosal vaccine delivery. *Pharm. Biotechnol.*, 1995, 6:473–493.

7. Payne, L.G., Jenkins, S.A., Woods, A.L., Grund, E.M., Geribo, W.E., Loebelenz, J.R., Andrianov, A.K., Roberts, B.E. Poly[di(carboxylatophenoxy)phosphazene] (PCPP) is a potent immunoadjuvant for an influenza vaccine. *Vaccine*, 1998, 16(1):92–98.

8. Lu, Y., Salvato, M.S., Pauza, C.D., Li, J., Sodroski, J., Manson, K., Wyand, M., Letvin, N., Jenkins, S., Touzjian, N., Chutkowski, C., Kushner, N., LeFaile, M., Payne, L.G., Roberts, B. Utility of SHIV for testing HIV-1 vaccine candidates in macaques. *J. Acquir. Immune Defic. Syndr. Hum. Retrovirol.*, 1996, 12(2):99–106.

9. Payne, L.G., Van Nest, G., Barchfeld, G.L., Siber, G.R., Gupta, R.K., Jenkins, S.A. PCPP as a parenteral adjuvant for diverse antigens. *Dev. Biol. Stand.*, 1998, 92:79–87.

10. Payne, L.G., Jenkins, S.A., Andrianov, A., Langer, R., Roberts, B.E. Xenobiotic polymers as vaccine vehicles. *Ad. Exp. Med. Biol.*, 1995, 371(B):1475–1480.

11. Andrianov, A.K. Polyphosphazenes as vaccine adjuvants. In Singh, M. ed. *Vaccine Adjuvants and Delivery Systems*. Wiley, Hoboken, NJ, 2007, p. 355–378.

12. Andrianov, A.K., Chen, J. Polyphosphazene microspheres: preparation by ionic complexation of phosphazene polyacids with spermine. *J. Appl. Polym. Sci*, 2006, 101(1):414–419.

13. Andrianov, A.K., Chen, J., Payne, L.G. Preparation of hydrogel microspheres by coacervation of aqueous polyphosphazene solutions. *Biomaterials*, 1998, 19(1–3):109–115.

14. Andrianov, A.K., Chen, J., Sule, S.S., Roberts, B.E. Ionically cross-linked polyphosphazene microspheres. *ACS Symp. Ser.*, 2000, 752:395–406.

15. Andrianov, A.K., Payne, L.G. Polymeric carriers for oral uptake of microparticulates. *Adv. Drug Deliv. Rev.*, 1998, 34(2–3):155–170.

16. Andrianov, A.K., Payne, L.G. Protein release from polyphosphazene matrices. *Adv. Drug Deliv. Rev.*, 1998, 31(3):185–196.

17. Andrianov, A.K., Marin, A., Peterson, P., Chen, J. Fluorinated polyphosphazene polyelectrolytes. *J. Appl. Polym. Sci.*, 2007, 103(1):53–58.

18. Andrianov, A.K., Chen, J., LeGolvan, M.P. Poly(dichlorophosphazene) as a precursor for biologically active polyphosphazenes: synthesis, characterization, and stabilization. *Macromolecules*, 2004, 37(2):414–420.

19. Andrianov, A.K., LeGolvan, M.P. Characterization of poly[di(carboxylatophenoxy)-phosphazene] by an aqueous gel permeation chromatography. *J. Appl. Polym. Sci.*, 1996, 60(12):2289–2295.

20. Andrianov, A.K., Svirkin, Y.Y., LeGolvan, M.P. Synthesis and biologically relevant properties of polyphosphazene polyacids. *Biomacromolecules*, 2004, 5(5):1999–2006.

21. Andrianov, A.K., Marin, A. Degradation of polyaminophosphazenes: Effects of hydrolytic environment and polymer processing. *Biomacromolecules*, 2006, 7(5): 1581–1586.

22. Andrianov, A.K., Marin, A., Peterson, P. Water-soluble biodegradable polyphosphazenes containing *N*-ethylpyrrolidone groups. *Macromolecules*, 2005, 38(19): 7972–7976.

23. Andrianov, A.K., Payne, L.G., Visscher, K.B., Allcock, H.R., Langer, R. Hydrolytic degradation of ionically cross-linked polyphosphazene microspheres. *J. Appl. Polym. Sci.*, 1994, 53(12):1573–1578.

24. Andrianov, A.K. Design and synthesis of functionalized polyphosphazenes with immune modulating activity. *PMSE Prepr.*, 2003, 88.

25. Andrianov, A.K. Water-soluble biodegradable polyphosphazenes: emerging systems for biomedical applications. *Polym. Prepr.*, 2005, 46(2):715.

26. Andrianov, A.K., Marin, A. Immunostimulating polyphosphazene compounds, us2006/0193820,2006.

27. Wu, J.Y., Wade, W.F., Taylor, R.K. Evaluation of cholera vaccines formulated with toxin-coregulated pilin peptide plus polymer adjuvant in mice. *Infecti. Immun.*, 2001, 69(12):7695–7702.

28. Mapletoft, J.W., Oumouna, M., Kovacs-Nolan, J., Latimer, L., Mutwiri, G., Babiuk, L.A., van Drunen Littel-van den Hurk, S. Intranasal immunization of mice with a formalin-inactivated bovine respiratory syncytial virus vaccine co-formulated with CpG oligodeoxynucleotides and polyphosphazenes results in enhanced protection. *J. Gen. Virol.*, 2008, 89:250–260.

29. McNeal, M.M., Rae, M.N., Ward, R.L. Effects of different adjuvants on rotavirus antibody responses and protection in mice following intramuscular immunization with inactivated rotavirus. *Vaccine*, 1999, 17(11–12):1573–1580.

30. Kabanov, A.V. From synthetic polyelectrolytes to polymer-subunit vaccines. *Pure Appl. Chem.*, 2004, 76(9):1659–1677.

31. Kabanov, A.V. Polymer genomics: an insight into pharmacology and toxicology of nanomedicines. *Adv. Drug Deliv. Rev.*, 2006, 58(15):1597–1621.

32. Kabanov, A.V., Batrakova, E.V., Sherman, S., Alakhov, V.Y. Polymer genomics. *Adv. Polym. Sci.*, 2006, 173–198.

33. Cairo, C.W., Gestwicki, J.E., Kanai, M., Kiessling, L.L. Control of multivalent interactions by binding epitope density. *J. Am. Chem. Soc.*, 2002, 124(8): 1615–1619.

34. Mutwiri, G., Gerdts, V., Lopez, M., Babiuk, L.A. Innate immunity and new adjuvants. *OIE Rev Sci Tech.*, 2007, 26(1):147–156.

35. Bouveret Le Cam, N.N., Ronco, J., Francon, A., Blondeau, C., Fanget, B. Adjuvants for influenza vaccine. *Res. Immunol.*, 1998, 149(1):19–23.

36. Gilbert, P.B., Chiu, Y.L., Allen, M., Lawrence, D.N., Chapdu, C., Israel, H., Holman, D., Keefer, M.C., Wolff, M., Frey, S.E. Long-term safety analysis of preventive HIV-1 vaccines evaluated in AIDS vaccine evaluation group NIAID-sponsored phase I and II clinical trials. *Vaccine*, 2003, 21(21–22):2933–2947.

37. Kim, J.H., Kirsch, E.A., Gilliam, B., Michael, N.L., VanCott, T.C., Ratto-Kim, S., Cox, J., Nielsen, R., Robb, M.L., Caudrelier, P., El Habib, R., McNeil, J. A phase I, open label, dose ranging trial of the Pasteur Merieux Connaught (PMC) oligomeric

HIV-1 Gp160mn/LAI-2 vaccine in HIV seronegative adults. *Abstracts of the 37th Annual Meeting of the Infectious Diseases Society of America*. Philadelphia, 1999.

38. Thongcharoen, P., Suriyanon, V., Paris, R.M., Khamboonruang, C., de Souza, M.S., Ratto-Kim, S., Karnasuta, C., Polonis, V.R., Baglyos, L., El Habib, R. A phase 1/2 comparative vaccine trial of the safety and immunogenicity of a CRF01_AE (subtype E) candidate vaccine: ALVAC-HIV (vCP1521) prime with oligomeric gp160 (92TH023/LAI-DID) or bivalent gp120 (CM235/SF2) boost. *J. Acquir. Immune Defic. Syndr.*, 2007, 46(1):48.

39. Final report on the safety assessment of methylparaben, ethylparaben, propylparaben, and butylparaben. *J. Am. Coll. Toxicol.*, 1984, 3(5): 145–209.

40. Soni, M.G., Burdock, G.A., Taylor, S.L., Greenberg, N.A. Safety assessment of propyl paraben: a review of the published literature. *Food Chem. Toxicol.*, 2001, 39(6):513–532.

41. Soni, M.G., Carabin, I.G., Burdock, G.A. Safety assessment of esters of *p*-hydroxybenzoic acid (parabens). *Food Chem. Toxicol.*, 2005, 43(7):985–1015.

42. Andrianov, A.K., LeGolvan, M.P., Sule, S.S., Payne, L.G. Degradation of poly [di(carboxylatophenoxy)phosphazene] in aqueous solution. In *Polymeric Materials Science and Engineering. Proceedings of the ACS Division of Polymeric Materials Science and Engineering*, San Francisco, 1997.

43. Allcock, H.R. *Chemistry and Applications of Polyphosphazenes*. Wiley, Hoboken, NJ, 2002, p. 725.

44. Andrianov, A.K., Marin, A., Chen, J., Sargent, J., Corbett, N. Novel route to sulfonated polyphosphazenes: single-step synthesis using noncovalent protection of sulfonic acid functionality. *Macromolecules*, 2004, 37(11):4075–4080.

45. Allcock, H.R., Kwon, S. An ionically cross-linkable polyphosphazene: poly[bis (carboxylatophenoxy)phosphazene] and its hydrogels and membranes. *Macromolecules*, 1989, 22(1):75–79.

4 Potential of Polyphosphazenes in Modulating Vaccine-Induced Immune Responses: I. Investigations in Mice

GEORGE MUTWIRI and PONN BENJAMIN

Vaccine and Infectious Disease Organization/International Vaccine Center, University of Saskatchewan, Saskatoon, Saskatchewan, Canada

ALEXANDER K. ANDRIANOV

Apogee Technology, Inc., Norwood, Massachusetts

LORNE A. BABIUK

Vaccine and Infectious Disease Organization/International Vaccine Center, University of Saskatchewan, Saskatoon, Saskatchewan, Canada; University of Alberta, Edmonton, Alberta, Canada

INTRODUCTION

Vaccines continue to play a critical role in the control of infectious diseases, with live as well as killed vaccines being available for a variety of infections. Live vaccines are often more efficacious, but there is concern that they could revert to virulence and cause disease, especially in immunocompromised hosts (Bowersock and Martin, 1999). These safety issues have led to increased interest in killed vaccines, which are generally regarded as safer. However, killed vaccines are often poorly immunogenic and require coadministration with adjuvants (Bowersock and Martin, 1999).

Adjuvants were first described by Ramon (Ramon, 1924) as substances that enhanced immune responses to a level higher than that enhanced by the antigen alone. Since then, they have been recognized as critical components of vaccines that consist of nonreplicating antigens. Over the years, numerous natural and synthetic substances have been evaluated as adjuvants. These include delivery systems such as alum, liposomes, microparticles, and oil–water emulsions

Polyphosphazenes for Biomedical Applications, Edited by Alexander K. Andrianov
Copyright © 2009 John Wiley & Sons, Inc.

(Singh and O'Hagan, 2002). However, few of these substances have been approved for clinical use in animals (Bowersock and Martin, 1999). In humans, alum is the most widely used adjuvant for human vaccines, and this is due partly to its long track record of safety and the fact that it promotes a Th2 type of immune response. However, alum does not appear to enhance Th1 immune response and is therefore not a rational choice for vaccines against pathogens for which Th1 or mixed Th1/Th2-type immune responses are required for protection. The need for safer and more effective adjuvants has attracted interest in research on new adjuvants. Immunostimulating adjuvants belong to a recently described class of immune-enhancing molecules, and these are primarily microbial components that act by stimulating innate immune responses (Singh and O'Hagan, 2002). In this regard, microbial components and their synthetic analog have been shown to stimulate innate immune responses and indeed have adjuvant activity (Davis et al., 1998; Merritt and Johnson, 1965; O'Hagan et al., 2001; Pasare and Medzhitov, 2005).

Polymers have been evaluated extensively as microparticle vaccine delivery vehicles (Bowersock and Martin, 1999; Mutwiri et al., 2005), but the adjuvant properties of synthetic polymers as aqueous formulations have not been explored systematically. Polyphosphazenes are synthetic polymers that consist of a backbone of alternating phosphorus and nitrogen atoms and organic side groups attached to each phosphorus atom (Payne and Andrianov, 1998). Structural modifications of such macromolecules can make them readily water soluble, biodegradable in vivo, and stable at room temperature. These properties make polyphosphazenes attractive candidates as vaccine adjuvants. One of the most investigated polyphosphazene polyelectrolytes, poly[di(carboxylatophenoxy)-phosphazene] (PCPP), was shown previously to have adjuvant activity in mice with a variety of viral and bacterial antigens (McNeal et al., 1999; Payne et al., 1998; Wu et al., 2001). We have recently reported that a new polyphosphazene derivative poly[di(sodium carboxylatoethylphenoxy)phosphazene] (PCEP), is a powerful adjuvant in mice immunized with influenza virus X:31 antigen (Mutwiri et al., 2007). The adjuvant effects of polyphosphazenes are thought to depend on the ability of the polymer to bind to the antigen (Andrianov et al., 2005). We reasoned that since this polymer–antigen interaction may vary from one antigen preparation to another, the adjuvant activity of polyphosphazenes may depend on the particular antigen preparation being evaluated. For this reason, we investigated whether PCEP and PCPP have any adjuvant activity when coadministered with hepatitis B virus surface antigen (HBsAg).

MATERIALS AND METHODS

Polyphosphazenes

PCEP and PCPP were synthesized using approaches described previously (Andrianov et al., 2004; Mutwiri et al., 2007) and were provided by Parallel

Solutions, Inc. (Cambridge, Massachusetts). Both polymers were dissolved at 2 mg/mL in PBS (phosphate-buffered saline, pH 7.2) and stored at room temperature in the dark, and were found to retain activity over a period of several months under these storage conditions. PCEP and PCPP vaccine formulation was achieved simply by mixing the antigen with an aqueous solution of polymer.

Animals and Immunization

All animal experiments were carried out according to the *Guide to the Care and Use of Experimental Animals*, provided by the Canadian Council on Animal Care. Experimental protocols were approved by the University of Saskatchewan Animal Care Committee. BALB/c mice were obtained from the Charles River Laboratories (North Franklin, Connecticut). Mice ($n = 5$ mice per group) were given a single subcutaneous (s.c.) immunization with either HBsAg (Biodesign International, Saco, Maine) at various doses (1.0, 0.2, or 0.04 µg per mouse) alone, or in combination with various adjuvants as follows: HbsAg + alum, HbsAg + PCEP, or HbsAg + PCPP. PCPP and PCEP were used at a dosage of 50 µg/mouse. Alum ($AlPO_4$, Cedarlane Laboratories, Hornby, Ontario, Canada) was used at 2.5 µg/µg of protein. All mice were observed for any signs of adverse reactions to immunization. Mice were bled prior to immunization and at weeks 2, 4, 8, 12, 16, 20, and 24 post-immunization.

Detection of Antigen-Specific Antibodies by ELISA

For detection of HBsAg-specific antibodies in the serum of mice, Immunolon II microtiter plates (Dynex Technology, Inc., Chantilly, Virginia) were coated overnight at 4°C with HBsAg at 1.0 µg/mL in carbonate coating buffer (15 mM Na_2CO_3, 35 mM $NaHCO_3$, pH 9.6), and 100 µL of the antigen was added to each well. Wells were washed six times with phosphate-buffered saline (pH 7.3) containing 0.05% Tween-20 (PBST). Diluted mouse serum samples were added to the wells at 100 µL/well and incubated for 2 h at room temperature. Wells were washed again with PBST. Biotinylated goat anti-mouse IgG, IgG1, and IgG2a antibodies (Caltag Laboratories, Carlsbad, California) were added to wells at 100 µl/well and plates were incubated for 1 h at room temperature. Wells were washed and alkaline phosphatase (AP) conjugated with streptavidin (Cedarlane Laboratories) was diluted 1/10,000, and 100 µL was added in each well, followed by 1 h of incubation at room temperature. Wells were washed eight times in double-distilled water (ddH$_2$O). Di(Tris) *p*-nitrophenyl phosphate (PNPP) (Sigma Chemical Company, St. Louis, Missouri) was diluted 1/100 in PNPP substrate buffer, and 100 µL/well was added. The reaction was allowed to develop for 15 min, and absorbance was read as optical density (OD) at 405 nm in a Microplate Reader (BioRad Laboratories, Hercules, California). Results are reported as titers, which are the reciprocal of the highest dilution that gave a positive OD reading. A positive titer was defined as an OD reading that was at least two times greater than the values for a negative sample. Negative samples were sera from naive, nonimmunized mice.

Statistical Analysis

All data on total IgG, IgG1, and IgG2a, antibody titers were analyzed as randomized design using the Proc GLM procedure of Statistical Analysis Software (SAS Institute, Inc., Cary, South Carolinia). Data that were not normally distributed were transformed logarithmically. The statistical model included the effects of adjuvant, time, and adjuvant × time. Mean comparisons were conducted to compare the magnitude of responses. Significant effects were declared at $p < 0.05$.

RESULTS

HBsAg-Specific Antibody Responses in Serum

High-Dose HBsAg   We tested whether PCEP or PCPP can enhance antibody responses of HBsAg in mice. Immunization of mice with a high dose of HBsAg (1 µg/mouse) combined with PCEP resulted in an increase in total IgG antibody titers, and this was significantly higher than in all other groups as early as 2 weeks post-immunization (Fig. 1a). Furthermore, the titers continued to increase over the 24-week experimental period. Mice immunized with HbsAg + PCPP also had increased total IgG titers, and these were significantly higher than those immunized with HBsAg + alum or HBsAg alone starting at 8 weeks post-immunization and maintained for the remainder of the experimental period (Fig. 1a). Alum had no significant effect on the total IgG titers compared to HBsAg alone.

When IgG isotypes were assessed, it was again observed that mice immunized with HbsAg + PCEP had significantly enhanced IgG1 titers compared to all other groups as early as 2 weeks post-immunization (Fig. 1b). Similarly, immunization with HbsAg + PCPP resulted in significantly increased IgG1 titers (Fig. 1b). Mice immunized with HBsAg + alum had a significant increase in IgG1 titers at week 2 but were not different from the HBsAg-alone group for the remainder of the 24-week experimental period (Fig. 1b).

FIGURE 1   Kinetics, amplitude, and duration of serum HBsAg-specific (a) total IgG, (b) IgG1, and (c) IgG2a antibody response in BALB/c mice given a single s.c. immunization with a high dose (1.0 µg) of HBsAg alone, with HBsAg + alum, HBsAg + PCPP, or HBsAg + PCEP. Each data point represents mean ± SE for titers of anti-HBsAg as determined by ELISA (enzyme-linked immunosorbent assay). Groups with different letters are significantly different ($p < 0.05$).

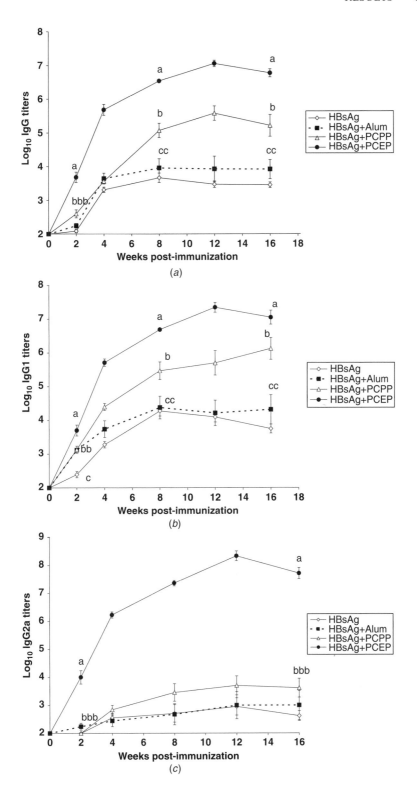

(a)

(b)

(c)

With regard to IgG2a, only PCEP increased IgG2a titers significantly starting at 2 weeks post-immunization, and by 16 weeks the titers in this group were at least 1000-fold higher than all the other groups (Fig. 1c). PCPP and alum did not induce any significant increase in IgG2a response (compared to HBsAg alone) at this dose of antigen.

Low-Dose HBsAg   Since the magnitude and quality of immune responses induced by an adjuvant can be influenced by the dose of antigen used, we then explored the adjuvant activity of PCPP, PCEP, and alum in mice immunized with a 25-fold lower dose of HBsAg (0.04 µg/mouse). As early as 2 weeks post-immunization, mice immunized with HBsAg + PCEP and HBsAg + PCPP had total IgG titers that were significantly higher than those immunized with HBsAg + alum and HBsAg alone (Fig. 2a). However, the HBsAg + PCEP titers continued to rise, and by week 12 post-immunization this group had titers that were significantly higher than those of the HBsAg + PCPP group (Fig. 2a). Alum did not enhance total IgG titers above those seen in mice immunized with HBsAg alone (Fig. 2a).

With regard to IgG1 titers, both HBsAg + PCEP and HBsAg + PCPP had significantly elevated titers, which were approximately 1000-fold higher than those of HBsAg + alum and HBsAg alone by week 8 (Fig. 2b). These titers were maintained for the remainder of the experimental period. Again, essentially no or very low levels of HBsAg-specific IgG1 antibodies were detected in mice immunized with HBsAg + alum and HBsAg alone.

PCEP increased IgG2a titers significantly, by at least 100-fold as early as 2 weeks post-immunization, and the titers were significantly higher than those of all the other groups (Fig. 2c). By 8 weeks post-immunization, PCEP titers were approximately 1000-fold higher than those of PCPP and alum groups (Fig. 2C). PCPP also increased antibody titers significantly above those in the HBsAg + alum and HBsAg-alone groups, and this was attributed to the fact that the latter two groups did not develop substantial IgG2a responses at the low-dose HBsAg (Fig. 2c).

Figure 3 summarizes the effect of HBsAg dose on antibody responses. Injection of mice with 0.04 µg of HBsAg alone or HBsAg + alum essentially induced no IgG1 and IgG2a antibody responses (Fig. 3). In contrast, there was significant induction of IgG1 and IgG2a antibody titers when PCEP or PCPP was added to 0.04 µg of HBsAg (Fig. 3), with PCEP having significantly higher IgG2a titers than the PCPP group (Fig. 3b). When the dose of HBsAg was

FIGURE 2   Kinetics, amplitude, and duration of serum HBsAg-specific (a) total IgG, (b) IgG1, and (c) IgG2a antibody response in BALB/c mice given a single s.c. immunization with a low dose (0.2 µg) of HBsAg alone, with HBsAg + alum, HBsAg + PCPP, or HBsAg + PCEP. Each data point represents mean ± SE for titers of anti-HBsAg as determined by ELISA. Groups with different letters are significantly different ($p < 0.05$).

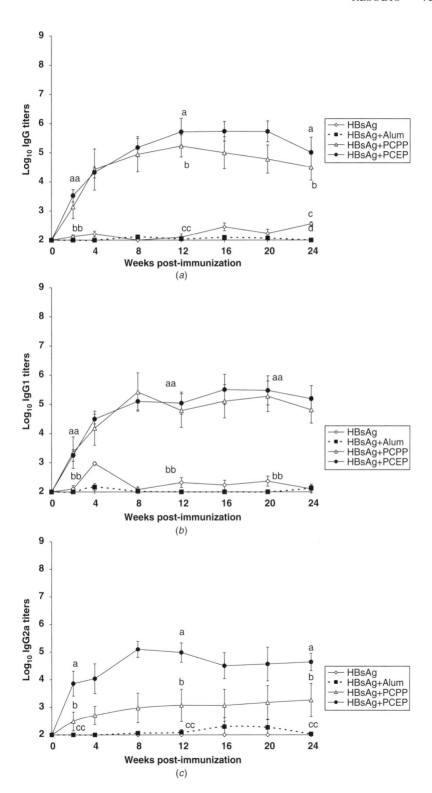

increased fivefold, to 0.2 µg/mouse, modest IgG1 and IgG2a titers were detected in all groups, including HBsAg alone, but at this time (12 weeks) only PCEP had significantly high titers of IgG2a compared to all other groups. Increasing the dose an additional fivefold to 1.0 µg/mouse had essentially no effect on antibody titers (Fig. 3). Thus, PCPP and PCEP induced significant IgG1 and IgG2 titers even when the dose of HBsAg was reduced 25-fold, whereas essentially no antibody titers were detected in mice immunized with HBsAg alone or HBsAg + alum at this low dose (Fig. 3).

We then used week 12 data from mice immunized with 0.2 µg of HBsAg with the various adjuvants to calculate the IgG1/IgG2a ratios. For HBsAg, HBsAg + alum, HBsAg + PCPP, and HBsAg + PCEP the ratios were 1.51,

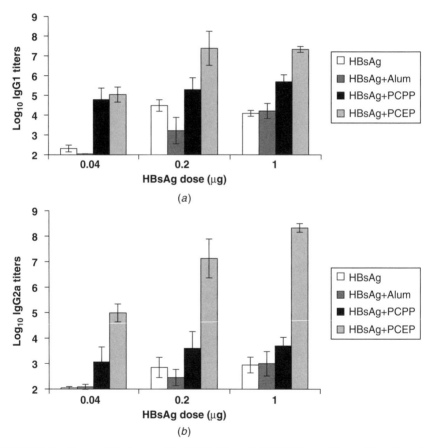

FIGURE 3 Serum HBsAg-specific (a) IgG1 and (b) IgG2a antibody titers in mice given a single s.c. immunization with a various doses of (0.04, 0.2, and 1.0 µg) of HBsAg alone, HBsAg + PCEP, HBsAg + PCPP, or HBsAg + alum. Each data point represents mean ± SEM of titers of anti-HBsAg as determined by ELISA. Groups with different letters are significantly different ($p < 0.05$).

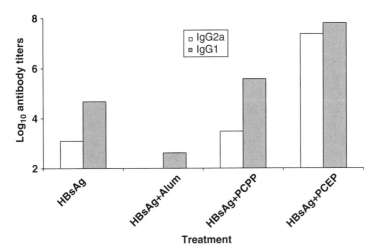

FIGURE 4 Serum IgG2a and IgG1 titers in mice 12 weeks post-immunization with 0.2 μg of HBsAg alone, HBsAg + PCEP, HBsAg + PCPP, and HBsAg + alum.

1.31, 1.60, and 1.06, respectively. These data indicated to us that the PCEP was associated with equivalent IgG1 and IgG2a antibody responses, while the HBsAg alone or with the other two adjuvants (alum and PCPP) were associated predominantly with a IgG1 response. This is demonstrated further in Figure 4 using the same week 12 data.

DISCUSSION

In the present investigation we have shown that the polyphosphazenes PCPP and PCEP enhance antigen-specific immune responses to HBsAg in mice. The adjuvant activity of PCEP was superior to that of PCPP and the conventional adjuvant alum. Polyphosphazenes are linear polymers that form noncovalent, water-soluble protein–polyphosphazene complexes that bind multiple protein antigens per polymer chain (Andrianov et al., 2005). It is thought that both the amount of antigen bound to the polymer and the ability to form the complex may be important in their ability to perform as adjuvants in vivo (Andrianov et al., 2005). It has been proposed that the antigen bound in such complexes may be more stable and protected from degradation, and also, complexes may be taken up by antigen-presenting cell (APC) more efficiently. We reasoned that the physicochemical interaction between polymer and antigen may depend on the nature of the antigen as well as its composition. Whereas PCPP has been tested widely with a variety of antigens (McNeal et al., 1999; Payne et al., 1998; Wu et al., 2001), PCEP has been tested only with X:31 antigen from influenza virus (Mutwiri et al., 2007). The present investigation confirms that PCEP and PCPP are both potent adjuvants with HBsAg in mice. The magnitude and quality (levels of IgG1 and IgG2a) of immune responses in the present studies are similar to those observed previously with influenza X:31 antigen (Mutwiri et al., 2007). An adjuvant that can enhance immune responses to multiple antigens would be suitable for enhancing immune responses to various antigens present in multivalent vaccines. While polymer binding to antigen may contribute to the adjuvant activity of PCPP and PCEP, other factors may be important in mediating the immune-enhancing activity of polyphosphazenes. In this regard, we have observed that PCEP and to a lesser extent PCPP induce cytokines IL-12 and IFN-γ in the absence of antigen, suggesting that stimulation of the innate immune system may contribute to the adjuvant activity of polyphosphazenes (Mutwiri, 2008).

The ability of PCPP and PCEP to reduce the dose of HBsAg without a proportional reduction in magnitude or compromising the quality of immune responses is a highly desirable attribute of an adjuvant because this can lead to reduced costs of the vaccine. In addition, polyphosphazenes did not seem to induce any severe tissue reaction at the site of injection, but this requires a more detailed investigation.

Evaluation of IgG1 and IgG2a antibody isotypes in response to immunization provides a relative measure of the contribution of Th2 and Th1 humoral immune responses, respectively (Finkelman et al., 1990). In our studies, PCEP induced similar levels of IgG1 and IgG2a antibody titers, suggesting that this polymer induced mixed Th1/Th2-type immune responses, whereas PCPP induced a predominantly Th2 type of response. However, to establish clearly that Th cell–derived cytokines were involved in the adjuvant activity of polyphosphazenes, it would be necessary to demonstrate the presence of HBsAg-specific Th1- and Th2-type IFN-γ- and IL-4-secreting cells in the splenocytes of immunized mice.

In conclusion, PCPP and PCEP are potent adjuvants that enhance antibody responses to hepatitis B virus antigen antigens. The present observations provide additional evidence that PCEP is a potent adjuvant and has the potential to improve the efficacy of killed vaccines.

Acknowledgments

Financial support for this work was provided by grants from the Natural Sciences and Engineering Council (NSERC), Canadian Agricultural and Rural Development Saskatchewan (CARDS), Alberta Beef Producers, and the Krembil Foundation. Published with permission from the director of VIDO as journal series 436.

REFERENCES

Andrianov, A.K., Svirkin, Y.Y., LeGolvan, M.P. 2004. Synthesis and biologically relevant properties of polyphosphazene polyacids. *Biomolecules*, 5:1999–2006.

Andrianov, A.K., Marin, A., Roberts, B.E. 2005. Polyphosphazene polyelectrolytes: a link between the formation of noncovalent complexes with antigenic proteins and immunostimulating activity. *Biomacromolecules*, 6:1375–1379.

Bowersock, T.L., Martin, S. 1999. Vaccine delivery to animals. *Adv. Drug Deliv. Rev.*, 38:167–194.

Davis, H.L., Weeratna, R., Waldschmidt, T.J., Tygrett, L., Schorr, J., Krieg, A.M. 1998. CpG DNA is a potent enhancer of specific immunity in mice immunized with recombinant hepatitis B surface antigen. *J. Immunol.*, 160:870–876.

Finkelman, F.D., Holmes, J., Katona, I.M., Urban, J.F., Jr., Beckmann, M.P., Park, L.S., Schooley, K.A., Coffman, R.L., Mosmann, T.R., Paul, W.E. 1990. Lymphokine control of in vivo immunoglobulin isotype selection. *Annu. Rev. Immunol.*, 8:303–333.

McNeal, M.M., Rae, M.N., Ward, R.L. 1999. Effects of different adjuvants on rotavirus antibody responses and protection in mice following intramuscular immunization with inactivated rotavirus. *Vaccine*, 17:1573–1580.

Merritt, K., Johnson, A.G. 1965. Studies on the adjuvant action of bacterial endotoxins on antibody formation: VI. Enhancement of antibody formation by nucleic acids. *J. Immunol.*, 94:416–422.

Mutwiri, G., Bowersock, T.L., Babiuk, L.A. 2005. Microparticles for oral delivery of vaccines. *Expert Opin. Drug Deliv.*, 2:791–806.

Mutwiri, G., Benjamin, P., Soita, H., Townsend, H., Yost, R., Roberts, B., Andrianov, A.K., Babiuk, L.A. 2007. Poly[di(sodium carboxylatoethylphenoxy)phosphazene] (PCEP) is a potent enhancer of mixed Th1/Th2 immune responses in mice immunized with influenza virus antigens. *Vaccine*, 25:1204–1213.

Mutwiri, G., Benjamin, P., Soita, H., Babiuk, L.A. 2008. Co-administration of CpG ODN strongly enhances immune responses in mice immunized with Hepatitis B surface antigen. *Vaccine*, 26:2680–2688.

O'Hagan, D.T., MacKichan, M.L., Singh, M. 2001. Recent developments in adjuvants for vaccines against infectious diseases. *Biomol. Eng.*, 18:69–85.

Pasare, C., Medzhitov, R. 2005. Toll-like receptors: linking innate and adaptive immunity. *Adv. Exp. Med. Biol.*, 560:11–18.

Payne, L.G., Andrianov, A.K. 1998. Protein release from polyphosphazene matrices. *Adv. Drug Deliv. Rev.*, 31:185–196.

Payne, L.G., Jenkins, S.A., Woods, A.L., Grund, E.M., Geribo, W.E., Loebelenz, J.R., Andrianov, A.K., Roberts, B.E. 1998. Poly[di(carboxylatophenoxy)phosphazene] (PCPP) is a potent immunoadjuvant for an influenza vaccine. *Vaccine*, 16:92–98.

Ramon, G. 1924. *Ann. Inst. Pasteur.*, 38:1.

Singh, M., O'Hagan, D.T. 2002. Recent advances in vaccine adjuvants. *Pharm. Res.*, 19:715–728.

Wu, J.Y., Wade, W.F., Taylor, R.K. 2001. Evaluation of cholera vaccines formulated with toxin-coregulated pilin peptide plus polymer adjuvant in mice. *Infect. Immun.*, 69:7695–7702.

5 Potential of Polyphosphazenes in Modulating Vaccine-Induced Immune Responses: II. Investigations in Large Animals

GEORGE MUTWIRI

Vaccine and Infectious Disease Organization/International Vaccine Center, University of Saskatchewan, Saskatoon, Saskatchewan, Canada

LORNE A. BABIUK

Vaccine and Infectious Disease Organization/International Vaccine Center, University of Saskatchewan, Saskatoon, Saskatchewan, Canada; University of Alberta, Edmonton, Alberta, Canada

INTRODUCTION

Killed or subunit vaccines are highly desirable, due primarily to their long track record of safety. Unfortunately, these vaccines are poorly immunogenic and their efficacy is critically dependent on adjuvants, compounds that enhance immune responses induced by vaccines (Ramon, 1924). Many natural and synthetic compounds have been evaluated for their adjuvant activity. Some of these have shown potential as adjuvants in laboratory animals, but few proceed to clinical trials and are subsequently used in commercial vaccines for a variety of reasons, but most fail because they are either not effective or have undesirable side effects in target hosts. In this regard, data obtained from mouse studies are not always predictive of efficacy in humans or large animals, and studies in target hosts are usually required, which is where most adjuvants fail. For example, when used alone, CpG DNA is a potent systemic and mucosal adjuvant in mice, but in large animals appropriate formulation with co-adjuvants or delivery systems is required for optimal adjuvant activity of CpG (Ioannou et al., 2002a; Mutwiri et al., 2004). Also, a safe adjuvant in mice may not necessarily reflect safety in humans and large animals. For these

Polyphosphazenes for Biomedical Applications, Edited by Alexander K. Andrianov
Copyright © 2009 John Wiley & Sons, Inc.

reasons, adjuvants have to be evaluated in the target host species in which their use is intended.

The adjuvant effects of polyphosphazenes have been well documented in mice by numerous investigators using a variety of candidate vaccine antigens (McNeal et al., 1999; Mutwiri et al., 2007; Payne et al., 1998; Wu et al., 2001). Although these data are encouraging, studies in target species are required to warrant clinical evaluation with commercial vaccines. In the present study we evaluated the adjuvant activity of the polyphosphazene poly[di(carboxylato-phenoxy)phosphazene] (PCPP) in large animals.

MATERIALS AND METHODS

Animals and Immunization

Suffolk sheep of either sex were obtained from the Department of Animal and Poultry Science (University of Saskatchewan, Saskatoon, Saskatchewan, Canada). The animals were housed at the Vaccine and Infectious Disease Organization animal facility and fed ad libitum on a diet of rolled barley and alfalfa hay. All experiments were carried out according to the *Guide to the Care and Use of Experimental Animals*, provided by the Canadian Council on Animal Care. All experimental protocols were approved by the Animal Care Committee of the University of Saskatchewan. Three separate immunization experiments were performed.

1. *Confirmation of adjuvant activity of PCPP in a large animal.* In a preliminary experiment, we sought to first determine whether PCPP has any adjuvant activity in sheep. Two groups of animals ($n = 3$ per group) were used in this experiment. The experimental group was immunized with a single subcutaneous (s.c.) injection of 50 µg of the model antigen, porcine serum albumin (PSA), plus 0.5 mg of PCPP (Parallel Solutions, Inc., Boston, Massachusetts), while the second (control) group was immunized with PSA alone. Animals were bled at 2-week intervals, and serum was separated and stored at −20°C until used. PSA-specific antibody responses were monitored in the serum by ELISA (enzyme-linked immunosorbent assay) for a period of 10 weeks.

2. *Antigen-dose titration study.* The second experiment was carried out to determine the lowest effective dose of PCPP that was required to induce adjuvant activity in sheep. For this experiment, four groups of lambs ($n = 7$ lambs per group) were immunized by a single s.c. injection with 50 µg of PSA alone or in combination with either 0.1, 0.5, or 1.0 mg PCPP. Serum samples were collected and antibody titers analyzed as above.

3. *Confirmation of adjuvant activity of PCPP with a viral antigen.* In the third experiment, lambs ($n = 5$ lambs per group) were immunized with a single s.c. injection of 20 µg of glycoprotein gD (tgD) alone, or in combination with either 0.5 mg of PCPP or 30% Emulsigen (a commercial adjuvant). Nonimmunized

control animals were injected with PBS. The site of injection was examined visually and palpated during the first 48 h following injection. Animals were bled before immunization and at 4 and 6 weeks after immunization. Serum samples were collected and stored at −20°C until used. tgD-specific antibody responses were assayed in serum by the ELISA technique.

Antigens

The model antigen PSA was purchased from Sigma Chemical Company (St. Louis, Missouri). tgD was prepared from bovine herpesvirus-1 transfected cells (BHV-1), as described previously (Ioannou et al., 2002b).

Assessment of Antibody Responses by ELISA

The ELISA technique to evaluate antigen-specific immune responses in sera from sheep was used as described previously (Ioannou et al., 2002b; Mutwiri et al., 2002) with minor modifications. Briefly, microtiter plates (Immunolon II, Dynatech Laboratories, Gaithersburg, Maryland) were coated overnight with antigen (PSA, tgD, or P1P5), washed, and then incubated with serially diluted sera. Alkaline phosphatase (AP)–conjugated rabbit anti-sheep IgG was used to detect immunoglobulin (Ig) captured by antigen. The assay was developed by using p-nitrophenylphenylphosphate (PNPP, Sigma) substrate. Absorbence was read as optical density (OD) at a wavelength of 405 nm on a microplate reader (BioRad Laboratories, Hercules, California). The titer was equal to the reciprocal of the highest dilution that gave a positive reading. A positive titer was defined as an OD reading that was at least two times greater than the values for a negative sample. Negative samples included sera from naive (nonimmunized) animals.

RESULTS

1. *PCPP has potent adjuvant activity in sheep.* Immunization of lambs with PSA alone induced antigen-specific antibody titers in serum as early as 2 weeks post-immunization (Fig. 1). The antibody titers did not increase further, but declined after 4 weeks and were down to baseline levels by 10 weeks post-immunization (Fig. 1). Lambs immunized with PSA + PCPP developed antibody titers as early as 2 weeks post-immunization, and these titers were almost 100-fold higher than those seen in lambs immunized with PSA alone (Fig. 1). Interestingly, antibody titers in lambs immunized with PSA + PCPP were sustained at peak levels for the entire 10-week experimental period (Fig. 1).

2. *Low doses of PCPP are sufficient to induce adjuvant activity in sheep.* The second experiment was carried out to determine the lowest effective dose of PCPP required to induce adjuvant activity in a large animal. Immunization of lambs with PSA alone induced a modest increase in antibody titers 2 weeks

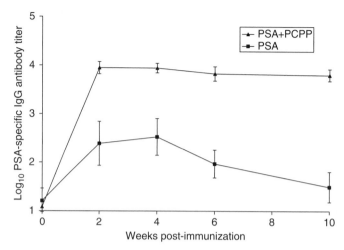

FIGURE 1   Adjuvant activity of PCPP in sheep. Young lambs were immunized with a single subcutaneous injection with either 50 µg of porcine serum albumin (PSA) alone (■) or with PSA + 500 µg PCPP (▲). PSA-specific antibody responses were monitored in serum over a period of 10 weeks. Each data point represents mean ± SEM of titers of anti-PSA antibodies as determined by ELISA.

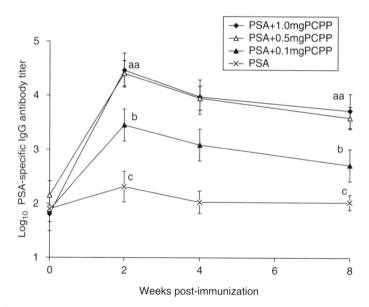

FIGURE 2   Determination of the optimal dose of PCPP required for adjuvant activity in sheep. Young lambs were immunized with a single subcutaneous injection with either PSA alone or with PSA + various doses of PCPP. Antibody responses were monitored in serum at intervals over a period of 6 weeks. Each data point represents mean ± SEM of titers of anti-PSA antibodies as determined by ELISA.

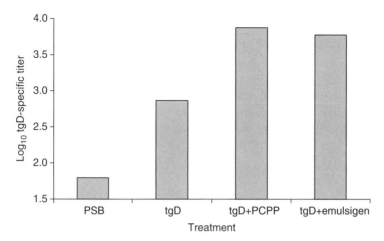

FIGURE 3 Adjuvant activity of PCPP with a viral antigen in sheep. Young lambs were immunized with a single subcutaneous injection with either the viral antigen tgD alone or with tgD + PCPP or tgD + Emulsigen. tgD-specific antibody responses were monitored in serum. Each data point represents mean $\pm$ SEM of titers of anti-PSA antibodies as determined by ELISA at 4 weeks post-immunization.

after immunization, but these were back to baseline levels by 4 weeks post-immunization (Fig. 2). In contrast, immunization of lambs with PSA in combination with as low as 0.1 mg of PCPP resulted in a significant increase in titers compared to those immunized with PSA alone, and the titers were sustained for the 8-week experimental period (Fig. 2). As expected, 0.5 mg of PCPP + PSA induced a significant increase in titers, and these were significantly higher than those in lambs immunized with the 0.1 mg dose of PCPP (Fig. 2). However, increasing the PCPP dose to 1.0 mg did not result in any additional increase in titers (Fig. 2), suggesting that a dose of 0.5 mg of PCPP was optimal.

3. *PCPP-enhanced antibody responds to a viral antigen in sheep.* We then tested adjuvant activity of PCPP with a viral antigen, tgD, from bovine herpesvirus-1. The PCPP induced an almost 10-fold increase in tgD-specific antibody titers in serum 4 weeks post-immunization (Fig. 3) compared to tgD alone, and the responses were maintained for at least 8 weeks (data not shown). The antibody response was similar to that observed with the conventional adjuvant Emulsigen. Although Emulsigen is known to cause severe tissue reaction, no tissue reaction was seen at the injection site in animals injected with antigen formulated in PCPP.

DISCUSSION

The present investigations demonstrate for the first time that PCPP has adjuvant activity in large animals, confirming that the adjuvant activity of polyphosphazenes is not limited to mice. This is significant given that some adjuvants have species-specific activity. Interestingly, only 0.1 mg of PCPP was sufficient to induce adjuvant activity in sheep. This is remarkable since doses of 0.05 to 0.1 mg of PCPP have previously been used in mouse studies (Mutwiri et al., 2007; Payne et al., 1998). These results show that only relatively small doses of PCPP are required for adjuvant activity, and this would be a desirable attribute for animal health applications since small doses of the adjuvant result in less cost for the adjuvant component in the vaccine. This, coupled with the fact that polyphosphazenes can be produced relatively inexpensively, makes these polymers attractive for animal vaccines from an economic standpoint. In addition, this study showed the optimal dose of PCPP to be 0.5 mg, confirming the need for dose selection studies for each species.

PCPP forms water-soluble complexes with antigens, and it has been suggested that the nature of the antigen determines the adjuvant activity of polyphosphazenes (Andrianov et al., 2005). This suggests that the adjuvant activity of polyphosphazenes may vary from one antigen to another. For this reason it was necessary to test the adjuvant activity of the PCPP with a variety of antigens. Our results show that PCPP is a versatile adjuvant in sheep, as indicated by its adjuvant activity with the model antigen PSA and the viral antigen tgD. An adjuvant that is compatible with many antigens is preferable, as it would have a broader application with different types of vaccines. The observation that PCPP enhanced antibody responses to tgD is particularly encouraging because when given in appropriate adjuvant formulation, this antigen was shown to protect cattle against infection with BHV-1 (Ioannou et al., 2002b), a virus involved in the pathogenesis of bovine respiratory disease complex ("shipping fever"). However, although encouraging, whether the antibodies induced by PCPP are functional needs to be confirmed, for example, in a virus neutralization assay, which is more predictive of protection than are ELISA titers.

Interestingly, PCPP performed as well as Emulsigen, an oil-in-water emulsion used widely in animal studies and in some vaccines (Ioannou et al., 2002a,b). Emulsigen has high efficacy and is considered a gold standard for veterinary vaccines. However, Emulsigen causes severe tissue damage at the site of injection. It should be noted that injection-site reactions have been associated with significant economic losses in the beef industry (Van Donkersgoed et al., 1997). Our observation that PCPP induced a similar level of antibody titers but no severe tissue reaction at the site of injection is encouraging.

In addition to enhancing the magnitude of the antibody responses, addition of PCPP to the antigen also induced an early onset of the responses, often as early as 2 weeks, and in some cases the peak response was achieved at this time. This is remarkable after a single primary immunization, where significant responses do not peak until at about 4 weeks. An early peak response is usually expected after multiple immunizations, or at least after a booster vaccination. The early onset of peak immune responses is a critical parameter because it suggests that animals are protected sooner following vaccination.

In summary, this study confirms that the polymer PCPP is an effective adjuvant with a viral vaccine antigen and performs as well as the conventional adjuvant emulsigen but has the added advantage of not causing injection-site tissue reaction.

Acknowledgments

Financial support for this work was supported by grants from the Alberta Beef Producers. The authors thank Ponn Benjamin for his technical assistance, and animal care personnel for their help with animal experiments. This article is printed with permission of the director of VIDO as journal series 493.

REFERENCES

1. Andrianov, A.K., Marin, A., Roberts, B.E. 2005. Polyphosphazene polyelectrolytes: a link between the formation of noncovalent complexes with antigenic proteins and immunostimulating activity. *Biomacromolecules*, 6:1375–1379.

2. Ioannou, X.P., Gomis, S.M., Karvonen, B., Hecker, R., Babiuk, L.A., van Drunen Little–van den Hurk, S. 2002a. CpG-containing oligodeoxynucleotides, in combination with conventional adjuvants, enhance the magnitude and change the bias of the immune responses to a herpesvirus glycoprotein. *Vaccine*, 21:127–137.

3. Ioannou, X.P., Griebel, P., Hecker, R., Babiuk, L.A., van Drunen Littel–van den Hurk, S. 2002b. The immunogenicity and protective efficacy of bovine herpesvirus 1 glycoprotein D plus Emulsigen are increased by formulation with CpG oligodeoxynucleotides. *J. Virol.*, 76:9002–9010.

4. McNeal, M.M., Rae, M.N., Ward, R.L. 1999. Effects of different adjuvants on rotavirus antibody responses and protection in mice following intramuscular immunization with inactivated rotavirus. *Vaccine*, 17:1573–1580.

5. Mutwiri, G., Bowersock, T., Kidane, A., Sanchez, M., Gerdts, V., Babiuk, L.A., Griebel, P. 2002. Induction of mucosal immune responses following enteric immunization with antigen delivered in alginate microspheres. *Vet. Immunol. Immunopathol.*, 87:269–276.

6. Mutwiri, G.K., Nichani, A.K., Babiuk, S., Babiuk, L.A. 2004. Strategies for enhancing the immunostimulatory effects of CpG oligodeoxynucleotides. *J. Control. Release*, 97:1–17.

7. Mutwiri, G., Benjamin, P., Soita, H., Townsend, H., Yost, R., Roberts, B., Andrianov, A.K., Babiuk, L.A. 2007. Poly[di(sodium carboxylatoethylphenoxy)-phosphazene] (PCEP) is a potent enhancer of mixed Th1/Th2 immune responses in mice immunized with influenza virus antigens. *Vaccine*, 25:1204–1213.

8. Payne, L.G., Jenkins, S.A., Woods, A.L., Grund, E.M., Geribo, W.E., Loebelenz, J.R., Andrianov, A.K., Roberts, B.E. 1998. Poly[di(carboxylatophenoxy)phosphazene] (PCPP) is a potent immunoadjuvant for an influenza vaccine. *Vaccine*, 16:92–98.

9. Ramon, G. 1924. *Ann. Inst. Pasteur*, 38:1.

10. Van Donkersgoed, J., Dixon, S., Brand, G., Vanderkop, M. 1997. A survey of injection site lesions in fed cattle in Canada. *Can. Vet. J.*, 12:767–772.

11. Wu, J.Y., Wade, W.F., Taylor, R.K. 2001. Evaluation of cholera vaccines formulated with toxin-coregulated pilin peptide plus polymer adjuvant in mice. *Infect. Immun.*, 69:7695–7702.

6 Polyphosphazenes as Adjuvants for Inactivated and Subunit Rotavirus Vaccines in Adult and Infant Mice

KARI JOHANSEN

Department of Virology, Swedish Institute for Infectious Disease Control, Solna, Sweden; Department of Microbiology, Tumor and Cell Biology, Karolinska Institute, Stockholm, Sweden

JORMA HINKULA

Department of Virology, Swedish Institute for Infectious Disease Control, Solna, Sweden; Department of Microbiology, Tumor and Cell Biology, Karolinska Institute, Stockholm, Sweden; Division of Molecular Virology, University of Linköping, Linköping, Sweden

CLAUDIA ISTRATE

Department of Virology, Swedish Institute for Infectious Disease Control, Solna, Sweden; Division of Molecular Virology, University of Linköping, Linköping, Sweden

ELIN JOHANSSON

Department of Virology, Swedish Institute for Infectious Disease Control, Solna, Sweden

DIDIER PONCET

Virologie Moléculaire et Structurale, CNRS-UMR 2472, INRA-UMR 1157, IFR 115, Gif-sur-Yvette, France

LENNART SVENSSON

Division of Molecular Virology, University of Linköping, Linköping, Sweden

INTRODUCTION

Rotavirus (RV) is the major etiological agent of severe dehydrating diarrhea in infants and neonate animals worldwide (Bishop, 1994). Annually, RV disease

Polyphosphazenes for Biomedical Applications, Edited by Alexander K. Andrianov
Copyright © 2009 John Wiley & Sons, Inc.

causes the death of approximately half a million (352,000 to 592,000) children, affecting mainly children in developing countries, and accounts for one-third of hospitalizations for diarrhea worldwide (Parashar et al., 2003; Parashar and Glass, 2006). A number of vaccine candidates have been tested in various animal models and human clinical trials (Barnes et al., 2002; Bernstein et al., 1998; Bertolotti-Ciarlet et al., 2003; Bhandari et al., 2006; Ciarlet et al., 1998; Conner et al., 1993, 1996; Coste et al., 2000; Crawford et al., 1999; Fernandez et al., 1996, 1998; Fromantin et al., 2001; Glass et al., 2005; Gonzalez et al., 2004; Jiang et al., 1999; Johansen et al., 2003; O'Neal et al., 1997; 1998; Parashar and Glass, 2006, Parez et al., 2006; Ruiz-Palacios et al., 2006; VanCott et al., 2006; Vesikari et al., 2006; Yuan et al., 2000; Wood, 2005). Most vaccine candidates are based on live-attenuated RV strains delivered orally to mimic natural infection (Parashar and Glass, 2006). Two of the vaccine candidates, Rotateq (Merck Vaccines, Whitehouse Station, New Jersey) and Rotarix (GlaxoSmithKline Biologicals, Rixensart, Belgium), both live-attenuated oral vaccines, were licensed in 2006 in Europe and several other countries throughout the world (Ruiz-Palacios et al., 2006; Vesikari et al., 2006).

Alternative, second-generation subunit or inactivated vaccines designed for parenteral or mucosal administration have been evaluated in animal models (Bertolotti-Ciarlet et al., 2003; Conner et al., 1993, 1996; Ciarlet et al., 1998; Johansen et al., 2003). Administration of subunit vaccines parenterally may offer advantages over live-attenuated vaccines administered orally, since live oral RV vaccines may be associated with (1) reduced intake by preexisting maternal antibodies transferred over the placenta and in breast milk, (2) development of new reassortments between vaccine strains and wild-type strains, and (3) development of chronic excretion in children with congenital or acquired immunodeficiency. Types of subunit vaccines considered for evaluation in animal models include formalin-inactivated rotavirus and rota-virus-like particles (RV-VLPs), as they are nonreplicating and are immuno-genically and antigenically similar to native infectious RVs.

In three animal models, the infant mouse model, the adult mouse model, and the rabbit model, partial or complete protective immunity has been induced with inactivated rotavirus using empty capsids of simian rotavirus SA-11 (Sheridan et al., 1984), β-propriolactone-, psoralen-, and ultraviolet- or formalin-inactivated rotavirus (Coffin et al., 1997; Conner et al., 1993; Johansen et al., 2003; McNeal et al., 1992; Offit and Dudzik, 1989). RV-VLPs expressing the main structural viral proteins in various combinations, such as 2/6, 2/6/7, 2/4/6/7, 8-2/6/7 have also been assessed for their efficacy in preventing rotavirus infection using a variety of animal models: mice (Bertolotti-Ciarlet et al., 2003; Coste et al., 2000; Crawford et al., 1999; Fromantin et al., 2001; Istrate et al., 2008; Jiang et al., 1999; Johansson et al., 2008) O'Neal et al., 1998), rabbits (Ciarlet et al., 1998; Conner et al., 1996), gnotobiotic piglets (Gonzalez et al., 2004; Yuan et al., 2000) and cows (Fernandez et al., 1996, 1998).

Different routes of administration for inactivated rotavirus particles and RV-VLPs—parenteral, intranasal, oral, and intrarectal—have also been investigated (Conner et al., 1996; Istrate et al., 2008; Johansson et al., 2008; O'Neal et al., 1997; Parez et al., 2006) and using the parenteral route, formalin-inactivated particles (Johansen, 1999; Johansen et al., 2003) and 2/4/6/7-VLPs were proven to confer homologous protection in rabbits (Conner et al., 1996) and heterologous protection in mice (Crawford et al., 1999; Istrate et al., 2008; Jiang et al., 1999; Johansson et al., 2008).

Potent adjuvants known to enhance the immune response to RV subunit vaccines such as alum, cholera toxin, LT(R192G), LT(R72), LT(K63), chimeric A1 subunit of cholera toxin (CTA1)-DD, Resiquimod, poly[di(carboxylato-phenoxy)phosphazene] (PCPP), and CpG ODN (Agnello et al., 2006; Choi et al., 2002; McNeal et al., 1999, 2006; Parez et al., 2006; VanCott et al., 2006) have been studied in various animal models. In a series of experiments we have shown that the potent adjuvant PCPP provides an adjuvant effect when given together with formalin-inactivated rotavirus or rotavirus VLPs parenterally (Istrate et al., 2008; Johansen, 1999; Johansson et al., 2008) and orally (Johansson et al., 2008), with the strongest adjuvant effect observed when used parenterally. Results from the parenteral studies are reviewed in this chapter.

MATERIAL AND METHODS

Production of Formalin-Inactivated Rotavirus and Rotavirus VLPs

Formalin-inactivated rotavirus particles were obtained as described previously after cultivation of rhesus rotavirus (I-RRV) on Ma 104 cells, gradient-purified on sucrose-CsCl, and inactivated with 0.01% formalin for 72 h (Johansen et al., 2003). Virus was titrated on Ma 104 cells before inactivation, and the titer obtained was used as for dose calculation. Remaining infectivity was excluded by virus titration on Ma 104 cells. Virus particles were examined by electron microscopy on the day of immunization to ensure that vaccine batches still contained more than 90% complete rotavirus particles.

Rotavirus-like particles RF 8-2/6/7, based on the bovine rotavirus strain RF, were produced in baculovirus-infected insect cells as described previously (Crawford et al., 1994; Labbe et al., 1991; Parez et al., 2006). Three recombinant baculoviruses were constructed to insert VP6, VP7, and VP8 fused to the VP2 protein. To produce chimeric RF 8-2/6/7 VLPs, Sf9 cells (*Spodoptera frugiperda* clone 9) were co-infected with each recombinant baculovirus at a multiplicity of infection of 5 plaque-forming units (PFU)/cell. Supernatants and cells were harvested 7 days post-infection, and VLPs were extracted with Freon 113 or VertrelXF and gradient purified in sucrose–CsCl (Istrate et al., 2008). The protein concentration in the VLP suspension was estimated by the method of Bradford (Pierce, Rockford, Illinois) with bovine serum albumin

(BSA) as standard. VLPs were inspected for retained integrity by electron microscopy on the day of immunization.

Adjuvant

The adjuvant, water-soluble polyphosphazene—poly[di(carboxylatophenoxy)-phosphazene] sodium salt (PCPP)—was provided by Parallel Solutions, Inc., Cambridge, Massachusetts (Andrianov et al., 2004, 2005; Mutwiri et al., 2007). The average molecular weight of the PCPP was approximately 800,000 g/mol. The PCPP was dissolved in Dulbecco's PBS (phosphate-buffered saline) before use. Various concentrations of PCPP were used in the different studies. In the first study, evaluating formalin-inactivated rotavirus, PCPP was used in a final concentration of 0.1% (100 μg/mouse). In subsequent studies evaluating rotavirus VLPs, 100 μg/mouse was used in adult mice, and newborn mice received a dose of 10 μg of PCPP.

Mice, Immunization, and Challenge

Six- to eight-week-old female BALB/c mice or 3- to 5-day-old infant mice were immunized once or twice in groups of six or seven mice. Individual adult mice were immunized intramuscularly with either formalin-inactivated rotavirus (2×10^5 to 10^8 PFU/dose) or rotavirus VLPs (2 or 20 μg) alone or in combination with PCPP. Infant mice were immunized subcutaneously with 10 μg RF 8-2/6/7 VLPs in a volume of 20 μL, neat or in combination with PCPP. Serum samples were obtained at 4 to 12 weeks post-vaccination in all sets of experiments. Fecal samples were obtained at 1 to 4 weeks post-vaccination from a set of adult mice immunized with formalin-inactivated rotavirus or rotavirus VLPs.

Detection of Rotavirus-Specific Antibodies in Sera and Stools

RV-specific IgG and IgA antibody titers were determined by ELISA (enzyme-linked immunosorkent assay) as described previously (Johansen et al., 2003). Briefly, 96-well plates (Costar, Cambridge, Massachusetts) coated with rabbit anti-rotavirus IgG (K224, SBL, Stockholm, Sweden), diluted 1/1000 in 0.05 M carbonate–bicarbonate buffer (pH 9.6) (Karolinska University Hospital, Department of Chemistry, Stockholm Sweden) were incubated overnight at 4°C. Plates were blocked with 0.1% bovine serum albumin (BSA) (Sigma-Aldrich, St. Louis, Missouri) in PBS for 1 h at 37°C. After blocking, RRV-infected cell lysate (MA 104 cells) diluted 1:100 was added to plates for 1 h at 37°C. Diluted samples were then added to plates and incubated for 2 h at 37°C. Serum samples were diluted in 10-fold dilution steps starting at 1:100, while fecal samples were diluted in twofold dilution steps starting at 1:2. Fecal samples were collected as described previously (VanCott et al., 1998; Lundholm et al., 1999). Briefly, the fecal pellets were weighed and suspended in 0.5 mL of PBS with protease inhibitors (1 µg/mL leuptin, 1.6 µg/mL aprotinin, Sigma-Aldrich, Steinheim, Germany). The debris was removed by centrifugation and the supernatants stored after filtration at −70°C. After each incubation step, the plates were washed five times with saline solution containing 0.05% Tween-20 (Sigma-Aldrich, St. Louis, Missouri). After washing, plates were incubated with horseradish peroxidase (HRP)-conjugated goat anti-mouse IgG diluted 1/40,000 (BioRad Laboratories, Richmond, Virginia) or with goat anti-mouse IgA (Southern Biotechnology Associates, Alabama) diluted 1:1000 for 60 min at 37°C. The substrate used was tetramethyl benzidine (ICN Biochemicals, Ohio) and the reaction developed at room temperature was stopped by addition of H_2SO_4 (2 M) (Fisher Chemicals, United Kingdom). Absorbance was measured at 450 nm with a Flow Titertech ELISA reader (Molecular Devices, Canada). A sample was considered positive when the optical density was greater or equal to twice the negative control sera or control fecal sample. Serum/fecal IgG and IgA RV-specific antibody titers were determined for individual dams and used to calculate the geometric mean titer (GMT) for each group.

Neutralizing antibodies to RRV or bovine rotavirus strain RF were analyzed by a virus peroxidase focus reduction assay as described previously (Johansen et al., 2003). Briefly, serum samples diluted in two-fold dilution steps starting at 1:2 were mixed with 200 µL of trypsin-activated (Sigma-Aldrich, St. Louis, Missouri) RRV or RF virus (titer 200 PFU/mL) and incubated for 1 h at room temperature. The mixture was added to an MA104 cell monolayer for 1 h at 37°C. Cells were then washed and incubated at 37°C. After 18 h of incubation, cells were fixed in 2% paraformaldehyde (Sigma) in PBS and permeabilized with 1% Triton X-100 (Sigma) in PBS for 15 min, and infected cells were stained with a monoclonal antibody diluted 1:100 directed against rotavirus VP6 (255) for 1 h at 37°C. A peroxidase-labeled goat-anti-mouse (Bio Rad) diluted 1:1000 was used as conjugate, and as substrate, 3-aminoethyl carbazole (1 mg/mL) (Sigma) in 0.05 M sodium acetate buffer (pH 5.2) (Karolinska University Hospital, Department of Chemistry, Stockholm Sweden) containing 0.01% H_2O_2 (Sigma) was used. The neutralizing titers were defined as the reciprocal of the serum dilution, showing a 60% reduction in the number of cells infected.

Statistical Analysis

Statistical comparisons between the groups were performed using the non-parametric Kruskal–Wallis test to analyze and compare antibody responses. A significant difference was considered when a p-value below 0.05 was obtained. A one-way analysis of variance nonparametric test was performed when comparing fecal IgA titers using GraphPad Prism version 4.0 for MacIntosh, OS 9, and Apple (GraphicPad Software), to compare medians between groups at the $p < 0.05$ and $p < 0.001$ levels.

RESULTS

1. *PCPP enhances rotavirus-specific serum IgG and IgA immune response to formalin-inactivated RRV and RF 8-2/6/7 VLPs administered intramuscularly in adult mice.* To evaluate whether administration of purified formalin-inactivated rotavirus in various doses (2×10^5 to 10^8 PFU/dose) combined with a constant dose of PCPP (100 μg) would induce a better humoral immune response than I-RRV alone, mice were immunized twice intramuscularly. In animals immunized with I-RRV-PCPP (2×10^6 to 10^8 PFU/dose), about 10-fold higher serum IgG antibody titers developed after two immunizations than in animals immunized with I-RRV (2×10^6 to 10^8 PFU/dose) alone (Table 1). While an excellent serum IgG response was noted, only low or undetectable serum IgA was seen. Except for the animals that received the lowest dose, all animals immunized with I-RRV-PCPP developed fecal IgG and IgA antibodies, which contrasts to animals immunized with I-RRV without adjuvant. Animals immunized with I-RRV-PCPP in doses from 2×10^6 to 10^8 PFU/dose also developed significantly higher neutralizing antibody titers (10- to 20-fold) than those for animals immunized with I-RRV without adjuvant (Table 1).

To further evaluate the specific systemic immune response developed after intramuscular immunization with RV-VLPs alone or in combination with PCPP as an adjuvant provided in a single-dose regimen, RV-specific serum IgG and IgA was determined by ELISA. Immunization with RF 8 2/6/7-VLPs, in a dose of 2 μg or 20 μg, induced RV-specific IgG antibodies with GMT titers that continued to rise until 8 weeks post-vaccination (pvac), when it leveled off (Fig. 1a). The highest dose tested, 20 μg of RV-VLPs, induced a good serum IgG response at 2, 4, 8, and 12 weeks. When the highest dose was combined with the adjuvant PCPP, specific IgG titers continued to rise until 12 weeks pvac, and the titers obtained were 10- to 90-fold higher at 2, 4, 8, and 12 weeks pvac (Fig. 1a). Animals immunized with 20 μg of RV-VLPs with adjuvant developed slightly but not statistically significant higher antibody responses than when 2 μg was administered, at weeks 2 and 4, whereas it was statistically significant at weeks 8 and 12. Dams from groups receiving RV-VLPs in the two vaccine doses evaluated, 2 μg and 20 μg, were also found to develop serum-neutralizing antibodies against bovine rotavirus strain RF during the 12-week period of observation. At week 12, a two- to fourfold increase in neutralizing antibodies was noted in groups that received the combined vaccine with the adjuvant PCPP compared to the RV-VLPs alone.

RV specific serum IgA developed at a lower level but followed the same pattern as IgG with a rise in the first 8 weeks when RV-VLPs were administered alone at 2, 4, 8, and 12 weeks pvac, and a rise for 12 weeks when RV-VLPs were administered in combination with adjuvant at 2, 4, 8, and 12 weeks pvac (Fig. 1b). The adjuvant effect obtained in groups that received adjuvant was 5- to 15-fold higher.

TABLE 1 Rotavirus Antibody Titers Following Two Parenteral Intramuscular Immunizations of Mice with Formalin-Inactivated RRV (I-RRV) with and Without PCPP[a]

Vaccine dose	Adjuvans 100 μg/mouse	Rotavirus-Specific Antibody Titers After Two Immunizations			
		Serum IgG[b] After 2nd Immunization	Neutralizing Antibodies[a] After 2nd Immunization	Fecal IgA	Fecal IgG
I-RRV 2×10^8	—	151,638±30,737 (6/6)[c]	28,735±6,400 (6/6)	2±0 (4/6)	9±11 (5/6)
I-RRV 2×10^7	—	29,182±8,100 (6/6)	14,367±7,010 (6/6)	2±0 (3/6)	2±0 (2/6)
I-RRV 2×10^6	—	11,682±4,276 (5/6)	3,591±9,893 (6/6)	<2 (0/6)	2±0 (2/6)
I-RRV 2×10^5	—	2,700±2,163 (2/6)	1,600±1,346 (3/6)	<2 (0/6)	<2 (0/6)
I-RRV 2×10^8	PCPP	1638,970±218,700 (6/6)	258,031±43,175 (6/6)	9±1 (6/6)	20±19 (6/6)
I-RRV 2×10^7	PCPP	546,323±72,900 (6/6)	204,800±57,115 (6/6)	5±1 (6/6)	6±5 (6/6)
I-RRV 2×10^6	PCPP	87,066±34,269 (5/6)	64,507±27,782 (6/6)	5±2 (6/6)	7±2 (6/6)
I-RRV 2×10^5	PCPP	8,100±0 (1/6)	1,795±1,145 (2/6)	ND[e]	<2 (0/2)
PBS	—	<100 (0/2)	<100 (0/2)	<2 (0/2)	<2 (0/2)

Source: Johansen (1999).

[a] Female BALB/c mice (six animals/group) were immunized intramuscularly at 0 and 3 months with formalin-inactivated RRV either with or without adjuvant. Serum samples were collected 2 weeks after last immunization and were analyzed for rotavirus-specific antibodies starting at a 1:100 dilution. Fecal samples were collected 2 weeks after the completed vaccination schedule and were analyzed for antibodies starting at a 1:2 dilution.

[b] The data are expressed as the reciprocal of serum and fecal antibody titers (GMT ± SEM).

[c] Number of responders in each group of six animals.

[d] The data are expressed as the reciprocal of the serum dilution, showing a 60% reduction in number of RRV-infected cells (GMT ± SEM).

[e] Not done, due to technical failure.

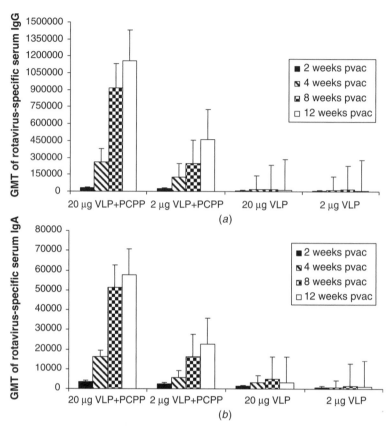

FIGURE 1 Rotavirus-specific serum (a) IgG and (b) IgA in adult mice immunized in a one-dose regimen with either 2 or 20 μg of RV-VLP with or without 100 μg PCPP. (From Istrate et al., 2008.)

2. *PCPP enhances serum rotavirus-specific fecal IgG and IgA immune response to formalin-inactivated RRV and RF 8-2/6/7 VLPs when administered intramuscularly in adult mice.* After vaccination with formalin-activated RRV, fecal IgG antibodies developed in a dose-dependent manner, but modest titers were noted overall (Table 1). Very low fecal IgA titers were also noted (Table 1). RV-specific fecal IgG and IgA were also determined by ELISA after vaccination with RV-VLPs to evaluate the specific fecal antibody levels after intramuscular immunization with RV-VLPs alone or in combination with the adjuvant. RV-specific fecal antibodies were already detectable 1 week pvac. Low levels of fecal IgA obtained in animals receiving 20 and 2 μg RV-VLPs at 1, 2, 3, and 4 weeks pvac were enhanced only slightly in combination with adjuvant (Fig. 2b). In addition, low levels of fecal IgG were obtained in groups that received 20 μg of RV-VLP at 1, 2, 3, and

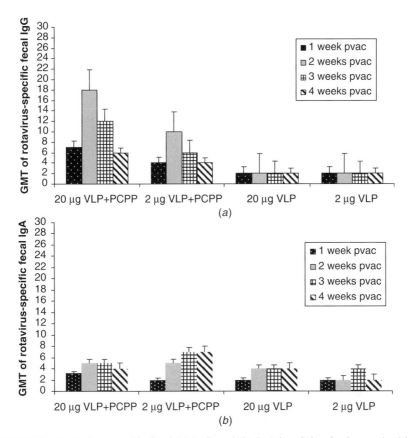

FIGURE 2   Rotavirus-specific fecal (a) IgG and (b) IgA in adult mice immunized in a one-dose regimen with either 2 or 20 µg of RV-VLP with or without 100 µg PCPP. (From Istrate et al., 2008.)

4 weeks pvac, but in both groups that received 2 µg and 20 µg of RV-VLP in combination with adjuvant, a three- to ninefold rise in IgG titers was observed at 1 to 4 weeks pvac (Fig. 2a).

3. *PCPP enhances rotavirus-specific serum IgG immune response to RF 8-2/6/7 VLPs administered subcutaneously to infant mice.* Offspring of previously rotavirus-uninfected dams responded to a single subcutaneous immunization with RF 8*-2/6/7 VLPs at days 3 to 5 of life, with an antibody response detected in all six individuals 4 to 8 weeks post-immunization (Table 2). Addition of the adjuvant PCPP, increased the antibody response only slightly at 4 and 8 weeks post-immunization (twofold, no statistical significance, Table 2). All individuals in these two groups showed a response with increased levels of IgG between 4 and 8 weeks post-vaccination.

TABLE 2   Rotavirus-Specific Serum IgG Titers in Offspring to Uninfected and Unimmunized Dams After Subcutaneous Immunization with Rf 8-2/6/7 VLP with or Without the Adjuvant PCPP at Day 3 of Age

Immunization	n	Time After Immunization[a]	
		4 Weeks; GMT (range)	8 Weeks; GMT (range)
Rf 8*-2/6/7 VLP	6	2,263 (800–6,400)	10,159 (3,200–12,800)
Rf 8*-2/6/7 VLP + PCPP	7	4,307 (400–12,800)	25,600 (6,400–51,200)

Source: Johansen (2008).
[a] GMT, geometric mean titer.

DISCUSSION

It is generally agreed that a rotavirus vaccine would improve child morbidity significantly in both developed and developing countries and would have a significant impact on child mortality, mainly in developing countries. Since rotaviruses most often infect the intestinal epithelium of the small intestine, initial development of an oral vaccine was natural. The protective efficacy after a natural rotavirus infection against subsequent symptomatic rotavirus infection has been estimated to be 58 to 75% (Mrukowicz et al., 1999; Velazquez et al., 1996; Ward and Bernstein, 1994).

Two new live oral RV vaccines have recently been licensed for use in children showing good efficacy (>90%) and safety profiles after two or three doses, respectively (Ruiz-Palacios et al., 2006; Vesikari et al., 2006). The two vaccines have, since licensure, been introduced in pediatric immunization programs in a number of countries throughout the world. However, results from clinical trials in developing countries, where the vaccine is most needed, still await.

Oral delivery of vaccines may be associated with certain limitations, such as poor long-term immunologic memory (John and Jayabal, 1972), inhibition of vaccines uptake by maternal antibodies in the youngest children (Steele et al., 2005), the possibilities of development of reassortments between vaccine strains and wild-type strains, limitations in providing a live vaccine to immunodeficient children, including HIV-infected and severely malnourished children, and competing pathogens in the intestinal tract (Brandzaeg, 2007). Parenteral immunization has therefore become a recognized potential vaccination strategy against rotavirus diarrhea.

Optimally, an RV vaccine should induce protective immunity after a single dose at a very early age. In the studies we performed, we started out with a two-dose schedule of formalin-inactivated rotaviruses combined with a potent adjuvant, PCPP. In this first study, we observed that the adjuvant PCPP induced a highly significant enhancement in the development of rotavirus-specific serum IgG, neutralizing serum antibodies, and fecal IgG. We have continued our studies with the novel RF 8-2/6/7-VLPs and shown systemic and fecal humoral immune responses in adult mice after intramuscular immunization in a single-dose regimen.

The possibility of using a single-dose RV vaccine has not been evaluated before and is attractive for inducing protective immunity in early life. RV-induced diarrhea in developed countries occurs most commonly in children between 6 and-24 months of age (Uhnoo et al., 1986), while the first RV illness in developing countries occurs in about 50% of children before 2 months of age (Espinoza et al., 1997). If a single-dose regimen would prove protective in other models and in clinical trials, the need for production of large quantities of vaccine would be reduced. The low dose of 2×10^6 PFU of RRV in combination with the adjuvant PCPP used in this study induced a good immune response and subsequently, protective immunity (Johansen, 1999). Similarly, a dose of $2\,\mu g$ of RV-VLPs in combination with the model adjuvant PCPP induced protective immunity. This dose includes significantly less VLP than used previously by other investigators (Bertolotti-Ciarlet et al., 2003; Coste et al., 2000; Crawford et al., 1999; Jiang et al., 1999; O'Neal et al., 1997; Parez et al., 2006).

For decades, parenteral administration has been used to stimulate protective immunity against pathogens whose entry or replication sites are limited to mucosal surfaces. This includes vaccines against measles, mumps, rubella, influenza, polio, hepatitis A, whooping cough, and more recently, human papillomavirus. The recently licensed oral live RV vaccines are a significant step forward toward reducing the morbidity caused by RV throughout the world. However, to reduce circulating rotavirus in larger geographical areas, a nonlive vaccine is probably needed.

Identifying new adjuvants for use in people of all ages has not been easy and is needed for several vaccines where a rapid immune response to a small amount of antigen is possible such as in the case of pandemic vaccines or where waning immunity is a concern (Siegrist, 2007). PCPP has been shown to be a potent adjuvant with many diverse antigens. The exact immunostimulatory effect of PCPP is not fully elucidated, but in a recent study it was shown that PCPP was associated predominantly with an IL-4 response (Mutwiri et al., 2007), although we have previously noted low IgG1/IgG2a ratios (<1), suggesting activation of both humoral and cell-mediated immune responses (Johansen, 1999). In all experiments with adult mice, a significant increase in the antibody response was observed. The less significant increase in infant mice could possibly be explained by the fact that newborn mice are less developed immunologically: for example, that the immune response is biased toward a Th2 direction in newborns compared to adults and by the fact that they were given a smaller dose of antigen and adjuvant for practical reasons related to the vaccine volume.

A measurable fecal IgG response was noted in all animals provided PCPP-adjuvanted vaccines, regardless of whether formalin-inactivated or VLP-based vaccine was used, whereas the total fecal IgG response, as well as the number of responders, were lower in the groups receiving the two vaccine types alone.

The fecal IgA response was in general low but was found to be higher in animals receiving PCPP-adjuvanted vaccines than in animals immunized with the vaccine without adjuvant.

We conclude that inactivated rotavirus vaccines provided intramuscularly in a two-dose regimen and subunit rotavirus vaccines provided intramuscularly in a single-dose regimen with a potent adjuvant such as PCPP stimulate an excellent immune response, including neutralizing antibodies in the adult mouse model. The response was somewhat diminished in the infant mouse model. This concept could be evaluated further in this and other animal models using different rotavirus G/P genotypes, evaluating the induction of possible heterotypic cross-reactive protection to all circulating genotypes in humans against further development of possible human vaccine candidates.

REFERENCES

1. Agnello, D., Hervé, C., et al. 2006. Intrarectal immunization with rotavirus 2/6 virus-like particles induces an anti-rotavirus immune response localized in the intestinal mucosa and protects against rotavirus infection in mice. *J. Virol.*, 80(8):3823–3832.

2. Andrianov, A., Svirkin, Y., et al. 2004. Synthesis and biologically relevant properties of polyphosphazene polyacids. *Biomacromolecules*, 5(5):1999–2006.

3. Andrianov, A., Marin, A., et al. 2005. Polyphosphazene polyelectrolytes: a link between the formation of non-covalent complexes with antigenic proteins and immunostimulating activity. *Biomacromolecules*, 6(3):1375–1379.

4. Barnes, G.L., Lund, J.S., et al. 2002. Early phase II trial of human rotavirus vaccine candidate RV3. *Vaccine*, 20:2950–2956.

5. Bernstein, D.I., Smith, V.E., et al. 1998. Safety and immunogenicity of live, attenuated human rotavirus vaccine 89-12. *Vaccine*, 16(4):381–387.

6. Bertolotti-Ciarlet, A., Ciarlet, M., et al. 2003. Immunogenicity and protective efficacy of rotavirus 2/6-virus-like particles produced by a dual baculovirus expression vector and administred intramuscularly, intranasally, or orally to mice. *Vaccine*, 21:3885–3900.

7. Bhandari, N., Sharma, P., et al. 2006. Safety and immunogenitcity of two live attenuated human rotavirus vaccines candidates, 116E and I321, in infants: results of a randomised controlled trial. *Vaccine*, 24(31-32):5817–5823.

8. Bishop, R. 1994. *Natural History of Human Rotavirus Infections*. Marcel Dekker, New York.

9. Brandzaeg, P. 2007. Induction of secretory immunity and memory at mucosal surfaces. *Vaccine*, 25:5467–5484.

10. Choi, A., McNeal, M., et al. 2002. Intranasal or oral immunization of inbred and outbred mice with murine of human rotavirus VP6 proteins protects against viral shedding after challenge with murine rotaviruses. *Vaccine*, 20:3310–3321.

11. Ciarlet, M., Crawford, S.E., et al. 1998. Subunit rotavirus vaccine administered parenterally to rabbits induces active protective immunity. *J. Virol.*, 72(11):9233–9246.

12. Coffin, S.E., Moser, C.A., et al. 1997. Immunologic correlates of protection against rotavirus challenge after intramuscular immunization of mice. *J. Virol.*, 71(10):7851–7856.

13. Conner, M.E., Crawford, S.E., et al. 1993. Rotavirus vaccine administered parenterally induces protective immunity. *J. Virol.*, 67(11):6633–6641.

14. Conner, M.E., Zarley, C.D., et al. 1996. Virus-like particles as a rotavirus subunit vaccine. *J. Infect. Dis.*, 174(suppl 1):S88–S92.

15. Coste, A., Sirard, J., et al. 2000. Nasal immunization of mice with virus-like particles protects offsprings against rotavirus diarrhea. *J. Virol.*, 74:8966–8971.

16. Crawford, S.E., Labbe, M., et al. 1994. Characterization of virus-like particles produced by the expression of rotavirus capsid proteins in insect cells. *J. Virol.*, 68(9):5945–5922.

17. Crawford, S.E., Estes, M.K., et al. 1999. Heterotypic protection and induction of a broad heterotypic neutralization response by rotavirus-like particles. *J. Virol.*, 73(6):4813–4822.

18. Espinoza, F., Paniagua, M., et al. 1997. Rotavirus infections in young Nicaraguan children. *Pediatri. Infecti. Dis. J.*, 16(6):564–571.

19. Fernandez, F.M., Conner, M.E., et al. 1996. Isotype-specific antibody responses to rotavirus and virus proteins in cows inoculated with subunit vaccines composed of recombinant SA11 rotavirus core-like particles (CLP) or virus-like particles (VLP). *Vaccine*, 14(14):1303–1312.

20. Fernandez, F.M., Conner, M.E., et al. 1998. Passive immunity to bovine rotavirus in newborn calves fed colostrum supplements from cows immunized with recombinant SA11 rotavirus core-like particle (CLP) or virus-like particle (VLP) vaccines. *Vaccine*, 16(5):507–516.

21. Fromantin, C., Jamot, B., et al. 2001. Rotavirus 2/6 virus-like particles administered intranasally in mice with or without the mucosal adjuvants cholera toxin and *Escherichia coli* heat-labile toxin, induce a Th1/Th2-like immune response. *J. Virol.*, 75:11010–11016.

22. Glass, R., Bhan, M., et al. 2005. Development of candidate rotavirus vaccines derived from neonatal strains in India. *J. Infecti. Dis.*, 192(suppl. 1):S30–S35.

23. Gonzalez, A., Nguyen, T., et al. 2004. Antibody responses to human rotavirus (HRV) in gnotobiotic pigs following a new prime/boost vaccine strategy using oral attenuated HRV priming and intranasal VP2/6 rotavirus-like particle (VLP) boosting with ISCOM. *Clini. Exp. Immunol.*, 135(3):361–372.

24. Istrate, C., Hinkula, J., et al. 2008. Parenteral administration of RF 8-2/6/7 rotavirus-like particles in a one-dose regimen induce protective immunity in mice. *Vaccine*, 26:4594–4601.

25. Jiang, B., Estes, M.K., et al. 1999. Heterotypic protection from rotavirus infection in mice vaccinated with virus-like particles. *Vaccine*, 17(7–8):1005–1013.

26. Johansen, K., 1999. Immune responses related to protection against rotavirus after natural infection and vaccination. Ph.D. dissertation, Microbiology and Tumorbiology Center and Department of Woman and Child Health, Karolinska Institute, Stockholm, Sweden.

27. Johansen, K., Schroder, U., et al. 2003. Immunogenicity and protective efficacy of a formalin-inactivated rotavirus vaccine combined with lipid adjuvants. *Vaccine*, 3565:1–8.

28. Johansson, E., Istrate, C., et al. 2008. Amount of maternal rotavirus-specific antibodies influence the outcome of rotavirus vaccination of newborn mice with virus-like particles. *Vaccine*, 26:778–785.

29. John, T., Jayabal, P. 1972. Oral polio vaccination in children in the tropics: the poor seroconversion rates and the absence of viral interference. *Am. J. Epidemiol.*, 96:263–269.

30. Labbe, M., Charpilienne, A., et al. 1991. Expression of rotavirus VP2 produces empty corelike particles. *J. Virol.*, 65(6):2946–2952.

31. Lundholm, P., Asakura, Y., et al. 1999. Induction of mucosal IgA by a novel jet delivery technique for HIV-1 DNA. *Vaccine*, 17(15–16):2036–2042.

32. McNeal, M.M., Sheridan, J.F., et al. 1992. Active protection against rotavirus infection of mice following intraperitoneal immunization. *Virology*, 191(1):150–157.

33. McNeal, M., Rae, M., et al. 1999. Effects of different adjuvants on rotavirus antibody responses and protection in mice following intramuscular immunization with inactivated rotavirus. *Vaccine*, 17:1573–1580.

34. McNeal, M., Stone, S., et al. 2006. Protection against rotavirus shedding after intranasal immunization of mice with a chimeric VP6 protein does not require intestinal IgA. *Virology*, 346:338–347.

35. Mrukowicz, J.Z., Thompson, J., et al. 1999. Epidemiology of rotavirus in infants and protection against symptomatic illness afforded by primary infection and vaccination. *Vaccine*, 17(7-8):745–753.

36. Mutwiri, G., Benjamin, P., et al. 2007. Poly[di(sodium carboxylatoethylphenoxy)-phosphazene] (PCEP) is a potent enhancer of mixed Th1/Th2 immune responses in mice immunized with influenza virus antigens. *Vaccine*, 25:1204–1213.

37. Offit, P.A., Dudzik, K.I. 1989. Noninfectious rotavirus (strain RRV) induces an immune response in mice which protects against rotavirus challenge. *J. Clini. Microbiol.*, 27(5):885–888.

38. O'Neal, C.M., Clements, J.D., et al. 1998. Rotavirus 2/6 viruslike particles administered intranasally with cholera toxin, *Escherichia coli* heat-labile toxin (LT), and LT-R192G induce protection from rotavirus challenge. *J. Virol.*, 72(4):3390–3393.

39. O'Neal, C.M., Crawford, S.E., et al. 1997. Rotavirus virus-like particles administered mucosally induce protective immunity. *J. Virol.*, 71(11):8707–8717.

40. Parashar, U., Glass, R. 2006. Public health. progress toward rotavirus vaccines. *Science*, 312(5775):851–852.

41. Parashar, U., Hummelman, E., et al. 2003. Global ilness and deaths caused by rotavirus disease in children. *Emerg. Infect. Dis.*, 9:565–572.

42. Parez, N., Fourgeux, C., et al. 2006. Rectal immunization with rotavirus virus-like particles induces systemic and mucosal humoral immune responses and protects mice against rotavirus infection. *J. Virol.*, 80(4):1752–1761.

43. Ruiz-Palacios, G., Pérz-Schael, I., et al. 2006. Safety and efficacy of an attenuated vaccine against severe rotavirus gastroenteritis. *N. Engl. J. Med.*, 354(1):11–22.

44. Sheridan, J.F., Smith, C.C., et al. 1984. Prevention of rotavirus-induced diarrhea in neonatal mice born to dams immunized with empty capsids of simian rotavirus SA-11. *J. Infect. Dis.*, 149(3):434–438.

45. Siegrist, C-A 2007. The challenges of vaccine responses in early life: selected examples. *J. Comp. Pathol.*, 137:S4–S9.

46. Steele, D., Tumbo, J., et al., 2005. *Concomitant administration of a live-attenuated oral rotavirus vaccin (RIX4414) with poliovirus vaccines in African infants.* Presented at the 23rd Annual Meeting of the European Society for Pediatric Infectious Diseases, Valencia, Spain.

47. Uhnoo, I., Olding-Stenkvist, E., et al. 1986. Clinical features of acute gastroenteritis associated with rotavirus, enteric adenoviruses, and bacteria. *Arch. Dis. Child.*, 61(8):732–738.

48. VanCott, T., Kaminski, R., et al. 1998. HIV-1 neutralizing anitbodies in the genital and respiratory tracts of mice intranasally immunized with oligomeric gp160. *J. Immunol.*, 160:2000–2012.

49. VanCott, J., Prada, A., et al. 2006. Mice develop effective but delayed protective immune responses when immunized as neonates either intranasally with nonliving VP6/LT(R192G) or orally with live rhesus rotavirus vaccine candidates. *J. Virol.*, 80(10):4949–4961.

50. Velazquez, F.R., Matson, D.O., et al. 1996. Rotavirus infections in infants as protection against subsequent infections. *N. Engl. J. Med.*, 335(14):1022–1028.

51. Vesikari, T., Matson, D., et al. 2006. Safety and efficacy of a pentavalent human–bovine (WC3) reassortant rotavirus vaccine. *N. Engl. J. Med.*, 354(1):23–33.

52. Ward, R.L., Bernstein, D.I. 1994. Protection against rotavirus disease after natural rotavirus infection. US Rotavirus Vaccine Efficacy Group. *J. Infect. Dis.*, 169(4):900–904.

53. Wood, D. 2005. WHO informal consultation on quality, safety and efficacy specifications for live attenuated rotavirus vaccines, Mexico City, 8–9 February. *Vaccine*, 23(48–49):5478–5487.

54. Yuan, L., Geyer, A., et al. 2000. Intranasal administration of 2/6-rotavirus-like particles with mutant *Escherichia coli* heat-labile toxin (LT-R192G) induces antibody-secreting cell responses but not protective immunity in gnotobiotic pigs. *J. Virol.*, 19(8843–8853).

7 Polyphosphazene Immunoadjuvants for Intradermal Vaccine Delivery

ALEXANDER K. ANDRIANOV, DANIEL P. DECOLLIBUS,
HELICE A. GILLIS, HENRY H. KHA, and
ALEXANDER MARIN

Apogee Technology, Inc., Norwood, Massachusetts

INTRODUCTION

Skin constitutes a natural barrier against infection and has a high density of dendritic cells, such as Langerhans cells, whose function it is to recognize foreign bacteria and viruses and initiate an effective immune response. This makes the immune system of the skin a desirable target for vaccination, and a number of reports have been published recently on the intradermal delivery of vaccines [1–5].

Intradermal immunization is, however, facing technical challenges that must be addressed to effectively administer vaccines to the skin. The approach requires either special personnel training [6] or the development of technologies that do not involve the use of conventional needles. To overcome the stratum corneum barrier and increase skin permeability, various alternative approaches have been explored, which include both chemical and physical techniques.

One of the most attractive methodologies is the use of microneedles, submillimeter structures capable of penetrating the stratum corneum and releasing the vaccine in the appropriate skin compartment, such as the epidermis or dermis [7,8] (Fig. 1). Two principal types of microneedles have been used for intradermal vaccine delivery: hollow microneedles and microneedles with solid-state vaccine formulation. Hollow microneedles allow infusion of liquid formulation into the skin. They generally do not require reformulation of vaccine, but are more expensive to produce and involve the added cost and complexity of integrating the injection device.

Microneedles that utilize solid vaccine formulations are either micro-fabricated from the formulation itself or have a metal support on which the

Polyphosphazenes for Biomedical Applications, Edited by Alexander K. Andrianov
Copyright © 2009 John Wiley & Sons, Inc.

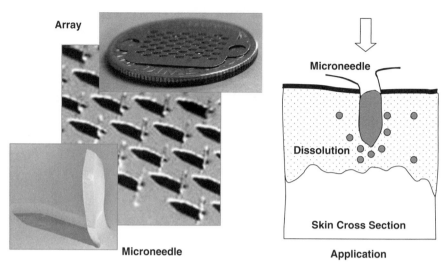

FIGURE 1 Scanning electron microscopy image of a single microneedle, optical microscopic image of the microneedle array, and schematic presentation of the application process.

formulation is deposited as a coating. In both cases, the use of binding polymer is critical to provide sufficient mechanical strength and to maintain the three-dimensional integrity of the formulation. Microneedles can be administered in the form of a patch that can be applied similar to a Band-Aid. After peeling off a protective release liner to expose the microneedles, the patch is pressed to the skin, which may be accomplished by hand or may require the use of a small device to assure correct microneedle insertion. When the microneedle is inserted in the skin, the formulation is dissolved to release vaccine antigen in a controlled fashion. This imposes further requirements to the binding polymer, such as fast dissolution or degradation profile in aqueous environments. The additional advantage of solid microneedles is in their potentially improved stability compared to conventional liquid formulations, which can lead to a better product shelf life.

Contemporary vaccine technologies, however, entail the use of immunoadjuvants to induce the desired immune responses [9]. Unfortunately, many vaccine adjuvants may not be compatible with an intradermal delivery approach. For example, alum, which is by far the most common adjuvant used in commercial vaccines, was shown to induce serious adverse effects, such as formation of granuloma, when administered intradermally [10]. Other potent adjuvants, such as emulsions or liposomes, are biphasic and may not be sufficiently stable to withstand the microneedle coating processes.

In a quest for an immunoadjuvant compatible with a solid-phase delivery technology, polyphosphazene polyelectrolytes appear to be positioned uniquely. Macromolecules combining ionic functionalities with a polyphosphazene

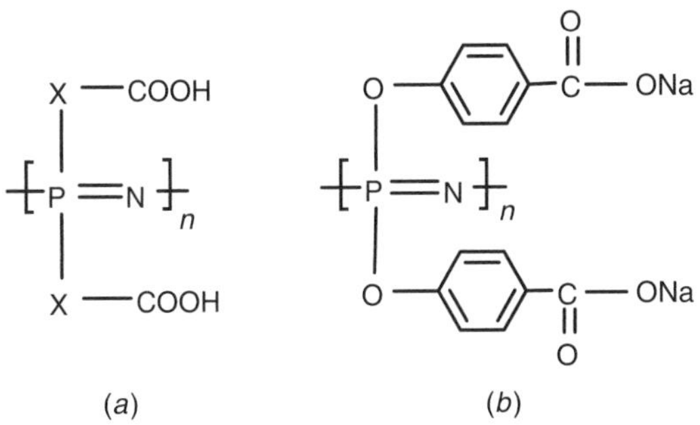

SCHEME 1   General structure of polyphosphazene vaccine adjuvants (a) and the lead compound, PCPP (b).

backbone have attracted considerable attention, due to their water solubility, degradability of the backbone, and microsphere-forming properties [11–13]. Most important, they demonstrated excellent immunostimulating characteristics, and a lead compound, poly[di(carboxylatophenoxy)phosphazene (PCPP) (Scheme 1), showed significant potential in multiple animal models and in clinical trials [14]. The macromolecular nature of ionic polyphosphazenes provides means for its use in two roles: as a coating-forming material for microneedles and as a vaccine adjuvant. This potential dual functionality is especially important for solid-formulation microneedle technology, since mechanical and dissolution requirements impose severe restrictions on the microneedle loading capacity. In this chapter we report on the fabrication of polyphosphazene-based microneedles, their general in vitro characteristics, and their performance in intradermal immunization studies.

FORMULATION ASPECTS OF MICRONEEDLE TECHNOLOGY

Fabrication of Microneedles with Solid Formulations

In our studies we used microneedles consisting of a metal support and a polymeric coating—a solid antigen containing formulation dissolvable upon administration to the skin. Microneedle arrays were produced in a two-stage process. First, arrays of 50 needlelike structures were produced by chemical etching of titanium foil and bent out of plane at a $90°$ angle. Each protrusion was $600\,\mu m$ long, and the arrays had dimensions of $1 \times 1\,cm$. A micro-dip-coating process was then used to coat the tips of these protrusions with a polymer–protein formulation to fabricate microneedles. The process involved the use of a reservoir containing individual microwells corresponding to each protrusion in the array, and consisted of contacting the protrusions with the formulation in the reservoir [15–17]. The procedure included drying the coating with anhydrous nitrogen and was repeated as needed to produce the desired dose of material in the microneedle. In our first experiments we used an aqueous solution of sodium salt of carboxymethyl cellulose (CMC) as a coating-forming polymer, formulated with a model protein, horseradish peroxidase (HRP). CMC is an excipient employed in many parenteral formulations and has also been a frequent choice in many microneedle formulations.

Antigen Dosage

Micro dip-coating is a relatively simple and scalable process, its main challenge for biological applications being precise antigen dosing. It is customary to correlate the amount of biological agent to be deposited on the microneedles to the number of coating cycles (dips) to which the microneedle was subjected. For example, it is often assumed that microneedles which undergo the same predetermined number of cycles will receive the same antigen dose. However, the evaporation process leads to changes in the concentration and viscosity of the formulation, resulting in the unequal deposition of antigen with every subsequent coating. To minimize these adverse effects, an approach based on measuring the actual volume of formulation supplied to the microneedles in the coating process has been developed. This volumetric method takes into account the actual consumption of formulation in the micro-dip-coating process rather than the numbers of cycles. Essentially, the microprotrusion is immersed in the coating formulation repeatedly until the formulation in the microwell of the coating apparatus is fully consumed (Fig. 2). Then the microwell is refilled and the process is repeated if necessary.

The correlation between the loading of the protein on the microneedle, as detected by high-performance liquid chromatography (HPLC), and the amount of the antigen and polymer supplied volumetrically to be consumed in the micro-dip-coating process is shown in Figure 3a. Results demonstrate the high accuracy of the approach and the linear dependence of the dose on the amount of protein supplied.

Efficiency of Protein Encapsulation

One of the critical parameters of any protein encapsulation technique, including microneedle coating, is how well the biological material maintains its functional integrity in the process. Although the micro-dip-coating method is carried out under mild conditions and ambient temperatures in an aqueous solution, it involves a drying process, which can be potentially detrimental to the protein. As shown in from Figure 3a, the dose of HRP deposited on the microneedles, as detected by HPLC, correlates well with the amounts supplied, which indicates no detectable loss of material in the coating process. The functional activity of the protein deposited on the microneedles, which in the case of HRP is enzymatic activity, was also evaluated. Figure 3b shows the activity of the enzyme coated on microneedles as a function of the amount of

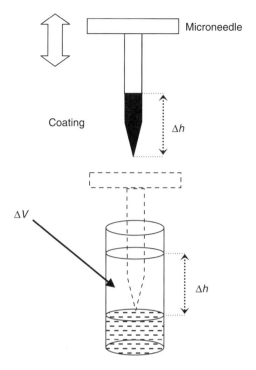

Microwell containing formulation solution

FIGURE 2   Fabrication of a microneedle using a micro-dip-coating process in which an antigen dose on the microneedle is controlled volumetrically. A precise volume of formulation solution is supplied to the microwell and then fully consumed in a process in which the microneedle repeatedly comes in to contact with the formulation.

enzyme used in the coating process. This activity ("experimental") was also compared to the activity of the enzyme calculated on the basis of HPLC quantitation and initial activity of the enzyme in solution ("theoretical"). The results show excellent correlation between theoretical and experimental curves, indicating practically no loss of activity in the coating process for this sensitive enzyme.

Protein Stability in Solid-State Microneedle Coatings

One of the most significant potential advantages of coated microneedle technology is the potentially better shelf life of solid-state formulations than that of their solution counterparts. The stability of coated microneedles containing HRP was compared to the same enzyme in aqueous formulation at 80°C. The accelerated degradation profiles show a dramatic improvement in the stability of a solid-state formulation (Fig. 4, curve 1) compared to that of HRP in solution (Fig. 4, curve 2).

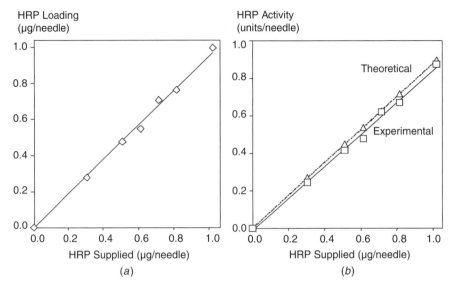

FIGURE 3 (a) Amount and (b) enzymatic activity of HRP loaded on microneedles versus the amount of HRP supplied for microneedle coating.

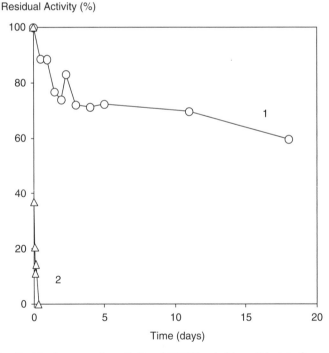

FIGURE 4 Residual enzymatic activity of HRP loaded in solid-state formulations on microneedles (1) and in aqueous solution (2) during accelerated thermal stability studies as a function of time (80°C, coating contained CMC as a binding polymer).

POLYPHOSPHAZENES AS MULTIFUNCTIONAL COMPOUNDS FOR MICRONEEDLE VACCINE DELIVERY

Formation of Microneedle Coating and Role of Polyphosphazenes

Coated microneedle technology imposes certain requirements on the properties of polymers, such as water solubility, biodegradability, adequate viscosity-enhancing characteristics, and ideally, a history of human use. In addition to these general prerequisites, it is also desirable that macromolecules improve process efficiency. For example, an accelerated construction of the coating can minimize production time and, even more important, decrease overall drying time: a phase that can potentially have an adverse effect on sensitive antigens. Therefore, polymeric film-forming agents should provide the minimal number of coating and drying cycles needed for formation of the coating.

PCPP, a lead polyphosphazene immunoadjuvant, appears to address most of the requirements noted above. It is water soluble, biodegradable, and has a history of use in clinical trials [18,19]. Furthermore, to afford strong immunostimulating activity, it is common to use PCPP with a molecular weight in excess of 500,000 g/mol [13,20], so that highly viscous solutions can be prepared.

To investigate the effect of polyphosphazene on the efficiency of the coating process, its film-forming properties were compared with those of CMC, which was used above as a coating-forming material. The performances of both macromolecules were evaluated based on their ability to facilitate the deposition on microneedles of the model antigen bovine serum albumin (BSA). The amount of protein in the microneedles was detected by rinsing the arrays and analyzing the solute using HPLC. To attain the same viscosity-enhancing characteristics [5.1 centipoise (cps) in protein-free solution], PCPP in the coating formulation was used at a concentration of 0.5% (w/v), while the concentration of CMC was somewhat higher, 0.8% (w/v). Unexpectedly, it was found that the rate of antigen deposition using BSA formulations was dramatically higher in the presence of PCPP (Fig. 5, curves 1 and 3). Furthermore, the increase in CMC concentration to 1.5% (w/v) (Fig. 5, curve 2) still did not bring the rate of coating formation to the levels achieved with PCPP. The advantage of PCPP formulations can be illustrated by the fact that the loading of 0.45 μg of BSA per microneedle can be achieved with only four coating cycles using PCPP containing formulation, whereas to accomplish the same result with the formulation based on CMC required 35 cycles.

This phenomenon can be explained potentially in light of previous studies on PCPP–protein complexes [21]. Noncovalent interactions between polyphosphazene and BSA, which can be expected in the formulation, can lead to a rise in the viscosity of the PCPP–BSA solution and even the formation of a physical network with protein acting as a cross-linker. Both of these circumstances result in the improved film-forming properties and superior coating performance of PCPP. Although the viscosities of the polymers were essentially

BSA Loading (µg/microneedle)

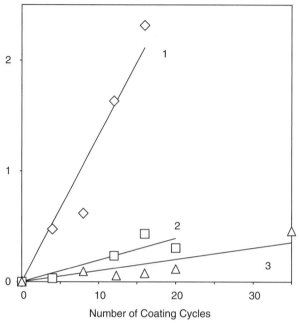

Number of Coating Cycles

FIGURE 5   BSA loading per microneedle as a function of a number of coating cycles in 5% (w/v) BSA formulations containing 0.5% (w/v) of PCPP (1), 1.5% (w/v) of CMC (2), and 0.8% (w/v) of CMC (3) in 0.1 × PBS.

the same in protein-free solutions, the addition of BSA led to a dramatic increase in the viscosity of PCPP–BSA solution compared to that of its CMC counterpart. Thus, based on the formulations tested, use of PCPP accelerates the coating process and results in the consumption of less material to deposit the protein dose desired. The latter is especially important since the amount of formulation (microneedle loading) can be limited because of dissolution and sharpness requirements.

Modulation of Antigen Release Using Polyphosphazenes

Prolongation of antigen release is often desirable to achieve advantageous pharmacological effects, such as the *depot effect*, which can result in more potent and persistent immune responses [9]. Although slow release can be in apparent contradiction with the highly desired short application time for the vaccination patch, there are approaches to achieving it without introducing any change in the treatment process. For example, sustained release can be accomplished when the formulation is detached from the needle upon administration in the form of a hydrogel, which can then release the antigen slowly.

Polyphosphazene polyelectrolytes can provide effective mechanisms for the modulation of protein release through the formation of hydrogels using a very mild and "protein-friendly" process of ionic complexation [12,22,23].

To investigate the validity of the approach for intradermal formulations, microneedles containing PCPP and fluorescein isothiocyanate conjugated bovine serum albumin (FITC-BSA) were treated with a solution of calcium chloride. Protein release was then monitored and compared to that for uncross-linked formulation in an aqueous medium containing 0.9% (w/v) of sodium chloride to mimic the ionic strength of the intradermal environment. As shown in Figure 6, formulation treated with calcium chloride showed slower dissolution (release of PCPP) and protein release rates. A good correlation between polymer and protein profiles indicates that the main mechanism for protein release is the dissolution of the coating, which generally takes place as calcium cross-linker is removed, due to the ion-exchange reactions with sodium. Although the release in solution is still fast, slower profiles can be expected for intradermal environments, where the rate of polymer swelling and the

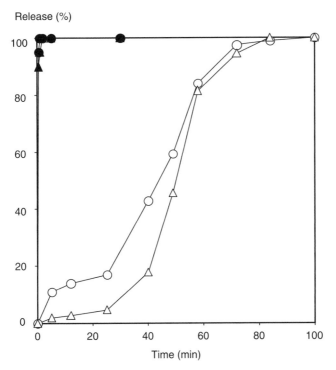

FIGURE 6 Release profiles of FITC–BSA (circles) and PCPP (triangles) from micro-needle formulation without cross-linking (solid markers) and after cross-linking with calcium chloride (open markers) in aqueous solutions [cross-linking solution contained 5% (w/v) of calcium chloride and 10% (w/v) of sodium chloride, release media, 0.9% (w/v) of sodium chloride].

kinetics of ion exchange can be somewhat retarded compared to solution under sink conditions used in vitro. Variations of cross-linking conditions can also bring about further reduction in the release rates.

Polyphosphazenes Are Potent Intradermal Immunoadjuvants

In vivo performance of microneedles containing PCPP formulations was evaluated in pigs using recombinant hepatitis B surface antigen (HBsAg). The study design included intradermal administration using PCPP-containing microneedles and microneedles in which formulation was integrated with inactive CMC. The results were also compared with those of intramuscular injections of HBsAg and HBsAg adjuvanted with the same dose of PCPP that was used on the microneedles.

To effect intradermal delivery, microneedle arrays were assembled in patches containing an adhesive layer to maintain close contact between the skin and the array. The application areas of the skin were shaved and cleaned, patches containing microneedle arrays were placed on the skin, and pressure was applied to the center of the patch for 1 min to facilitate microneedle insertion. The patches were then allowed to remain in place for a total of 30 min. Experiments conducted both in vivo and in vitro using porcine cadaver skin indicated that this treatment was generally sufficient for the formulation to be dissolved in the skin. This was confirmed by microscopic examination of needles before and after application, as well as HPLC studies comparing the amounts of protein and polymer on the needles after treatment with representative samples.

The results of a single immunization study demonstrate that PCPP-containing microneedles were superior to those in all other groups in the induction of immune responses (Fig. 7). Serum IgG-specific HBsAg titers for PCPP containing microneedles were approximately 10 times higher than those induced with a formulation of HBsAg and PCPP administered intramuscularly. They were also up to three orders of magnitude higher than those induced with PCPP-free microneedle formulations containing the same dose of HBsAg or through intramuscular administration of nonadjuvanted antigen. In addition, intradermal administration of PCPP-adjuvanted HBsAg demonstrated dose-sparing potential. Immunization with 10 μg of HBsAg using PCPP micronee-dles induced approximately a 10-fold higher response than intramuscular injection of 20 μg of HBsAg adjuvanted with PCPP (data not shown).

Although it has been known that PCPP is a potent adjuvant for parenteral immunization, it appears that its immunostimulating activity can be improved dramatically if the adjuvant is delivered intradermally. In fact, it is difficult to anticipate the results for "PCPP-enhanced" intradermal delivery based on data for microneedle administration and the adjuvant effect of PCPP in intramuscular immunization (Fig. 7). Such a "nonadditive" or synergistic effect of intradermal administration and PCPP adjuvant suggests some differences in immunostimulatory mechanisms between these routes of administration.

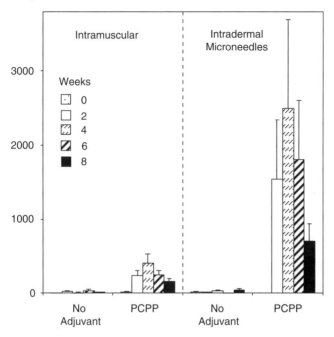

FIGURE 7   Antibody titers after single-dose intradermal and intramuscular immunizations of pigs with HBsAg formulated with or without PCPP (microneedles were used for intradermal studies; microneedles without PCPP were prepared using CMC as a binding polymer; seven animals per group; 20 µg of HbsAg; PCPP formulations contained 66 µg of the polymer).

For example, it is known that the formation of water-soluble complexes between PCPP and the antigen and physicochemical characteristics of such macromolecular structures can have a significant effect on the potency of polyphosphazene adjuvant [21]. Intradermal delivery of PCPP using microneedles results in a delivery of solid-state PCPP, in which case the properties of the antigen–polymer assembly can be affected significantly. Thus, it is not inconceivable to suggest that solid-state delivery of PCPP can result in modified antigen–polymer interactions, leading to a more effective antigen presentation.

EXPERIMENT

Materials

Poly[di(carboxylatophenoxy)phosphazene] (PCPP) (Sigma-Aldrich, St. Louis, Missouri) purified by multiple precipitation using sodium chloride [13] and sodium carboxymethyl cellulose (CMC) (Aqualon, USP/NF grade, low viscosity, Hercules, Wilmington, Delaware) without additional purification were employed. Solutions of PCPP and CMC were filtered through γ-irradiated, 0.45-μm Millex syringe filters before use. Fluorescein isothiocyanate conjugated bovine serum albumin (FITC-BSA) (Sigma-Aldrich), hepatitis B surface antigen, recombinant (HBsAg) (Fitzgerald Industries International, Inc., Concord, Massachusetts), polyoxyethylene (20) sorbitan monolaurate (Tween-20) (TCI America, Portland, Oregon), Dulbecco's phosphate-buffered saline (PBS) (sterile, without calcium or magnesium, Lonza, Walkersville, Maryland), calcium chloride dihydrate, USP, FCC grade (Mallinckrodt Baker Inc., Phillipsburg, New Jersey), sodium chloride (EMD Chemicals Inc., Gibbstown, New Jersey and Red 40 (Allura Red AC, TCI America) were used. Titanium arrays, each containing 50 needlelike protrusions approximately 600 μm in length, were prepared by chemical etching and then washed in an ultrasonic bath.

The coating formulation was fed to a 50-microwell reservoir using a Genie Plus syringe pump (Kent Scientific, Torrington, Connecticut). A microneedle array was secured on an array holder and then attached to an X–Y–Z micropositioning system using alignment pins and holders. Using the micropositioning system, the coating procedure was performed by submerging the microneedles into the wells in the coating reservoir and then removing them immediately, followed by a drying step in which the arrays were purged with anhydrous nitrogen gas. The needles were dipped until the wells were depleted of formulation, at which time they were replenished. A stereo zoom microscope (STZ-45-BS-FR), with a 2.0-megapixel 1616×1216 digital camera (Caltex Scientific, Irvine, California), was used to monitor the process.

Analysis of the Coating

Quantitative analysis of the coating was performed using ultraviolet–visible spectrophotometry (Hitachi U-2810 Spectrophotomer, Hitachi High Technologies America, Inc., San Jose, California) and size-exclusion HPLC (Hitachi LaChrom Elite HPLC system, equipped with an L-2I3OHTA pump and degasser, an L-2200 autosampler, an L-2455 diode array detector, and an L-2490 refractive index detector, Hitachi High Technologies America). Each coated array was placed in an individual plastic weigh boat, along with 1 mL of $0.1 \times$ PBS, to dissolve the coating. Standards were prepared and calibration curves set up for both UV/vis and HPLC. The optical density was measured at 235 nm (PCPP) and 495 nm (FITC-BSA). HPLC analysis was conducted using an Ultrahydrogel 250 size-exclusion column (Waters Corporation, Milford, Massachusetts) for separation. The mobile phase consisted of $0.1 \times$ PBS with 10% acetonitrile, and the flow rate was set to 0.75 mL/min. Data were processed using EZChrom Elite software (Hitachi High Technologies America).

Cross-Linking and Release Study

The arrays were coated with a formulation containing 1% (w/v) of PCPP, 0.3% (w/v) of FITC-BSA, and 0.1% (v/v) of Tween-20 in 0.5 × PBS, resulting in a coating that contained 30 μg of BSA and 100 μg of PCPP. Cross-linking with calcium chloride was performed by first placing the arrays in individual plastic weigh boats along with 0.5 mL of calcium chloride and sodium chloride in deionized water. After 10 min they were dried with anhydrous nitrogen for 1 min and then left in a desiccator overnight.

A release study was performed at ambient temperature, with arrays in 0.5 mL of 0.9% (w/v) sodium chloride in deionized water. The solution was refreshed after each time point was recorded. The amount of FITC-BSA and PCPP released from the microneedles was analyzed via UV/vis spectrophotometry by obtaining the optical densities at 495 nm and 235 nm, respectively.

Stability Studies

Accelerated thermal stability studies of solid-state microneedle formulations containing HRP were performed at 80°C. Coated microneedles were prepared using formulation containing 2% (w/v) of CMC, 0.5% (w/v) of HRP, and 0.2% (w/v) of Tween-20 in 1 × PBS. HRP solutions containing 5 mg/mL of HRP in 1 × PBS were prepared for comparative purposes. Samples were placed in a MaxQ Mini 4450 shaker-incubator (Barnstead International, Dubuque, Iowa) for a predetermined length of time. Microneedle samples were then rinsed with 1 mL of PBS to dissolve the coating and analyzed for protein content using HPLC and for enzymatic activity using 2,2′-azino-bis(3-ethyl-benzthiazoline-6-sulfonic acid) as a substrate (enzymatic assay of peroxidase from horseradish, EC 1.11.1.7, Sigma product P-6782, Sigma-Aldrich). The residual activity was then calculated as an experimentally observed enzymatic activity of the sample divided by the amount of enzyme as determined by HPLC.

In Vivo Immunization Experiments

In vivo immunization experiments were conducted in Land Race Cross pigs, which were divided into groups each containing seven animals. The pigs were 3–4 weeks old at the start of the study and weighed 5 to 8 kg each. The application sites for intradermal administration were clipped of all hair and then shaved to further ensure a smooth surface. The sites were then washed with water and allowed to air dry. The patches containing microneedle arrays were put on the skin manually and pressure was applied on the center of the patch for 1 min to facilitate microneedle insertion. The patch was allowed to remain in place, undisturbed, for 29 additional minutes before removal. Each animal immunized via intramuscular route received a 1-mL injection of liquid formulation in the neck, right behind the ears. Intradermal injection using a hypodermal syringe was carried out using four 50-µL injections of liquid formulation in four different spots on the ear, for a total of 200 µL of 100 µg/mL HBsAg solution (i.e. 20-µg dose). All subjects were anesthetized during the immunization with a combination of Xylazine and Ketamine. The blood samples were collected prior to being immunized (0 weeks) and then at 2, 4, 6, and 8 weeks post-immunization.

Detection of IgG-specific HBsAg Titers

Antigen-specific antibodies (IgG) in pig serum were determined by ELISA (enzyme-linked immunosorbent assay in 96-well Immunolon II plates coated with HbsAg in sodium carbonate buffer, pH 9.6. The plates were washed six times with PBS containing 0.05% Tween-20 (PBST). Twofold serial dilutions of sera in PBST containing 0.5% gelatin were added to the wells and the plate was incubated 2 h at ambient temperature. Unbound serum was removed by washing the plates six times with PBST. Biotinylated goat anti-mouse IgG (Caltag Laboratories) was added and the plates were incubated for 1 h at ambient temperature. The plates were washed six times with PBST, and alkaline phosphatase–conjugated streptavidin (BioCan Scientific) was added and plates were incubated for 1 h at ambient temperature. Unbound conjugate was removed by washing eight times with deionized water and serum antibodies were detected by adding 1 mg/mL of p-nitrophenyl phosphate di(Tris) salt in 1% diethanolamine–0.5 mM magnesium chloride buffer, pH 9.8. The reaction was allowed to run for 15 min and the absorbance was measured at 405 nm using a Benchmark microplate reader (BioRad Laboratories, Hercules, California). The endpoint titers were the reciprocal of the highest sample dilution, producing a signal identical to that of an antibody-negative sample at the same dilution plus three times standard deviation. The average antibody titers for a group of mice were expressed as mean titers.

REFERENCES

1. Alarcon, J.B., Hartley, A.W., Harvey, N.G., Mikszta, J.A. Preclinical evaluation of microneedle technology for intradermal delivery of influenza vaccines. *Clin. Vaccine Immunol.*, 2007, 14(4):375–381.

2. Belshe, R.B., Newman, F.K., Cannon, J., Duane, C., Treanor, J., Van Hoecke, C., Howe, B.J., Dubin, G. Serum antibody responses after intradermal vaccination against influenza. *N. Engl. J. Med.*, 2004, 351(22):2286–2294.

3. Glenn, G.M., Kenney, R.T., Hammond, S.A., Ellingsworth, L.R. Transcutaneous immunization and immunostimulant strategies. *Immunol. Allergy Clin. of North Am.*, 2003, 23(4):787–813.

4. Hooper, J.W., Golden, J.W., Ferro, A.M., King, A.D. Smallpox DNA vaccine delivered by novel skin electroporation device protects mice against intranasal poxvirus challenge. *Vaccine*, 2007, 25(10):1814–1823.

5. Kenney, R.T., Frech, S.A., Muenz, L.R., Villar, C.P., Glenn, G.M. Dose sparing with intradermal injection of influenza vaccine. *N. Engl. J. Med.*, 2004, 351(22):2295–2301.

6. La Montagne, J.R., Fauci, A.S. Intradermal influenza vaccination: Can less be more? *N. Engl. J. Med.*, 2004, 351(22):2295–2301.

7. Prausnitz, M.R. Microneedles for transdermal drug delivery. *Adv. Drug Deliv. Revi.*, 2004, 56(5):581–587.

8. Prausnitz, M.R., McAllister, D.V., Kaushik, S., Patel, P.N., Mayberry, J.L., Allen, M.G. Microfabricated microneedles for transdermal drug delivery. *Am. Soci. Mech. Engi., Bioeng. Div. (Publ.)*, 1999, 42, 89–90.

9. Singh, M, ed. *Vaccine Adjuvants and Delivery Systems*. Wiley-Interscience, Hoboken, NJ, 2006.

10. Vogelbruch, M., Nuss, B., KÃrner, M., Kapp, A., Kiehl, P., Bohm, W. Aluminium-induced granulomas after inaccurate intradermal hyposensitization injections of aluminium-adsorbed depot preparations. *Allergy*, 2000, 55(9):883–887.

11. Andrianov, A.K. Water-soluble polyphosphazenes for biomedical applications. *J. Inorg. Organomet. Polym. Mater.*, 2006, 16(4):397–406.

12. Andrianov, A.K., Chen, J. Polyphosphazene microspheres: Preparation by ionic complexation of phosphazene polyacids with spermine. *J. Appl. Polym. Sci.*, 2006, 101(1):414–419.

13. Andrianov, A.K., Svirkin, Y.Y., LeGolvan, M.P. Synthesis and biologically relevant properties of polyphosphazene polyacids. *Biomacromolecules*, 2004, 5(5):1999–2006.

14. Andrianov, A.K. Polyphosphazenes as vaccine adjuvants. In Vaccine Adjuvants and Delivery Systems. Singh, M., ed., Wiley, Hoboken, NJ, 2007, pp. 355–378.

15. Gill, H.S., Prausnitz, M.R. Transdermal drug delivery via coated microneedles. AIChE Annual Meeting, Conference Proceedings. 2005.

16. Gill, H.S., Prausnitz, M.R. Coating formulations for microneedles. *Pharm. Res.*, 2007, 24(7):1369–1380.

17. Gill, H.S., Prausnitz, M.R. Coated microneedles for transdermal delivery. *J. Controll. Release*, 2007, 117(2):227–237.

18. Bouveret Le Cam, N.N., Ronco, J., Francon, A., Blondeau, C., Fanget, B. Adjuvants for influenza vaccine. *Res. Immunol.*, 1998, 149(1):19–23.

19. Thongcharoen, P., Suriyanon, V., Paris, R.M., Khamboonruang, C., de Souza, M.S., Ratto-Kim, S., Karnasuta, C., Polonis, V.R., Baglyos, L., El Habib, R. A phase 1/2 comparative vaccine trial of the safety and immunogenicity of a CRF01_AE (subtype E) candidate vaccine: ALVAC-HIV (vCP1521) prime with oligomeric gp160 (92TH023/LAI-DID) or bivalent gp120 (CM235/SF2) boost. *J. Acquir. Immune Defici. Syndro.*, 2007, 46(1):48.

20. Payne, L.G., Jenkins, S.A., Woods, A.L., Grund, E.M., Geribo, W.E., Loebelenz, J.R., Andrianov, A.K., Roberts, B.E. Poly[di(carboxylatophenoxy)phosphazene] (PCPP) is a potent immunoadjuvant for an influenza vaccine. *Vaccine*, 1998, 16(1):92–98.

21. Andrianov, A.K., Marin, A., Roberts, B.E. Polyphosphazene polyelectrolytes: a link between the formation of noncovalent complexes with antigenic proteins and immunostimulating activity. *Biomacromolecules*, 2005, 6(3):1375–1379.

22. Andrianov, A.K., Chen, J., Payne, L.G. Preparation of hydrogel microspheres by coacervation of aqueous polyphosphazene solutions. *Biomaterials*, 1998, 19(1–3): 109–115.

23. Andrianov, A.K., Payne, L.G. Protein release from polyphosphazene matrices. *Adv. Drug Deliv. Rev.*, 1998, 31(3):185–196.

PART III
Biomaterials

8 Biodegradable Polyphosphazene Scaffolds for Tissue Engineering

SYAM P. NUKAVARAPU and SANGAMESH G. KUMBAR

Department of Orthopaedic Surgery, University of Connecticut, Farmington, Connecticut; Department of Chemical, Materials and Biomolecular Engineering, University of Connecticut, Storrs, Connecticut

HARRY R. ALLCOCK

Department of Chemistry, The Pennsylvania State University, University Park, Pennsylvania

CATO T. LAURENCIN

Department of Orthopaedic Surgery, University of Connecticut, Farmington, Connecticut; Department of Chemical, Materials and Biomolecular Engineering, University of Connecticut, Storrs, Connecticut

INTRODUCTION

Tissue engineering is the interplay between biodegradable materials, cells, and signaling molecules (Langer and Vacanti, 1993; Laurencin et al., 1999). In the last few years considerable work has been devoted to develop three-dimensional (3D) porous scaffolds from novel biocompatible and biodegradable polymeric materials for the repair, restoration, and regeneration of various tissues, such as skin, ligament, tendon, and bone (Cooper et al., 2006; Kumbar et al., 2008; Laurencin et al., 2006). The synthetic polymers polylactide (PLA), polyglycolide (PGA), poly(lactide-*co*-glycolide acid) (PLAGA), polycaprolactone, polyanhydrides, and polyphosphazenes have attracted great attention as biomaterials for tissue engineering scaffolding applications (Allcock, 2006; Langer et al., 2004; Levenberg and Langer, 2004; Nair et al., 2006a).

Polyphosphazenes are inorganic–organic hybrid polymers with a backbone consisting of alternating phosphorus and nitrogen atoms (Allcock, 2003). Each phosphorus atom in the backbone is substituted for by two organic side groups that dictate polymer properties to a great extent. Synthetic flexibility associated with a broad range of physical and chemical properties has been a unique

Polyphosphazenes for Biomedical Applications, Edited by Alexander K. Andrianov
Copyright © 2009 John Wiley & Sons, Inc.

feature of this class of polymers. A variety of polyphosphazenes from biostable to biodegradable have been reported so far for various biomedical applications (Allcock, 2006; Heyde and Schacht, 2004; Laurencin and Ambrosio, 2003). Biodegradable phosphazenes with hydrolytically sensitive side groups such as amino acid ester, glucosyl, glycerol, lactide, and glycolide esters have raised broad interest as biomaterials for tissue engineering and drug delivery applications (Allcock, 2001; Kumbar et al., 2006; Laurencin et al., 2005; Nair et al., 2003). Controlled degradation and the degradation behavior modulation (surface vs. bulk) associated with nontoxic degradation products caused biodegradable polyphosphazenes to become a potential class of biomaterials for tissue engineering.

Tissues are of 3D origin, and cells alone cannot form complex tissue structures without 3D support during regeneration. This support, referred to as a *scaffold*, is often carved out of synthetic or natural polymers with interconnected pores and mechanical properties similar to those of the regenerating tissue. In the beginning, the scaffold at the defect site acts as a negative template and allows cell ingrowth, nutrient supply, and waste removal through its pore structure. Over time the scaffold degradation and neotissue formation is achieved to regenerate tissues successfully with the porosity required. The ideal scaffold should exhibit biocompatibility, biodegradability, mechanical compatibility, interconnected porosity, and the nontoxic nature of the degradation products (Hollister, 2005; Hutmacher, 2001). Biodegradable polyphosphazenes were proven to be biocompatible in vitro and in vivo with nontoxic degradation products (Ambrosio et al., 2002; Kumbar et al., 2006; Laurencin et al., 1993) In addition, the mechanical properties are tunable to match a tissue of interest. Therefore, biodegradable polyphosphazenes are a promising class of biomaterials for scaffold-based tissue engineering.

So far, reviews in the literature have dealt only with the synthetic flexibility and biocompatibility of various biodegradable polyphosphazenes (Allcock, 2001; Allcock et al., 2002; Heyde and Schacht, 2004; Kumbar et al., 2006; Laurencin et al., 2003; Luten et al., 2003; Nair et al., 2003; Singh et al., 2007). However, in this chapter we cover biodegradable polyphosphazenes as scaffold materials for tissue engineering. Scaffold fabrication, characterization, and their suitability for tissue regeneration are discussed further.

BIODEGRADABLE POLYPHOSPHAZENES

Polyphosphazenes with hydrolytically labile side groups such as glucosyl, glycerol, imadazolyl and lactide, glycolide, and amino acid esters form a subclass of polymers known as biodegradable polyphosphazenes (Allcock and Kwon, 1988; Allcock et al., 1977, 1994b, 1994c, 2003). These important classes of polyphosphazenes are used as biomaterials in both tissue engineering and controlled drug delivery (Allcock et al., 1994a; Ibim et al., 1998; Laurencin and Ambrosio, 2003; Nair et al., 2003; Passi et al., 2000). Hydrolytically sensitive side groups sensitize the phosphorus–nitrogen backbone and degrade into various by-products, as shown in Scheme 1 (Allcock, 2003; Allcock et al., 1982; Andrianov and Marin, 2006). The polymer degradation rate can be varied from a few hours to years based on the side groups chosen for polymer backbone substitution. Various side-group combinations resulted in numerous biodegradable polyphosphazenes with a wide range of degradation time and physicochemical properties suitable for a variety of biomedical applications (Allcock, 2001; Heyde and Schacht, 2004; Kumbar et al., 2006). For example,

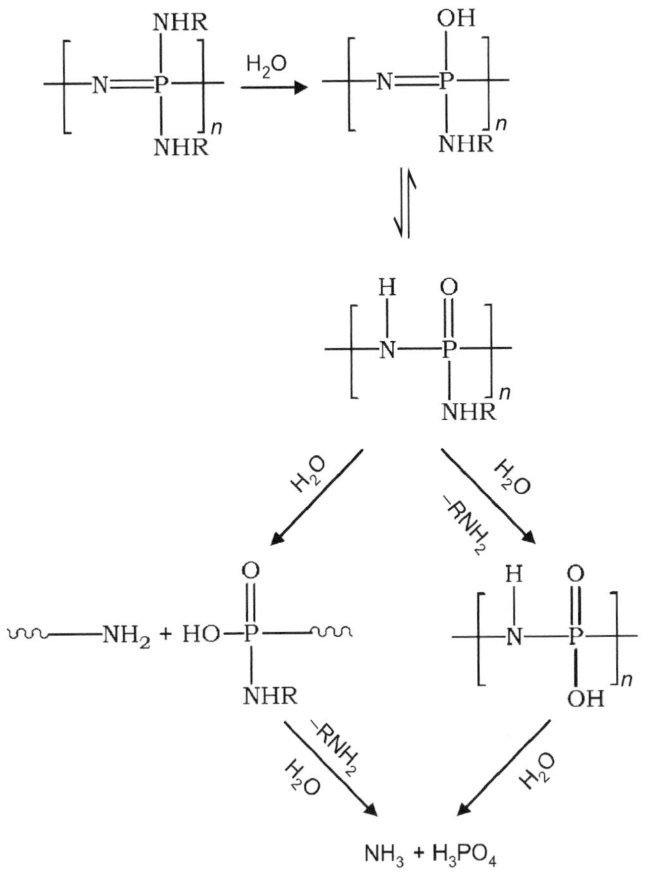

SCHEME 1 Generalized degradation pathway for amino acid ester–substituted polyphosphazenes. The degradation is initiated by displacement of the side group with hydroxyl ion and protonating the backbone nitrogen leading to phosphazene formation. These polymers can also degrade via ester functionalization (not shown here). In both ways, polyphosphazenes produce body-friendly degradation products (amino acids and ammonium phosphate buffer). This ensures improved biocompatibility and makes them attractive candidate materials in scaffold-based tissue engineering.

amino acid ester polyphosphazenes have shown great potential as scaffold materials for tissue repair and regeneration (Laurencin et al., 1993; Nair et al., 2006b; Sethuraman et al., 2006), while biodegradable poly[di(carboxylatophenoxy)phosphazene] (PCPP) is currently under clinical trail as a polyelectrolyte vaccine immunoadjuvant (Allcock and Kwon, 1989; Andrianov et al., 2004, 2005; Mutwiri et al., 2007).

POLYPHOSPHAZENES AND "TUNABILITY"

During tissue repair and regeneration, scaffolds are also expected to support mechanical functionality of the tissue until regeneration is complete. This is possible with scaffolds that are mechanically compatible with the regenerating tissues. For example, scaffolds for bone tissue engineering should possess bone-matching mechanical strength and a compressive modulus, while vascular and skin replacement scaffolds should be able to sustain larger strains without permanent deformation. These requirements have been achieved primarily by the selection of a suitable polymer for the required application (Blan and Birla, 2008; Khan et al., 2008; Ng and Hutmacher, 2006; Stegemann et al., 2007).

In this respect, polyphosphazenes with the required scaffold properties and degradation rates can be fabricated through proper side-group selection. For example, a large family of amino acid ester side-group polyphosphazenes has shown great promise for a variety of biomedical applications (Allcock et al., 1994b; El-Amin et al., 2006; Heyde and Schacht, 2004; Kumbar et al., 2006; Nair et al., 2006b). Polyphosphazenes bearing smaller amino acid moieties such as glycine and alanine are mechanically soft and fast degrading. In contrast, bulkier groups such as phenylalanine and lysine result in slow-degrading, mechanically tough polymers. Also, co-substituted polyphosphazenes were developed to have polymers with desirable mechanical and degradation performance. Using this approach, Singh et al. have developed a series of polyphosphazenes with alanine and other side groups, including glycine, p-methylphenoxy, or p-phenylphenoxy groups (Singh et al., 2005). The resulting polymers showed glass transition temperatures from -10 to $35°C$ with tensile modulus in the range 31.4 to 456 MPa. Water contact angles for these polymers varied from 63 to 107°, and hence different levels of cell response were observed. Combinations of different side groups, composition, and mixed substitution approaches resulted in a large family of biodegradable polyphosphazenes with desired physicochemical properties for a variety of biomedical applications (Singh et al., 2006, pp. 914–918). Furthermore, side-group modification with cell adhesive groups such as galactose using various spacers has resulted in a new class of polymers with increased protein interaction and biocompatibility (Heyde et al., 2007, 2008). This ability to create a large group of polymers with the required properties renders polyphosphazenes more versatile than some of the polymer systems available for tissue engineering applications.

POLYPHOSPHAZENE BONE CEMENTS

Hydroxyapatite, a major component of bone, has traditionally been used as a bone-filling material. Forming hydroxyapatite (HAp) through a cementing reaction at physiological temperature is always advantageous because it offers an easy way to form bone implants at injury sites. Osteoconductive bone grafts such as calcium phosphate, calcium sulfate, and tricalcium phosphate (TCP) have been used traditionally to form hydroxyapatite-based bone implants in vivo. For example, the reaction between TCP and dicalcium phosphate (DCP) or dibasic calcium phosphate anhydrate (DCPA) results in hydroxyapatite, and this formulation is currently used as a calcium phosphate bone cement under various trade names: Hydroset, Norion SRS, Norion CRS, and others. Since HAp alone is brittle, formation of its composites with biodegradable polymers is favorable for developing tough, bone-mimicking synthetic composite cements.

The necessary requirement to mimic bone tissue closely is to develop a composite wherein the ceramic and polymer phases interdigitate at the molecular level. This has been achieved by reacting TCP and DCPA in the presence of acid-PCPP or its salts with Na/K (Greish et al., 2005, 2006; Ten Huisen et al., 1996). In both cases, HAp formation was associated with calcium cross-linked polymer salt formation. Calcium bridging acted as a nucleation site for hydroxyapatite formation. It has also provided an opportunity to form a true composite wherein the ceramic and polymer phases interact at the molecular level to attain the mechanical properties required. Furthermore, composite formation has been demonstrated using amino acid ester polyphosphazenes where Ca salt formation was observed through ester hydrolysis (Greish et al., 2008). So far, studies suggest the true composite formation in the presence of polyphosphazenes; however, higher proportions of polymer inhibited HAp formation by coating the calcium precursors with a viscous polymer layer. Polyphosphazene-assisted composite formation through a cementlike reaction is a big step toward developing composite implants in vivo for bone repair and regeneration applications.

APPLICATIONS IN TISSUE ENGINEERING

Due to their synthetic flexibility and physicochemical property tunability, polyphosphazenes have long been considered for tissue engineering applications. Biodegradable polyphosphazenes have been processed into three-dimensional porous scaffolds via salt leaching, microsphere sintering, and electrospinning. Salt leaching and microsphere sintering methodologies resulted in scaffolds that are implantable at load-bearing sites for bone tissue engineering. Flexible membranes generated by electrospinning have been studied for non-load-bearing bone regeneration. Scaffold physical properties, along with their ability to regenerate bone tissue, are described below.

Polyphosphazenes for Bone Scaffolding

Laurencin and co-workers were the first to use polyphosphazenes for bone tissue engineering (Laurencin et al., 1992). Since then, various biodegradable polyphosphazenes have been developed for bone scaffolding applications (Deng et al., 2008; Nair et al., 2006). The apatite layer formation and the in vitro, in vivo osteocompatibility observed proved their ability as potential materials for bone scaffolding (Brown et al., 2007; Deng et al., 2008; Sethuraman et al., 2006). The requirement for a polymer to be used for bone regeneration is to have its glass transition temperature higher than the physiological temperature. Scaffolds fabricated using such polymers can withstand the mechanical load without deformation at the load-bearing sites. Polyphosphazenes with small side groups such as glycine and alanine esters showed T_g values lower than 0°C (see Table 1). However, polyphosphazenes with bulkier groups, such as valine, leucine, and phenyl alanine, showed higher

TABLE 1   Polyphosphazene Single-Substituent Polymers and Their Glass Transition Temperatures[a]

R (Amino Acid Ester Group)	Corresponding Amino Acid	Glass Transition Temperature, T_g (°C)
$NH-CH_2-COOC_2H_5$	Glycine	−40
$\begin{array}{c} CH_3 \\ \vert \\ NH-CH-COOC_2H_5 \end{array}$	Alanine	−10
$\begin{array}{c} H_3C\diagdown \diagup CH_3 \\ CH \\ \vert \\ CH_2 \\ \vert \\ NH-CH-COOC_2H_5 \end{array}$	Leucine	15
$\begin{array}{c} H_3C\diagdown \diagup CH_3 \\ CH \\ \vert \\ NH-CH-COOC_2H_5 \end{array}$	Valine	25
$\begin{array}{c} C_6H_5 \\ \vert \\ CH_2 \\ \vert \\ NH-CH-COOC_2H_5 \end{array}$	Phenylalanine	42

[a] Increase in the size of amino acid ester side groups resulted in polyphosphazenes with higher glass transition temperatures. Polymers with glass transition temperatures close to or higher than body temperature are suitable as scaffold materials for load-bearing tissue engineering.

T_g values, due to the sterically limited motion of the polymer backbone. Of these, phenylalanine ethyl ester phosphazene showed a T_g value of 42°C, which is one of the highest reported so far in the class of biodegradable polyphosphazenes (Nukavarapu et al., 2008b). This may be due to the possible π–π interactions between the phenyl rings, causing the side groups to become ordered. The increase in side group order because of the bulkier side group and phenyl group π–π interactions limit the flexibility of the backbone and raise the glass transition temperature.

Scaffolds for Load-Bearing Bone Tissue Engineering

Laurencin et al. have used a salt leaching method for polyphosphazenes and achieved a 3D scaffold from biodegradable poly[(methylphenoxy)(ethyl glycinato)phosphazene] (Ambrosio et al., 2003; Laurencin et al., 1996). The scaffold showed interconnected porosity with an average pore diameter of 165 μm that supports osteoblast cell infiltration. In contrast to 2D films, these matrices showed progressive osteoblast proliferation for the 21-day study period in vitro. This study has confirmed osteoblast growth not only on the surface but also within the pores, to achieve cellularized 3D constructs for bone tissue repair and regeneration.

Polyphosphazene Microsphere Scaffolds Microsphere sintering is often achieved by heating microspheres closer to the glass transition temperature of the particular polymer (Borden et al., 2002a,b). The polyphosphazenes studied so far exhibited chemical decomposition close to the T_g value and hence were sintered using an alternative method of solvent–nonsolvent sintering (Brown et al., 2008; Nukavarapu et al., 2008b). A solvent–nonsolvent sintering approach is presented in Figure 1. A range of solvent–nonsolvent compositions with the same fractional solubility parameters as polymer parameters were employed in this method. Differential evaporation (nonsolvent > solvent) drives the entire composition from soluble to insoluble over the process time to achieve fusion between adjacent microspheres. This method has been used effectively to fabricate 3D porous scaffolds implantable at load-bearing sites for bone tissue engineering.

Solvent–nonsolvent sintering has been applied successfully to various biodegradable polyphosphazenes: poly[bis(ethyl alaninato)phosphazene] (PNEA), poly[bis(methyl valinato)phosphazene] (PNMV), and poly[bis(ethyl phenylalaninato) phosphazene] (PNEPhA) (Brown et al., 2008). For PNEA the solvent and nonsolvent selected were acetone and hexane, respectively; for both PNMV and PNEPhA the solvent was tetrahydrofuran (THF) and the nonsolvent was again hexane. Microspheres of the aforementioned polymers were sintered with a range of solvent–nonsolvent compositions to obtain scaffolds with tunable porosity, pore diameter, and the mechanical properties. Matrices thus obtained exhibited optimal scaffold properties for their use in bone regeneration.

Microspheres   Microsphere Slurry   Sintered Microspheres

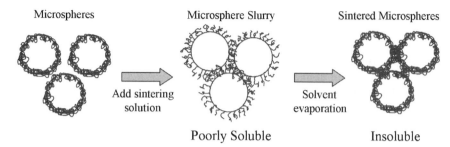

Add sintering   Solvent
solution   evaporation

Poorly Soluble   Insoluble

FIGURE 1   Dynamic solvent approach for sintering PNEPhA microspheres into 3D scaffolds. Initially, microsphere slurry was formed with a THF (solvent) and hexane (nonsolvent) combination. The solution fractional solubility parameters are such that microsphere surfaces swell and open up the polymer chains. Dynamic solvent transition from poor solvent to nonsolvent (because THF is more volatile than hexane) results in polymer chain locking and hence permanent bonding between the adjacent microspheres. In this method bonding between the microspheres and the ultimate scaffold properties (porosity and mechanical strength) can be fine-tuned through proper selection of the THF/hexane ratio.

Polyphosphazene–Nanohydroxyapatite Composite Scaffolds   Bone is a composite comprising of nanohydroxyapatite (nHAp) dispersed in a collagen matrix. A scaffold is very efficient in bone regeneration when it mimics the bone in structure and composition. Various studies to date have focused on achieving 3D scaffolds from biodegradable polymer–HAp composites which are compositionally similar to bone (Christenson et al., 2007; Kim et al., 2006; Wang et al., 2008). On this basis polyphosphazene–nHAp composite microsphere scaffolds were fabricated for orthopedic applications (Nukavarapu et al., 2008b). Following our discussion above, polymer PNEPhA is the obvious choice as a candidate material for load-bearing tissue engineering. The PNEPhA was further combined with 20% of nHAp to fabricate composite microspheres. Composite microsphere scaffolds were fabricated using dynamic solvents composed of different ratios of THF and hexane.

Porosity and mechanical strength are the two key parameters that determine the scaffold efficacy for any tissue engineering applications. In bone regeneration, scaffolds should be able to bear the mechanical load at the defect site while allowing cell infiltration, nutrient supply, and waste removal through the interconnected pores. To meet these requirements, a range of dynamic solvent compositions (THF compositions 17.5, 20, and 22.5 vol%, with the remaining volume of hexane) were utilized to obtain PNEPhA–20nHAp composite microsphere scaffolds (Fig. 2). It is clear from Figure 2b that increasing the THF content in the dynamic solvent increases the amount of bonding between the microspheres. The resulting scaffold mechanical strength and porosity are listed in Table 2. Scaffolds showed a compressive modulus (46 to 81 MPa) and compressive strength (6.5 to 13 MPa) that are comparable with the values

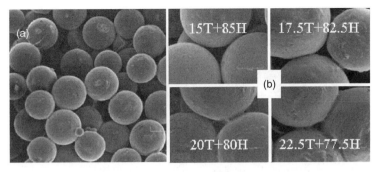

FIGURE 2 Scanning electron micrographs (SEMs) showing the typical morphology of polyphosphazene scaffolds (a) sintered with a solvent mixture of 20% THF + 80% hexane, and (b) close-up images of scaffolds sintered with various THF–hexane combinations, showing different degrees of microsphere bonding.

reported previously for scaffolds used for bone regeneration, and these properties are also in the lower range of human cancellous bone (Borden et al., 2002b, 2004; Jiang et al., 2006). The median pore diameter (145 to 86 μm) and the pore volume (24 to 15%) support bone cell migration into these 3D microsphere scaffolds. Three-dimensional microsphere scaffolds with porosity around 20 to 30% cause a negative template to form around 70 to 80% of the porous structure that matches the native bone tissue following complete degradation of the scaffold.

The increased presence of cytoskeletal actin on the microsphere scaffolds indicates the progressive growth of primary rat osteoblasts. As shown in Figure 3a microsphere adjoining areas attract a higher number of osteoblasts than do nonadjoining regions. This was because microsphere adjoining areas with high surface area per unit volume attract a larger amount of matrix proteins and hence result in more cell adhesions. By day 12, cells migrated to microsphere surfaces and formed a semiconfluent cellularized 3D construct. The quantitative cell proliferation (Fig. 3b) and alkaline phosphatase (ALP) expression showed a trend similar to that observed for PLAGA composite scaffolds.

TABLE 2   Effect of Dynamic Solvent Composition on Scaffold Physical Properties[a]

Dynamic Solvent Composition	Compressive Modulus	Compressive Strength	Pore Diameter	Percent Porosity
17.5T–82.5H	45.9 ± 3.7	6.5 ± 1.6	145.1 ± 4.3	23.7 ± 0.8
20T–80H	69.5 ± 4.9	11.2 ± 1.8	117.7 ± 6.9	19.9 ± 1
22.5T–77.5H	80.6 ± 10	12.9 ± 2.5	86.2 ± 4.6	15.1 ± 0.8

[a] Increasing the THF (or decreasing hexane) content in sintering solution resulted in increased mechanical performance with reduced pore diameter and porosity.

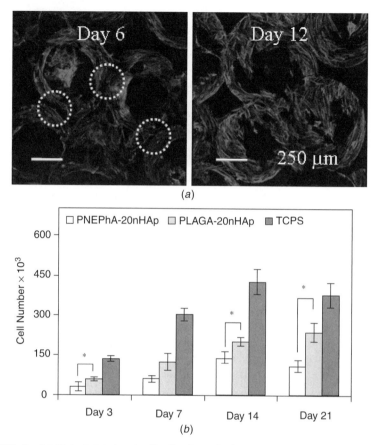

FIGURE 3   (a) Cytoskeletal actin distribution of primary rat osteoblast cells grown on polyphosphazene–nHAp composite microsphere scaffold. The circled regions show higher initial cell proliferation at the microsphere adjoining areas; the entire scaffold became confluent by day 12. (b) Primary rat osteoblast cell proliferation on polyphosphazene–nHAp scaffolds compared with PLAGA–nHAp scaffolds and planar TCPS. *, Denotes significant difference with $p < 0.05$.

However, at early time points, both cell proliferation and phenotypic ALP expression of osteoblasts on PNEPhA 3D composite scaffolds were significantly lower than those of PLAGA composite scaffolds. These observed differences were attributed to the difference in the hydrophobicity of the PNEPhA and PLAGA polymers. Phenyl alanine is known to be the second-highest hydrophobic amino acid among the natural amino acids, and hence the PNEPhA polymer. The higher hydrophobic nature allows stronger interactions between the surface and the adhered proteins and would inhibit the adhered cell reorientation, migration, and cellular phenotypic expression (Wilson et al., 2005). Efforts including various side groups, compositional variation (Singh et al., 2005), and plasma treatment (Allcock et al., 2007) are in progress to further improve the polyphosphazene scaffolds suitable for the repair, restoration, and regeneration of bone.

Scaffolds for Non-Load-Bearing Bone Tissue Engineering

Electrospinning has been a very popular technique to create nanofiber scaffolds for tissue-regeneration applications (Nair et al., 2004b; Nukavarapu et al., 2008a). Nanofiber scaffolds not only provide structural support but also offer features to closely mimic the native tissue extracellular matrix for enhanced tissue repair and regeneration (Kumbar et al., 2007; Nukavarapu et al., 2007). Ongoing research activities in the Laurencin laboratory have focused on PNEA, PNmPh, and their copolymers to fabricate nanofiber scaffolds via electrospinning (Laurencin et al., 2005). PNmPh component was used as a modulator to alter the copolymer degradation rate, surface property (hydrophilicity/hydrophobicity), and mechanical performance. For example a higher PNmPH content resulted in a prolonged nanofiber degradation rate and improved mechanical properties over the PNEA nanofiber matrices. Electrospun nanofiber matrices of poly[bis(p-methylphenoxy)phosphazene] (PNmPh) supported the adhesion of bovine coronary artery endothelial cells (BCAECs) as well as promoting the adhesion and proliferation of osteoblasts such as MC3T3-E1 cells (Fig. 4c) (Nair et al., 2004b).

PNEA was electrospun to produce nanofiber scaffolds for tissue engineering and drug delivery applications. A PNEA solution concentration of 9% (w/v) in THF, electrospun at 15 kV at a distance of 30 cm with a 2-mL/h flow rate, resulted in nanofibers 331 ± 108 nm in diameter (Bhattacharyya et al., 2006). Electrospinning was used further to fabricate PNEA–nHAp composite nanofiber scaffolds for orthopedic applications (Bhattacharyya et al., 2008). The nanofibers had a rougher surface and the nodules observed along the length of the fibers suggest the nHAp encapsulation. Electron-microscopic investigation confirmed the encapsulation of 20- to 40-nm nHAp crystals into the nanofibers

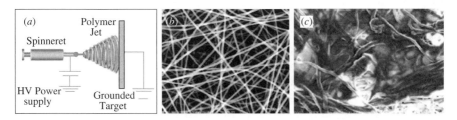

FIGURE 4   (a) Schematic showing the electrospinning process to generate 3D nanofiber scaffolds, (b) SEM of electrospun PNmPh nanofiber scaffold, and (c) MC3T3-E1 osteoblast cell adhesion and spreading on these polyphosphazene nanofiber scaffolds. (From Nair et al., 2004a, printed with permission from ACS.).

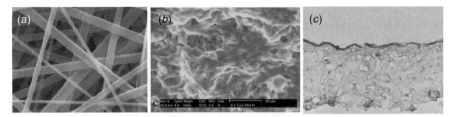

FIGURE 5 (a) SEM of an electrospun polyphospazene 3D scaffold; (b) NEC culture resulting in a confluent monolayer by day 16; (c) NECs failed to migrate through a nanofiber matrix, and the inner part of the scaffold remained acellular, as evident from the hematoxylin-stained cellular construct (400 ×). (From Carampin et al., 2007, with permission from Wiley Periodicals.)

(diameters ranging from 100 to 310 nm). These scaffolds are of great interest because they provide cells with a natural bonelike environment and a fibrillar structure that can enhance bone tissue regeneration and repair. Preliminary studies using nanofiber matrices of these polymers and composites have shown encouraging osteoblast performance, and detailed in vitro and in vivo investigations are needed to further confirm their suitability as scaffolds for regenerating tissues.

Carampin et al. (2007) synthesized a biodegradable poly[(ethyl phenylala-nato)$_{1.4}$(ethyl glycinato)$_{0.6}$ phosphazene] (PPhe–GlyP) that had an intrinsic viscosity of $[\eta] = 3.10$ dL/g at 25°C. This polymer was optimized for various electrospinning parameters to obtain bead-free electrospun nanofiber matrices and tubular scaffolds. A polymer solution of 4% (w/v) in THF, 0.8-mm needle diameter, 10-cm collector-tip distance, and 9 kV accelerating voltage resulted in random fibers with a fiber diameter of 850 ± 150 nm and porosity of $85\% \pm 2\%$ (Carampin et al., 2007). Tubular constructs were fabricated by making use of a 2-mm stainless steel rod as a target with manual rotation. Thus, fabricated nanofiber matrices were characterized for primary rat neuromicrovascular endothelial cell (NEC) adhesion and proliferation at different time intervals. Nanofiber matrices showed significantly higher NEC growth than the control tissue culture polystyrene (TCPS) plates after 4 days in culture. Endothelial cells formed a confluent monolayer on the outer surface of the tubular construct after 16 days and were unable to migrate through the wall thickness, and the inner part of the scaffold remained acellular (Fig. 5b and c). Amino acid–substituted polyphosphazene nanofibers matrices closely mimic the structure of natural ECM and are suitable materials to fabricate in vitro small-caliber vessel substitutes. However, efforts are needed to improve the pore volume and diameter to facilitate cell migration within the scaffold structure. Cell seeding both inside and outside tubular structures can potentially increase the endothelialization of the scaffold.

Polyphosphazene–PLAGA Blend Scaffolds

A blend approach has been widely accepted to generate polymeric biomaterials with tunable properties. This is a popular way to design a biomaterial with controlled physical, biological, and degradation properties (Coutinho et al., 2008; Li et al., 2006; Sell et al., 2006). Laurencin, Allcock, and co-workers have applied this concept to biodegradable polyphosphazenes and developed blend systems with PLA, PGA, and PLAGA (Ambrosio et al., 2002; Ibim et al., 1997; Krogman et al., 2007; Laurencin et al., 2007). This synergistic approach has resulted in various miscible blends with optimal mechanical, degradation, and enhanced biocompatibility properties. In vitro and in vivo biocompatibility studies further confirmed their potentiality as scaffold materials (Deng et al., 2008; Nair et al., 2005; 2006; Qiu, 2002). Selected formulations were further fabricated into 3D scaffolds via microsphere sintering and electrospinning.

Recently, authors have successfully fabricated microspheres from various polyphosphazene–PLAGA 85:15 blends. Microspheres formed between PNEA and PLAGA 85:15 showed a single glass transition temperature, confirming their miscible nature (data not reported). Microsphere sintering was achieved by a solvent–nonsolvent method utilizing a suitable solvent for one of the polymers (acetone or CH_2Cl_2) and a nonsolvent (ethanol or hexane) for both polymers. Even though both methods resulted in microsphere binding, scaffold characterization is required to identify the efficient sintering process for blend microspheres. It is also interesting to note that some of the blend systems resulted in nanophase separation on the microsphere surface. Further studies are needed to sinter these microspheres, without loosing the nanofeatures, to develop nanofeatured 3D scaffolds for tissue engineering.

Nonwoven fiber scaffolds of PNEA, PLA, and their blends were fabricated via conventional wet spinning (Conconi et al., 2006). In brief, polymer solutions in dichloromethane were slowly injected into n-heptane (nonsolvent) using a hypodermic syringe fitted with a 26G needle at 25°C to obtain fiber–filament scaffolds. These scaffolds were tested for the performance of primary rat osteoblasts obtained from bone marrow. Fiber scaffolds of amino acid–substituted polyphosphazene and their blends resulted in enhanced cell adhesion and growth. Fiber scaffolds resulted in decreased ALP compared to TCPS after 4 days, which might be due to osteoblasts being in the proliferation phase. Such fiber scaffolds of polyphosphazenes and their blends could be useful as fillers where limited mechanical loading occurs.

CONCLUSIONS

Biodegradable polyphosphazenes are a unique class of hybrid inorganic–organic polymers suitable for various biomedical applications. Amino acid ester polyphosphazenes have attracted particular attention as candidate biomaterials because of their biocompatibility, controllable hydrolytic degradation rates, and nontoxic degradation products. Therefore, various amino acid ester–based polyphosphazenes were developed and characterized for scaffold-based tissue engineering applications. Some of these amino acid ester polyphosphazenes were fabricated successfully into 3D scaffolds using a variety of techniques, such as salt leaching, microsphere sintering, and electrospinning. Among these, polyphosphazene–nHAp composite microsphere scaffolds showed appropriate mechanical, porosity, and osteoblast cell affinity properties to be used for bone tissue engineering. Electrospun nanofiber matrices with the ability to mimic the structure and function of native tissue extracellular matrix can better serve as scaffolds for enhanced tissue regeneration. The fine-tuning option available at the material level can be used effectively to design polyphosphazene scaffolds with the required mechanical, biological, and degradation characteristics suitable for any scaffolding applications. Although biodegradable polyphosphazenes are considered increasingly for scaffolding applications, there is a strong need to investigate these polymers further (for incomplete degradation and imunoadjuvant behavior) before using them for tissue engineering applications.

Acknowledgments

The authors acknowledge funding from the National Institutes of Health (R01 EB004051 and R01 AR052536). Dr. Laurencin is the recipient of a Presidential Faculty Fellow award from the National Science Foundation.

REFERENCES

Allcock, H.R. 2001. Rational design and synthesis of polyphosphazenes for tissue engineering. In Atala, A. Lanza, R., eds., *Methods of Tissue Engineering*. Academic Press, San Diego, CA, pp. 597–607.

Allcock, H.R. 2003. *Chemistry and Applications of Polyphosphazenes*. Wiley, Hoboken, NJ.

Allcock, H.R. 2006. Design of new biomedical materials with targeted properties. *Abstracts of Papers, 232nd ACS National Meeting*; pp. 10–14.

Allcock, H.R., Kwon, S. 1988. Glyceryl polyphosphazenes: synthesis, properties, and hydrolysis. *Macromolecules*, 21:1980–1985.

Allcock, H.R., Kwon, S. 1989. An ionically crosslinkable polyphosphazene: poly[bis (carboxylatophenoxy)phosphazene] and its hydrogels and membranes. *Macromolecules*, 22:75–79.

Allcock, H.R., Fuller, T.J., Mack, D.P., Matsumura, K., Smeltz, K.M. 1977. Synthesis of poly[(amino acid alkyl ester)phosphazenes]. *Macromolecules*, 10:824–830.

Allcock, H.R., Fuller, T.J., Matsumura, K. 1982. Hydrolysis pathways for aminophosphazenes. *Inorg. Chem.*, 21:515–521.

Allcock, H.R., Pucher, S.R., Scopelianos, A.G. 1994a. Poly[(amino acid ester)pho-sphazenes] as substrates for the controlled release of small molecules. *Biomaterials*, 15:563–569.

Allcock, H.R., Pucher, S.R., Scopelianos, A.G. 1994b. Poly[(amino acid ester)pho-sphazenes]: synthesis, crystallinity, and hydrolytic sensitivity in solution and the solid state. *Macromolecules*, 27:1071–1075.

Allcock, H.R., Pucher, S.R., Scopelianos, A.G. 1994c. Synthesis of poly(orgnaopho-sphazenes) with glycolic acid ester and lactic acid ester side groups: prototypes for new bioerodible polymers. *Macromolecules*, 27:1–4.

Allcock, H.R., Laredo, W.R, Bender, J.D. Bender, J.D., Ambrosio, A.M. 2002. Bioerodiable polymers and responoive hydrogels for tissue engineering and biome-dical devices. *Polym. Prepri.*, 43:456.

Allcock, H.R., Singh, A., Ambrosio, A.M., Laredo, W.R. 2003. Tyrosine-bearing polyphosphazenes. *Biomacromolecules*, 4:1646–1653.

Allcock, H.R., Steely, L.B., Kim, S.H., Kim, J.H., Kang, B.K. 2007. Plasma surface functionalization of poly[bis(2,2,2-trifluoroethoxy)phosphazene] films and nanofi-bers. *Langmuir*, 23:8103–8107.

Ambrosio, A.M., Allcock, H.R., Katti, D.S., Laurencin, C.T. 2002. Degradable polyphosphazene/poly(alpha-hydroxyester) blends: degradation studies. *Biomater-ials*, 23:1667–1672.

Ambrosio, A.M., Sahota, J.S., Runge, C., Kurtz, S.M., Nair, L.S., Allcock, H.R., Laurencin, C.T. 2003. Novel polyphosphazene–hydroxyapatite composites as bio-materials. *IEEE Eng. Med. Biol. Mag.*, 22:18–26.

Andrianov, A.K., Marin, A. 2006. Degradation of polyaminophosphazenes: effects of hydrolytic environment and polymer processing. *Biomacromolecules*, 7:1581–1586.

Andrianov, A.K., Svirkin, Y.Y., LeGolvan, M.P. 2004. Synthesis and biologically relevant properties of polyphosphazene polyacids. *Biomacromolecules*, 5:1999–2006.

Andrianov, A.K., Marin, A., Roberts, B.E. 2005. Polyphosphazene polyelectrolytes: a link between the formation of noncovalent complexes with antigenic proteins and immunostimulating activity. *Biomacromolecules*, 6:1375–1379.

Bhattacharyya, S., Nair, L.S., Singh, A., Krogman, N.R., Greish, Y.E., Brown, P.W., Allcock, H.R., Laurencin, C.T. 2006. Electrospinning of poly[bis(ethyl alanato) phosphazene] nanofibers. *J. Biomed. Nanotech.*, 2:1–10.

Bhattacharyya, S., Kumbar, S.G., Khan, Y.M., Nair, L.S., Singh, A., Krogman, N.R., Brown, P.W., Allcock, H.R., Laurencin, C.T. 2008. Biodegradable polyphospha-zene–nanohydroxyapatite composite nanofibers: scaffolds for bone tissue engineer-ing. *J. Biomed. Nanotech.*, 4:1–7.

Blan, N.R., Birla, R.K. 2008. Design and fabrication of heart muscle using scaffold-based tissue engineering. *J. Biomed. Mater. Res.*, A86:195–208.

Borden, M., Attawia, M., Laurencin, C.T. 2002a. The sintered microsphere matrix for bone tissue engineering: in vitro osteoconductivity studies. *J. Biomed. Mater. Res.*, 61:421–429.

Borden, M., Attawia, M., Khan, Y.M., Laurencin, C.T. 2002b. Tissue engineered microsphere-based matrices for bone repair: design and evaluation. *Biomaterials*, 23:551–559.

Borden, M., Attawia, M., Khan, Y.M., El-Amin, S.F., Laurencin, C.T. 2004. Tissue-engineered bone formation in vivo using a novel sintered polymeric microsphere matrix. *J. Bone Joint Surg. Am.*, 86:1200–1208.

Brown, J.L., Nair, L.S., Bender, J.D., Allcock, H.R., Laurencin, C.T. 2007. The formation of an apatite coating on carboxylated polyphosphazenes via a biomimetic process. *Mater. Lett.*, 61:3692–3695.

Brown, J.L., Nair, L.S., Laurencin, C.T. 2008. Solvent/non-solvent sintering: a novel route to create porous microsphere scaffolds for tissue regeneration. *J. Biomed. Mater. Res.*, 86B:396–406.

Carampin, P., Conconi, M.T., Lora, S., Menti, A.M., Baiguera, S., Bellini, S., Grandi, C., Parnigotto, P.P. 2007. Electrospun polyphosphazene nanofibers for in vitro rat endothelial cells proliferation. *J. Biomed. Mater. Res.*, A80:661–668.

Christenson, E.M., Anseth, K.S., van den Beucken, J.J., Chan, C.K., Ercan, B., Jansen, J.A., Laurencin, C.T. 2007. Nanobiomaterial applications in orthopedics. *J. Orthop. Res.*, 25:11–22.

Conconi, M.T., Lora, S., Menti, A.M., Carampin, P., Parnigotto, P.P. 2006. In vitro evaluation of poly[bis(ethyl alanato)phosphazene] as a scaffold for bone tissue engineering. *Tissue Eng.*, 12:811–819.

Cooper, J.A., Bailey, L.O., Carter, J.N., Castiglioni, C.E., Kofron, M.D., Ko, F.K., Laurencin, C.T. 2006. Evaluation of the anterior cruciate ligament, medial collateral ligament, achilles tendon and patellar tendon as cell sources for tissue-engineered ligament. *Biomaterials*, 27:2747–2754.

Coutinho, D.F., Pashkuleva, I.H., Alves, C.M., Marques, A.P., Neves, N.M., Reis, R.L. 2008. The effect of chitosan on the in vitro biological performance of chitosan–poly(butylene succinate) blends. *Biomacromolecules*, 9:1139–1145.

Deng, M., Nair, L.S., Nukavarapu, S.P., Kumbar, S.G., Jiang, T., Krogman, N.R., Singh, A., Allcock, H.R., Laurencin, C.T. 2008. Miscibility and in vitro osteocompatibility of biodegradable blends of poly[(ethyl alanato)(P-phenyl phenoxy)phosphazene] and poly(lactic acid–glycolic acid). *Biomaterials*, 29:337–349.

El-Amin, S.F., Kwon, M.S., Starnes, T., Allcock, H.R., Laurencin, C.T. 2006. The biocompatibility of biodegradable glycine containing polyphosphazenes: a comparative study in bone. *J. Inorg. Organomet. Polym.*, 16:387–396.

Greish, Y.E., Bender, J.D., Nair, L.S., Brown, P.W., Allcock, H.R., Laurencin, C.T. 2005. Composite formation from hydroxyapatite with sodium and potassium salts of polyphosphazene. *J. Mater. Sci. Mater. Med.*, 16:613–620.

Greish, Y.E., Bender, J.D., Nair, L.S., Brown, P.W., Allcock, H.R., Laurencin, C.T. 2006. Formation of hydroxyapatite–polyphosphazene polymer composites at physiologic temperature. *J. Biomed. Mater. Res.*, A77:416–425.

Greish, Y.E., Sturgeon, J.L., Singh, A., Krogman, N.R., Touny, A.H., Sethuraman, S., Nair, L.S., Laurencin, C.T., Allcock, H.R., Brown, P.W. 2008. Formation and properties of composites comprised of calcium-deficient hydroxyapatites and ethyl alanate polyphosphazenes. *J. Mater. Sci. Mater. Med.*, 19:3153–3160.

Heyde, M., Schacht, E. 2004. Biodegradable polyphosphazenes for biomedical applications. In Gleria, M., De Jaeger, R., eds., *Applicative Aspects of Poly(organophosphazenes)*. Nova Science, Hauppauge, NY, pp. 1–32.

Heyde, M., Moens, M., Van Vaeck, L., Shakesheff, K.M., Davies, M.C., Schacht, E.H. 2007. Synthesis and characterization of novel poly[(organo)phosphazenes] with cell-adhesive side groups. *Biomacromolecules*, 8:1436–1445.

Heyde, M., Claeyssens, M., Schacht, E.H. 2008. Interaction between proteins and polyphosphazene derivatives having a galactose moiety. *Biomacromolecules*, 9:672–677.

Hollister, S.J. 2005. Porous scaffold design for tissue engineering. *Nat. Mater.*, 4:518–524.

Hutmacher, D.W. 2001. Scaffold design and fabrication technologies for engineering tissues: state of the art and future perspectives. *J. Biomater. Sci. Polym. Ed.*, 12:107–124.

Ibim, S.E., Ambrosio, A.M., Kwon, M.S., El-Amin, S.F., Allcock, H.R., Laurencin, C.T. 1997. Novel polyphosphazene/poly(lactide-*co*-glycolide) blends: miscibility and degradation studies. *Biomaterials*, 18:1565–1569.

Ibim, S.E., El-Amin, S.F., Goad, M.E., Ambrosio, A.M., Allcock, H.R., Laurencin, C.T. 1998. In vitro release of colchicine using poly(phosphazenes): the development of delivery systems for musculoskeletal use. *Pharm. Dev. Tech.*, 3:55–62.

Jiang, T., Abdel-Fattah, W.I., Laurencin, C.T. 2006. In vitro evaluation of chitosan/poly(lactic acid–glycolic acid) sintered microsphere scaffolds for bone tissue engineering. *Biomaterials*, 27:4894–4903.

Khan, Y.M., Yaszemski, M.J., Mikos, A.G., Laurencin, C.T. 2008. Tissue engineering of bone: material and matrix considerations. *J. Bone Joint Surg. Am.*, 90:36–42.

Kim, S.S., Sun Park, M., Jeon, O., Yong Choi, C., Kim, B.S. 2006. Poly(lactide-*co*-glycolide)/hydroxyapatite composite scaffolds for bone tissue engineering. *Biomaterials*, 27:1399–1409.

Krogman, N.R., Singh, A., Nair, L.S., Laurencin, C.T., Allcock, H.R. 2007. Miscibility of bioerodible polyphosphazene/poly(lactide-*co*-glycolide) blends. *Biomacromolecules*, 8:1306–1312.

Kumbar, S.G., Bhattacharyya, S., Nukavarapu, S.P., Khan, Y.M., Nair, L.S., Laurencin, C.T. 2006. In vitro and in vivo characterization of biodegradable poly(organophosphazenes) for biomedical applications. *J. Inorg. Organomet. Polym. Mater.*, 16:365–385.

Kumbar, S.G., James, R., Nukavarapu, S.P., Laurencin, C.T. 2007. Electrospun nanofiber scaffolds: engineering soft tissues. *Biomed. Mater.*, 3:1–15.

Kumbar, S.G., Nukavarapu, S.P., James, R., Nair, L.S., Laurencin, C.T. 2008. Electrospun poly(lactic acid-*co*-glycolic acid) scaffolds for skin tissue engineering. *Biomaterials*, 29:4100–4107.

Langer, R., Tirrell, D.A. 2004. Designing materials for biology and medicine. *Nature*, 428:487–492.

Langer, R., Vacanti, J.P. 1993. Tissue engineering. *Science*, 260:920–926.

Laurencin, C.T., Ambrosio, A.M. 2003. Biodegradable polyphosphazenes for biomedical applications. In Arshady, R., eds., *Biodegradable Polymers*. PBM Series, London, pp. 153–173.

Laurencin, C.T., Morris, C.D., Pierre-Jacques, H., Schwartz, E.R., Keaton, A.R., Zou, L. 1992. Osteoblast culture on bioerodible polymers: studies of initial cell adhesion and spread. *Polym. Adv. Tech.*, 3:359–364.

Laurencin, C.T., Norman, M.E., Elgendy, H.M., El-Amin, S.F., Allcock, H.R., Pucher, S.R., Ambrosio, A.M. 1993. Use of polyphosphazenes for skeletal tissue regeneration. *J. Biomed. Mater. Res.*, 27:963–973.

Laurencin, C.T., El-Amin, S.F., Ibim, S.E., Willoughby, D.A., Attawia, M., Allcock, H.R., Ambrosio, A.M. 1996. A highly porous 3-dimensional polyphosphazene polymer matrix for skeletal tissue regeneration. *J. Biomed. Mater. Res.*, 30:133–138.

Laurencin, C.T., Ambrosio, A.M., Borden, M.D., Cooper, J.A. 1999. Tissue engineering: orthopedic applications. *Annu. Rev. Biomed. Eng.*, 1:19–46.

Laurencin, C.T., Nair, L.S., Battacharyya, S., Allcock, H.R., Bender, J.D., Brown, P.W. Greish, Y.E. 2005. Polymeric nanofibers for tissue engineering and drug delivery. U.S. patent 2005025630.

Laurencin, C.T., Khan, Y.M., Kofron, M.D., El-Amin, S.F., Botchwey, E., Yu, X., Cooper J.A. Jr., 2006. The ABJS Nicolas Andry award: tissue engineering of bone and ligament: a 15-year perspective. *Clin. Orthop. Relat. Res.*, 447:221–236.

Laurencin, C.T., Nair, L.S., Deng, M. 2007. Polyphosphazene poly(lactide-*co*-glycolide) blends: the development of a novel biomedical material. Presented at the 233rd ACS National Meeting, Chicago, 2007; PMSE-097.

Levenberg, S., Langer, R. 2004. Advances in tissue engineering. *Curr. Top. Dev. Biol.*, 61:113–134.

Li, M., Mondrinos, M.J., Chen, X., Gandhi, M.R., Ko, F.K., Lelkes, P.I. 2006. Co-electrospun poly(lactide-*co*-glycolide), gelatin, and elastin blends for tissue engineering scaffolds. *J. Biomed. Mater. Res.* A, 79:963–973.

Luten, J., van Steenis, J.H., van Someren, R., Kemmink, J., Schuurmans-Nieuwenbroek, N.M., Koning, G.A., Crommelin, D.J., van Nostrum, C.F., Hennink, W.E. 2003. Water-soluble biodegradable cationic polyphosphazenes for gene delivery. *J. Control. Release*, 89:483–497.

Mutwiri, G., Benjamin, P., Soita, H., Townsend, H., Yost, R., Roberts, B., Andrianov, A.K., Babiuk, L.A. 2007. Poly[di(sodium carboxylatoethylphenoxy)phosphazene] (PCEP) is a potent enhancer of mixed Th1/Th2 immune responses in mice immunized with influenza virus antigens. *Vaccine*, 25:1204–1213.

Nair, L.S., Katti, D.S., Laurencin, C.T. 2003. Biodegradable polyphosphazenes for drug delivery applications. *Adv. Drug Deliv. Rev.*, 55:467–482.

Nair, L.S., Laurencin, C.T. 2006a. Polymers as biomaterials for tissue engineering and controlled drug delivery. *Adv. Biochem. Eng. Biotech.*, 102:47–90.

Nair, L.S., Bhattacharyya, S., Bender, J.D., Greish, Y.E., Brown, P.W., Allcock, H.R., Laurencin, C.T. 2004a. Fabrication and optimization of methylphenoxy substituted polyphosphazene nanofibers for biomedical applications. *Biomacromolecules*, 5:2212–2220.

Nair, L.S., Bhattacharyya, S., Laurencin, C.T. 2004b. Development of novel tissue engineering scaffolds via electrospinning. *Expert Opin. Biol. Ther.*, 4:659–668.

Nair, L.S., Bender, J.D., Singh, A., Sethuraman, S., Greish, Y.E., Brown, P.W., Allcock, H.R., Laurencin, C.T. 2005. Biodegradable poly[bis(ethyl alanato)phosphazene]–poly(lactide-*co*-glycolide) blends: miscibility and osteocompatibility evaluations. *Mater. Res. Soc. Symp. Proc.*, 844:319–325.

Nair, L.S., Lee, D.A., Bender, J.D., Barrett, E.W., Greish, Y.E., Brown, P.W., Allcock, H.R., Laurencin, C.T. 2006b. Synthesis, characterization, and osteocompatibility evaluation of novel alanine-based polyphosphazenes. *J. Biomed. Mater. Res. A* 76:206–213.

Ng, K.W., Hutmacher, D.W. 2006. Reduced contraction of skin equivalent engineered using cell sheets cultured in 3D matrices. *Biomaterials*, 27:4591–4598.

Nukavarapu, S.P., Kumbar, S.G., Nair, L.S., Laurencin, C.T. 2007. Nanostructures for tissue engineering regenerative medicine. In Gonsalves, K.E., Halberstadt, C.R., Laurencin, C.T., Nair, L.S., eds., *Biomedical Nanostructures*. NJ, Hoboken, Wiley, pp. 377–407.

Nukavarapu, S.P., Kumbar, S.G., Merrell, J.G., Laurencin, C.T. 2008a. Electrospun polymeric nanofiber scaffolds for tissue regeneration. In Laurencin, C.T., Nair, L.S., eds., *Nanotechnology and Tissue Engineering: The Scaffold.* CRC Press, Boca Raton, FL, pp. 199–219.

Nukavarapu, S.P., Kumbar, S.G., Brown, J.L., Krogman, N.R., Weikel, A.L., Hindenlang, M.D., Nair, L.S., Allcock, H.R., Laurencin, C.T. 2008b. Polyphosphazene/nano-hydroxyapatite composite microsphere scaffolds for bone tissue engineering. *Biomacromolecules*, 9:1818–1825.

Passi, P., Zadro, A., Marsilio, F., Lora, S., Caliceti, P., Veronese, F.M. 2000. Plain and drug loaded polyphosphazene membranes and microspheres in the treatment of rabbit bone defects. *J. Mater. Sci. Mater. Med.*, 11:643–654.

Qiu, L. 2002. In vivo degradation and tissue compatibility of polyphosphazene blend films. *Sheng Wu Yi Xue Gong Cheng Xue Za Zhi*, 19:191–195.

Sell, S.A., McClure, M.J., Barnes, C.P., Knapp, D.C., Walpoth, B.H., Simpson, D.G., Bowlin, G.L. 2006. Electrospun polydioxanone-elastin blends: potential for bioresorbable vascular grafts. *Biomed. Mater.*, 1:72–80.

Sethuraman, S., Nair, L.S., El-Amin, S.F., Farrar, R., Nguyen, M.T., Singh, A., Allcock, H.R., Greish, Y.E., Brown, P.W., Laurencin, C.T. 2006. In vivo biodegradability and biocompatibility evaluation of novel alanine ester based polyphosphazenes in a rat model. *J. Biomed. Mater. Res. A* 77:679–687.

Singh, A., Krogman, N.R., Sethuraman, S., Nair, L.S., Sturgeon, J.L., Brown, P.W., Laurencin, C.T., Allcock, H.R., 2005. Synthesis, characterization and in vitro degradation of L-alanine co-substituted polyphosphazenes. Presented at the 230th ACS National Meeting, Washington, DC, 2005; POLY 340.

Singh, A., Krogman, N.R., Sethuraman, S., Nair, L.S., Sturgeon, J.L., Brown, P.W., Laurencin, C.T., Allcock, H.R. 2006. Effect of side group chemistry on the properties of biodegradable L-alanine cosubstituted polyphosphazenes. *Biomacromolecules*, 7:914–918.

Singh, A., Steely, L., Krogman, N.R., Allcock, H.R. 2007. Development of polyphosphazenes for surface and biomedical applications. Presented at the 233rd ACS National Meeting, Chicago, 2007; PMSE-095.

Stegemann, J.P., Kaszuba, S.N., Rowe, S.L. 2007. Advances in vascular tissue engineering using protein-based biomaterials. *Tissue Eng.*, 13:2601–2613.

Ten Huisen, K.S., Brown, P.W., Reed, C.S., Allcock, H.R. 1996. Low temperature synthesis of a self-assembling composite: hydroxyapatite poly[bis(sodium carboxylatophenoxy)phosphazene]. *J. Mater. Sci. Mater. Med.*, 7:673.

Wang, X., Wang, X., Tan, Y., Zhang, B., Gu, Z., Li, X. 2008. Synthesis and evaluation of collagen–chitosan–hydroxyapatite nanocomposites for bone grafting. *J. Biomed. Mater. Res. A* (May 13, Epub ahead of print).

Wilson, C.J., Clegg, R.E., Leavesley, D.I., Pearcy, M.J. 2005. Mediation of biomaterial–cell interactions by adsorbed proteins: A review. *Tissue Eng.*, 11:1–18.

9 Biodegradable Polyphosphazene Blends for Biomedical Applications

MENG DENG

Department of Chemical Engineering, University of Virginia, Charlottesville, Virginia; Department of Orthopaedic Surgery, University of Connecticut, Farmington, Connecticut

LAKSHMI S. NAIR

Department of Orthopaedic Surgery, University of Connecticut, Farmington, Connecticut; Department of Chemical, Materials and Biomolecular Engineering, University of Connecticut, Storrs, Connecticut

NICHOLAS R. KROGMAN and HARRY R. ALLCOCK

Department of Chemistry, The Pennsylvania State University, University Park, Pennsylvania

CATO T. LAURENCIN

Department of Orthopaedic Surgery, University of Connecticut, Farmington, Connecticut; Department of Chemical, Materials and Biomolecular Engineering, University of Connecticut, Storrs, Connecticut

BACKGROUND

A *biomaterial* is defined as "a nonviable material used in a medical device, intended to interact with biological systems" [1]. In other words, biomaterials are generally "substances other than food or drugs contained in therapeutic or diagnostic systems that are in contact with tissue or biological fluids" [2]. Such definitions of interfacing synthetic or natural materials with biology illustrate well their use as temporary substrates for a number of applications, including temporary prostheses, tissue engineering scaffolds, and drug delivery devices.

Over the last 30 years, the trend toward developing appropriate biomaterials to match complex requirements for different biomedical applications has motivated numerous research endeavors and has experienced consistent growth. Biomaterials such as polymers, ceramics, and metals are widely used

Polyphosphazenes for Biomedical Applications, Edited by Alexander K. Andrianov
Copyright © 2009 John Wiley & Sons, Inc.

in many pharmaceutical preparations, in medical devices ranging from contact lenses to kidney dialyzers, and in implants from vascular grafts to cardiac pacemakers. They are found in more than 8000 different types of medical devices, 2500 separate diagnostic products, and 40,000 different pharmaceutical preparations [2]. The industry and markets for biomaterials and medical devices have evolved to be worth $100 billion [3]. Even though biomaterials have already contributed greatly to the improvement of human health, the need for better biomaterial systems continues to increase.

Among the existing biomaterials, polymeric materials are very attractive, due to the ease of tailoring their chemical, physical, and biological properties for target applications. It is possible to impart specific hydrophilic or hydrophobic entities, biodegradable repeating units, or multifunctional motifs [4]. Diverse properties can be obtained by directly incorporating desirable functional groups or indirectly modifying existing structures with different functional groups [5]. Due to their versatility, they are rapidly replacing other biomaterial classes, such as metals and ceramics. As a result, the annual sales of polymeric biomaterials amount to $7 billion, which is attributed to about 88% of the total market of biomaterials in 2003. It was expected that the biocompatible materials market would reach almost $11.9 billion by 2008, implying significant growth for polymeric biomaterials [6]. More specifically, biodegradable polymeric materials are preferred for biomedical applications, due to their biodegradability and inherent biocompatibility [7].

Biodegradable Polymers

Biodegradable polymers, including both natural and synthetic polymers, have been investigated extensively for a number of biomedical applications. These polymeric biomaterials are degraded into low-molecular-weight fragments via hydrolysis or enzymolysis under physiological conditions [8]. Therefore, biodegradable polymers can be classified as hydrolytically degradable polymers or enzymatically degradable polymers, based on the type of degradation. The inherent bioactivity associated with biological polymers has made them very attractive materials for clinical uses. However, most of them undergo enzymatic degradation. The degradation rate depends on the availability and concentrations of the enzymes. It is difficult to predict or control their degradation in vivo. Moreover, the disadvantages, such as microbial contamination, strong immunogenic responses, and possible disease transmission, limit their applications. On the other hand, synthetic biodegradable polymers are generally biologically inert. They have unique advantages, such as more predictable properties and ease of tailoring for specific applications, devoid of many of the disadvantages of natural polymers.

Since the success of the first poly(glycolic acid) (PGA) suture system, there has been a great increase in using synthetic biodegradable polymers as transient implants for tissue engineering and drug delivery applications. In 1995, the market of commercial products made from absorbable polymers in the United

States was over $300 million [9]. Total health care costs for Americans involving tissue damage and end-stage organ failure were estimated to be around $400 billion [10]. Tissue engineering technologies were developed to treat these problems with synthetic biodegradable polymers. Since then, interest in technologies that use biodegradable polymers has increased dramatically, as was showed by the Pubmed and Medline databases. The market worldwide for biodegradable polymers was more than $215 million and is expected to increase at a rate of 12.6%, to $388 million annually by 2010 [11].

Currently, some major classes of synthetic biodegradable polymers that have been investigated for biomedical applications include polyesters, poly(α-amino acids), and polyanhydrides [7]. Among them, aliphatic polyesters such as polymers of lactic acid (PLA) and glycolic acid (PGA) or their co-polymers poly(lactic acid–glycolic acid) (PLAGA) are the most commonly used bio-degradable polymers for biomedical applications, due to their commercial availability, their approval by the U.S. Food and Drug Administration (FDA) for certain clinical applications, their established biocompatibility, their controlled degradation rate, and their excellent mechanical properties. However, most biodegradable polymers suffer from distinct drawbacks. For example, the bulk erosion of aliphatic polyesters often leads to uncontrollable release kinetics, which is not desirable for controlled drug delivery applications. These polymers degrade via unstable backbone ester hydrolysis into the corresponding acids, which can compromise the structural integrity and potentially affect biocompatibility both in vitro and in vivo [12–15]. It has been of concern that in anatomical sites such as articular cartilage, a significant accumulation of these acid degradation products could occur and affect the cells and the tissues surrounding the implants [13,15]. Furthermore, it is possible that these acids can inactivate sensitive molecules such as the proteins used for drug delivery applications [16]. To overcome these limits, PLA or PLAGA with a high lactide content has been investigated for various medical applications. However, several clinical complications have also been reported [13]. These studies have propelled researchers to search for and develop novel biodegradable polymeric materials with nontoxic and neutral degradation products, appropriate degra-dation rates, and mechanical integrity that can appropriately match the specific and unique requirements of each specific medical application.

The development of novel biodegradable polymers focuses primarily on the design of novel synthetic polymers with unique chemistries and the adoption of novel synthesis methods or a combination of both. Considering the inherent complexity and the wide range of targeted applications, it is very difficult for one polymeric system to satisfy all the requirements and be considered as an ideal biomaterial.

Polymer Blends

Blending of two or more polymers is a practical way to develop novel biomaterials by synergistically combining the advantages of the parent polymers. It has advantages such as the ease of blend preparation and efficient control of blend properties via compositional changes in the parent components [17]. Polymer blends are of great commercial interest and occupy over 30% of the polymer market [17].

More recently, efforts have been made to improve the properties of polyesters by blending them with a variety of other polymers. For example, PLAGA was blended with 2-methacryoloxyethyl phosphoroylcholine (PMPC) to reduce the inflammatory reaction of adherent cells on the surface [18]. Meanwhile, PLAGA was also blended with hydrophilic polymers such as poly(vinyl alcohol) [19] as well as natural polymers such as chitin [20] to control the degradation rate of PLAGA and the delivery of proteins. Similarly, blends of PLLA with PDLLA (or PCL) and pluronic surfactants were prepared to make PLLA tougher and more suitable for orthopedic applications [21]. The triblock copolymers of poly(ethylene oxide) (PEO) and poly(propylene oxide) (PPO), PEO–PPO–PEO, were blended with PLLA to minimize the initial protein burst and extend protein release [22]. The effect of enantiomeric polymer blending (PLA/PDLLA) on the degradation rate of PLLA has been investigated extensively [23,24].

A Unique Class of Polymers: Polyphosphazenes

Polyphosphazenes (PPHOs) are linear polymers with alternating nitrogen and phosphorus atoms in the backbone and with two organic side groups attached to each phosphorus atom, as depicted in Scheme 1 [25]. The synthetic flexibility provided by altering the side groups attached to the backbone enables a wide array of PPHOs with diverse physical, chemical, and biological properties. Among them, biodegradable PPHOs are a unique class of polymers suitable for

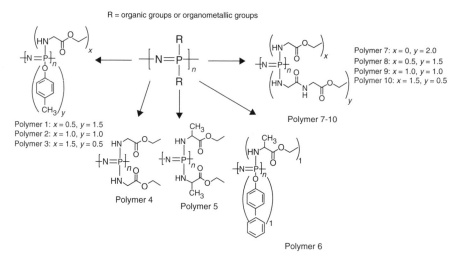

SCHEME 1   Structures of various PPHOs.

a wide range of biomedical applications. They undergo hydrolytic degradation, yielding an ammonium phosphate buffer solution and side-group degradation by-products. In recent years, amino acid ester–substituted PPHOs have attracted much attention, due to their synthetic versatility, nontoxic and buffering degradation products composed of phosphate and ammonia, and recognized biocompatibility both in vitro and in vivo [26–32]. However, the synthesis of biodegradable PPHOs is a complex process that involves the initial synthesis of a highly reactive prepolymer poly(dichlorophosphazene) followed by simultaneous or sequential nucleophilic substitution reactions. So far, this nontraditional process is still under commercial development.

Recently, a great deal of research has focused on developing composite materials as novel biomaterials for biomedical applications. Specifically, during the past two decades, significant advances have been made in the development of polyphosphazene-blend materials for biomedical applications [31,33–39]. Each of these applications requires biomaterials with specific physical, chemical, and biological properties. Therefore, a wide range of PPHOs blend with degradable or nondegradable polymers are being developed.

POLYPHOSPHAZENE-BASED BLENDS

Polymeric blending is a simple method to develop novel biomaterials while offering controllable properties and degradation rates. Both biodegradable and nonbiodegradable blend materials have been developed based on PPHOs. Among them, degradable PPHO blends are preferred candidate materials for developing medical devices such as temporary prostheses, three-dimensional tissue-engineered scaffolds, and controlled drug delivery vehicles. In the meantime, the nondegradable PPHO blends have been used primarily as implants or prostheses.

In general, PPHO blends have several unique advantages compared to other current polymer blend systems. Blending or combining PPHOs with other biodegradable polymers, such as PLAGA, serves as a practical method to develop novel materials that combine the beneficial features of PLAGA, such as biocompatibility and wide applicability with the well-tuned degradability and buffering degradation products of PPHOs. In the blend systems, the synthetic flexibility of PPHOs allows us to design and develop specific polymers that can have strong hydrogen-bonding interactions with PLAGA to achieve miscibility. More interestingly, the degradable blends composed of biodegradable PPHOs and PLAGA make it possible to have self-neutralizing materials for biomedical applications where the ammonium phosphate buffering solution of PPHOs degradation products could neutralize the acidic degradation products of PLAGA. Furthermore, specific functional groups can be introduced easily to the blends through the regulation of side groups in PPHOs and thus make the blend materials more attractive. For example, carboxyl-substituted PPHOs such as poly[bis(carboxylate phenoxy)phosphazene]

(PCPP) have been shown to interact with simulated body fluid (SBF) and form a bonelike apatite layer on the surface [40]. Such bioactivity would greatly improve the osteointegrativity of scaffolds and implants for bone tissue engineering [41]. Therefore, the PPHOs blends are very attractive candidate materials for biomedical applications.

PPHOs/Synthetic Nonbiodegradable Polymer Blends

PPHOs of nonbioerodible character have been blended with other polymers to fabricate polymer composite blends [42]. This work outlines the utility of polyphosphazene used in polymer blends and lays the groundwork for bioerodible polymer blends, which we discuss further later.

PPHOs/Synthetic Biodegradable Polymer Blends

Different blend systems of PPHOs with a variety of classical synthetic biodegradable polymers have been reported [31,33–39]. These classical polymers include polyesters such as PLA, PLAGA, and polyanhydrides, which are commonly used materials for scaffold-based tissue engineering or controlled drug delivery applications. Through either surface erosion or bulk erosion, depending on the side-group chemistry, PPHOs can degrade into neutral and buffering degradation products, which are composed mainly of an ammonium phosphate-buffering solution and side-group by-products. These nontoxic degradation products can act as a buffering solution to the acids, including lactic acid and glycolic acid produced by polyesters such as PLA, PGA, and PLAGA. Besides the self-neutralizing ability, such blend systems allow efficient control of materials properties by adjusting the component ratios and can easily be tailored as temporary substrates for different biomedical applications.

The concept and feasibility of fabricating various miscible blends of PPHOs with PLAGA and the ease of tailoring the blend degradation rate was first demonstrated by Ibim et al. [33]. In the study, three different PPHOs containing different molar ratios of ethyl glycinato to p-methylphenoxy (polymer 1, 2, or 3 in Scheme 1) were synthesized and blended with PLAGA (50:50) at a weight ratio of 50:50 [33]. All blends showed homogeneous morphologies, and their miscibility was confirmed by differential scanning calorimetry (DSC). The degradation studies demonstrated that blends degraded at different rates, implying that the blend properties can easily be tuned for a variety of different biomedical settings. The follow-up study by Ambrosio et al. demonstrated the self-neutralizing ability of such blend systems and the possibility of maximizing the neutralization effect by optimizing the blend composition [34]. Three types of samples, including PLAGA (50:50), polymer 2, and their blends were subjected to degradation in nonbuffered solutions at pH 7.4 over a period of 40 weeks. The degradation of the matrices was evaluated by monitoring mass loss as well as molecular-weight decline, while the degradation medium was analyzed for the amount of acid released in nonbuffered solutions (Fig. 1).

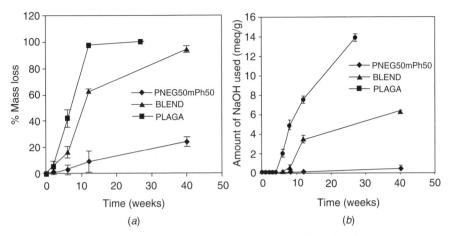

FIGURE 1 (a) Percent mass loss of a PLAGA–polymer 2 blend and parent polymers in pH 7.4 distilled water at 37°C over 12 weeks. (b) Amount of NaOH (mEq/g) used to neutralize the acid released during the degradation of the blend and its parent polymers. (From ref. 34, with permission. Copyright © 2002 Elsevier Ltd.)

The amounts of neutralizing base (0.01 N solution of sodium hydroxide) for titrating the medium back to pH 7.4 were measured for each type of sample throughout degradation. Results showed that the blend required significantly lower amounts of neutralizing base than PLAGA, which indicated that the blend released fewer acidic degradation products, due to the buffering effect of PPHOs degradation products (Fig. 1b). In addition, the blend showed an intermediate degradation rate between these of the parent polymers, and the hydrolysis of PPHOs could have been catalyzed by the acidic degradation products of PLAGA (Fig. 1a). Hence, these two earlier studies demonstrated the advantages of self-neutralizing blends as novel biomaterials and established the ground for developing PPHO-based blends as next-generation materials for controlled drug delivery and other biomedical applications.

Meanwhile, Qiu et al. have investigated a series of degradable blends comprising poly[bis(glycine ethyl ester)phosphazene] (polymer 4, Scheme 1) and polyesters or polyanhydrides for both degradation characteristics and tissue biocompatibility [37–39]. In the study, different blends of polymer 4 with PLA, PLAGA (80:20), poly(sebacic anhydride) (PSA), and poly(sebacic anhydride-*co*-trimellitylimidoglycine)-*block*-poly(ethylene glycol) (30:50:20 by mole) (PSTP) were fabricated in various ratios using a solvent-mixing technique [37]. The polymer 4–PLA blends were completely immiscible, whereas the polymer 4–PLAGA and polymer 4–PSTP blends were found to be partially miscible as evaluated by DSC, Fourier transfer infrared [FT-IR], and phase-contrast microscopy. Slabs of polymer 4–PLAGA and polymer 4–PSTP blends were subjected to degradation studies in distilled water with or without *Rhizopus delemer* lipase at 37°C [38]. The blend of polymer 4–PLAGA

(70 : 30 w/w) took 120 days to disappear completely, whereas polymer 4–PSTP (70 : 30 by weight) slabs needed only 20 days. The degradation rate of polymer 4–PLAGA blends was strongly accelerated by the enzyme, and the degree of enzymatic degradation depended on the weight percentage of PLAGA in the blend. In vivo degradation of the polymer blends was carried out by implanting them subcutaneously in the back of mice, and some differences in degradation were observed compared to in vitro degradation. Such variation was attributed to more complex physiological environment, which indicated that the degradation mechanism of polymer 4–PLAGA was a combination of hydrolysis and enzymolysis, while that of polymer 4–PSTP was only hydrolysis. In addition, both in vitro and in vivo studies showed that the degradation rate of the blends could be controlled easily by adjusting the blend composition. The tissue biocompatibility of polymer 4–PLAGA was found to be better than that of polymer 4–PSTP, and the tissue biocompatibility of polymer 4–PSTP blends was improved by increasing the weight percentage of polymer 4 in blends [39]. These findings suggested that the blends of polymer 4 and PLAGA or PSTP may be used as drug delivery matrices or for other potential biomedical applications.

Thereafter, a series of biodegradable blends have also been developed for bone tissue engineering applications. Nair et al. have developed the biocompatible blends of polymer 5 (Scheme 1) and PLAGA (85 : 15) at two weight ratios of 25 : 75 and 50 : 50 [36]. The two blends were characterized for miscibility, degradation, mechanical performance, and in vitro osteocompatibility. The blends were found to be partially miscible, as confirmed by scanning electron microscopy (SEM), DSC, and FT-IR. The weak hydrogen-bonding interactions are attributed to the steric hindrance from the α-CH$_3$ groups in the alanine and lactide units. Furthermore, osteocompatibility studies using primary rat osteoblasts showed significantly higher cell adhesion and proliferation for both blends than those of the parent polymers. This study demonstrated the superior osteocompatibility of alanine-based PPHO blends and supported their potential use in bone tissue repair and regeneration.

More recently, Deng et al. have developed high-strength blends with a polyphosphazene that contains a 1 : 1 ratio of the side groups alanine ethyl ester and phenylphenoxy (polymer 6, Scheme 1) and PLAGA (85 : 15) for bone tissue engineering applications [31]. The solvent effect on blend miscibility was investigated and tetrahydrofuran (THF) was used to produce different blends at three different weight ratios: 25 : 75 (BLEND25), 50 : 50 (BLEND50), and 75 : 25 (BLEND75). As confirmed by SEM, DSC, and FT-IR, BLEND25 was miscible, whereas BLEND50 and BLEND75 were partially miscible (Fig. 2), indicating the limitation of the solvent effect on blend miscibility. The blends were able to nucleate bonelike apatite via a biomimetic process (Fig. 3), which could induce bioactivity in vivo. All blends showed comparable osteoblast cell adhesion and proliferation to PLAGA. After 21 days of cell culture, multilayers of osteoblasts covered the blend surfaces as shown in Figure 4. Furthermore, the polymer 6 component in blends resulted in an increase in osteoblast

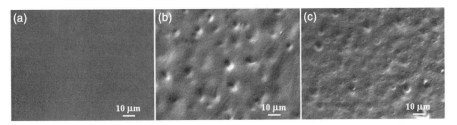

FIGURE 2 SEM micrographs of various blends: (a) BLEND25; (b) BLEND50; (c) BLEND75. BLEND25 showed uniform morphology, indicating blend miscibility, whereas the other two blends showed visible phase separation. (From ref. 31, with permission. Copyright © 2007 Elsevier Ltd.)

phenotypic expression and mineralized matrix synthesis in vitro. Therefore, the high-strength polymer 6 and PLAGA blends are attractive candidate materials for a variety of musculoskeletal applications.

Blend miscibility is one of the prerequisites in applications that require uniform and predictable properties. Efforts have also been made to improve blend miscibility by introducing different side groups containing multiple hydrogen-bonding sites in PPHOs. Krogman et al. have synthesized five different glycyl-glycine ethyl ester dipeptide-based PPHOs (polymer 7–10, Scheme 1) and investigated their blends with both PLAGA (50:50) and PLAGA (85:15) [35]. Among them, polymer 8 was found to form completely miscible blends with both PLAGA blends a fact characterized by SEM, DSC, and attenuated total reflection–IR. It was found that the T_g value for each blend was lower than for each parent polymer, implying that PPHOs and PLAGA in the blend acted as plasticizers for each other. Further, hydrolysis studies showed the blends degraded at a slower rate than did both parent

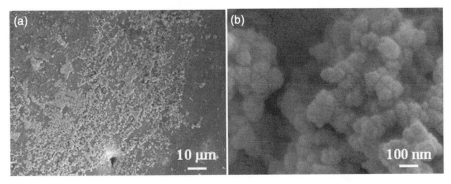

FIGURE 3 SEM micrographs of BLEND25 substrate after 21 days of incubation in 1.5 × SBF at 37°C at a magnification of (a) 1000; (b) 100,000. (From ref. 31, with permission. Copyright © 2007 Elsevier Ltd.)

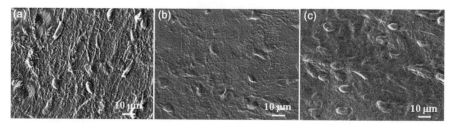

FIGURE 4   SEM micrographs of primary rat osteoblasts cultured on different blends for a period of 21 days: (a) BLEND25; (b) BLEND50; (c) BLEND75. Blend surfaces were covered completely by PRO multilayers. (From ref. 31, with permission. Copyright © 2007 Elsevier Ltd.)

polymers (Fig. 5a). As evident in Figure 5b, the pH of the degradation medium possessing the blends increased from 2.5 to 4, which was attributed to the buffering capacity of PPHO degradation products. This study has demonstrated the feasibility of fine-tuning the PLAGA degradation profile by either adjusting the side-group ratios of PPHOs or by compositional changes in the blend. It also showed that the neutralization effect could be maximized by regulating the PPHO side groups. Such novel dipeptide-based blends of PPHOs with PLAGA are believed to be beneficial for a variety of biomedical applications.

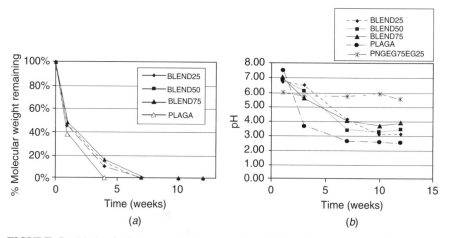

FIGURE 5   (a) Hydrolytic degradation profile of PLAGA–polymer 8 blends and PLAGA in pH 7.4 distilled water at 37°C over 12 weeks. (b) pH change of aqueous media where the PLAGA–PNGEG50EG50 blends and their parent polymers were degraded over 12 weeks. (From ref. 35, with permission. Copyright © 2007 American Chemical Society.)

Blends of PPHOs Alone

The degradability of PPHO materials can be controlled by the content of side groups. An alternative approach is the blending of a more hydrolysis-sensitive depsipeptide-containing polymer with poly[(amino acid ester)phosphazenes]. A variety of materials with diverse degrees of degradability can be obtained by combining the parent polymers [43–45]. Crommen et al. have demonstrated that the biodegradability of PPHO materials can be controlled not only by varying the content of depsipeptide side groups but also by blending poly [(amino acid ester)phosphazenes] with a hydrolysis-sensitive depsipeptide-containing a PPHO derivative [43]. Amino acid ester–substituted PPHOs were blended with depsipeptide-containing polymers in different weight ratios. The rate of degradation can be adjusted by blending (50 : 50 w/w) poly[(ethyl glycinate)-co-(ethyl 2-(O-glycyl lactate))phosphazene] (3% gly-lact-OEt) with poly[(ethyl alanate)phosphazene] [43]. However, it was found that the presence of the depsipeptide-substituted polymer only promoted the release of glycine (ethyl ester); it did not affect release of the alanine side groups. Lemmouchi et al. observed a similar phenomenon for a blend (50 : 50 w/w) of poly[(ethyl glycinate)phosphazene] with poly[(ethyl alanate)-co-(ethyl 2-(O-glycyl lactate))-phosphazene] (3% gly lact-OEt) [44]. Only the release of alanine and alanine ethyl ester was affected significantly. Since the blend materials showed a single T_g value, such interesting degradation behavior of depsipeptide-containing PPHO blends might be attributed to the intromolecular catalysis of the polymer by the pendant carboxylic acid.

In recent years, thermosensitive polymer hydrogels have attracted much attention as injected delivery systems of drugs, factors, or cells in the field of drug delivery and tissue engineering. Such polymer hydrogels show sol–gel transitions under certain stimuli, such as temperature and pH. Different physical cross-links can be used to develop such gel systems. Most of them are formed via miscelle packing or hydrophobic interactions between polymer chains and contain both hydrophilic and hydrophobic regions. Earlier studies have reported several synthetic polymer hydrogels, including N-isopropylacrylamide (NiPAAM) copolymers poly(lactide-co-glycolide)–poly(ethylene glycol)–poly(lactide-co-glycolide) (PLAGA–PEG–PLAGA) and poly(ethylene oxide)–poly(propylene oxide)–poly(ethylene oxide) (PEO–PPO–PEO) [46–49]. More recently, Lee et al. demonstrated the formation of thermosensitive poly(organophosphazene) gels bearing α-amino-ω-methoxy poly(ethylene glycol) (AMPEG) and hydrophobic amino acid esters such as isoleucine ethyl ester, leucine ethyl ester, and valine ethyl ester [50]. The nature and compositional ratios of the hydrophilic and hydrophobic moieties directly affect the gelation properties of the polymers. For example, a PPHO gel that contains a greater hydrophobic component, such as isoleucine ethyl ester (IleOEt), generally shows a higher gel viscosity and lower gelation temperature. On the other hand, a PPHO polymer containing less hydrophobic or hydrolysis-sensitive groups such as ethyl-2-(O-glycyl) lactate (GlyLacOEt) tends to exhibit lower gel viscosities and higher gelation temperatures. Hence, to have the desired gelation properties, precise control over the balance between hydrophilicity and hydrophobicity is required in order. However, complexity and many unexpected factors involved in polymer synthesis make such control unpredictable. Kang et al. demonstrated the efficacy of controlling the gelation behavior of hydrogels via compositional regulation over the hydrophobic and hydrophilic components in polymer blends [51]. In the study, both hard and soft PPHOs were synthesized. A hard polymer including $[NP(IleOEt)1.16(AMPEG550)0.84]_n$ and $[NP(IleOEt)1.13(GlyLacOEt)0.03(AMPEG550)0.84]_n$ exhibits a low gelation temperature and a high viscosity, whereas a soft one such as $[NP(IleOEt)1.07(GlyLacOEt)0.02(AMPEG550)0.91]_n$ shows a high gelation temperature and a low viscosity. By blending these two types of PPHOs at an appropriate ratio, the blended aqueous solution was able to change into a transparent hydrogel that showed great strength at 37°C. According to DSC and IR, the two polymers blended homogenously and formed a thermosensitive injectable hydrogel with a T_{max} value of 37 to 38°C.

CONCLUSIONS

Given unsurpassable synthetic flexibility, controlled degradability, excellent biocompatibility, and buffering ability, PPHO-blend materials have vast potential in tissue engineering and drug delivery. A great number of researchers have investigated extensively from the design and development of appropriate PPHO blends to the in vitro and in vivo evaluation of the biological performance. In particular, PPHO blends with polyesters are very promising. The buffering capacity of PPHO degradation products can neutralize the acidic degradation products of PLAGA, while the degradation rates of blends can easily be regulated with compositional change. This will bring great benefits to medical therapies, since the acidic degradation products of PLAGA have been reported to have some adverse effects. The synthetic flexibility of PPHOs allows us to maximize the buffering effect and develop self-neutralizing polymer blends with appropriate properties for a wide range of biomedical applications.

FUTURE TRENDS

In recent years there has been tremendous interest in biodegradable materials that can degrade into nontoxic products under physiological conditions. Among them, synthetic biodegradable polymers such as polyesters and polyanhydrides are being widely investigated for scaffold-based tissue engineering or controlled drug delivery because of their degradability, suitable mechanical properties, and tunable degradation rates. However, most of them are associated with certain drawbacks.

Blending PPHOs with these classical biodegradable polymers provides an efficient approach to combine the beneficial features of both parent polymers. PPHO blends have already showed great potential in the biomedical field. Therefore, it is a very promising area for future research to explore. This will include the following aspects. First, as for blend fabrication, most of the research effort utilizes the solvent casting method. Several studies have demonstrated the negative effect of solvents on the biological performance of blends. Other alternative methods that eliminate the solvent effect should be investigated. Second, the miscibility of PPHO blends with PLAGA, especially PLAGA 85:15, remains a challenge because of the high degree of crystallinity caused by lactide units. To achieve superior properties compared to parent polymers, complete miscibility needs to be achieved. Our recent studies have demonstrated the success of achieving miscibility through the manipulation of polymer chemistries with different solvents or PPHO side groups, which provide important guidance in that direction. Third, three-dimensional structures should be developed to take advantage of the unique properties of PPHO blends for medical therapies.

Acknowledgments

The authors gratefully acknowledge financial support from the National Institutes of Health (RO1 EB004051) and the National Science Foundation (EFRI-0736002).

REFERENCES

1. Williams, D. Definitions in biomaterials. In *Progress in Biomedical Engineering*, 4th ed. Elsevier, Amsterdam, 1987.

2. Langer, R., Peppas, N.A. Advances in biomaterials, drug delivery, and bionanotechnology. *AIChEJ.*, 2003, 49:2990–3006.

3. Ratner, B.D., Bryant, S.J. Biomaterials: where we have been and where we are going. *Annu. Rev. Biomed. Eng.*, 2004, 6:41–75.

4. Peppas, N.A., Huang, Y., Torres-Lugo, M., Ward, J.H., Zhang, J. Physicochemical foundations and structural design of hydrogels in medicine and biology. *Annu. Rev. Biomed. Eng.*, 2000, 2:9–29.

5. Bures, P., Huang, Y., Oral, E., Peppas, N.A. Surface modifications and molecular imprinting of polymers in medical and pharmaceutical applications. *J. Control. Release*, 2001, 72:25–33.

6. RB-072N biocompatible materials for the human body [online]. Business Communications Company, Inc., Norwalk, CT, July 2003.

7. Nair, L.S., Laurencin, C.T. Biodegradable polymers as biomaterials. *Prog. Polym. Sci. Polym. Biomed. Appl.*, 2007, 32:762–798.

8. Katti, D.S., Lakshmi, S., Langer, R., Laurencin, C.T. Toxicity, biodegradation and elimination of polyanhydrides. *Adv. Drug Deliv. Rev. Polyanhydrides Poly(ortho esters)*, 2002, 54:933–961.

9. *U.S. Absorbable and Erodible Biomaterials Products Markets.* Frost & Sullivan, Mountain View, CA, 1995, Chap. 10.

10. Atala, A., Mooney, D.J., eds. *Synthetic Biodegradable Polymer Scaffolds.* Springer, Boston, 1997.

11. Schlechter, M. Business Communications Company Market Study, 2005.

12. Fu, K., Pack, D.W., Klibanov, A.M., Langer, R. Visual evidence of acidic environment within degrading poly(lactic-*co*-glycolic acid) (PLGA) microspheres. *Pharm. Res.*, 2000, 17:100–106.

13. Bostman, O.M. Osteoarthritis of the ankle after foreign-body reaction to absorbable pins and screws: a three- to nine-year follow-up study. *J. Bone Joint Surg. Br.*, 1998, 80-B:333–338.

14. Agrawal, C.M., Athanasiou, K.A. Technique to control pH in vicinity of biodegrading PLA–PGA implants. *J. Biomed. Mater. Res.*, 1997, 38:105–114.

15. Taylor, M.S., Daniels, A.U., Andriano, K.P., Heller, J. Six bioabsorbable polymers: in vitro acute toxicity of accumulated degradation products. *J. Appl. Biomater.*, 1994, 5:151–157.

16. Yang, L., Alexandridis, P. Physicochemical aspects of drug delivery and release from polymer-based colloids. *Curr. Opin. Colloid Interface Sci.*, 2000, 5: 132–143.

17. Utracki, L.A. *Polymer Alloys and Blends.* Hanser Verlag, Munich, Germany, 1989.

18. Iwasaki, Y., Sawada, S.-I., Ishihara, K., Khang, G., Lee, H.B. Reduction of surface-induced inflammatory reaction on PLGA/MPC polymer blend. *Biomaterials*, 2002, 23:3897–3903.

19. Pitt, G.G., Cha, Y., Shah, S.S., Zhu, K.J. Blends of PVA and PGLA: control of the permeability and degradability of hydrogels by blending. *J. Control. Release*, 1992, 19:189–199.

20. Mi, F.-L., Shyu, S.-S., Lin, Y.-M., Wu, Y.-B., Peng, C.-K., Tsai, Y.-H. Chitin/PLGA blend microspheres as a biodegradable drug delivery system: a new delivery system for protein. *Biomaterials*, 2003, 24:5023–5036.

21. Chen, C.-C., Chueh, J.-Y., Tseng, H., Huang, H.-M., Lee, S.-Y. Preparation and characterization of biodegradable PLA polymeric blends. *Biomaterials*, 2003, 24:1167–1173.

22. Park, T.G., Cohen, S., Langer, R. Poly(L-lactic acid)/pluronic blends: characterization of phase separation behavior, degradation, and morphology and use as protein-releasing matrixes. *Macromolecules*, 1992, 25:116–122.

23. Ikada, Y., Tsuji, H. Biodegradable polyesters for medical and ecological applications. *Macromol. Rapid Commun.*, 2000, 21:117–132.

24. Tsuji, H. In vitro hydrolysis of blends from enantiomeric poly(lactide)s: 4. well-homo-crystallized blend and nonblended films. *Biomaterials*, 2003, 24:537–547.

25. Allcock, H.R. *Chemistry and Applications of Polyphosphazenes.* Wiley, Hoboken, NJ, 2003.

26. Kumbar, S., Bhattacharyya, S., Nukavarapu, S., Khan, Y., Nair, L., Laurencin, C. In vitro and in vivo characterization of biodegradable poly(organophosphazenes) for biomedical applications. *J. Inorg. Organomet. Polym. Mater.*, 2006, 16:365–385.

27. Lakshmi, S., Katti, D.S., Laurencin, C.T. Biodegradable polyphosphazenes for drug delivery applications. *Adv. Drug Deliv. Rev.*, 2003, 55:467–482.

28. Singh, A., Krogman, N.R., Sethuraman, S., et al. Effect of side group chemistry on the properties of biodegradable ʟ-alanine cosubstituted polyphosphazenes. *Biomacromolecules*, 2006, 7:914–918.

29. Laurencin, C.T., Norman, M.E., Elgendy, H.M., et al. Use of polyphosphazenes for skeletal tissue regeneration. *J. Biomed. Mater. Res.*, 1993, 27:963–973.

30. Laurencin, C.T., El-Amin, S.F., Ibim, S.E., et al. A highly porous 3-dimensional polyphosphazene polymer matrix for skeletal tissue regeneration. *J. Biomed. Mater. Res.*, 1996, 30:133–138.

31. Deng, M., Nair, L.S., Nukavarapu, S.P., et al. Miscibility and in vitro osteocompatibility of biodegradable blends of poly[(ethyl alanato) (*p*-phenyl phenoxy)phosphazene] and poly(lactic acid–glycolic acid). *Biomaterials*, 2008, 29:337–349.

32. Allcock, H.R., Pucher, S.R., Scopelianos, A.G. Poly[(amino acid ester)phosphazenes]: synthesis, crystallinity, and hydrolytic sensitivity in solution and the solid state. *Macromolecules*, 1994, 27:1071–1075.

33. Ibim, S.E.M., Ambrosio, A.M.A., Kwon, M.S., El-Amin, S.F., Allcock, H.R., Laurencin, C.T. Novel polyphosphazene/poly(lactide-*co*-glycolide) blends: miscibility and degradation studies. *Biomaterials*, 1997, 18:1565–1569.

34. Ambrosio, A.M.A., Allcock, H.R., Katti, D.S., Laurencin, C.T. Degradable polyphosphazene/poly([alpha]-hydroxyester) blends: degradation studies. *Biomaterials*, 2002, 23:1667–1672.

35. Krogman, N.R., Singh, A., Nair, L.S., Laurencin, C.T., Allcock, H.R. Miscibility of bioerodible polyphosphazene/poly(lactide-*co*-glycolide) blends. *Biomacromolecules*, 2007, 8:1306–1312.

36. Nair, L.S., Allcock, H.R., Laurencin, C.T. Biodegradable poly[bis(ethyl alanato)-phosphazene]–poly(lactide-*co*-glycolide) blends: miscibility and osteocompatibility evaluations. *Mater. Res. Soc. Symp. Proc.*, 2005.

37. Qiu, L.Y., Zhu, K.J. Novel blends of poly[bis(glycine ethyl ester) phosphazene] and polyesters or polyanhydrides: compatibility and degradation characteristics in vitro. *Polym. Int.*, 2000, 49:1283–1288.

38. Qiu, L.Y. In vitro and in vivo degradation study on novel blends composed of polyphosphazene and polyester or polyanhydride. *Polym. Int.*, 2002, 51:481–487.

39. Qiu, L.Y. Degradation and tissue compatibility of polyphosphazene blend films in vivo. *Shengwu Yixue Gongchengxue Zazhi*, 2002, 19:191–195.

40. Brown, J.L., Nair, L.S., Bender, J., Allcock, H.R., Laurencin, C.T. The formation of an apatite coating on carboxylated polyphosphazenes via a biomimetic process. *Mater. Lett.*, 2007, 61:3692–3695.

41. Hench, L.L., Polak, J.M. Third-generation biomedical materials. *Science*, 2002, 295:1014–1017.

42. Allcock, H.R., Visscher, K.B. Preparation and characterization of poly(organophosphazene) blends. *Chem. Mater.*, 1992, 4:1182–1187.

43. Crommen, J., Vandorpe, J., Schacht, E. Degradable polyphosphazenes for biomedical applications. *J. Control. Release*, 1993, 24:167–180.

44. Lemmouchi, Y., Schacht, E., Dejardin, S. Biodegradable poly[(amino acid ester)phosphazenes] for biomedical applications. *J. Bioact. Compat. Polym.*, 1998, 13:4–18.

45. Schacht, E., Vandorpe, J., Dejardin, S., Lemmouchi, Y., Seymour, L. Biomedical applications of degradable polyphosphazenes. *Biotechnol. Bioeng.*, 1996, 52: 102–108.

46. Chen, G., Hoffman, A.S. Graft copolymers that exhibit temperature-induced phase transitions over a wide range of pH. *Nature*, 1995, 373:49–52.

47. Jeong, B., Bae, Y.H., Kim, S.W. Thermoreversible gelation of PEG-PLGA-PEG triblock copolymer aqueous solutions. *Macromolecules*, 1999, 32:7064–7069.

48. Zentner, G.M., Rathi, R., Shih, C., et al. Biodegradable block copolymers for delivery of proteins and water-insoluble drugs. *J. Control. Release*, 2001, 72:203–215.

49. Glatter, O., Scherf, G., Schillen, K., Brown, W. Characterization of a poly(ethylene oxide)–poly(propylene oxide) triblock copolymer (EO27-PO39-EO27) in aqueous solution. *Macromolecules*, 1994, 27:6046–6054.

50. Lee, B.H., Song, S.-C. Synthesis and characterization of biodegradable thermosensitive poly(organophosphazene) gels. *Macromolecules*, 2004, 37:4533–4537.

51. Kang, G.D., Heo, J.-Y., Jung, S.B., Song, S.-C. Controlling the thermosensitive gelation properties of poly(organophosphazenes) by blending. *Macromol. Rapid Commun.*, 2005, 26:1615–1618.

10 Polyphosphazenes from Condensation Polymerization

PATTY WISIAN-NEILSON

Department of Chemistry, Southern Methodist University, Dallas, Texas

INTRODUCTION

As other chapters is this book have clearly indicated, the polyphosphazenes are unique among polymers in their structural diversity and broad range of properties. Since the first well-characterized polyphosphazenes were prepared in the mid-1960s [1], both the number of new polyphosphazenes and the synthetic pathways for producing these polymers have continued to expand [2]. The traditional preparation begins with the formation of poly(dichlorophosphazene), $[Cl_2PN]_n$, by ring-opening polymerization of hexachlorocyclotriphosphazene, $[Cl_2PN]_3$, followed by nucleophilic substitution reactions to replace the halogen atoms with organic side groups. Although ring-opening-substitution is the most popular way to prepare many polyphosphazenes, condensation polymerization processes have garnered increasing attention in the past three decades [2,3]. In addition to offering potential for controlling molecular weight, condensation reactions supplement the ring-opening process by offering access to different types of polyphosphazenes. In this chapter the preparation of polyphosphazenes by condensation polymerization, recent advances in the process, and the chemistry and properties of some of the unique polymers accessible by this approach are discussed.

CONDENSATION POLYMERIZATION METHODS

In 1980 [4] the first synthesis of high-molecular-weight polyphosphazenes by condensation polymerization was reported. In these initial experiments, an N-silylphosphoranimine, $Me_3SiN{=}P(OCH_2CF_3)Me_2$, was heated in a sealed

Polyphosphazenes for Biomedical Applications, Edited by Alexander K. Andrianov
Copyright © 2009 John Wiley & Sons, Inc.

glass ampoule for several days at 160°C [eq. (1)] producing a volatile condensation product, $Me_3SiOCH_2CF_3$, and an essentially quantitative yield of poly (dimethylphosphazene), $[Me_2PN]_n$, **1**. The opaque, film-forming polymer was soluble in chlorinated hydrocarbons and ethanol but insoluble in water and tetrahydrofuran (THF) and had a molecular weight of 50,000, or approximately 650 repeat units. This polymer may be considered an isoelectronic analog of the well-known silicone, poly(dimethylsiloxane), and it was the first reported polyphosphazene in which all the substituents were attached to the PN backbone by direct P–C bonds.

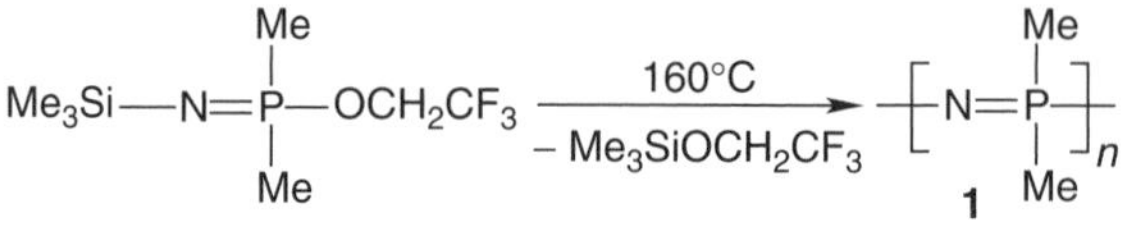

$$Me_3Si-N{=}P(\text{Me})(\text{Me})-OCH_2CF_3 \xrightarrow[- Me_3SiOCH_2CF_3]{160°C} \left[N{=}P(\text{Me})(\text{Me})\right]_n \quad \mathbf{1} \tag{1}$$

The thermal condensation polymerization of other phosphoranimines containing simple alkyl and aryl groups also proceeds smoothly to produce P–C-substituted polymers [i.e., poly(alkyl/arylphosphazene)s], including non-geminally substituted polymers with two different groups on each phosphorus [5]: for example, $[Me(Ph)PN]_n$ (**2**) and $[Me(alkyl)PN]_n$ (**3**) [6]; simple random copolymers: for example, $[Me(Ph)PN]_m[Me_2PN]_n$ (**4**) [5,7]; and an unusual system in which the backbone phosphorus is part of a five-membered carbon ring, **5** [8] (Chart 1). Typically, molecular weights are approximately 10^5 and polydispersities are small (1.2 to 3) relative to polyphosphazenes obtained by ring opening. Thus far, fully P–C-substituted poly(alkyl/arylphosphazene)s are not accessible by either the ring-opening-substitution of $[Cl_2PN]_n$ or direct ring opening of fully alkylated/arylated cyclic phosphazenes. Hence, condensation polymerization can be viewed as complementary to ring-opening substitution methods which provide access primarily to P–N- and P–O-substituted polymers. In general, the P–C bond is more hydrolytically and chemically stable than P–O and P–N bonds, so the poly(alkyl/arylphosphazene)s offer a significantly different dimension to the applications of this polymer system. For example, most poly(alkyl/arylphosphazene)s do not degrade under strongly basic conditions even under prolonged exposure, so they are not likely candidates for drug release. On the other hand, their stability could potentially make them useful biostructural materials.

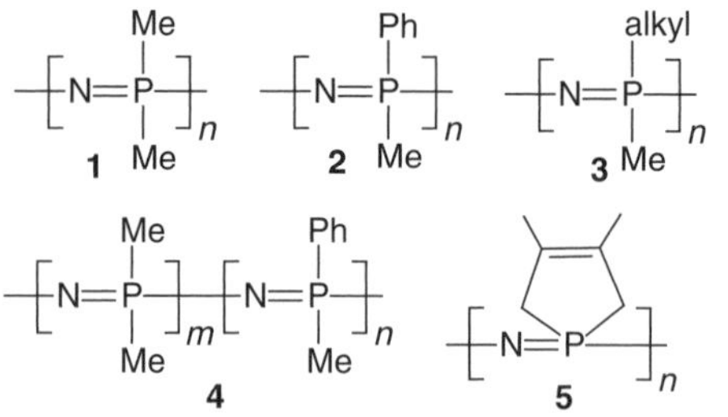

CHART 1

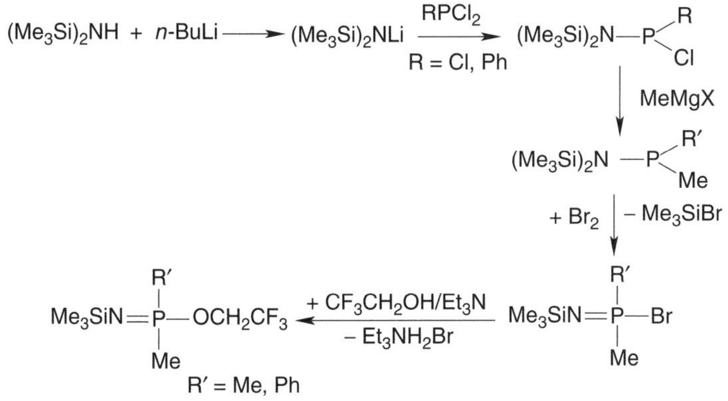

$$(Me_3Si)_2NH + \textit{n-BuLi} \longrightarrow (Me_3Si)_2NLi \xrightarrow[R = Cl, Ph]{RPCl_2} (Me_3Si)_2N-P\overset{R}{\underset{Cl}{<}}$$

$$\downarrow MeMgX$$

$$(Me_3Si)_2N-P\overset{R'}{\underset{Me}{<}}$$

$$+ Br_2 \; \left| \; - Me_3SiBr \right.$$

$$\underset{\underset{Me}{|}}{\overset{\overset{R'}{|}}{Me_3SiN=P}}-OCH_2CF_3 \xleftarrow[- Et_3NH_2Br]{+ CF_3CH_2OH/Et_3N} \underset{\underset{Me}{|}}{\overset{\overset{R'}{|}}{Me_3SiN=P}}-Br$$

R' = Me, Ph

SCHEME 1

The synthesis of the *N*-silylphosphoranimine precursors for condensation polymerization utilizes well-established, relatively straightforward reactions [7,9] (Scheme 1). This facilitates the incorporation of P–C-bonded organic groups on phosphorus in the small molecule precursors and avoids problems such as incomplete substitution and chain degradation that occur with attempts to substitute preformed $[Cl_2PN]_n$ with alkyl or aryl moieties using organometallic reagents. An additional feature of the condensation polymerization is that the diversity of the organic substituents can be enhanced by select modification reactions on the pendant phosphorus groups in the *N*-silylphosphoranimines [eq. (2)] [10].

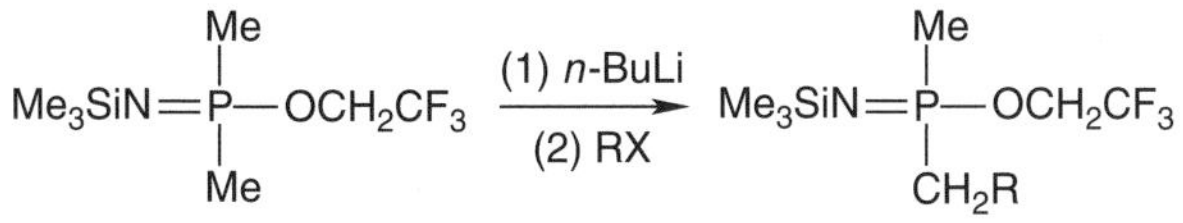

$$\underset{\underset{Me}{|}}{\overset{\overset{Me}{|}}{Me_3SiN=P}}-OCH_2CF_3 \xrightarrow[\text{(2) RX}]{\text{(1) } \textit{n-BuLi}} \underset{\underset{CH_2R}{|}}{\overset{\overset{Me}{|}}{Me_3SiN=P}}-OCH_2CF_3 \qquad (2)$$

R = Me, CH_2Ph, $CH_2CH=CH_2$, PPh_2, $P(NMe_2)_2$, Br,
$SiMe_2R'$ [R' = Me, Ph, H, $CH=CH_2$, $CH_2(CH_2)_2CN$]

X = Cl, Br, I

A number of variations in the condensation polymerization of *N*-silylphosphoranimines have been developed since the initial report. In 1990, fluoride ions were used to catalyze the polymerization of a closely related trialkoxy-substituted *N*-silylphosphoranimine [eq. (3)] [11]. The process was carried out in solution at milder temperatures than those used in the original report. However, the method was not reported to work with alkyl- and aryl-substituted phosphoranimines.

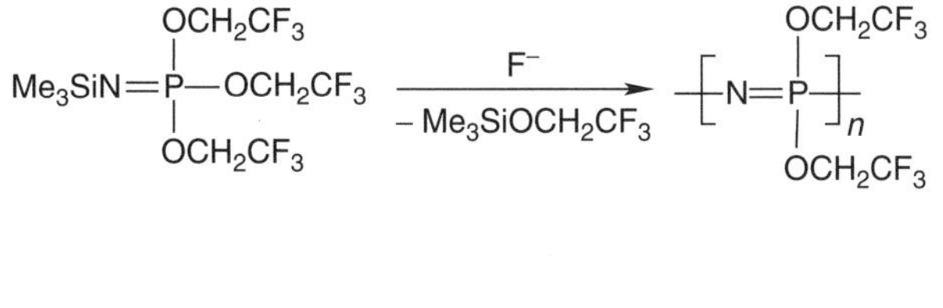

$$\underset{\underset{OCH_2CF_3}{|}}{\overset{\overset{OCH_2CF_3}{|}}{Me_3SiN=P}}-OCH_2CF_3 \xrightarrow[- Me_3SiOCH_2CF_3]{F^-} \underset{\underset{OCH_2CF_3}{|}}{\overset{\overset{OCH_2CF_3}{|}}{\left[N=P \right]_n}} \qquad (3)$$

Since then, we have found that the phenoxy leaving group is a less expensive alternative to the trifluoroethoxy group, and in the presence of NaOPh, several hundred grams of *P*-phenoxy-*N*-silylphosphoranimines can be polymerized simply by heating in a flask at temperatures of about 130°C for 2 to 3 days [12]. This is further facilitated by the fact that many of the precursors, in particular the *P*-phenyl-*P*-methylphosphoranimine, can be prepared on a large scale (at least 1 to 2 mol) in a one-pot sequence (Scheme 2). Yet another variation in the process is the use of hexachloroethane in place of bromine for the oxidation–halogenation of the (silylamino)phosphines. The latter is generally easier to handle on a large scale, produces more easily removed by-products (i.e., Me_3SiCl and LiCl), and avoids the contamination of HBr often found in reagent-grade bromine. This improved precursor synthesis and phenoxide ion catalysis is the procedures currently used in our labs to prepare 50- to 100-g quantities of poly(methylphenylphosphazene), $[Me(Ph)PN]_n$ (**2**), the polymer discussed at length below.

In the mid-1990s, a room-temperature living cationic polymerization of the *N*-silylphosphoranimine $Me_3SiN=PCl_3$ was first reported [13]. This occurs in the presence of small amounts of PCl_5 at room temperature in either solution or bulk phases. The poly(dichlorophosphazene) produced in this process has narrow molecular-weight distributions (<1.3) and molecular weights on the order of 10^4 [eq. (4)]. In a short time, the method was extended to the synthesis of polymers with directly P–C-bonded groups [14], the "living" cationic ends of the polymer were used to prepare a variety of block copolymers [15], and studies showed that there is a relationship between initiator–monomer stoichiometry [16]. The capability of obtaining polymers with controllable narrow molecular weights is ultimately useful for biological applications since polymer properties are related to molecular weight [16].

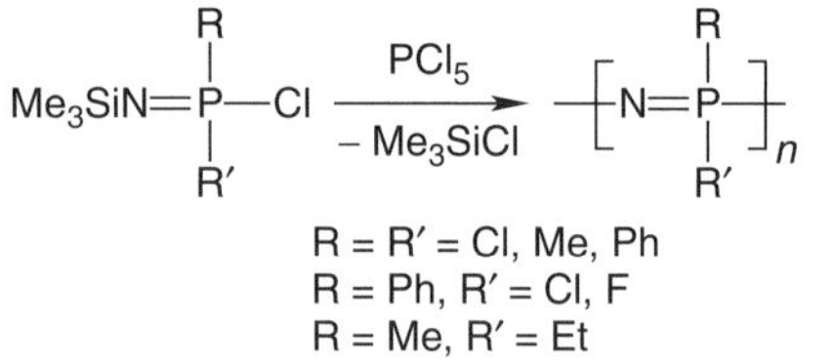

$$Me_3SiN=\overset{\overset{\displaystyle R}{|}}{\underset{\underset{\displaystyle R'}{|}}{P}}-Cl \xrightarrow[-\,Me_3SiCl]{PCl_5} \left[N=\overset{\overset{\displaystyle R}{|}}{\underset{\underset{\displaystyle R'}{|}}{P}} \right]_n$$

$$R = R' = Cl, \, Me, \, Ph$$
$$R = Ph, \, R' = Cl, \, F$$
$$R = Me, \, R' = Et$$

$$(4)$$

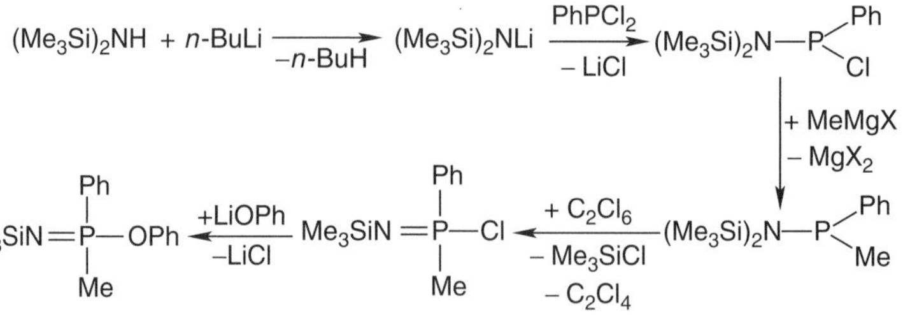

SCHEME 2 One-pot reaction sequence for preparation of a precursor to $[(Me)(Ph)PN]_n$ (**2**).

The most recent advance in the synthesis of polyphosphazenes from *N*-silylphosphoranimines is a result of efforts to form phosphoranimine cations stabilized by phosphine ligands [17]. While treatment of the *P*-bromo-*N*-silylphosphoranimine $Me_3SiN=P(Br)Me_2$ with simple trialkylphosphines such as Me_3P formed stable phosphoranimine salts (Scheme 3), the analogous phosphite, $(MeO)_3P$, produced quantitative yields of high-molecular-weight poly(dimethylphosphazene), $[Me_2PN]_n$ (**1**). The reaction was carried out on a small scale in chloroform and was complete after only a few hours at room temperature. Similarly, poly(methylphenylphosphazene) (**2**) was prepared under the same conditions from the appropriate P–Br phosphoranimine. Preliminary studies show that the molecular weights are high (10^5), polydispersities are below 2, and the molecular weights appear to be controllable via both time and initiator stoichiometry. Clearly, the ability to form polyphosphazenes with controllable molecular weights from *N*-silylphosphoranimines at room temperature in solution using such reagents as PCl_5 and $(MeO)_3P$ is an important advance. In addition, these newer methods utilize phosphoranimines with P-halo groups, thus eliminating the additional synthetic step needed to incorporate trifluorethoxy or phenoxy leaving groups.

Two other types of condensation processes have been used to prepare polyphosphazenes. One of these involves phosphine azide intermediates that eliminate nitrogen on heating [eq. (5)]. This process was reported for the synthesis of diaryl-substituted phosphazenes, including amorphous and, hence, soluble, poly(phenyl-*p*-tolulylphosphazene) [18]. However, the phosphine azides and related azide intermediates have a strong tendency to detonate on heating, especially on larger scales, thus limiting the viability for obtaining synthetically useful quantities of polymers.

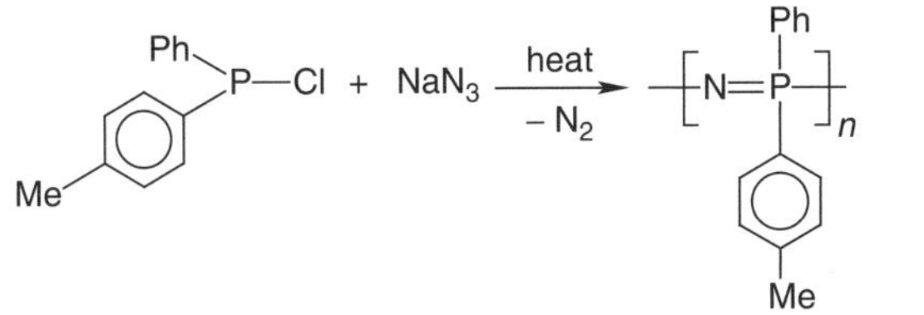

$$\text{(5)}$$

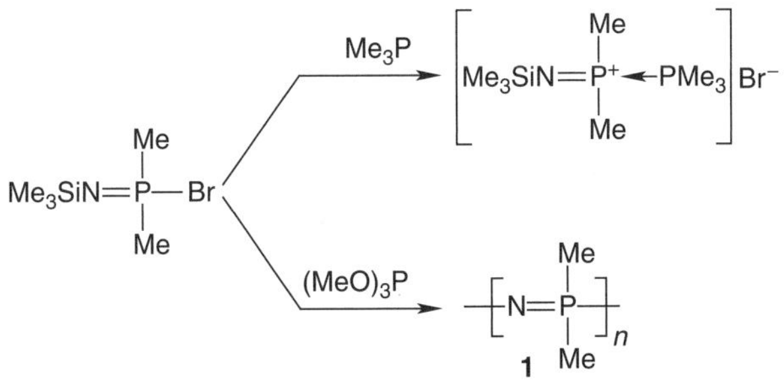

SCHEME 3

A somewhat different condensation polymerization process has been used to prepare poly(dichlorophosphazene) [eq. (6)] [19]. The process involves loss of $O=PCl_3$ in bulk polymerizations at about 200°C. The chlorine on the resulting polymer, $[Cl_2PN]_n$, is readily replaced with simple alkoxy and aryloxy groups to make stable, usable polymers, as is typically done with $[Cl_2PN]_n$ from ring opening of hexachlorocyclophosphazene. This process was commercialized in France in the 1980s.

$$O=\overset{\overset{\displaystyle Cl}{|}}{\underset{\underset{\displaystyle Cl}{|}}{P}}-N=\overset{\overset{\displaystyle Cl}{|}}{\underset{\underset{\displaystyle Cl}{|}}{P}}-Cl \xrightarrow[- POCl_3]{ca.250°C} \left[N=\overset{\overset{\displaystyle Cl}{|}}{\underset{\underset{\displaystyle Cl}{|}}{P}} \right]_n \qquad (6)$$

VARIATIONS IN POLY(ALKYL/ARYLPHOSPHAZENES)

One of the most important outcomes of condensation polymerization is access to fully P–C-substituted polyphosphazenes. This is possible because most simple alkyl and aryl groups are incorporated into the N-silylphosphorani-mines in a relatively straightforward manner, as discussed above (Schemes 1 and 2). Once the simple N-silylphosphoranimines are formed, some variation in the substituents at phosphorus can be introduced by relatively straightforward reactions [eq. (2)]. Various combinations of copolymers, polymers with long-chain side groups [6,20] and more recently, large P–C-bonded aromatic groups [21] allow for additional control of properties such as glass transition tempera-tures and hydrophobicity.

There are, however, some limitations in the types of functionalities that can be incorporated into polymers at the precursor level. This arises because the reactivity of the P–C-bonded functional groups must be considered in light of the high reactivity of the Si–N and P–O or P–X bonds in the precursor phosphoranimines as well as the stability of these functional groups under condensation reaction conditions (i.e., heat, PCl_5, or phosphite). One way to circumvent such problems is to prepare polymers initially with simple groups and then pursue macromolecular substitution reactions. Among the best candidates for this approach is $[(Me)(Ph)PN]_n$ (**2**). Synthetically, preparation of this polymer is the most straightforward and is readily accomplished on scales of 50 to 100 g. Unlike semicrystalline polymers such as $[Me_2PN]_n$ (**1**) and $[Et_2PN]_n$, which display solubility in only a few solvents, such as chloroform, ethanol, and dichloromethane, $[Me(Ph)PN]_n$ (**2**), is a white, amorphous, film-forming material that is soluble not only in chlorinated hydrocarbons but also in THF [7], the latter of which facilitates a broader range of reactions, especially reactions at the alkyl groups. In addition, $[(Me)(Ph)PN]_n$ offers potential reactivity at the phenyl group, and as in all poly(alkyl/arylphosphazene)s, coordination reactions occur readily at the relatively basic nitrogen atom. Each of these three types of reactivity (Scheme 4) and many of the polymers derived from such reactions are discussed below.

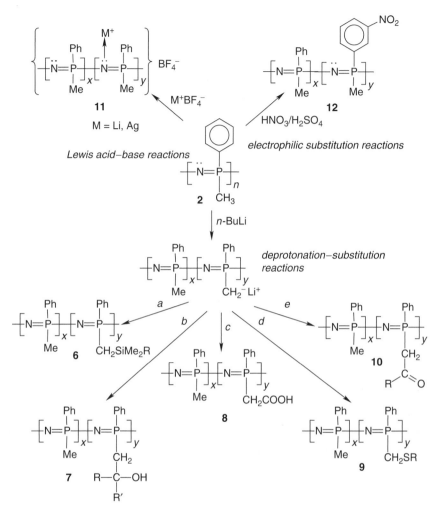

SCHEME 4 Three reactive sites in poly(methylphenylphosphazene), $[(Me)(Ph)PN]_n$ (**2**). Reagents: (a) RMe_2SiCl, where $R = Me$, H, $CH=CH_2$, $CH_2(CH_2)_2CN$, $CH_2(CH_2)_xCH_3$, $(CH_2)_2(CF_2)_yCF_3$; (b) $RR'C(=O)$, where R, $R' =$ alkyl, aryl, ferrocenyl, thiophenyl, etc.; (c) CO_2 followed by H^+; (d) RSSR, where $R = Me$, Ph; (e) $RC(=O)OEt$, where $R =$ alkyl, aryl.

The richest source of new polyphosphazenes with P–C-bonded groups results from the relative acidity of methyl groups attached to phosphorus, thus facilitating deprotonation to form intermediate carbanions and subsequent reaction with a variety of electrophiles (Scheme 4) [22]. These *deprotonation–substitution* reactions are carried out in THF, and the number of sites that are substituted is controlled by the stoichiometry of *n*-BuLi and the electrophile. Complete deprotonation can be achieved to form an intermediate anion

[23], and up to 95% of the methyl groups can be substituted successfully, with the steric size of the electrophile limiting the degree of substitution [23,24]. The lower portion of Scheme 4 shows some of the functional groups that have been incorporated by deprotonation–substitution. For example, polymers with a variety of silyl groups (pathway *a*) were prepared where R is alkyl $(CH_2)_xCH_3$ or fluoroalkyl $(CH_2)_2(CF_2)_nCF_3$ chains. Both the surface hydrophobicity, as measured by contact angles [24], and the gas permeability [25] of these polymers varied with the chain length and the amount of fluorine incorporated into the side groups. Reactions of the intermediate polymer anion with aldehydes and ketones (pathway b) afforded polymers with more hydrophobic fluorine groups [26], with electroactive ferrocene and thiophene moieties [27], and with several ester groups [28]. The electron transport properties of the ferrocene polymers were studied. Enhanced water solubility as well as the formation of hydrogels was reported for the carboxylate polymers that resulted from treatment of the polymer anion with carbon dioxide (pathway c) [29]. More recently, sulfide groups have been attached (pathway d) and subsequently oxidized to sulfone groups, which imparted more thermal stability [30]. The incorporation of sulfur groups suggests that there is potential for linking sulfur-containing amino acids to the polyphosphazenes. Finally, adding ketone moieties [31] (pathway e) provides yet another reactive site, the carbonyl group, which could be useful for attaching bioactive groups to the polymers.

Grafted copolymers of the poly(alkyl/arylphosphazenes) are accessible directly from the anion intermediate of **2** via anionic addition polymerization (polyphosphazene-*graft*-polystyrene) [32] or ring-opening polymerization (polyphosphazene-*graft*-polydimethylsiloxane) [33] (Scheme 5). An alcohol derivative, **7**, from the deprotonation–substitution (pathway b) has also served to provide initiation sites for atom transfer radical polymerization (ATRP) and subsequent formation of polyphosphazene-*graft*-poly(methyl methacrylate) [34] (Scheme 6).

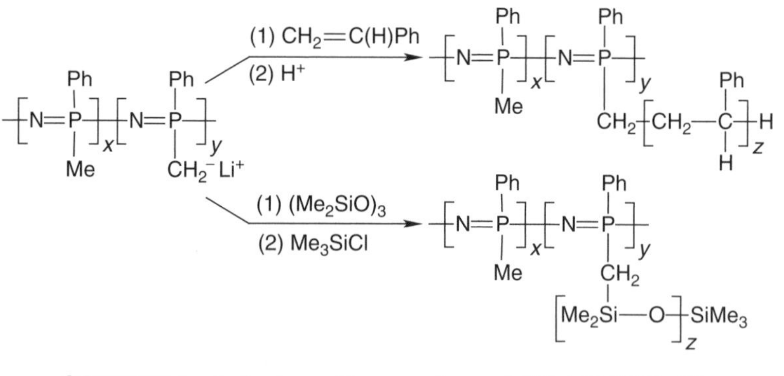

SCHEME 5 Anionic grafting reactions of $[(Me)(Ph)PN]_n$ (**2**).

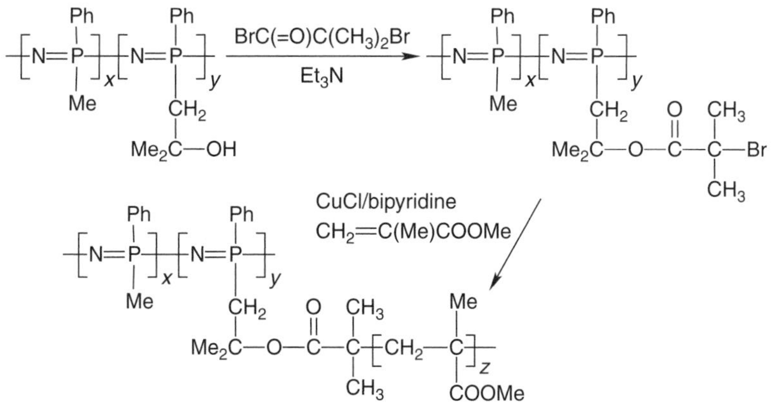

SCHEME 6   ATRP grafting of $[(Me)(Ph)PN]_n$ (**2**).

A second site of reactivity in polymer **2** is the aromatic group. *Electrophilic aromatic substitution* reactions (upper right in Scheme 4) are hindered by the deactivating effect of the directly attached phosphorus atom. Thus far, only nitration reactions have been reported on **2**, and the nitro groups were subsequently converted to amines and amides [35]. Similar reactions on the surface of aryloxy-substituted polyphosphazenes have been used to attach enzymes to polyphosphazenes [36]. In addition, preliminary work [21] indicates that sulfonation of **2** occurs, thus introducing sites for ionic conduction, a common process used to prepare ion-conducting membranes.

Finally, poly(alkyl/arylphosphazene)s are somewhat unique among polyphosphazenes because of the enhanced basicity of the backbone nitrogen that results from the electron releasing alkyl and aryl groups on the adjacent phosphorus atoms in the backbone. Treatment of polymer solutions with reagents such as $MgSO_4$ or $PtCl_2$ indicate that divalent metals facilitate cross-linking of these polymers. However, soluble polymers, **11**, were isolated with coordination to Li^+ and Ag^+ ions [37] (Scheme 4, upper left). The basicity and coordinating abilities of $[(Me)(Ph)PN]_n$ (**2**) are also evident in the stabilization of gold nanoparticles [38]. In the preparation of these metal: polymer nanocomposites, it was not necessary to use alkylammonium salts as phase transfer reagents to move gold ions from aqueous solutions into an organic phase when $[(Me)(Ph)PN]_n$ (**2**) was present. This is probably due to the amphiphilic character of this polymer.

It should be noted that most of the modification reactions discussed in earlier paragraphs are readily adaptable to related poly(alkyl/arylphosphazene)s with either longer alkyl chains, such as butyl and hexyl groups (polymers **3**) and with larger aromatic groups. As reported for functionalized polymers derived from $[(Hex)(Me)PN]_n$ and $[(Bu)(Me)PN]_n$, the presence of such groups significantly alters properties such as solubility and glass transition temperature [20]. Recent work shows that these reactions may also be applied to very hydrophobic systems with biphenyl and naphthyl groups instead of simple phenyl groups [21].

POTENTIAL FOR BIOMEDICAL APPLICATIONS

As noted above, two significant advantages of the preparation of polyphosphazenes by condensation polymerization are access to fully P–C-substituted polymers and, more recently, promising potential for control of molecular weight. Given the stringent requirements and the diverse applications of biomaterials, the ability to prepare polymers reproducibly with well-defined molecular weight and narrow distributions of molecular weight is vital. The chemical properties required for biomaterials are also important and are strongly related to the actual application. For example, for drug release, slow degradation of the polymer can be used for controlled release of encapsulated drugs, while more biologically stable materials are needed for longer-term applications to replace structural materials such as bones and blood vessels [39].

In general, the poly(alkyl/arylphosphazene)s have good thermal stability with the simplest polymers [i.e., $[Me_2PN]_n$ (**1**) and $[(Me)(Ph)PN]_n$ (**2**)], showing onsets of decomposition well above 350°C. The addition of various organofunctional groups (e.g., alcohols and ketones) usually lowers the onset of decomposition by as much as 50°C. The glass transition temperatures (T_g) of these polymers vary widely depending on the nature of the side groups. Semicrystalline $[Me_2PN]_n$ (**1**) has a T_g of −40°C, and the T_g of amorphous $[(Me)(Ph)PN]_n$ (**2**) is +40°C. Adding polar OH [27] or COOH [29] groups or larger biphenyl groups [21] raises these values to as high as 100°C, while adding long flexible groups decreases the T_g to as low as −69°C for $[(Me)(n\text{-hexyl})PN]_n$ [6].

The solubility of the poly(alkyl/arylphosphazene)s is also very dependent on the polarity of the side groups. Although most of the polymers discussed in this chapter are soluble in various organic solvents, it should be noted that the carboxylated salt derivative, $[(NaOOCCH_2)(Ph)PN]_x[(Me)(Ph)PN]_y$ (**8**), becomes soluble in water with increasing amounts of monomer unit, x (i.e., over 50% substitution of carboxylate groups) [29]. This is more pronounced with terpolymer derivatives that also include $[Me_2PN]$ units in the polymer backbone. In fact, the homopolymer $[Me_2PN]_n$ (**2**) is very hygroscopic and it becomes soluble in water when only small amounts of acid are added.

Although no direct studies related to biological applications have been reported for the poly(alky/arylphosphazene)s or their derivatives, some preliminary toxicity tests of samples from our labs have been conducted [40]. Simple cytotoxicity agarose overlay and MEM (minimum essential medium) elution tests on powders of $[Me_2PN]_n$ (**1**), $[(Me)(Ph)PN]_n$ (**2**), and the alcohol derivative of polymer **2**, $[(HOCMe_2CH_2)(Ph)PN]_x[(Me)(Ph)PN]_y$ (**7**), showed no toxicity. The same three polymers were also found to be nonhemolytic in a direct-contact in vitro hemolysis test. On the other hand, the carboxylic acid derivative of $[(Me)(Ph)PN]_n$ (**2**), $[(HOOCCH_2)(Ph)PN]_x[(Me)(Ph)PN]_y$ (**8**), was found to be toxic and hemolytic in the same tests. It is likely that both the enhanced solubility in aqueous solutions and the high coordinating ability of the carboxylate group account for these results. Some other preliminary studies of samples from our group indicated that films of **2** initially showed good adhesion of proteins, but the proteins were also eliminated after as little as 24 h. This is probably due to reorientation of the polymer, causing an inversion of the surface groups, thus changing the hydrophobicity, a common phenomenon in many polymers.

SUMMARY

The use of condensation polymerization in the preparation of polyphosphazenes is an increasingly important process. The recent reports of improvements in the condensation process that facilitate molecular-weight control as well as narrow molecular-weight distributions underscore the vast potential for this type of polymerization. The method also offers promise for lowering the cost of production of polyphosphazenes, since room-temperature polymerizations have now been achieved. Significant cost-effectiveness could be enhanced with improved methods for synthesis of the phosphoranimine precursors. Although condensation polymerization is a potential method for preparing most polyphosphazenes, it is currently the only way to obtain polymers with directly P–C-bonded substituents. Hence, it complements the better developed ring-opening substitution synthesis of polymers with side groups attached by P–O and P–N bonds. Like the latter type of polymer, macromolecular substitution reactions of simple poly(alkyl/arylphosphazene)s provide access to a variety of polyphosphazenes. This reactivity and the potential to attach bioactive groups, as well as the lack of toxicity of the simplest poly(alkyl/arylphosphazene)s, suggest potential applications of this class of compounds as biomaterials.

REFERENCES

1. Allcock, H.R., Kugel, R.L. *J. Am. Chem. Soc.*, 1965, 87:4216.
2. Allcock, H.R. *Chemistry and Applications of Polyphosphazenes*. Wiley-Interscience, Hoboken, NJ, 2003.
3. Mark, J.E., Allcock, H.R., West, R. *Inorganic Polymers*, 2nd ed. Oxford University Press, New York, 2005, Chap. 3.
4. Wisian-Neilson, P., Neilson, R.H. *J. Am. Chem. Soc.*, 1980, 102:2848.
5. Neilson, R.H., Hani, R., Wisian-Neilson, P., Meister, J.J., Roy, A.K., Hagnauer, G.L. *Macromolecules*, 1987, 20:910.
6. (a) Neilson, R.H., Jinkerson, J.L., Kucera, W.R., Longlet, J.J., Samuel, R.C., Wood, C.E. *ACS Symp. Ser.*, 1994, 572:232; (b) Jinkerson, D.L. Ph.D. dissertation, Texas Christian University, 1989.

7. Wisian-Neilson, P., Neilson, R.H. *Inorg. Synth.*, 1989, 25:69.

8. Gruneich, J.A., Wisian-Neilson, P. *Macromolecules*, 1996, 29:5511.

9. Wisian-Neilson, P., Neilson, R.H. *Inorg. Chem.*, 1980, 19:1875.

10. (a) Roy, A.K., Hani, R., Neilson, R.H., Wisian-Neilson, P. *Organometallics*, 1987, 6:378; (b) Wettermark, U.G., Wisian-Neilson, P., Scheide, G.M., Neilson, R.H. *Organometallics*, 1987, 6:959; (c) Roy, A.K., Wettermark, U.G., Scheide, G.M., Wisian-Neilson, P., Neilson, R.H. *Phosphorus Sulfur*, 1987, 33:147.

11. Montague, R.A., Matyjaszewski, K. *J. Am. Chem. Soc.*, 1990, 112:6721.

12. Neilson, R.H., Wisian-Neilson, P. *Chem. Rev.*, 1988, 88:541.

13. (a) Honeyman, C.H., Manners, I., Morrissey, C.T., Allcock, H.R. *J. Am. Chem. Soc.*, 1995, 117:7035; (b) Allcock, H.R., Crane, C.A., Morrissey, C.T., Nelson, J.M., Reeves, S.D., Honeyman, C.H., Manners, I. *Macromolecules*, 1996, 29:7740.

14. (a) Allcock, H.R., Nelson, J.M., Reeves, S.D., Honeyman, C.H., Manners, I. *Macromolecules*, 1997, 30:50; (b) Allcock, H.R., Reeves, S.D., Nelson, J.M., Manners, I. *Macromolecules*, 2000, 33:3999.

15. (a) Allcock, H.R., Reeves, S.D., Nelson, J.M., Crane, C.A., Manners, I. *Macromolecules*, 1997, 30:2213; (b) Prange, R., Allcock, H.R. *Macromolecules*, 1999, 32:6390; (c) Allcock, H.R., Reeves, S.D., Nelson, J.M., Manners, I. *Macromolecules*, 2000, 33:5763.

16. Allcock, H.R., Scott, S.D., deDenus, C.R., Crane, C.K. *Macromolecules*, 2001, 34:748.

17. Huynh, K., Lough, A.J., Manners, I. *J. Am. Chem. Soc.*, 2006, 128:14002.

18. (a) Matyjaszewski, K., Franz, U., Montague, R.A., White, M.L. *Polymer*, 1994, 35:5005; (b) Franz, U., Nuyken, O., Matyjaszewski, K. *Macromolecules*, 1993, 26:3723.

19. (a) Helioui, M., De Jaeger, R., Puskaric, E., Heubel, J. *Makromol. Chem*, 1982, 183:1137; (b) D'Hallum, G., De Jaeger, R., Chambrette, J.P., Potin, P. *Macromolecules*, 1992, 25:1254.

20. Wisian-Neilson, P., Koch, K.A., Zhang, C. *Macromolecules*, 1998, 31:1808.

21. Wisian-Neilson, P., Zhang, C., Jung, J.-H. Unpublished results.

22. Wisian-Neilson, P., Jung, J.-H., Potluri, S.K. *ACS Symp. Ser.*, 2006, 917:335.

23. Wisian-Neilson, P., Claypool, C.L., Bahadur, M. *Macromolecules*, 1994, 27:7494.

24. Wisian-Neilson, P., Bailey, L., Bahadur, M. *Macromolecules*, 1994, 27:7713.

25. Wisian-Neilson, P., Xu, G.-F. *Macromolecules*, 1996, 29:3457.

26. Wisian-Neilson, P., Xu, G.-F., Wang, T. *Macromolecules*, 1995, 28:8657.

27. (a) Wisian-Neilson, P., Ford, R.R. *Organometallics*, 1987, 6:2258; (b) Crumbliss, A.L., Cooke, D., Castillo, J., Wisian-Neilson, P. *Inorg. Chem.*, 1993, 32:6088.

28. Wisian-Neilson, P., Huang, L., Islam, M.Q., Crane, R.A. *Polymer*, 1994, 35:4985.

29. Wisian-Neilson, P., Islam, M.S., Ganapathiappan, S., Scott, D.L., Raghuveer, K.S., Ford, R.R. *Macromolecules*, 1989, 22:4382.

30. Jung, J.-H., Kmecko, T., Claypool, C.L., Zhang, H., Wisian-Neilson, P. *Macromolecules*, 2005, 38:2122.

31. Wisian-Neilson, P., Zhang, C. *Macromolecules*, 1998, 31:9084.

32. Wisian-Neilson, P., Schaefer, M.A. *Macromolecules*, 1989, 22:2003.

33. Wisian-Neilson, P., Islam, M.S. *Macromolecules*, 1989, 22:2026.

34. Cambre, J.N., Wisian-Neilson, P. *J. Inorg. Organomet. Polym. Mater.*, 2006, 16:311.

35. Wisian-Neilson, P., Bahadur, M., Iriarte, J.M., Wood, C.E., Ford, R.R. *Macromolecules*, 1994, 27:4471.

36. Allcock, H.R., Kwon, S. *Macromolecules*, 1986, 19:1502.

37. Wisian-Neilson, P., García-Alonso, F.J. *Macromolecules*, 1993, 26:7156.

38. Walker, C.H., St. John, J.V., Wisian-Neilson, P. *J. Am. Chem. Soc.*, 2001, 123:3846.

39. Allcock, H.R. *Contemporary Polymer Chemistry*, 2nd ed. Prentice Hall, Englewood Cliffs, NJ, 1990, Chap. 24.

40. Tests were done by North American Science Associates, Inc., Northwood, OH, 1992.

11 Electrospun Polyphosphazene Nanofibers for In Vitro Osteoblast Culture

MARIA TERESA CONCONI

Dipartimento di Scienze Farmaceutiche, Università di Padova, Padova, Italy

PAOLO CARAMPIN

Institute of Organic Synthesis and Photoreactivity, C.N.R., Bologna, Italy; School of Pharmacy and Pharmaceutical Sciences, University of Manchester, Manchester, United Kingdom

SILVANO LORA

Institute of Organic Synthesis and Photoreactivity, C.N.R., Bologna, Italy

CLAUDIO GRANDI and PIER PAOLO PARNIGOTTO

Dipartimento di Scienze Farmaceutiche, Università di Padova, Padova, Italy

INTRODUCTION

In the past 20 years, both natural and synthetic materials had been studied and applied to bone tissue engineering (Laurencin et al., 1999). Synthetic materials such as expanded poly(tetrafluoroethylene) (e-PTFE) (Walters et al., 2003), titanium (Fedorowicz et al., 2007), and ethyl cellulose (Dowell et al., 1991) often required secondary surgical procedure for retrieval, whereas natural materials such as chitosan (Di Martino et al., 2005), poly(lactic-*co*-glycolic acid) (PLGA) (Peltoniemi et al., 2002), and silk fibroin nanofiber (Altman et al., 2003) possess insufficient mechanical strength and unstable properties. In this context, poly(organophosphazene)s, high-molecular-weight polymers with a backbone of alternating phosphorus and nitrogen atoms and two organic side groups bonded to each phosphorus atom, can represent an attractive

Polyphosphazenes for Biomedical Applications, Edited by Alexander K. Andrianov
Copyright © 2009 John Wiley & Sons, Inc.

alternative to the materials used currently. Starting from the chloride parent polymer, poly(dichlorophosphazene), a wide variety of amino, alkyloxy, aryloxy, or organometallic side groups can substitute for the chlorine atoms by nucleophilic reaction to create polymers with tailored chemical and physical properties. The introduction of amino acid esters moieties confers hydrolytic instability to the backbone of the polymer, and a number of polyphosphazenes substituted by these labile side groups have been reported (Allcock et al., 1977). The polymers degrade in aqueous medium to nontoxic products, including ammonia, phosphate, amino acids, and the corresponding alcohol. Polyphosphazenes with ethyl or benzyl glycolic ester and lactic ester as substituents were also found to undergo hydrolytic degradation, yielding, in this case, phosphate and the corresponding acid and alcohol (Allcock and Kwon, 1988). The amino acid ester polyphosphazenes are perhaps the most extensively studied biodegradable polyphosphazenes, particularly as controlled drug delivery systems. Indeed, polyphosphazene-based implantable devices were used for controlled delivery of succinylsulfathiazole (an antibacterial agent) and naproxen (an anti-inflammatory agent) (Grolleman et al., 1986; Veronese et al., 1998) and of melphalan in local antitumoral treatment (Goedemoed et al., 1991) or for release of the cytostatic agent mitomycin C (Schacht et al., 1996). The use of degradable polyphosphazenes as matrices for controlled release of macromolecular drugs was also reported (Caliceti et al., 2000; Ibim et al., 1996). For more on the subject, we refer the reader to recently published reviews (Allcock, 2003; Heyde and Schacht, 2004; Lakshmi et al., 2003) and to other chapters in this book. Collectively, all these studies indicate that polyphosphazenes with amino acid ester substituents or mixed side groups (amino acid ester and imidazole, amino acid ester and 4-methylphenoxy, and so on) meet important properties: (1) the polymers are prone to hydrolytic breakdown; (2) their hydrolysis rate can be tuned by appropriate choice of the nature and the ratio of the two side moieties; and (3) the degradation products are harmless at the physiological conditions. Despite these properties, there are few works related to the use of biodegradable polyphophazenes as materials in tissue engineering applications. Laurencin and co-workers reported that polyphosphazenes promote the growth, adhesion, and spreading of osteoblast cells (Laurencin et al., 1993, 1996; Nair et al., 2004, 2006). Kajiwara studied the culture of Bowes and Chinese hamster ovary cell lines onto a polyphosphazene membrane (Kajiwara, 1992). Alanine ethyl ester–substituted polyphosphazene membranes was used as a temporary barrier with release of trimethoprin or succinylsulfathiazole for tissue regeneration in the treatment of periodontal diseases (Veronese et al., 1999) and as guides for peripheral nerve regeneration in an experimental animal model (Langone et al., 1995). In our previous work we have demonstrated that scaffolds composed of poly[(ethyl phenylalanato)$_{1.4}$(ethyl glycinato)$_{0.6}$phosphazene] (PPhe–GlyP) and prepared by an electrospinning method support in vitro adhesion and growth of endothelial cells and may be a useful tool for the construction of small blood vessels (Carampin et al., 2007). Starting from these findings, in this work we have verified whether PPhe–GlyP scaffolds obtained from either solvent casting or electrospinning methods could represent a suitable material for bone tissue engineering purposes. We have also evaluated the effects on osteoblast attachment and proliferation of PPhe–GlyP blends with two widely used polymers: poly(lactic acid) (PLA) and poly(caprolactone) (PCL) (Frenot and Cronakis, 2003; Reneker and Chun, 1996).

MATERIALS AND METHODS

Materials

Adult BALB-c male mice (20 to 25 g body weight) were purchased from Charles-River (Como, Italy). The study protocol was approved by the local Ethics Committee for Animal Studies, and all procedures were carried out according to Italian laws for animal care. Zoletil was obtained from Laboratories Virbac (Carros, France) and Rompun from Bayer (Leverkusen, Germany). Hexachlorocyclotriphosphazene was purchased from Nissho Iwai Corporation (Tokyo, Japan) and used after three careful sublimations. L-Glycine ethyl ester hydrochloride 99% and L-phenylalanine ethyl ester hydrochloride 99% from Aldrich (Steinheim, Germany) were dried in a vacuum cabinet at 30°C for 24 h. Tetrahydrofuran anhydrous 99.9% (THF) and poly(caprolactone) (PCL, average molecular weight ca. 65,000) were from Aldrich. Poly(lactic acid), Resomer L 210 S (PLA) (i.v. 3.3 to 4.3 dL/g) was from Boehringer Ingelheim (Ingelheim, Germany). Triethylamine puriss. p.a. was a Fluka product and dried over molecular sieves before use. All the other chemicals and reagents were obtained from Carlo Erba (Milan, Italy) and used as received. The following cell culture media and reagents were purchased from Sigma Chemical Company (St. Louis, Missouri): Dulbecco's modified Eagle's medium (DMEM), glutamine, ascorbic acid, dexametasone, β-glycerophosphate, ethylenediaminetetraacetic acid (EDTA), trypsin, trypan blue, and antibiotic–antimicotic solution. Fetal bovine serum (FBS) was from Seromed (Berlin, Germany), α-minimum essential medium (α-MEM) from Life Technologies (Paisley, Scotland, UK), and Ultroser from Invitrogen Corporation (Carlsbad, CA). Cresyl violet was provided from Fluka (Buchs, Switzerland). For cell proliferation assays, Cell Titer 96 Aqueous One Solution reagent was purchased from Promega Italia (Milan, Italy).

Apparatus

A Spellman high-voltage dc supply was used as a high-voltage power supply for electrospinning. The scanning electron microscopy (SEM) of the polymeric samples was performed with a Philips ESEM XL30 (Cambridge, UK) operating at 20-kV accelerating voltage. Microplate Autoreader EL 13 was purchased from Bio-Tek Instruments (Winooski, Vermont).

Synthesis of Poly[(ethyl phenylalanato)$_{1.4}$(ethyl glycinato)$_{0.6}$phosphazene]

Preparation and characterization of PPhe–GlyP has recently been described with the following procedure variations (Carampin et al., 2007). Briefly, the poly(dichlorophospazene) was obtained from thermal ring-opening polymerization of hexachlorocyclotriphosphazene. The polymer was dissolved in anhydrous THF and a partial substitution of the chloride atoms (70%) was obtained by reaction with L-phenylalanine ethyl ester using anhydrous triethylamine as an acceptor of the by-product hydrochloric acid. A new solution with a large excess of L-glycine ethyl ester was added, under constant stirring, at the reaction mixture to complete the chloride substitution. After addition, stirring was continued at 0°C for 6 h and at 40°C for a further 36 h. All the manipulations were performed under a nitrogen atmosphere. The insoluble triethylamine hydrochloride salt was removed by centrifugation and the polymer was obtained by precipitation into *n*-heptane. The crude polymer was further purified by successive dissolution in THF, centrifugation, and precipitation with *n*-heptane. Finally, the polyphosphazene was dried until a constant weight was achieved and was then stored in vacuum to avoid exposure to moisture. The elemental analysis for C, H, and N, the infrared, and ^{1}H, ^{31}P, and ^{13}C nuclear magnetic resonance spectra were consistent with polymer composition poly[(ethyl phenylalanato)$_{1.4}$(ethyl glycinato)$_{0.6}$ phosphazene]. The intrinsic viscosity of the polymer in THF at 25°C was $[\eta] = 3.0\,dL/g$.

Production of Implants

Polymer film was obtained by a solvent casting method: 500 mg of PPhe–GlyP was dissolved in 4 mL of dichloromethane in a closed recipient at room temperature. The mixture was poured in a glass Petri disk (30 cm^2), and the solvent was evaporated. After 24 h, the membrane was removed from the plates and vacuum dried for 24 h to remove any residual traces of solvent. Disks were bored out from the polymer film and used for in vitro and in vivo tests as described below. All operations were performed under sterile conditions. The disks used for in vivo tests were sterilized by ultraviolet (UV) light irradiation.

In Vitro Degradation Assays

Disks of PPhe–GlyP (15 mm in diameter, 0.6 mm thickness, 45.0 mg) were used for the time-dependent in vitro degradation studies. The films were immersed in distilled water and incubated at 37°C and pH 7.4 without stirring. The water uptake of the polymer was determined by the weight of the wet disk, and the polymer degradation by mass loss of the dried sample. The determinations were carried out at scheduled times for up to 100 days, and the results were reported over time. The studies were carried out in triplicate.

In Vivo Assays

Under general anaesthesia (35 mg/kg Zoletil and 2 mg/kg Rompun), PPhe–GlyP disks (6 mm in diameter, 0.6 mm thickness, 7.2 mg) were implanted into subcutaneous pockets of 18 BALB-c mice. The animals were housed at 22°C on a 12-h light/dark cycle and received food and water *ad libitum*. Their health and behavior were assessed daily until polymeric samples were collected at 2, 7, 14, 21, 28, and 60 days post-implantation and fixed with 10% formalin in phosphate-buffered saline (PBS) for 24 h. The samples were dehydrated and embedded in paraffin. Five-micrometer sections were stained with hematoxylin and eosin. Twelve slices from each samples were observed and photographed with an optical microscope.

Electrospinning Process

The physical setup employed in this study for the electrospinning process consisted of a variable high-voltage power supply (voltage range 0 to 50 kV), a 5-mL glass syringe with a needle (0.8 to 1.0 mm internal diameter and 3 cm long) and an aluminum disk (5 cm diameter) as collector, placed at a variable distance (9 to 10 cm) from the capillary tip. The syringe was fixed at the support and tilted at approximately 45° from horizontal. The positive electrode of the high-voltage power supply was attached to the needle of the syringe and the negative electrode to the grounded collector. The polymer solution (3.3 to 6% w/v) was poured in the syringe and a falling drop was formed at the tip of the needle by the force of gravity. When an electric field was created between the capillary tip and the collector, an electrostatic charge was induced on the surface of the drop. Once a critical voltage is exceeded, a positive-charged jet was ejected from the drop and splayed to the negative-charged target. During the jet travel, the solvent gradually evaporated, leaving a continuous fiber that accumulated on the collector target. The process results in the production of nonwoven fibrous matrices that were used as scaffolds for cell cultures. The matrices were removed from the collector, cut in disks (12 mm in diameter), which were rinsed in *n*-heptane to draw out the residual traces of solvent, dried, then placed into 24-well culture plates and sterilized by ultraviolet light irradiation for 60 min per side. All the operations were carried out in an airflow cabinet. The nanoscale features and the diameter of the fibers depend on different parameters, such as solvent, polymer concentration, capillary diameter, electric potential, and tip–collector distance. We prepared various matrices of PPhe–GlyP, alone or mixed with PLA or PCL, by electrospinning, changing the processing parameters in order to obtain scaffolds with similar nanofiber diameter. The thickness of polymer fibers was determined by SEM measuring their diameter on 50 images.

Porosity of the Electrospun Scaffolds

The thickness of the scaffolds was measured by a micrometer knowing the bulk density of PPhe–GlyP (Carampin et al., 2007), PLA, and PLC: The porosity of the scaffolds was calculated using the corresponding apparent densities according to Ma et al. [25].

Cell Cultures

Osteoblasts were obtained from femurs of adult Sprague-Dowley rats according to the method proposed by Richard and co-workers (Richard et al., 1994). Cells were collected from bone marrow by repeated washing with DMEM supplemented with FBS and antibiotic. Cell suspension was centrifuged (5 min, 1500 rpm) and the cells extracted were resuspended in α-MEM containing 1% antibiotic–antimytotic solution and supplemented with the following factors: FBS (10%), glutamine (2 mM), ascorbic acid (50 µM), dexametasone (10^{-8} M), and Ultroser (2%). Five days after seeding, the culture medium was replaced and added with 10 mM β-glicerophosphate. Subconfluent cultures were removed with 0.002% EDTA and 0.25% trypsin in PBS, and centrifuged (5 min, 1500 rpm).

Adhesion and Proliferation Assays

Osteoblasts from the second to the fourth passage were used to study cell adhesion and proliferation. Cells were harvested and seeded on polymeric sheets in a 24-well microtiter plate and cultured under standard conditions. For adhesion assay, 3.0×10^4 cells/cm^2 were plated on the polymeric matrices and allowed to adhere for 3 h. Cultures were washed twice with PBS to remove unattached cells and fixed overnight with 10% paraformaldehyde in phosphate buffer. The fixed cells were stained with 0.04% cresyl violet in 20% methanol for 30 min. The dye was extracted by 0.1 M citric acid in 50% ethanol, and absorbance was determined at 600 nm. The results, means of four experiments, were expressed as percent change from cultures seeded onto tissue culture–treated polystyrene plates. To study the cell proliferation, osteoblasts (7.5×10^3 cells/cm^2) were cultured for 96 h, and then their viability was checked using the Cell Titer 96 Aqueous One Solution reagent. Briefly, a mixture containing Cell Titer 96 Aqueous One Solution reagent and culture medium without FBS and phenol red was added into each well and plates were incubated for 3 h. Metabolically active cells were able to reduce the tetrazolium compound (MTS) into a purple formazan product, whose absorbance was measured at 490 nm. The results, means of four experiments, were expressed as percent change from cultures seeded onto tissue culture–treated polystyrene plates.

Statistical analysis

All results were expressed as mean $\pm$ SD of four separate experiments. Their statistical comparison was performed by analysis of variance, followed by Student's t-test. The level of significance chosen was $p < 0.05$.

RESULTS AND DISCUSSION

In Vitro and In Vivo Degradation Studies of PPhe–GlyP Membranes Obtained by the Solvent Casting Method

As shown in Figure 1, PPhe–GlyP disks obtained by the solvent casting method presented a smooth surface with several holes whose diameter ranged from 0.5 to 2 μm. The in vitro degradation assays for PPhe–GlyP disks were carried out over a 100-day period in 0.1 M sodium phosphate buffer at 37°C and pH 7.4 without stirring, and the medium was changed daily. The water uptake and the mass loss percentages were determined by weight of the wet and dry samples, respectively, at specific time intervals, and the results are reported in Figure 2. In the first week, the water uptake profile displayed a quick increase due to the matrix properties microporosity and its hydrophilicity, whereas the mass loss due to degradation was negligible. Successively, the wet polymer sample increases its weight slowly up to 100 days as a result of two degradation process: hydrolysis of the ester linkages, which generate pendant carboxylic acid groups, and hydrolytic cleavage of an external P–N bond to yield a P–OH derivate, which make the polymer more and more hydrophilic. However, the mass loss was balanced by the higher hydrophilicity of the sample. The degradation process was evident after the first week for the dried sample. In fact, in the first week the weight loss of the dried sample was negligible; after this period the polymer matrix displayed a nearly constant degradation kinetic, losing approximately 20% of its mass in 100 days. The mass loss of the dried sample can be ascribed to the two previous degradation processes and to hydrolysis of the skeleton of the polyphosphazene chain by the deprotected carboxylic acid function of the substituent units or attack by H$^+$ ions on skeletal nitrogen atoms according to the mechanism of

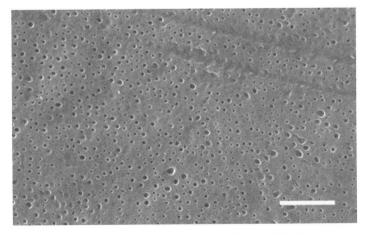

FIGURE 1   SEM micrographs (scale bar = 10 μm) of PPhe–GlyP disks prepared by the solvent casting method.

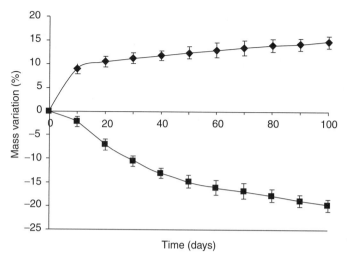

FIGURE 2   Percentage mass loss (■) and water uptake (◆) vs. time (days) of PPhe–GlyP disks in 0.1 M sodium phosphate buffer at 37°C and pH 7.4.

hydrolytic breakdown for amino acid ester polyphosphazene (Allcock and Kwon, 1988). This degradation reaction of phosphorus–nitrogen chains is slower that the two previous mechanisms and is the most important process for the weight loss of the polymer because it leads to a molecular-weight decrease (Carampin et al., 2007; Ruiz et al., 1993.) It is interesting to observe that the disk retains its dimensions and that the weight of the wet film is always higher than the initial weight: The water uptake due to the increase in polymer hydrophilicity hides the weight loss during the 100-day test period. After this term the wet film will start to lose its weight because of the progressive breakage in the polymeric main chain, as already observed for poly[(ethyl alanato)$_2$phosphazene] (Carenza et al., 2004).

To verify the in vivo biocompatibility of scaffolds, PPhe–GlyP disks were inserted into a subcutaneous pocket of BALB/c mice, and three animals were sacrificed at scheduled times. The polymeric implants and the adjacent tissues were withdrawn, and the macroscopic appearance of the samples was examined: No signs of inflammation, such as edema, blush, or necrosis, were seen during tissue sampling. The implants were well integrated in the host tissue without lacerations at the borders. Figure 3 shows the polymeric film before and 60 days after the implantation. The implant appeared opaque for the water uptake, but it maintained the initial dimensions.

The histological analysis of the polymeric implants and adjacent tissues revealed that 2 days after implantation (Fig. 4a), the samples showed the typical features of acute inflammatory response, probably due to the implant surgery. The inflammatory infiltrate present in the surrounding tissues decreased at 7 days (Fig. 4b) and disappeared completely at 14 days (Fig. 4c). However, a thin

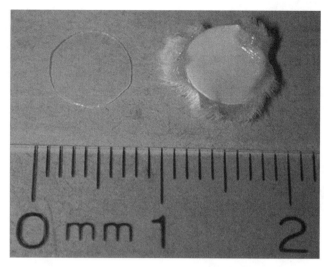

FIGURE 3 PPhe–GlyP disks before (on the left) and 60 days after (on the right) implantation into dorsal subcutaneous pockets of BALB-C mice.

fibrous capsule around the polymeric disk was still present until 60 days, and no cells were visible inside the implants (Fig. 4d).

Collectively, these findings confirm the in vivo biocompatibility of PPhe–GlyP, but also suggest that the solvent casting method is not suitable to obtain

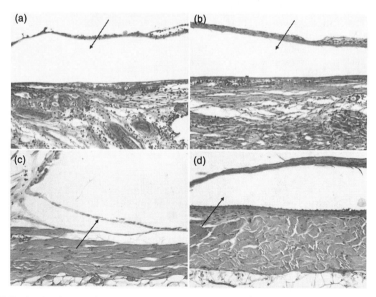

FIGURE 4 Sections of implants after 2 (a), 7 (b), 14 (c), and 60 (d) days from implantation. Arrows indicate PPhe–GlyP disks (magnification × 100).

PPhe–GlyP scaffolds able to allow host cell ingrowth. We can suppose that the dense and compact microstructure of disks could lead to an insufficient and slow degradation rate of the polymer, and its smooth surface could represent a poor stimulus for cell adhesion. It has been demonstrated that (1) highly porous scaffolds enhance cell adhesion and proliferation (Liu and Ma, 2004), and (2) the fiber diameter of nanofibrous scaffold has a significant influence on cell behavior (Laurencin et al., 1999). Indeed, in tissue regeneration the scaffold should provide, at a microscopic level, physical structural features (i.e., fiber dimension and porosity of the scaffolds) that mimic the nanoscale dimensions of the natural extracellular matrix. Starting from these considerations, using the electrospinning method, we have obtained PPhe–GlyP scaffolds that assure the formation of ultrathin fibers generating a high surface area for cell adhesion and growth.

In Vitro Studies of PPhe–GlyP Scaffolds Obtained by Electrospinning

Based on our previous work (Carampin et al., 2007), the electrospinning process was carried out using the following experimental conditions: 4% PPhe–GlyP (w/v) in THF, an 0.8-mm needle diameter, a 10-cm collector–tip distance, and a 9-kV accelerating voltage. So PPhe–GlyP scaffolds composed of submicrometer or nanometer fibers were obtained (Fig. 5). Under these experimental conditions, the spraying process was not formed, so that fibers were deposited onto a stationary collecting disk. The fibers of the matrix were randomly oriented, smooth, without bead defects, and their average diameters were 600 ± 150 nm. The scaffolds presented a three-dimensional structure with large interconnected voids; their porosity, determined according to Ma et al. (2005), was $85 \pm 2\%$.

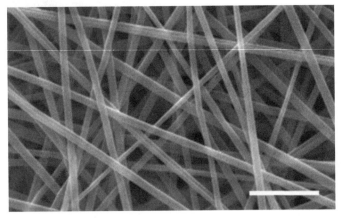

FIGURE 5   SEM micrographs (scale bar $= 5 \, \mu m$) of PPhe–GlyP scaffolds prepared by the electrospinning method.

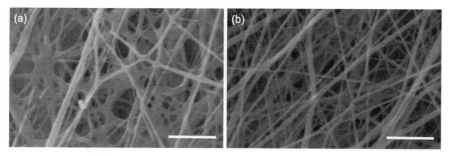

FIGURE 6   SEM micrographs (scale bar = 5 μm) of PLA (a) and PLA/PPhe–GlyP 75/25 (b) scaffolds prepared by the electrospinning method using 3.3% (w/v) polymeric solution in DCM (a,b), tip–collector distance 9 cm (a,b), diameter of needle 1 mm (a,b), applied voltage 13 kV (a), 15 kV (b).

We also prepared electrospun scaffolds composed of PLA, PLC alone, or as blends with PPhe–GlyP [PLA/PPhe–GlyP 75:25 (w/w) and PCL/PPhe–GlyP 75:25 (w/w)]. The electrospinning processing parameters were set to obtain scaffolds of similar nanofiber diameter. Figure 6 shows representative SEM images of electrospun fibers obtained from solutions in dichloromethane (DCM) of PLA (a) and PLA/PPhe–GlyP 75/25 (b), and Figure 7 shows the matrices of PLC (a) and PCL/PPhe–GlyP 75:25 (b). The experimental conditions are reported in the captions. All these scaffolds contained randomly oriented fibers whose average diameters were 600 ± 300 nm. Large interconnected voids were clearly visible and the porosities were higher ($85 \pm 5\%$) than that observed for PPhe–GlyP. In both PLA and PLC scaffolds, some fibers appeared fused or linked to each other.

When osteoblasts were seeded on electrospun scaffolds of PPhe–GlyP or PCL and its blend, cell adhesion and proliferation were slightly but significantly ($p < 0.05$) lower than when determined in control cultures seeded on tissue

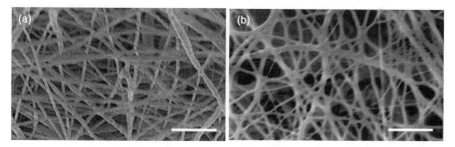

FIGURE 7   SEM micrographs (scale bar = 5 μm) of PCL (a) and PCL/PPhe–GlyP 75:25 (b) scaffolds prepared by the electrospinning method using 6% (w/v) polymeric solution in DCM (a,b), tip–collector distance 9 cm (a,b), diameter of needle 0.8 mm (a,b), applied voltage 13 kV (a), 18 kV (b).

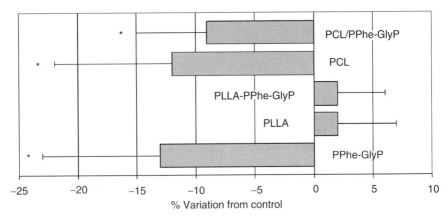

FIGURE 8 Adhesion of osteoblasts cultured for 3 h on scaffolds obtained by the electrospinning method. Bars are means ± SD of four separate experiments. *, $p < 0.05$ vs. control cultures grown on tissue culture–treated polystyrene plates; Student's t-test.

culture–treated polystyrene plates (Figs. 8 and 9). On the contrary, we have already observed that electrospun scaffolds of PPhe–GlyP (presenting the same porosity and fiber diameter of the scaffold here used) were able to improve both adhesion and proliferation of neuromicrovascular endothelial cells (Carampin et al., 2007). Moreover, poly[bis(ethyl alanato)phosphazene] (PAlaP) enhanced osteoblast attachment and growth compared with that observed on tissue culture polystyrene plates (Conconi et al., 2006). Taken together, these findings

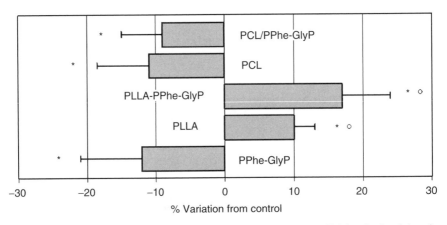

FIGURE 9 Viability of osteoblasts cultured for 96 h on scaffolds obtained by the electrospinning method. Bars are means ± SD of four separate experiments. *, $p < 0.05$ vs. control cultures grown on tissue culture–treated polystyrene plates; °, $p < 0.05$ vs. PPhe–GlyP; Student's t-test.

indicate that the cellular response to materials may depend on cell type. Here, the presence inside the polymer skeleton of glycine probably modifies the hydrophilicity properties of the polymer, making it less attractive for osteoblast but not endothelial cells.

No variations in cell adhesion were observed using PLA and its blend, whereas cell proliferation was significantly improved on PLA and its blend compared with that on PPhe–GlyP or tissue culture–treated polystyrene plates. Moreover, PLA/PPhe–GlyP 75:25 enhanced osteoblast growth at a greater rate than that determined using PLA scaffolds (Fig. 9). Although many studies show good biocompatibility of devices based on the polymers of lactic acid and their copolymers, concerns could rise regarding their acidic degradation products (i.e., lactic acid and glycolic acid) which could have important implications for cell growth when these polymers are used for long-term-applications. It has been shown that the degradation products of the poly(amino acid ester)phosphazene neutralized the acid degradation products of poly(lactic acid–glycolic acid) copolymers and that the buffering action of polyphosphazenes may be due to the release of ammonia and phosphates produced by hydrolysis of the polymeric skeleton (Ambrosio et al., 2002). Thus, the presence of PPhe–GlyP polymer inside the scaffold fibers could improve the long-term in vitro biocompatibility of PLA.

CONCLUSIONS

Collectively, our data indicate that PPhe–GlyP, bearing phenylalanine ethyl ester and glycine ethyl ester as side groups, possesses in vivo biocompatibility. Nevertheless, membranes of PPhe–GlyP obtained by the solvent casting method display a smooth surface and a slow degradation rate, losing approximately 20% of the initial mass in 100 days. In vivo they did not allow host cell ingrowth and were not remodeled into living tissue until 60 days from implantation.

On the contrary, PPhe–GlyP scaffolds by the electrospinning method showed good porosity and fiber dimensions resembling those of the natural extracellular matrix. Although PPhe–GlyP supports osteoblast adhesion and growth to a lesser degree than that observed for electrospun PLA, a synergic effect on cell proliferation was noted when osteoblasts were cultured on PLA/PPhe–GlyP 75:25. Since polyphosphazenes can exert a buffering effect on acidic degradation products of PLA, electrospun PPhe–GlyP may represent an interesting material to use together with PLA for bone tissue engineering. Finally, it must be noted that the poor mechanical properties of nanofibrous scaffolds make these materials useful only to repair defects whereby limited mechanical loading occurs, such as some cranial and maxillofacial bone defects.

REFERENCES

Allcock, H.R. 2003. *Chemistry and Application of Polyphosphazenes*. Wiley, Hoboken, NJ.

Allcock, H.R., Kwon, S. 1988. Glyceryl polyphosphazenes: synthesis, properties, and hydrolysis. *Macromolecules*, 21:1980–1985.

Allcock, H.R., Fuller, T.J., Mack, D.P., Matsumura, K., Smeltz, K.M. 1977. Synthesis of poly[(amino acid alkyl ester) phosphazenes]. *Macromolecules*, 10:824–830.

Altman, G.H., Diaz, F., Jakuba, C., Calabro, T., Horan, R.L., Chen, J., Lu, H., Richmond, J., Kaplan, D.L. 2003. Silk-based biomaterials. *Biomaterials*, 24(3): 401–416.

Ambrosio, A.M., Allcock, H.R., Katti, D.S., Laurencin, C.T. 2002. Degradable polyphosphazene/poly(α-hydroxyester) blends: degradation studies. *Biomaterials*, 23(7):1667–1672.

Caliceti, P., Veronese, F.M., Lora, S. 2000. Polyphosphazenes microspheres for insulin delivery. *Int. J. Pharm.*, 211(1–2):57–65.

Carampin, P., Conconi, M.T., Lora, S., Menti, A.M., Baiguera, S., Bellini, S., Grandi, C., Parnigotto, P.P. 2007. Electrospun polyphosphazene nanofibers for in vitro rat endothelial cells proliferation. *J. Biomed. Mater. Res.*, 80A:661–668.

Carenza, M., Lora, S., Fambri, L. 2004. Degradable phosphazene polymers and blends for biomedical applications. *Adv. Exp. Med. Biol.*, 553:113–122.

Conconi, M.T., Lora, S., Menti, A.M., Carampin, P., Parnigotto, P.P. 2006. In vitro evaluation of poly[bis(ethyl alanato)phosphazene] as a scaffold for bone tissue engineering. *Tissue Eng.*, 12(4):811–819.

Di Martino, A., Sittinger, M., Risbud, M.V. 2005. Chitosan: a versatile biopolymer for orthopaedic tissue-engineering. *Biomaterials*, 26(30):5983–5990.

Dowell, P., Moran, J., Quteish, D. 1991. Guided tissue regeneration. *Br. Dent. J.*, 171(5):125–127.

Fedorowicz, Z., Nasser, M., Newton, J.T., Oliver, R.J. 2007. Resorbable versus titanium plates for orthognathic surgery. *Cochrane Database Syst. Rev.*, 2:CD006204.

Frenot, A., Cronakis, I.S. 2003. Polymer nanofibers assembled by electrospinning. *Curr. Opin. Colloid Interface*, 8:64–75.

Goedemoed, J.H., De Groot, K., Claessen, A.M.E., Scheper, R.J. 1991. Development of implantable antitumor devices based on polyphosphazenes. II. *J. Control. Release*, 17:235–244.

Grolleman, C.W.J., de Visser, A.C., Wolke, J.G.C., van der Goot, H., Timmerman, H. 1986. Studies on a bioerodible drug carrier system based on a polyphosphazene: II. Experiment in vitro. *J. Control. Release*, 4:119–131.

Heyde, M., Schacht, E. 2004. Biodegradable polyphosphazenes for biomedical applications. In Gleria, M., De Jaeger, R., eds., *Phosphazenes: A Worldwide Insight*. Nova Science, Hauppauge, NY, pp. 365–398.

Ibim, S.M., Ambrosio, A.M., Larrier, D., Allcock, H.R., Laurencin, C.T. 1996. Controlled macromolecule release from poly(phophazene) matrices. *J. Control. Release*, 40:31–39.

Kajiwara, M. 1992. The study of the cultivation of Chinese hamster ovary and Bowes cell lines with poly(organophosphazene) membrane. *Appl. Biochem. Biotechnol.*, 33:43–50.

Lakshmi, S., Katti, D.S., Laurencin, C.T. 2003. Biodegradable polyphosphazenes for drug delivery applications. *Adv. Drug. Deliv. Rev.*, 55:467–482.

Langone, F., Lora, S., Veronese, F.M., Caliceti, P., Parnigotto, P.P., Valenti, F., Palma, G. 1995. Peripheral nerve repair using a poly(organo)phosphazene tubular prosthesis. *Biomaterials*, 16:347–353.

Laurencin, C.T., Norman, M.E., Elgendy, H.M., El-Amin, S.F., Allcock, H.R., Pucher, S.R., Ambrosio, A.M. 1993. Use of polyphosphazenes for skeletal tissue regeneration. *J. Biomed. Mater. Res.*, 27(7):963–967.

Laurencin, C.T., El-Amin, S.F., Ibim, S.E., Willoughby, D.W., Attawia, M., Allcock, H.R., Ambrosio, A.M. 1996. A highly porous 3-dimensional polyphosphazene polymer matrix for skeletal tissue. *J. Biomed. Mater. Res.*, 30:133–138.

Laurencin, C.T., Ambrosio, A.M., Borden, M.D., Cooper J.A. Jr., 1999. Tissue engineering: orthopedic applications. *Annu. Rev. Biomed. Eng.*, 1:19–46.

Liu, X., Ma, P.X. 2004. Polymeric scaffolds for bone tissue engineering. *Ann. Biomed. Eng.*, 32:477–486.

Ma, Z.W., Kotaki, M., Yong, T., He, W., Ramakrishna, S. 2005. Surface engineering of electrospun polyethylene terephthalate (PET) nanofibers towards development of a new material for blood vessel engineering. *Biomaterials*, 26:2527–2536.

Nair, L.S., Bhattacharyya, S., Bender, J.D., Greish, Y.E., Brown, P.W., Allcock, H.R., Laurencin, C.T. 2004. Fabrication and optimization of methylphenoxy substituted polyphosphazene nanofibers for biomedical application. *Biomacromolecules*, 5(6):2212–2220.

Nair, L.S., Lee, D.A., Bender, J.D., Barrett, E.W., Greish, Y.E., Brown, P.W., Allcock, H.R., Laurencin, C.T. 2006. Synthesis, characterization, and osteocompatibility evaluation of novel alanine-based polyphosphazenes. *J. Biomed. Mater. Res. A*, 76(1):206–213.

Peltoniemi, H., Ashammakhi, N., Kontio, R., Waris, T., Salo, A., Lindqvist, C., Grätz, K., Suuronen, R. 2002. The use of bioabsorbable osteofixation devices in craniomaxillofacial surgery. *Oral Surg. Oral Med. Oral Pathol. Oral Radiol. Endo.*, 94(1): 5–14.

Reneker, D.H., Chun, I. 1996. Nanometer diameter fibres of polymers produced by electrospinning. *Nanotechnology*, 7:216–223.

Richard, D.J., Sullivan, T.A., Sherker, B.J., Leboy, P.S., Kazhdan, I. 1994. Induction of rapid osteoblast differentiation in rat bone stromal cell cultures by dexamethasone and BMP-2. *Dev. Biol.*, 161:218–228.

Ruiz, E.M., Ramirez, C.A., Aponte, M.A., Barbosa-Canovas, G.V. 1993. Degradation of poly[(glycine ethyl ester)phosphazene] in aqueous media. *Biomaterials*, 14:491–496.

Schacht, E., Vandorpe, J., Dejardin, S., Lemmouchi, Y., Seymour, L. 1996. Biomedical application of degradable polyphosphazenes. *Biotechnol. Bioeng.*, 52:102–108.

Veronese, F.M., Marsilio, F., Caliceti, P., De Filippis, P., Giunchedi, P., Lora, S. 1998. Polyorganophosphazene microspheres for drug release: polymer synthesis, microsphere preparation, in vitro and in vivo naproxen release. *J. Control. Release*, 52:227–237.

Veronese, F.M., Marsilio, F., Lora, S., Caliceti, P., Passi, P., Orsolini, P. 1999. Polyphosphazene membranes and microspheres in periodontal diseases and implant surgery: pheripheral nerve repair using poly(organo)phosphazene tubular prosthesis. *Biomaterials*, 20:91–98.

Walters, S.P., Greenwell, H., Hill, M., Drisko, C., Pickman, K., Scheetz, J.P. 2003. Comparison of porous and non-porous Teflon membranes plus a xenograft in the treatment of vertical osseous defects: a clinical reentry study. *J. Periodontol.*, 74(8):1161–1168.

12 Phosphazenes and Surfaces

MARIO GLERIA

Dipartimento di Processi Chimici dell'Ingegneria, Università di Padova, Padova, Italy

ROBERTO MILANI

Dipartimento di Scienze Chimiche, Università di Padova, Padova, Italy

ROBERTA BERTANI

Dipartimento di Processi Chimici dell'Ingegneria, Università di Padova, Padova, Italy

ANGELO BOSCOLO BOSCOLETTO

Polimeri Europa, Tecnologia Chimica di Base, Venezia, Italy

ROGER DE JAEGER

Laboratoire de Spectrochimie Infrarouge et Raman, UMR-CNRS 8516, Université des Sciences et Technologies de Lille, Villeneuve d'Ascq, France

INTRODUCTION

The study of surfaces, their preservation, their modification through a variety of techniques, and their utilization is of both fundamental and technological interest [1–4]. In fact, tailored functionalization of material surfaces plays an important role in adapting external characteristics of different substrates to a broad range of applications. Thus, the control of material surface features can be reached by film-coating processes and by physical or chemical methods to tune valuable experimental parameters such as friction characteristics and tribology, wear reduction, surface lubrication, surface conductivity, biocompatibility, corrosion, adhesion, flame and heat resistance, and intumescence for specific industrial applications.

It is now quite a widely held opinion that surface-modification processes should lead to the preparation of materials showing completely different surface properties while maintaining the bulk physical and mechanical features

Polyphosphazenes for Biomedical Applications, Edited by Alexander K. Andrianov
Copyright © 2009 John Wiley & Sons, Inc.

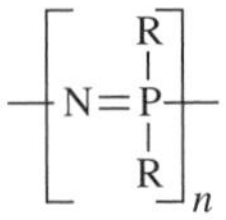

$$\left[N = \underset{\underset{R}{|}}{\overset{\overset{R}{|}}{P}} \right]_n$$

SCHEME 1   General structure of poly(organophosphazenes).

of the pristine substrates. This would avoid research on completely new materials having totally unknown general characteristics, saving effort, time, and money. Moreover, surface modifications introduced in a substrate should involve only superficial layers of the material (3 to 10 Å), as thicker layers could modify bulk properties while thinner ones could be removed easily by mechanical erosion or by environmental factors.

Surface chemical modifications can be carried out easily on substrates already containing functional groups (e.g., –OH or –NH₂) susceptible to undergoing successive functionalization reactions according to well-known and exploited organic chemistry reactions [5]. As examples of these materials we may mention such products as oxides [6], silicon [7–9], and glass [10,11], all of them containing free hydroxylic groups in their chemical structure. As far as intrinsically apolar substrates are concerned (e.g., polyolefins [12], polybutadiene [13,14], polystyrene [15,16]), a variety of physical (thermal [17,18], photochemical [19,20], γ-radiolytic [21,22], plasma [23,24], etc.) and chemical (oxidation [25–27], sulfonation [28,29], nitration [30,31], chloromethylation [32,33], grafting [34–36], etc.) techniques are available to introduce polar functionalities (CHO, CO, COOR, COOH, NH₂, NR₂, OH, NO₂, SO₃H, etc.) on the surface of these materials. Post-functionalization reactions are also exploited to tailor surfaces in a specific way [37,38].

Despite the great number of practical applications already explored [39], the utilization of poly(organophosphazene)s, POPs (Scheme 1), for surface property modification can be considered as a relatively underserved area, with a reduced number of papers and patents published in this field. Looking through the literature, it can be realized that surface problems in phosphazene chemistry have been faced over time according to two main strategies: (1) modification of the surface of solid poly(organophosphazene) films through a variety of experimental techniques, and (2) modification of the surface of conventional, carbon-backboned macromolecules by exploiting different polyphosphazene derivatives.

In this chapter we survey the utilization of phosphazene substrates in surface chemistry together with their importance and implication in both academic studies and technological applications. The utilization of selected cyclophosphazenes for lubrication and antiwear applications has been the object of past review articles [40–42] and is not discussed here.

SURFACE MODIFICATION OF PHOSPHAZENE FILMS OR FIBERS

Despite the great importance of POPs from both a scientific and a technological point of view [43–46], research on the surface properties of these materials started only about 15 years after their original synthesis, put forth by H.R. Allcock in 1965 [47–50], when the first series of reports on this argument started to appear in the literature. Since then, the importance of surface studies on POPs has increased rapidly [51–54], and now a number of strategies have been individuated that illustrate the main tools exploited in this field:

1. Ultraviolet–visible, γ-rays, and electron beam irradiation
2. Specially tailored chemical reactions, including hydrolysis, oxidation, sulfonation, nitration, hydrosilylation, methatetical exchange, and grafting
3. Physical techniques such as plasma or electrospinning

In the first part of this chapter we highlight work that has been carried out over time on the surface modification of poly(organophosphazene)s, putting emphasis on the practical application of surface-functionalized polyphosphazenes as hydrophilic and/or hydrophobic materials in the biomedical field, environmentally and chemically resistant coatings, films or fibers, self-cleaning surfaces, and tribology materials, among others [55].

Surface Modification of POP Films Through Electromagnetic Radiation

Research on the surface functionalization of phosphazene polymer films or coatings originated with the need in macromolecular technology of polymers both to resist surface modifications and/or to undergo surface functionalization in a predictable way. The first point could be accomplished rather easily and simply by selecting phosphazene polymers bearing substituents that possess spectroscopic features similar to those of the inorganic P=N backbone. It is well known that the skeleton of these materials does not absorb light at wavelengths longer than 220 to 230 nm [39,56–59], so aliphatic and linear fluorinated alcohols can be considered ideal candidates for the preparation of phosphazene macromolecules that are stable under ultraviolet (UV) and visible photochemical irradiation.

In fact, early studies on the light-induced surface modification of POPs concerned mostly [poly(bis(trifluoroethoxy)phosphazene] [60–62] ([NP$(OCH_2CF_3)_2]_n$; PTFEP), a material that possesses a weak UV absorption coefficient in the near-UV range of the spectrum ($\varepsilon_{254} \sim 10$) [60], the wavelength of light usually selected for the irradiation of phosphazene films.

According to the first studies, PTFEP films irradiated under vacuum proved to be moderately resistant to radiation damage without skeletal cleavage of the material being observed. In contrast, the occurrence of a light cross-linking effect could be observed, as could be inferred by the increased solution viscosity of the material. Moreover, the formation of different low-molecular-weight products, such as CO_2, CH_3CF_3, HCF_3, $[NP(CH_2CF_3)_2]_3$, and $N_3P_3(OCH_2CF_3)_4Cl_2$, could be detected by mass spectrometry, whose origin was explained according to Scheme 2. In this scheme, the formation of carbon

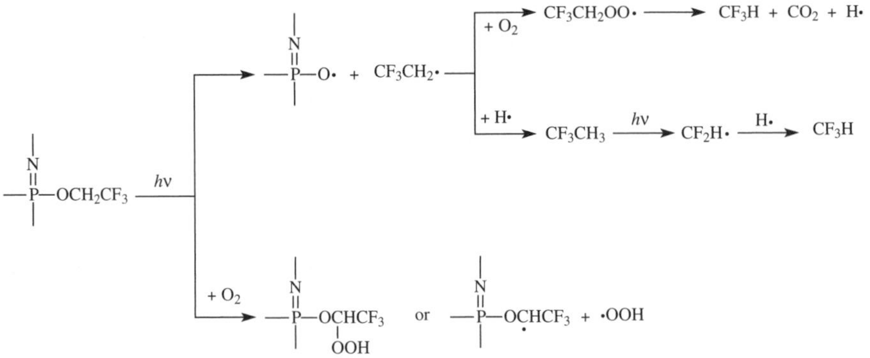

SCHEME 2   Surface modification of poly[bis(trifluoroethoxy)phosphazene] films irradiated at 254 nm in vacuum.

oxide was accounted for on the basis of the presence of molecular oxygen trapped inside the polymer, while the formation of cyclophosphazenes was explained by considering the occurrence of skeletal unzipping reactions during irradiation. The fact that one of the cyclophosphazenes formed contained chlorine atoms in its chemical structure was attributed to the involvement in the PTFEP degradation process of residual P–Cl units coming from an incomplete substitution of polydichlorophosphazene with trifluoroethoxy groups during synthetic reaction. The feasibility of the attack of oxygen atoms on the trifluoroethoxy substituent shown in scheme 2 was also supported by x-ray photoelectron spectroscopy (XPS) data reported by Hiraoka et al. [61].

Starting from these studies, it was eventually shown by Reichert et al. [62] that film morphology, surface phosphorus distribution, and contact angles do not undergo appreciable variations in nonirradiated and irradiated polymer samples, supporting the substantial inertness of PTFEP and copolymer (poly [bis(trifluoroethoxy)(fluoroalkoxy telomers)phosphazene]) films. The only experimental variations observed were in dielectric measurements, which evidenced an initial enhancement of side-chain mobility under irradiation. This fact was attributed to the detachment of some phosphorus substituents that facilitated motion of the vicinal lateral chains. This enhanced mobility was, however, followed immediately by a restriction of the polymer movements, due to the onset of cross-linking phenomena originated by radical coupling processes.

On the contrary, to observe the surface modification of phosphazene films under UV–visible irradiation in an almost predictable way, it was necessary to consider another class of phosphazene materials, poly(aryloxyphosphazene)s, as these polymers contain aromatic phenoxy or naphthoxy groups in their chemical structures that shift their absorption spectra to higher wavelengths, in some cases up to the limit of the visible range [63]. For example, the surface

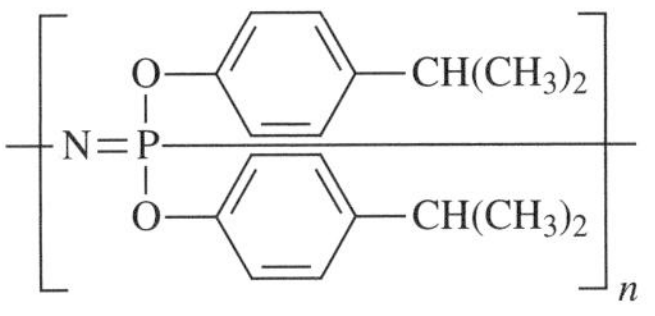

SCHEME 3   General structure of poly[bis(4-*iso*-propylphenoxy)phosphazene].

modification of poly[bis(4-*iso*-propylphenoxy)phosphazene], PiPP (see Scheme 3) could be obtained easily by irradiating films of this polymer with UV light in air [19] and analyzing the corresponding surface using secondary ion mass spectrometry (SIMS) [64]. According to the chemistry of PiPP, the tertiary hydrogen present in the isopropyl residue of the side phenoxy substituent can be removed quite easily under irradiation, to form surface hydroperoxydes by oxygen uptake that evolve successively to hydroxylic, carbonyl, carboxyl, or other functions according to Scheme 4. This process can be used to increase surface energy, wettability, and adhesion of the PiPP surface film or to exploit the surface-introduced chemical groups as starting points for further functionalization reactions. Electromagnetic radiations could, moreover, be used for surface functionalization purposes to induce grafting reactions of different monomers onto the surface of polyphosphazene films.

In fact, thermally, photochemically, and/or γ-radiolytically induced grafting reactions of organic conventional macromolecules onto the surface of polyphosphazene films have been investigated extensively to change surface polarity and wettability of phosphazene materials or to induce enhanced biocompatibility.

The overall process (Scheme 5) consists of:

1. An initial step in which mobile hydrogen atoms (usually present as alkyl radicals in the *para* position of the phenoxy substituents of poly(aryloxyphosphazene)s are removed by a variety of processes to induce the formation of phosphazene macroradicals; this initial process can also be induced by suitable photosensitizers (e.g., benzophenone [65,66]).

2. A propagation step in which the radical polymerization of vinyl monomers starts from the phosphazene macromolecule and eventually

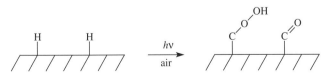

SCHEME 4   Surface chemical modification of [bis(4-*iso*-propylphenoxy)phosphazene] films under UV irradiation in air.

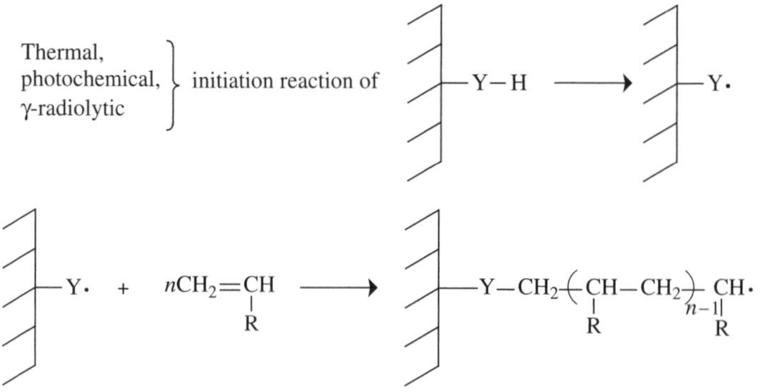

SCHEME 5 General scheme for thermally, photochemically, and γ-radiolytically induced grafting processes of organic macromolecules onto polyphosphazene films.

originates organic polymers grafted onto the surface of polyphosphazene films.

3. A termination step (omitted in Scheme 5 for simplicity) in which the final organic macroradicals coming from the growing chains are scavenged by hydrogen atoms derived from solvents, impurities, and so on, eventually stopping the growth of grafted chains.

As this argument has already been reviewed [39,59,67–71], it is not considered further in this chapter.

Phosphazene Surface Modification Through Nitration Reactions

The utilization of nitro group–containing polyphosphazenes is interesting for surface functionalization reactions and for the preparation of biologically important substrates [72–75]. Thus, poly[bis(phenoxy)phosphazene] was deposited as a coating on the porous surface of alumina particles with the aim of maximizing the exposed surface area and treated with 90% fuming nitric acid to introduce $-NO_2$ groups in the aromatic ring of the phenoxy substituents. These groups were successively reduced to $-NH_2$ by the action of sodium dithionite in H_2O at 100°C (see Scheme 6) and then reacted with glutaraldehyde, cyanogen

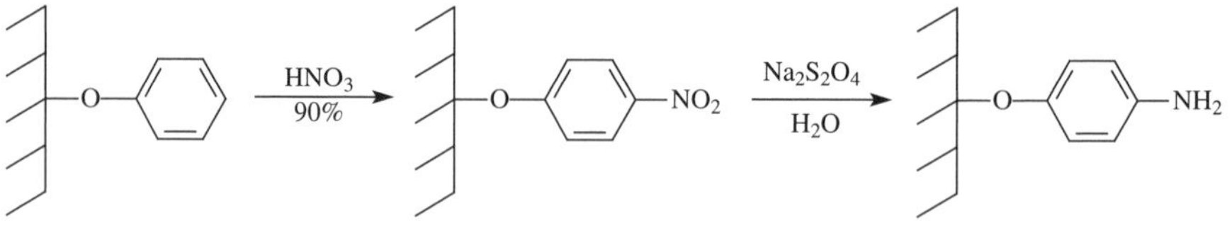

SCHEME 6 General nitration process of poly[bis(phenoxy)phosphazene] followed by reduction of the $-NO_2$ group to $-NH_2$.

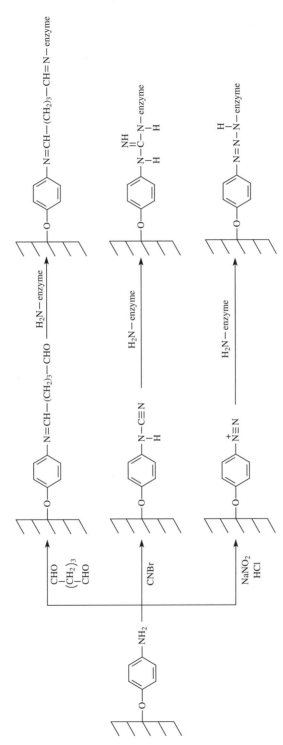

SCHEME 7 Surface functionalization of aryloxy-substituted phosphazene films with enzymes.

bromide, or nitrous acid and eventually used to graft enzymes (e.g., glucose-6-phosphate dehydrogenase or trypsin) according to Scheme 7. The activity of the immobilized surface enzymes was also investigated.

As a variant of this method, 2-(2-aminoethoxy)ethanol was attached to [poly(bis(trifluoroethoxy)phosphazene] by methathetical exchange and the pending free amino groups were reacted with glutaraldehyde and enzymes following the procedure described in Scheme 7 [73,75] to functionalize the surface of phosphazene films with trypsin.

Functionalization of POP Surfaces via Sulfonation Reactions

Surface sulfonation reactions of aryloxy-substituted poly(organophosphazene)s were carried out by Allcock et al. [76], who exploited poly[bis(phenoxy) phosphazene] together with phosphazene copolymers containing variable amounts of 3-ethylphenoxy residues by depositing films of these polymers onto the surface of clean glasses and cross-linking them by irradiation with γ-rays. This was necessary both to prevent dissolution of the phosphazene samples during sulfonation processes and to stabilize the surface structure. After this treatment, the samples were exposed to a series of sulfonating agents, such as SO_3, fuming H_2SO_4, and $SOCl_2$, according to Scheme 8.

The sulfonated phosphazene films were characterized by attenuated total reflection–infrared (ATR-IR) spectroscopy (onset of new peaks at 1160 and 1040 cm^{-1} assigned to the S=O and –SO$_3$Na groups, respectively), by XPS (a new peak at 168.4 eV indicated the presence of highly oxidized sulfate groups), optical microscopy (which made possible determination of the depth of the sulfonated layer between about 3 and 12 μm from the surface), and contact-angle measurements (the initial contact angles of unsulfonated polymers decreased from 70–90° to 5–40° for the sulfonated materials, to assess the enhanced hydrophylicity of the films obtained). These films proved to be able to form surface hydrogels by absorbing water, to exchange Na$^+$ with Mg^{2+} ions, and to immobilize polypeptides (e.g., protamine) ionically.

It was also reported [76] that sulfonated phosphazene polymers prepared using chlorosulfonic acid as a sulfonating agent always bear a considerable percentage (32%) of –SO$_2$Cl moieties that are able to undergo successive functionalization reactions according to Scheme 9 to obtain sulfonamide groups attached to the phosphazene surface. In the same paper it was also

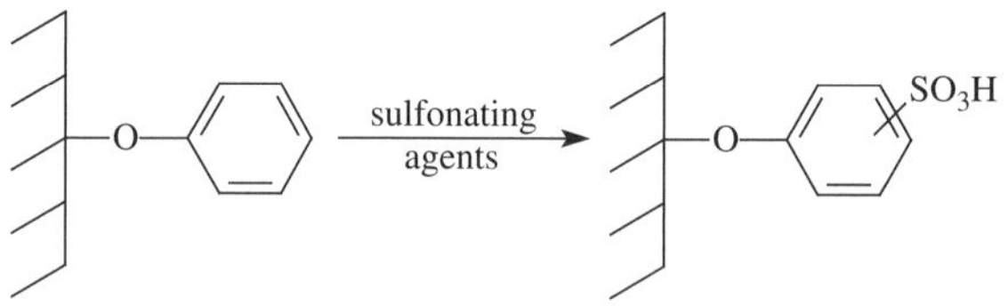

SCHEME 8 Sulfonation process of poly[bis(phenoxy)phosphazene].

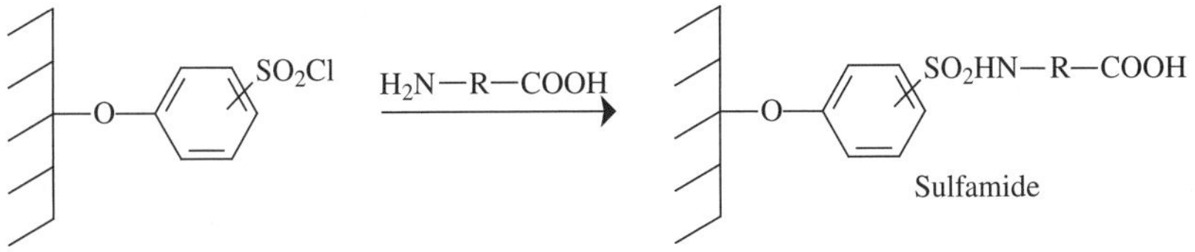

SCHEME 9 Surface functionalization of aryloxy-substituted phosphazenes with sulfonamide moieties.

shown that sulfonated phosphazene films exhibited enhanced biocompatibility features as revealed by preliminary Lindholm tests.

Bromomethylation Reactions of Polyphosphazene Surfaces

Bromomethylation reactions have been exploited by Neenan and Allcock [77,78] to prepare surface polyphosphazene–heparin complexes for nonthrombogenic purposes. The phosphazene macromolecule considered for this research was poly[bis(4-methylphenoxy)phosphazene] (PMPP), which was treated with N-bromosuccinimide (NBS) to brominate the $-CH_3$ group present in the phenoxy substituent of this polymer. The resulting $-CH_2Br$ functions were then treated with triethylamine to quaternize the polymer, which was eventually able to complex heparin molecules. The overall reaction sequence is reported in Scheme 10. The resulting polyphosphazene surface functionalized with heparin groups was subjected to a bovine blood-clotting test and was shown to have appreciable nonthrombogenic features.

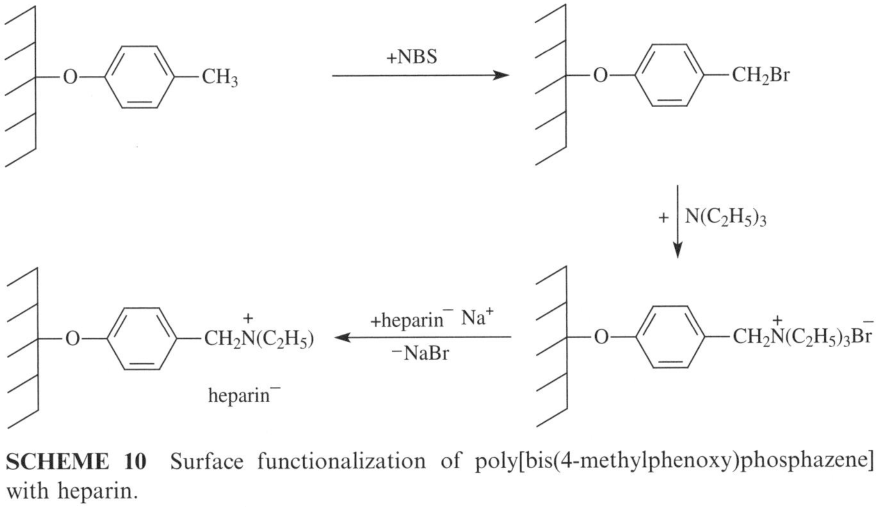

SCHEME 10 Surface functionalization of poly[bis(4-methylphenoxy)phosphazene] with heparin.

Oxidation and Hydrolysis Reactions on POP Surfaces

Surface oxidation [79,80] of 4-methylphenoxy substituents in PMPP and in phosphazene copolymers containing variable percentages of the same substituent, or hydrolysis reactions of propyl-4-carboxylatophenoxy moieties [80] in poly[bis(4-carboxylatophenoxy propylester)phosphazene], PCPP, are also useful reactions to introduce free carboxylic groups on the surface of poly(organophosphazene) films. The general procedure adopted for these reactions is reported in Scheme 11.

Oxidation reactions are particularly interesting because only one type of surface functional group (i.e., –COOH [79]) was obtained by reacting the polymer surface with the $KMnO_4$–NaOH system. Moreover, the structure of the final oxidized films could be tuned either by carefully controlling oxidation conditions (i.e., reaction times and temperatures) or by tuning the chemical composition of the starting 4-methylphenoxy/phenoxy copolymers [80].

The resulting oxidized films were characterized by XPS [onset of a new C(1 s) peak at 288.7 eV assigned to the carbon atom of the carboxylic group], ATR-IR spectroscopy (detection of a new peak at $1710 \, cm^{-1}$ assigned to the C=O

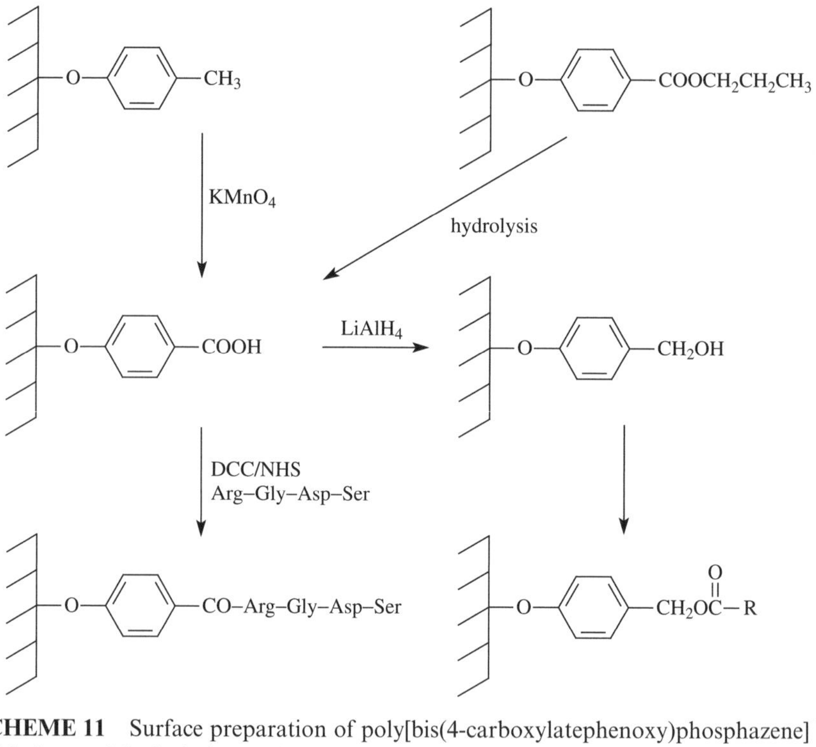

SCHEME 11 Surface preparation of poly[bis(4-carboxylatephenoxy)phosphazene] by oxidation and hydrolysis reactions of suitable polyphosphazenes and successive functionalization processes.

groups of the carboxylic function), scanning electron micrographs (no variations in the surface morphology could be observed for films prepared at 25 and 50°C, while those obtained at 80°C showed rougher surfaces, especially for longer reaction times), and contact angle measurements (in general, the initial contact angle of 92° for unoxidized films of PMPP decreased considerably after the oxidation treatment and was also found to be pH dependent [79]).

Hydrolysis reactions were also exploited to produce surface carboxylic groups by immersing films of poly[bis(4-carboxylatophenoxy propylester)phosphazene] in aqueous or methanol 1 M potassium hydroxide solutions at 50–80°C. Even in this case the characterization of the surfaces obtained was carried out by contact angle measurements (decrease in contact angles with an increase in the degree of hydrolysis and pH dependence), XPS spectroscopy [decrease in C/O ratio with an increase in the hydrolysis process, and an O(1s) region of the spectra consistent with the formation of carboxylate species].

Once obtained, surface carboxylic groups proved to be excellent starting points for further surface functionalization reactions, as shown in Scheme 11. Thus, the carboxylic groups present on the surface of the polyphosphazene films were able to exchange Na^+ with Ca^{2+} ions when treated with $CaCl_2$ solutions, could link a tetrapeptide Arg–Gly–Asp–Ser through the action of N-hydroxysuccinimide (NHS) and dicyclohexylcarbodiimide (DCC), and could be reduced to the corresponding benzyl alcohol groups by the action of 1 M $LiAlH_4$ solution in ethyl ether. In turn, this $-CH_2OH$ function was eventually used to link 4-nitrofuroic acid through an esterification reaction.

Surface Modification of POPs Through Methatetical Exchange Reactions

Another very useful synthetic strategy to modify the surface features of POPs is based on the utilization of methatetical exchange reactions of trifluoroalkoxy substituents in PTFEP with diethylene glycol monomethyl ether [81,82] (pathway a), different fluorine-containing alcoholates [83] (pathway b), or with other alcoholates containing OH [84] (pathway c), NH_2 [73–75,84] (pathway d), CN [84] (pathway e) functions, or hydroxylic groups [85] (pathway f), according to Scheme 12.

These modifications have been achieved by immersing films of poly[bis(trifluoroethoxy)phosphazene] in solutions of the sodium salts of the different nucleophiles at temperatures ranging between room temperature and 100°C (pathways a to e), or in an 8 M aqueous solution of NaOH at 80°C in the presence of tetrabutylammonium bromide as the phase transfer catalyst (pathway f) for variable periods of time. The resulting modified films were characterized by Fourier transform infrared spectroscopy, XPS, and contact-angle measurements. These functionalizations have been carried out to improve surface biological compatibility [82], adhesion properties [85], and resistance to polar solvents [83] of PTFEP while retaining bulk physical features or to exploit this polymer for immobilizing biologically active substances [73–75,81].

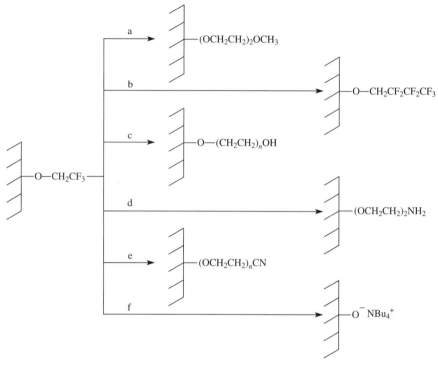

SCHEME 12 Methatetical exchange reactions of poly[bis(trifluoroethoxy)phosphazene] with various nucleophiles.

Surface Properties of Metal-Containing POPs

Despite the fact that many cyclo- and polyphosphazenes exist that bear different types of metals [86–91], the surface metallation of phosphazene polymer films seems to be comparatively less frequent. One example of metallophosphazene has been reported by Allcock et al., [92] who described the covalent attachment of an iron complex on the aryloxy-side substituent of a phosphazene macromolecule cross-linked by γ-rays. For the synthesis of this material, metal–halogen exchange reaction was attempted between n-BuLi and poly[bis(4-bromophenoxy)phosphazene], PBrPP, at low temperature, followed by the addition of cyclopentadienyliron dicarbonyl iodide [CpFe(CO)$_2$I] to a lithiated phosphazene intermediate. The overall reaction is described in Scheme 13.

The iron-containing polyphosphazene was characterized by contact angle measurements (θ decreases from the initial 120° up to 54° for the treated polymer), and by ATR-IR spectroscopy (band at 1410 cm^{-1} attributed to the C–C stretching of the cyclopentadienyl moieties). A total of 30 μm of the reaction depth was also estimated by scanning electron microscopy. More recently, in our group we could succeed in supporting different metal catalysts

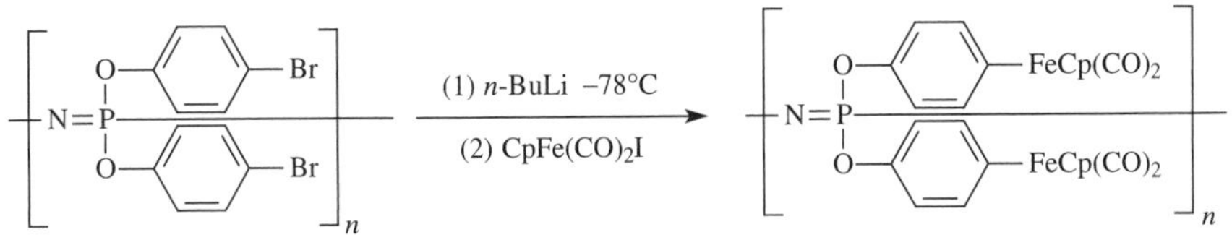

SCHEME 13 Surface functionalization of poly[bis(4-bromophenoxy)phosphazene] with iron complexes.

on the surface of several polyphosphazenes. Two synthetic procedures have been exploited to support metals: the utilization of relatively unstable complexes (pathway A) and the metal vapor synthesis (MVS) (pathway B) technique [93,94], according to Scheme 14.

Ru clusters have been supported on the surface of the polyphosphazene films using the first synthetic procedure, while the MVS approach has been exploited to prepare Rh, Pd, and Pt metal clusters on phosphazenes. As far as poly(organophosphazene)s are concerned, the macromolecules shown in Scheme 15 have been utilized. HRTEM analysis [95,96] showed that the metallic particles are monodispersed on the polymer surface, with average dimensions of 1.5 to 2.5 nm.

Ru/poly(dimethylphosphazene) composites (Ru/PDMP) proved to be particularly efficient in hydrogenation reactions of olefins, carbonyl-containing compounds and aromatics, operating in both homogeneous and heterogeneous phases [95]. Rh and Pt on poly(dimethylphosphazene) (Ru/PDMP and Pt/PDMP), were used in the selective hydrogenation reaction of cynnamaldehyde, a compound considered as a prototype for α,β-unsaturated carbonyl products [97]. Eventually, Pd/PDMP catalyst proved to be particularly efficient

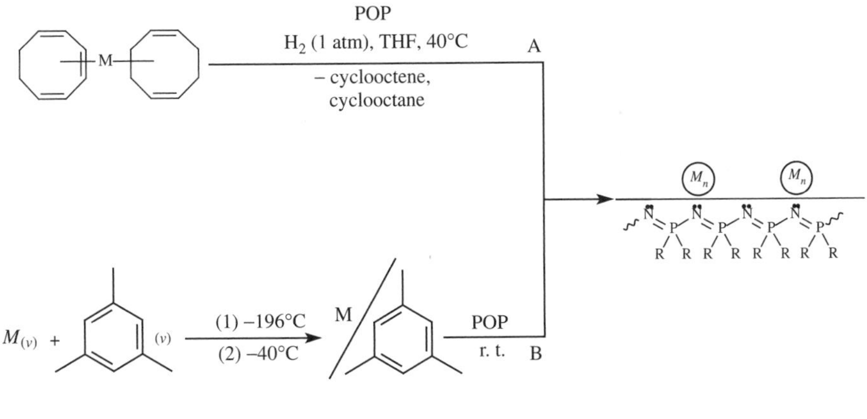

where M = Ru, Rh, Pd, Pt

SCHEME 14 Surface functionalization of polyphosphazene films with metal clusters.

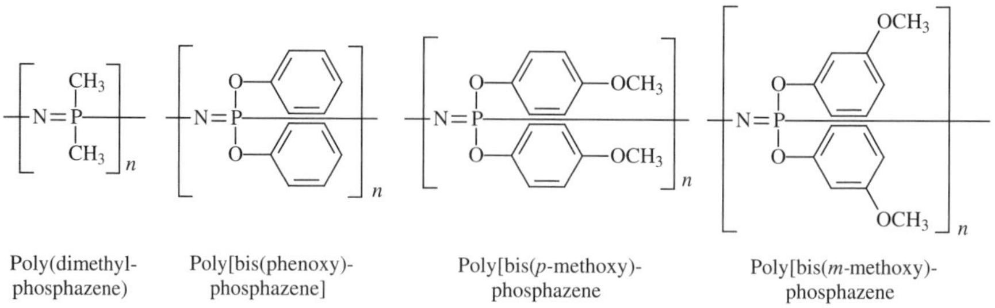

| Poly(dimethyl-phosphazene) | Poly[bis(phenoxy)-phosphazene] | Poly[bis(p-methoxy)-phosphazene | Poly[bis(m-methoxy)-phosphazene |

SCHEME 15   General structure of the phosphazene polymers exploited for metal surface functionalization processes.

in promoting Heck-type reactions between iodobenzene and methyl methacrylate in N-methylpyrrolidone [97]. The great advantage of these types of catalysts is that they can easily be isolated from the reaction mixture and reused without appreciable loss of activity.

Surface Features of Phosphazene Polymers Substituted with Silicon-Containing Groups

The synthesis of phosphazene polymers and copolymers substituted with silicon-containing groups could be achieved by preparing phosphazene copolymers partly substituted with trifluoroethoxy, phenoxy, or methoxy–ethoxy–ethoxy moieties, the residual chlorines eventually reacting with different types of aminosiloxane substituents [98]. The general structure of these materials is reported in Scheme 16. Surface characterization of these new materials was carried out by contact angle measurements that evidenced the hydrophobic nature of these compounds (θ ranging between 111 and 98°). No direct correlation between the polymer structure and contact angles could be detected.

Silicon-containing phosphazene macromolecules could also be prepared by surface hydrosilylation processes [99] between allyl group–containing polyphosphazenes and heptamethylhydrotrisiloxane derivatives in the presence of platinum-based catalysts, according to Scheme 17. The resulting polymers were characterized by XPS and contact-angle techniques. Possible applications for these new materials could be as low-temperature elastomers, thermally

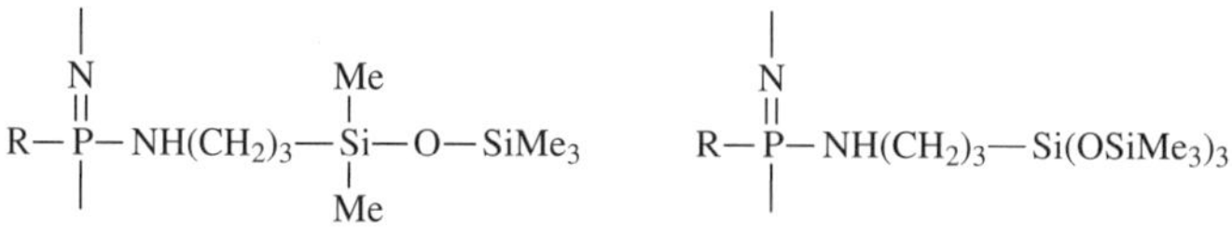

where $R = CF_3CH_2O$, C_6H_5O or $CH_3OCH_2CH_2OCH_2CH_2O$

SCHEME 16   General structure of phosphazene copolymers containing pending aminosiloxane substituents.

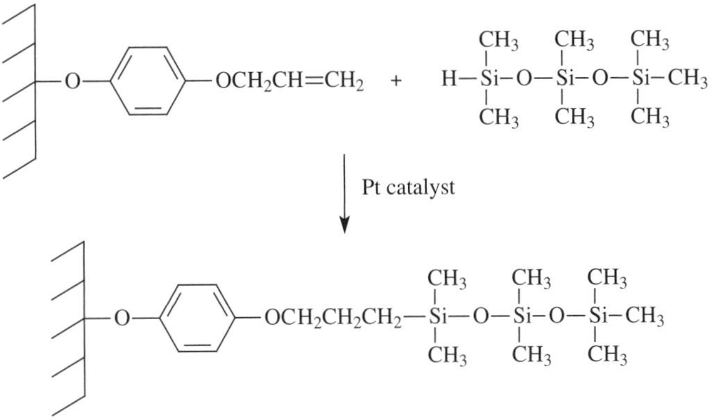

SCHEME 17 Surface hydrosilylation of allyl group–containing aryloxy phosphazene polymers.

stable and flame-retardant substrates, or membranes with improved gas permeability.

POP Surface Modification by Various Techniques

The behavior of a series of polyphosphazenes [i.e., poly(dichlorophosphazene), PDCP; poly[bis(aniline)phosphazene], PAP; and poly[bis(piperidino)phosphazene], PPP] under pulsed (1064 nm) or continuous (488 nm) laser irradiation has been investigated by Exarhos [100]. The materials, deposited on silica or silicon substrates by dip-coating or spin-casting techniques, showed extensive morphological damages, cratering and delamination under pulsed laser irradiation, with the formation of large ablated regions and debris from the ablate material distributed randomly over the damage area (PAP), the formation of cratered regions (PPP), and the obtainment of a spheroidal surface morphology indicative of melting processes followed by condensation (PDCP).

Investigations on atomic oxygen–induced surface change in phosphazene polymer films and coatings have been carried out by Fewell [101] in the attempt to develop macromolecules suitable for use in planned long-duration space missions. The polymers considered for this research are PTFEP and poly[bis (3- and 4-chlorophenoxy)phosphazene]s (P3ClPP and P4ClPP, respectively). XPS investigations on the treated films of these materials evidenced rearrangements in all the polymers investigated, which took place mostly at the skeletal nitrogen, leading to branching and/or cross-linking phenomena (=N–N=groups at 400.3 eV) and to the formation of oxidized species (N→O, 402 eV). Very low mass removal were found in the case of PTFEP, which increased in the case of P3ClPP and P4ClPP, possibly due to the loss of chlorine atoms in these materials. On the basis of these results it was concluded that the aforementioned structural changes in the POPs examined brought about an enhancement of resistivity under the experimental conditions selected.

Significant modifications of PTFEP surface properties were obtained by Singh et al. [55] using the electrospinning technique to prepare nanofibers of this polymer. The polymer showed a shifting of the water contact angle value for films from the hydrophobic value of 104° up to the superhydrophobic range 135–159° for electrospun nanofibers of different diameters and surface morphology.

Finally, surface modification of PTFEP films has been obtained by Allcock et al. by plasma techniques [102] using different types of plasma treatments, involving the use of O_2, N_2, CH_4, or CF_4/H_2 as reactive gases. It was found that the initial treatment of the polyphosphazene material with oxygen plasma induced a contact angle of 5°, which is highly hydrophilic; this value was enhanced successively to 39°, 68°, and 151° by using nitrogen, methane, or carbon tetrafluoride materials, respectively, in the plasma equipment.

SURFACE MODIFICATION OF ORGANIC AND INORGANIC SUBSTRATES WITH PHOSPHAZENES

Surface modification of conventional polymer films, inorganic oxide, beads or surfaces, human tissues, and so on, with phosphazene substrates has been carried out using both cyclic and polymeric phosphazene materials, according to a variety of different experimental approaches which imply utilization of:

1. Electromagnetic radiation (e.g., UV–visible and γ-rays)
2. Chemical reactions (e.g., silanization, oxidation, or grafting processes)
3. Surface reactions with chlorophosphazenes

General aims for these investigations could be:

- A change in the surface features of polymeric films for biomedical applications (biocompatibilization, implantation, etc.)
- The preparation of burn- or heat-resistant coatings
- Improvement in insulation or adhesion characteristics

In this section we provide examples of the way in which these modifications could be obtained by exploiting phosphazene materials.

Surface Modification with Phosphazenes by Means of Electromagnetic Irradiation

The surface functionalization of polymeric carbon-backboned materials can be reached quite easily by using UV–visible light or γ-rays to cross-link reactive

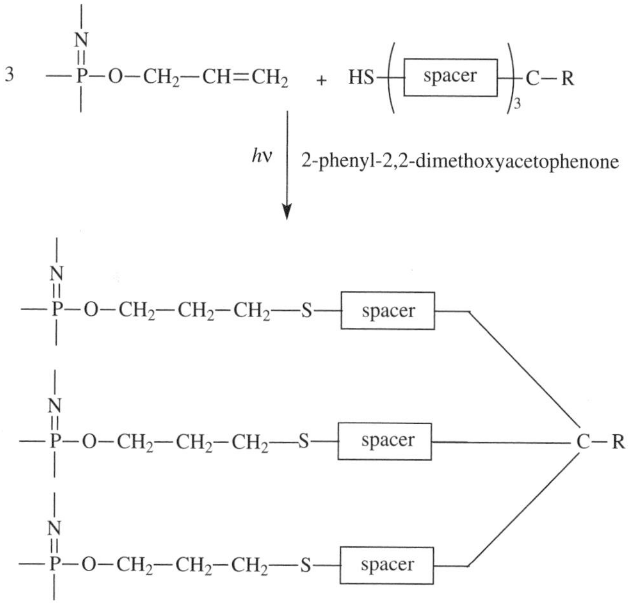

SCHEME 18 Surface modification of organic substrates by light-induced ene–thiol reactions of allyl group–containing phosphazene substrates.

cyclophosphazenes or selected polyphosphazenes with the aim of changing surface features or improving valuable characteristics. Thus, coating compositions suitable as burn-resistant coatings for thermoplastic flooring sheet or tile products have been prepared by UV light–induced ene/thiol reactions of cyclophosphazene and polyphosphazene mixtures containing unsaturated allyl substituents and tri- or tetrathiolic compounds in the presence of free-radical initiators [103]. The overall reaction is reported in Scheme 18.

Similarly, the surfaces of poly(propylene) [104], poly(vinyl chloride) [104], poly(ethylene terephthalate) [104], poly(bisphenol A carbonate) [104], and poly(methyl methacrylates) [104] could be modified by the γ-ray-induced grafting reaction of poly[bis(methoxyethoxymethoxy)phosphazene], MEEP (a) and poly[bis(methylamino)phosphazene], PMAP (b) (Scheme 19) to enhance surface hydrophilicity and to form surface hydrogels. Both these polyphosphazenes, in fact, are known to be water-soluble macromolecules [105–107] susceptible to undergoing clean γ-induced [108–110] cross-linking processes. The entire sequence of synthetic steps is reported in Scheme 20, in which an organic polymer film is immersed in a solution of the POP selected, cross-linked by γ-irradiation and allowed to absorb water until a surface hydrogel is formed. Polymer surfaces prepared according to this method showed enhanced hydrophilic features and potential applications in biomedical field [104].

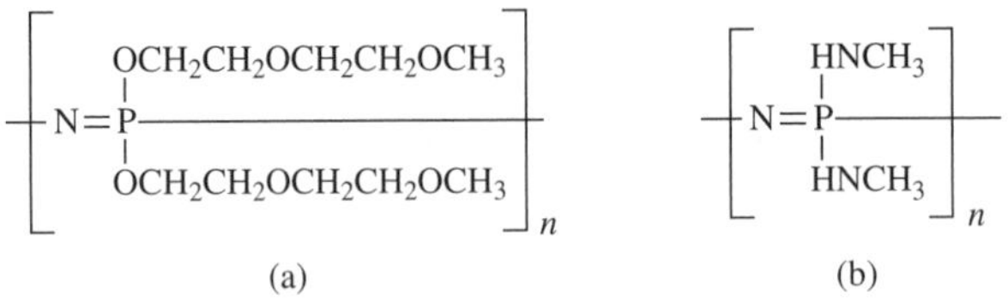

$$\left[\begin{array}{c} OCH_2CH_2OCH_2CH_2OCH_3 \\ | \\ N=P \\ | \\ OCH_2CH_2OCH_2CH_2OCH_3 \end{array} \right]_n \qquad \left[\begin{array}{c} HNCH_3 \\ | \\ N=P \\ | \\ HNCH_3 \end{array} \right]_n$$

(a)   (b)

SCHEME 19   General structure of poly[bis(methoxyethoxyethoxy)phosphazene] and poly[bis(methylamino)phosphazene].

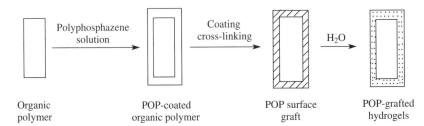

| Organic | POP-coated | POP surface | POP-grafted |
| polymer | organic polymer | graft | hydrogels |

SCHEME 20   Surface functionalization of organic polymers by grafting reactions of water-soluble polyphosphazenes.

Surface Modification with Phosphazenes Through Chemical Reactions

Classical reactions of organic chemistry are considered to be a valuable tool to reach the modification of surface properties of different substrates [5]. In this section we provide examples of how this surface functionalization approach was exploited to modify the surface of both organic and inorganic materials by means of free-radical, oxidation, silanization, and plasma processes. In addition, the surface modification reaction of silicon-based materials [e.g., silica beads, soda-lime glasses, crystalline silicon wafers (100)] will also be described by surface reaction with hexachlorocyclophosphazene (HCCP).

Free-Radical Reactions   Free-radical reactions have been used for the preparation of inert coatings for natural or synthetic rubber articles of biomedical interest, such as catheters, by surface-curing a fluorinated phosphazene copolymer (PNF 200 [111]) by using unsaturated products [112–114]. Thus, the surface of these catheters was first treated with a solution of di- or polyvinyl compounds (e.g., divinylbenzene) containing organic peroxides as a source of free radicals (e.g., benzoylperoxide), and then immersed in a solution containing a phosphazene copolymer, equimolecularly substituted with trifluoroethoxy and heptafluorobutoxy substituents (PNF 200 [111]), in the presence of α,α′-bis((*t*-butylperoxy)diisopropyl benzene) and of triisobutylaluminum as a new accelerating catalyst. The final mixture was eventually heated at 60°C for 15 h. The overall sequence is reported in Scheme 21. The resulting coated catheters

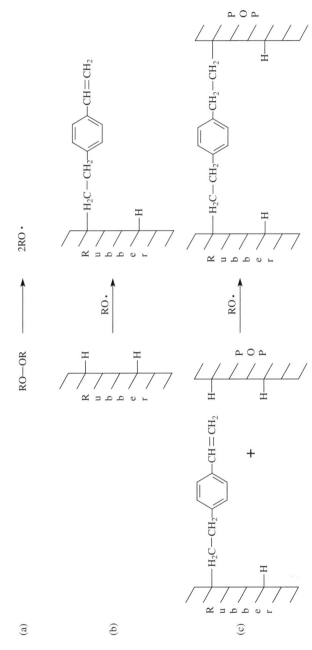

SCHEME 21 General scheme for the preparation of inert phosphazene coatings for natural or synthetic rubber articles through free-radical reactions.

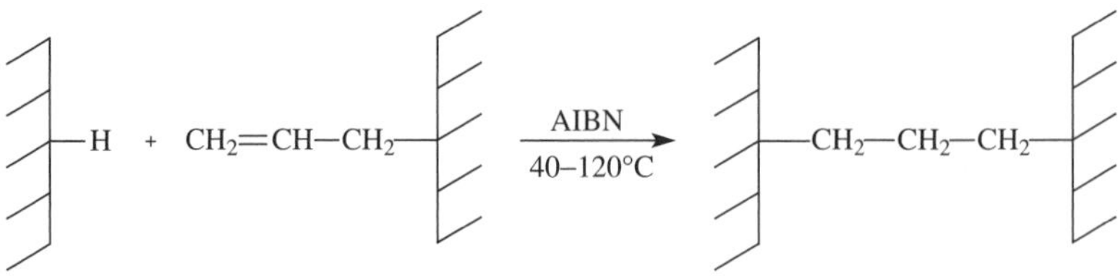

Polypropylene   Eypel   Polypropylene coated
                       (POP)   with POP

SCHEME 22   Thermally induced grafting reaction of allyl group–containing phosphazene copolymers on the surface of polypropylene hollow fibers.

showed tissue-reaction characteristics very close to those of silicone-based catheters.

Similarly, Eypel gum by Ethyl Co. could be grafted on the surface of hollow fibers of polypropylene in the presence of 2,2′-azobisisobutyronitrile (AIBN) as a thermal radical initiator [115], due to the thermal lability of the tertiary hydrogens on the polymer and to the presence of allylic double bonds in the polyphosphazene substrate (Scheme 22). These materials were exploited for electrophoresis and electroosmotic flow control for small molecule (e.g., pyridines and DNA bases) separation.

Oxidation Reactions   Oxidation of organic and/or inorganic substrates is considered as a suitable tool to modify the surface of a variety of materials through the introduction of free hydroxylic groups [5]. Once these functions are available, they can be used to change surface features by reacting with a great many products, including chlorinated cyclo- and polyphosphazenes and amino-substituted trialkoxy silanes. Free hydroxylic groups have been introduced onto the surface of different materials in a variety of ways that are discussed below.

Surface Oxidation Through Caro's Acid   Major work in this area has been carried out during the last 10 years by Grunze and co-workers [116–151], who could succeed in preparing surfaces of biomedical devices (e.g., stents or catheters) covered by PTFEP. In fact, preliminary observations by Tur et al. [152] demonstrated that PTFEP possesses good blood and tissue compatibility both in bulk and as a coating material. This important property of the phosphazene polymer was reported [118–120,123] to be due to the selective absorption of high quantities of albumin from blood plasma and selective rejection of some of the proteins that stimulate coagulation (typically, fibrogen). The surface of PTFEP films or coatings therefore appeared to be passivated against platelet adhesion. Moreover, absorption of these proteins was found to be low and reversible, thus indicating a low tendency to denaturation. As a consequence, it was thought that PTFEP has some potential applications in biomedicine as a material for blood-contacting devices, such as stents, protheses, and dental implants.

Films of PTFEP were deposited on the surface of biomedical devices using a variety of methods, all of them shown in Scheme 23. According to this scheme, the first step in the coating procedure implies the chemically induced oxidative cleaning of the surface by treatment with Caro's acid (1:3 mixture of H_2O_2 30% and concentrated H_2SO_4 [128]) and the introduction of a considerable number of free hydroxylic groups (pathway A). These functions can be reacted in a variety of methods, according to the following procedures:

1. Methatetical exchange of the functionalized surfaces with PTFEP and grafting of this polymer on the surface of the biomedical device (pathway B).
2. Reaction with PDCP, and successive grafting of PTFEP on the surface of the functionalized films by methatetical exchange (pathway C).
3. Reaction with γ-aminopropyltrimethoxy silane (APTMS), to produce surfaces containing free aminic functionalities (pathway D), from which the grafting of PTFEP can be achieved by successive treatment with PDCP (pathway E) and nucleophilic substitution of the residual un-reacted chlorines with trifluoroethoxy moieties (pathway F) or by direct methatetical exchange with PTFEP (pathway G).

The new surfaces obtained by the use of PTFEP coatings showed out-standing mechanical features, antithrombogenic characteristics, and good biocompatibility [52,53]. They also possessed the ability to prevent or reduce secondary injuries following implantation and the uncontrolled cellular growth that causes re-stenosis in implanted stents and inflammatory reactions follow-ing introduction of medical devices in the human body [125,135,146,149,150] (the same result could be obtained by random insertion along the phosphazene skeleton of ^{32}P radioisotopes, which emit β radiation during their radioactive decay [121,122,128,130]); the possibility of preventing or reducing inflamma-tions and autoimmune reactions following incorporation of foreign implants in the organism, thus imparting bacterial resistance and avoiding inflammatory responses [124,126,137,148]; microstructured surfaces, produced with a variety of techniques, in order to affect the flow behavior of liquids in catheters or stents in a particularly favorable way (e.g., sharkskin or lotus effect) [131–133,143,145]; pharmacologically active substances able to allow synthetic implants to act as drug delivery systems in a controlled way [153–156]; the capability to be used as a coating for plastic articles, to obtain wound coverage showing excellent biological and pharmacological properties [129,134]; the capability to be used for the preparation of artificial implants (e.g., heart valves) [116]; the possibility to be exploited as films or coatings for medical devices (e.g., stents or catheters) [127]; a basic structure formed by Nitinol (an intermetallic compound based on nickel and titanium metals that possesses

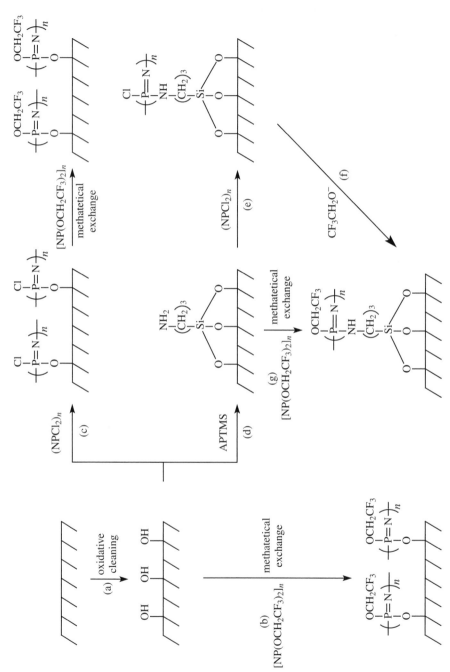

SCHEME 23 Surface functionalization of oxidized biomedical devices with poly[bis(trifluoroethoxy)phosphazene].

206

thermal and mechanical shape memory and shows very good body compatibility) covered by a film of PTFEP, suitable for use in the preparation of medical devices for artificial implants [136,139–141,151]; and bacterial resistance for dental implants [138]. Use of these materials, as obtained by surface modification with PTFEP, is mainly in the biomedical field.

Surface Oxidation Through Plasma Technique   Surface oxidation reactions were also carried out in our group using argon cold plasma treatments, with chlorinated phosphazenes as coupling agents. High-density polyethylene (HDPE) and polyamide-6 (PA6) were selected as examples of apolar and polar materials, respectively, and treated with argon plasma at room temperature to form free radicals on the surface [157–160]. This surface was then exposed to the atmosphere to form a series of oxidized functionalities (e.g., carbonyl, ester, ether, and hydroxylic groups) [159,161]. These functions were reacted successively with chlorinated phosphazenes in order to graft these materials on the surface of the polymers through the formation of strong C–O–P–Cl covalent bonds. The residual chlorines left on the grafted phosphazene materials were eventually substituted for by a variety of nucleophilic substituents [43–45], thus deeply modifying the surface properties of the original polymers. The overall reaction is shown in Scheme 24. Among the possible substituents that could be used to functionalize HDPE and PA6 surfaces, fluorinated alcohols of variable lengths and 4-hydroxyazobenzene were of particular importance because they were able to impart high surface hydrophobicity features and photochromic properties to the materials, respectively.

Functionalization of Compounds Containing Hydroxylated Surfaces   The surface of substrates containing free hydroxylic groups is particularly suitable for

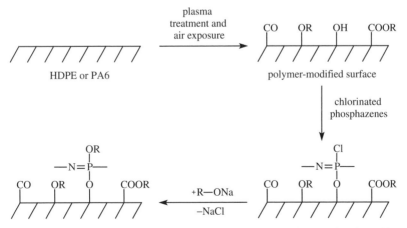

SCHEME 24   Surface modification of high-density polyethylene and polyamide-6 by the plasma technique using chlorinated phosphazenes as coupling agents.

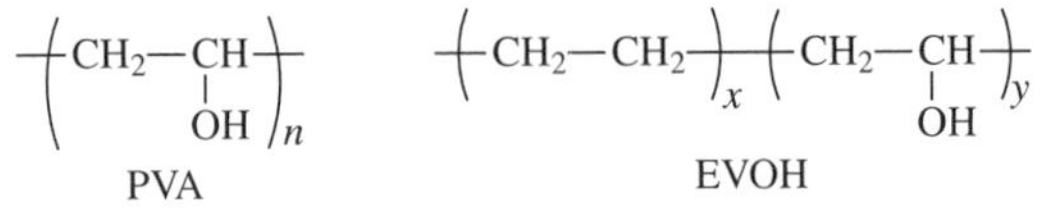

$$\left(CH_2{-}\underset{\underset{OH}{|}}{CH}\right)_n \qquad \left(CH_2{-}CH_2\right)_x\left(CH_2{-}\underset{\underset{OH}{|}}{CH}\right)_y$$

PVA   EVOH

SCHEME 25   General structure of poly(vinyl alcohol) and of a vinyl alcohol–ethylene copolymer.

undergoing modification reactions because of the chemical reactivity of these functions with different organic and inorganic reactants [5]. Two series of materials have been considered over time: polymers already containing free hydroxylic groups in their chemical structure, such as poly(vinyl alcohol) (PVA) and ethylene–vinyl alcohol copolymers (EVOH) (Scheme 25), and the surface of crystalline (100) silicon wafers, SiO_2 beads, and soda-lime glasses.

Surface Functionalization of PVA and EVOH Polymer Films   Basic investigations of the surface functionalization of PVA and EVOH films carried out by De Jaeger and colleagues [162–166], who considered these materials because of the excellent gas barrier properties of PVA in its dry state, which become very poor under moister conditions [163]. For this reason the surfaces of PVA films were treated with a variety of products to reach the grafting of some phosphazene copolymers containing different functional groups, all of them showing good hydrophobic features. The overall reaction for the preparation of these materials is shown in Scheme 26. Thus, poly[(phenoxy)(4-ethylphenoxy)(2-methyl-4-allylphenoxy)phosphazene] (Scheme 27), a phosphazene copolymer containing free allylic moieties, was grafted onto the surface of PVA by treating this substrate with a toluene solution of the polyphosphazene at 80–110°C in the presence of benzoylperoxide as a radical initiator [162] (pathway a, Scheme 26). The resulting materials showed enhanced hydrophobicity with respect to the pristine macromolecule, which is expected to be helpful in improving the oxygen-barrier features of PVA under high-humidity conditions.

Similarly, PVA films were immersed in a toluene solution of a methyl methacrylate polymer having pending cyclophosphazene units of the chemical structure shown in Scheme 28 in the presence of benzoyl peroxide as a radical initiator at 135°C (pathway b, Scheme 26). Under these conditions the ethyl group present in the *para* position of the phenoxy substituent in the cyclophosphazene is able to react with the PVA surface and induce grafting of the cyclophosphazene–methacrylate polymer onto the surface of PVA [164], thus forming a coating layer on this polymer.

In a third case (pathway c, Scheme 26), the surface functionalization of PVA macromolecules could be obtained by reacting a phosphazene copolymer containing free succinic anhydride residues of the structure shown in Scheme 29 with the hydroxylic functions present on the surface of the PVA film. This polymer, in turn, could be prepared by reaction of poly[(phenoxy)(4-ethylphenoxy)phosphazene] with maleic anhydride [167–169]. The opening of the

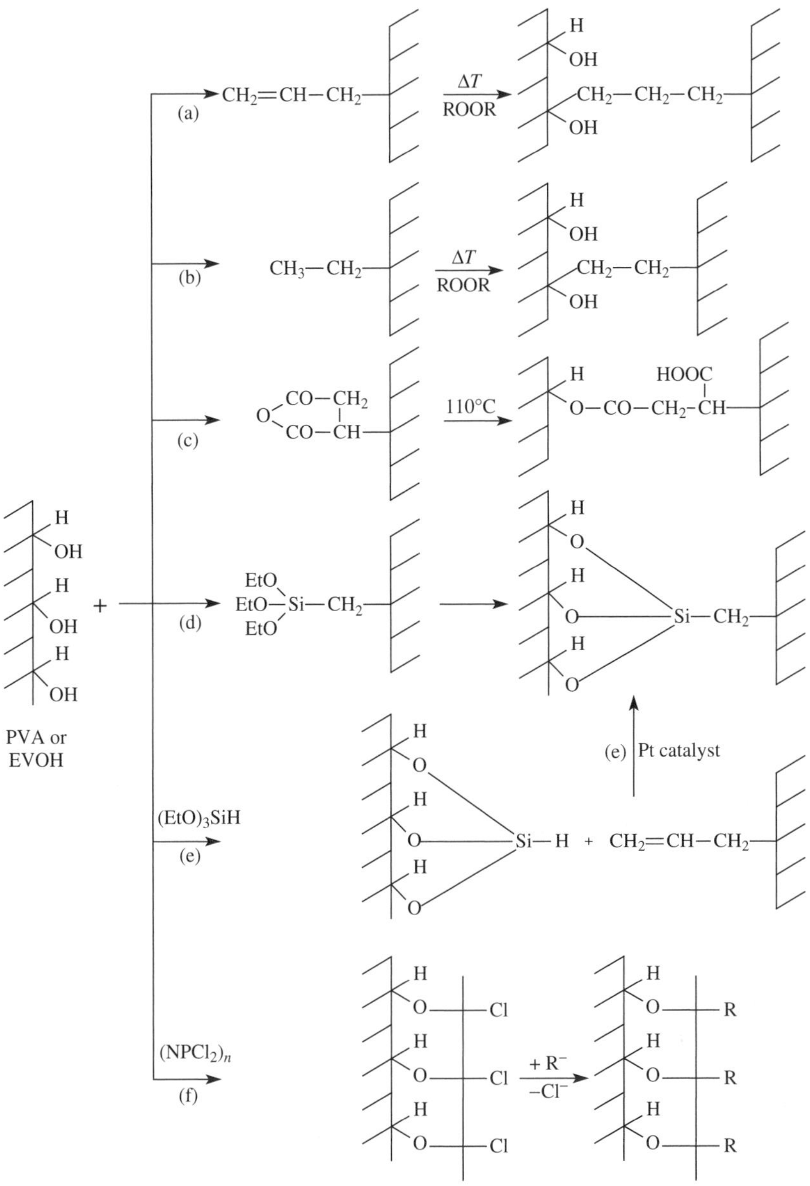

SCHEME 26 Grafting of phosphazene polymers and copolymers on the surface of poly(vinyl alcohol) and a vinyl alcohol–ethylene copolymer.

anhydride moieties resulting from this reaction was able to induce grafting of the polyphosphazene onto the surface of the organic substrate [166]. The resulting materials were proved to have enhanced surface hydrophobicity with respect to that of the starting PVA.

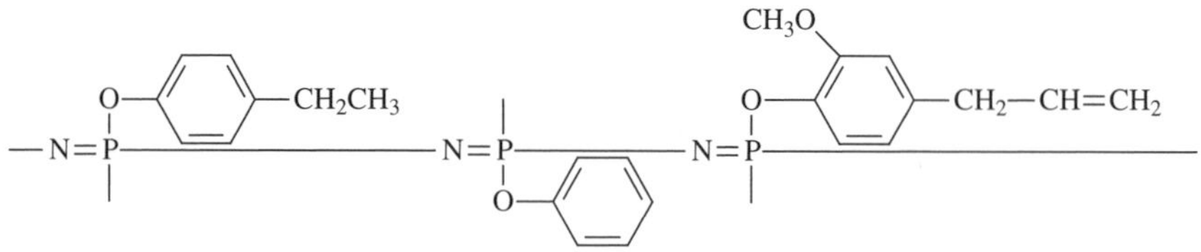

SCHEME 27 General structure of a (phenoxy)(4-ethylphenoxy)(2-methoxy-4-allyl-phenoxy) phosphazene copolymer.

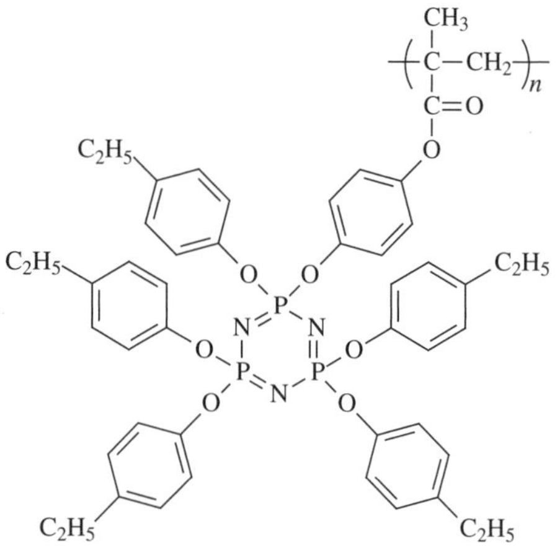

SCHEME 28 General structure of a methyl methacrylate copolymer having pending cyclophosphazene units substituted for by 4-ethylphenoxy groups.

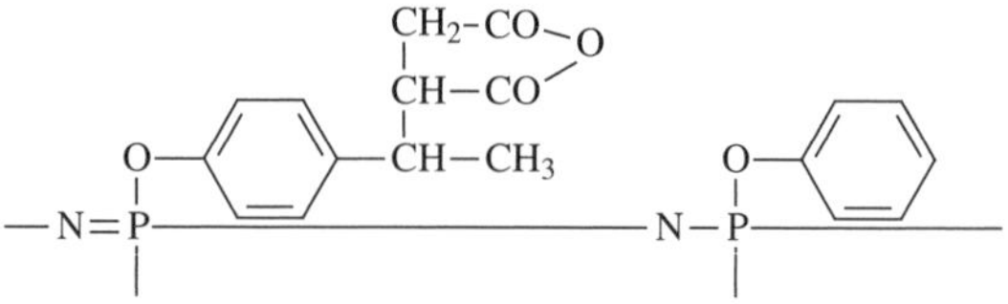

SCHEME 29 General structure of a phosphazene copolymer containing free succinic anhydride residues.

Hydrosilylation reactions have also been used to modify the surface characteristics of PVA and EVOH with phosphazene materials [165]. In fact, a phosphazene copolymer containing free triethoxysilane moieties [obtained by reacting the allylic functions present in the polyphosphazene with HSi(OEt)$_3$, tri(ethoxy alkyl)silane] was reacted with PVA or EVOH according to pathway d of Scheme 26, forming a phosphazene coating on the surface of the organic substrates. The same materials could be obtained by reversing this procedure (pathway e, Scheme 26), first grafting HSi(OEt)$_3$ onto the PVA surface, followed by a hydrosilylation reaction with the allylic residue on the polyphosphazene. In both cases, surface hydrophobicity of the pristine PVA could be enhanced remarkably.

Finally, functionalizing the surface of EVOH films with a variety of interesting organic molecules by treating EVOH with poly(dichlorophosphazene) or hexachlorocyclophosphazene according to the classical reactivity of the chlorinated phosphazene substrates has recently succeeded [47,49]. This reaction is illustrated in pathway f of Scheme 26. The successive reaction of the residual chlorines in the phosphazenes with suitable nucleophiles allowed the preparation of EVOH surfaces functionalized with fluorinated alcohols showing enhanced hydrophobic features and surfaces containing 4-hydroxyazobenzene moieties showing interesting photochromic processes [170].

Surface Functionalization of Hydroxylated Silicon-Based Materials   Basic investigations on the surface functionalization of hydroxylated silicon-based materials have been carried out in our groups [171–175], taking into consideration silica beads, crystalline (100) silicon wafers, and soda-lime glasses. These studies have been carried out by means of two different projects: first investigating the interactions that take place between hexachlorocyclophosphazene (HCCP) and the free–OH groups on the surface of soda-lime glass and (100) silicon wafers, and successively by studying the surface functionalization of silica gel beads and silicon wafers with specially tailored cyclophosphazenes.

Research on the surface functionalization of soda-lime glasses and (1 0 0) crystalline silicon wafers with HCCP was originated by the very ancient observation that the thermally induced ring-opening polymerization of HCCP to PDCP in sealed ampoules under vacuum [43,58] was influenced markedly by the glass walls of the flask where the polymerization process was run [176,177], strongly affecting the reproducibility of the overall process. This fact could indicate that chemical interactions could take place between the glass surface and the chlorinated phosphazene material [172,178,179]. It was found that by immersing a soda-lime slide in a THF solution with 0.1 M HCCP for variable periods of time and analyzing the resulting surface by the XPS technique, the initial ratio Cl:N:P=2:1:1 changed to Cl:N:P=1:1:1. In this way, three chlorine atoms of HCCP could be involved in a reaction with the surface –OH groups of the glass to graft HCCP on the surface of the material. The hypothesized reaction mechanism for this process is reported in Scheme 30.

Theoretical *ab initio* calculations carried out on this reaction [172], however, showed that this process, although thermodynamically favored ($\Delta E = 16.28$ kcal/mol), presents a considerably high kinetic energy barrier (108 kcal/mol), indicating intrinsic difficulty in this reaction taking place according to Scheme 30. The same calculations, moreover, made it possible to individuate several possible alternative *chemisorbed* configurations for the cyclophosphazene and

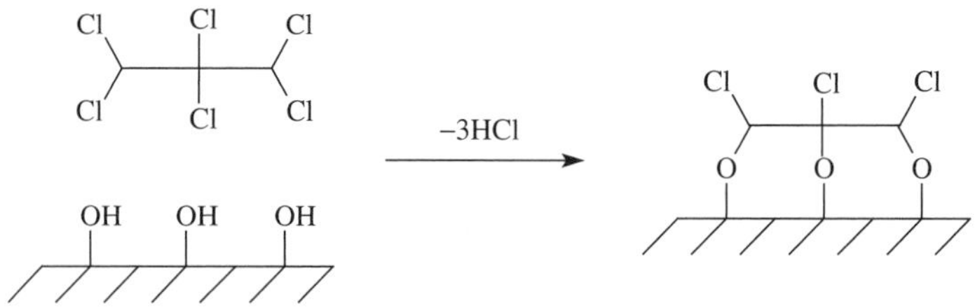

SCHEME 30 Hypothesized mechanism for the reaction between soda-lime glass surfaces and hexachlorocyclophosphazene.

evidenced the crucial role played by water molecules present on the glass surface and by different solvents used for the functionalization process.

Furthermore, additional investigations were carried out [175] aimed to study the effect of several experimental parameters, such as reaction temperature and time, the solvent employed, the effect of a final washing with THF, and a variety of post-reaction drying processes. Substrates were dried under vacuum at different temperatures to verify the influence of the hydration degree of the surface to be functionalized.

As it turned out, the use of THF as a solvent yielded higher HCCP deposits than when toluene was used, probably due to the higher polarity of the former. Drying the substrate under vacuum before immersion into the HCCP solution seemed to reduce the hydrolysis of P–Cl functions of the phosphazene, thus confirming the role of surface *physisorbed* water, although no clear trend depending on the dehydration temperature could be found; conversely, different drying procedures after the reaction, performed through temperature variation and duration and vacuum conditions did not appear to lead to significant differences. When functionalization was carried out at higher temperatures (e.g., at THF reflux), exceptionally low values of the Cl/P atomic percentage ratio were found from XPS (between 0.1 and 0.4), and the N/P ratio dropped to values ranging from 0.4 to 0.9, with such effects being stronger as reaction time increased. Analyses of the N_{1s} XPS peak shape and comparison with literature data [180–182] supported the hypothesis of the occurrence of degradative processes, probably generated by the hydrolysis of P–Cl moieties to P–OH and subsequent phosphazene–phosphazane isomerization, as illustrated in Scheme 31. Therefore, this procedure appeared to allow the deposition of HCCP-derived films with relatively strong covalent (P–O–Si) or hydrogen (P–OH⋯HO–Si or P=O⋯HO–Si) bonds that made them at least partially resistant to washing in THF, but these films would be hydrolytically unstable and undergo degradative phenomena, especially under high-temperature conditions.

As an extension of this research, we also succeeded in the surface functionalization of crystalline (100) silicon wafers through glow discharge–induced sublimation (GDS) processes [174].

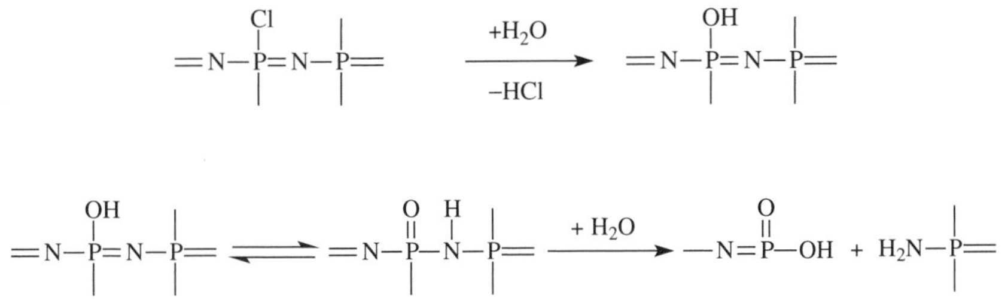

SCHEME 31 Hydrolysis reaction of chlorophosphazene units, followed by phosphazene–phosphazane processes and degradation.

In the second case the surface functionalization of silica gels beads and (100) crystalline silicon wafers with HCCP was obtained by reacting the cyclophosphazene first with 3 equivalents of four different nucleophilic substituents: 4-cyanophenol (4CNP), poly(ethylene glycol)monomethyl ether, average molecular weight 750 Da, (PEG-750-ME), tetrafluoropropanol (TFP), and 4-hydroxyazobenzene (AzB), to prepare partially substituted phosphazene cycles according to Scheme 32, where the average structures of the compounds have been reported.

The reactive chlorines still present in the partially substituted phosphazene derivatives were reacted successively with γ-aminopropyltriethoxysilane (APTES) according to the literature [183,184], to produce cyclophosphazenes having trialkoxy silane groups, according to Scheme 33. The cyclophosphazene derivatives were eventually exploited for surface functionalization reactions of silica gel beads by reaction with activated silica particles in anhydrous toluene solutions (Scheme 34) or were intermolecularly condensed, in both the presence or the absence of tetraethoxysilane (TEOS), to form monoliths or thin films, depending on the experimental conditions selected. An approximate structure

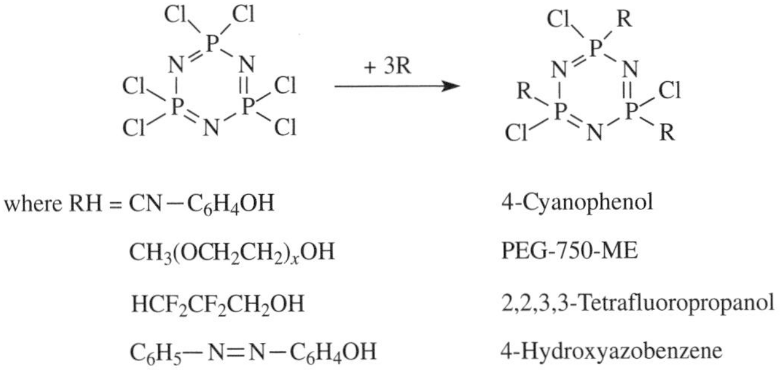

where RH = $CN-C_6H_4OH$ 4-Cyanophenol

$CH_3(OCH_2CH_2)_xOH$ PEG-750-ME

$HCF_2CF_2CH_2OH$ 2,2,3,3-Tetrafluoropropanol

$C_6H_5-N=N-C_6H_4OH$ 4-Hydroxyazobenzene

SCHEME 32 Synthesis of partially (ca. 50%) substituted cyclophosphazenes with 4-cyanophenol, poly(ethylene glycol monomethyl ether) (average molecular weight 750), tetrafluoropropanol, and 4-hydroxyazobenzene.

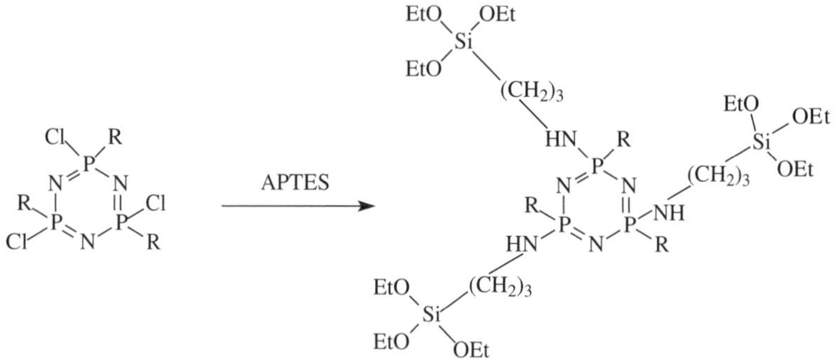

where R has the same meaning as in Scheme 32

SCHEME 33 Saturation reaction of the cyclophosphazenes reported in Scheme 32 with γ-aminopropyltriethoxysilane.

of the condensed phosphazenes during sol–gel experiments is illustrated in Scheme 35.

The materials prepared have been characterized according to standard spectroscopic techniques (infrared and ultraviolet, ^{1}H, ^{13}C, ^{31}P, and ^{29}Si

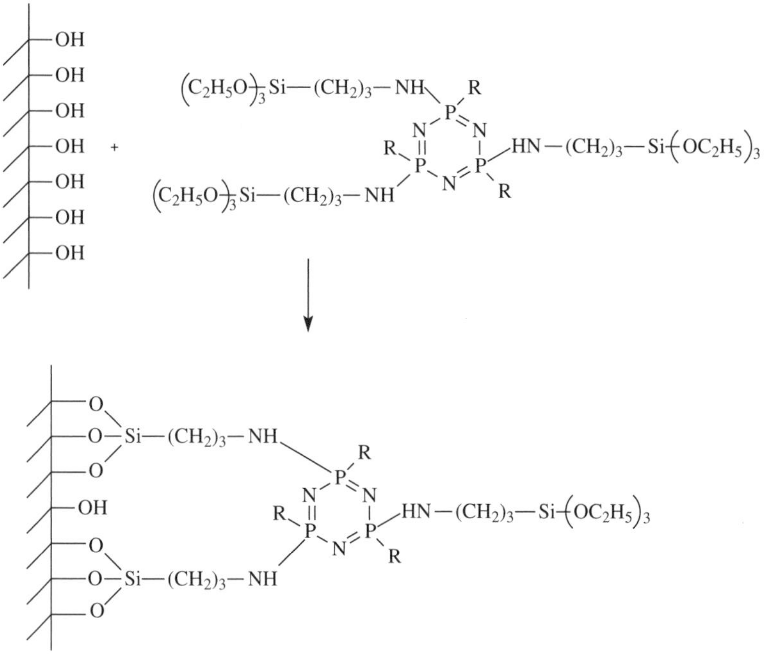

SCHEME 34 Surface functionalization of silica gel beads with cyclophosphazenes containing APTES groups.

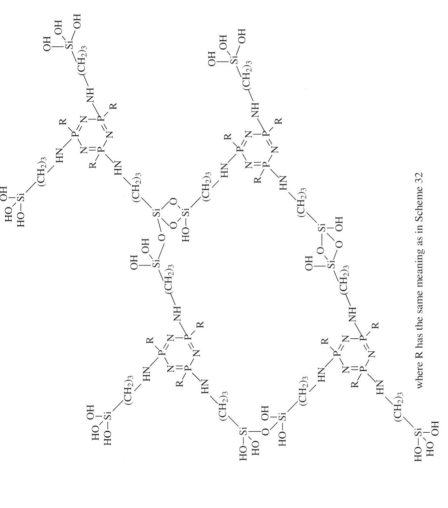

SCHEME 35 Approximated structure of monoliths and thin films prepared by condensation reaction of APTES-containing cyclophosphazenes through the sol–gel technique.

where R has the same meaning as in Scheme 32

215

solution nuclear magnetic resonance (NMR), and ^{29}Si solid-state NMR), and by thermal analysis (differential scanning calorimetry and DMTA) to prove the onset of photochromic features in thin films containing cyclophosphazenes functionalized with AzB and tunable mechanical properties in monoliths containing AzB, TFP, and PEG-750-ME functionalized trimers in the dependence of the phosphazene substituent group [173].

Surface Modification of Carbon Black with Phosphazenes An alternative method to modify the surface of organic substrates is based on the grafting of PDCP onto carbon black, followed by the substitution of the residual chlorines present in this polymer with alkoxy, aryloxy, or amino groups [185]. The overall reaction sequence is shown in Scheme 36. According to which the original phenol-like groups present on the surface of carbon black were treated with sodium hydroxide in methanol [186] and the resulting phenolate moieties were reacted with PDCP already partially substituted for by phenoxide groups to reduce its intrinsic high reactivity. The chlorine atoms of the grafted PDCP were eventually saturated with phenoxide, ethoxide, or aniline substituents, to graft the corresponding polyphosphazenes on the surface of carbon black. Thermal analysis tests on the grafted materials obtained evidenced an improvement in the thermal stability of PDCP with respect to that of the pristine material.

Polyphosphazenes as Metal Adhesives Poly(organophosphazene)s substituted for by alkyl phenoxy groups containing tertiary hydrogen atoms groups were exploited in adhesion problems in combination with metals (e.g., aluminum and copper). Thus, poly[bis(4-*sec*-butylphenoxy)phosphazene], PsBPP, was reacted with maleic anhydride to graft on the substrate succinic anhydride residues [187]. The resulting polymer (Scheme 37) was deposited on the surface of aluminum or copper plates or bars and tested for adhesion. The results indicated that the adhesion strength in the functionalized polyphosphazenes is

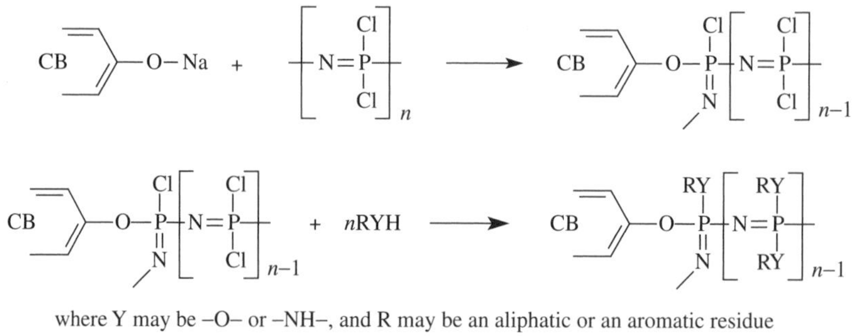

where Y may be −O− or −NH−, and R may be an aliphatic or an aromatic residue

SCHEME 36 Surface functionalization of carbon black with various phosphazene derivatives.

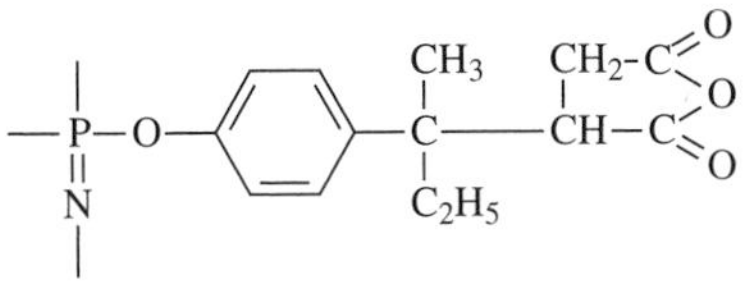

SCHEME 37   General structure of a (4-*sec*-butylphenoxy)phosphazene copolymer functionalized with succinic anhydride groups for metal adhesion processes.

related strictly to the amount of grafted anhydride groups and increased considerably for high quantities of this group [188].

CONCLUSIONS

In this chapter we highlighted the work that has been carried out over time concerning surface properties of poly(organophosphazene)s and their modification. Polyphosphazenes appear to be very versatile substrates for surface functionalization reactions considering either the intrinsic modification of surface properties of phosphazene films, coatings, and fibers, or the surface modifications that have been induced by phosphazene polymers and copolymers once deposited on the surface of organic and/or inorganic materials. Possible practical utilization of the modified materials described in this chapter are hydrophilic, hydrophobic, and/or superhydrophobic materials, biomedical substrates, surface metal–containing polymers suitable for catalysis, silicon-containing hybrid substrates, adhesives, photochromic products, and others.

REFERENCES

1. Garbassi, F., Morra, M., Occhiello, E. *Polymer Surfaces: From Physics to Technology*. Wiley, Chichester, UK, 1998.
2. Jones, R.A.L., Richards, R.W. *Polymers at Surfaces and Interfaces*. Cambridge University Press, Cambridge, UK, 1999.
3. Penn, L.S., Wang, H. *Polym. Adv. Technol.*, 1994, 5:809.
4. Totten, G.E., Liang, H., eds. *Surface Modification and Mechanisms: Friction, Stress, and Reaction Engineering*. Marcel Dekker, New York, 2004.
5. Smith, M.B., March, J. *March's Advanced Organic Chemistry: Reactions, Mechanisms, and Structure*. Wiley, New York, 2001.
6. Lin, J., Siddiqui, J.A., Ottenbrite, R.M. *Polym. Adv. Technol.*, 2001, 12:285.
7. Anderson, R.C., Muller, R.S., Tobias, C.W. *J. Electrochem. Soc.*, 1993, 140:1393.
8. Neergaard Waltenburg, H., Yates, J.T. *Chem. Rev.*, 1995, 95:1589.
9. Buriak, J.M. *Chem. Rev.*, 2002, 105:1271.
10. Taga, Y. *J. Non-cryst. Solids*, 1997, 218:335.

11. Sharma, A., Jain, H., Miller, A.C. *Surf. Interface Anal.*, 2001, 31:369.

12. Chung, M.T.C. *Functionalization of Polyolefins*. Academic Press, London, 2002.

13. Carey, D.H., Ferguson, G.S. *Macromolecules*, 1994, 27:7254.

14. Smith, L., Doyle, C., Gregonis, D.E., Andrade, J.D. *J. Appl. Polym. Sci.*, 1992, 27:1269.

15. Idage, S.B., Badrinarayanan, S., *Langmuir*, 1998, 14:2780.

16. Clément, F., Held, B., Soulem, N., Guimon, C. *Eur. Phys. J. App. Phys.*, 2002, 18:135.

17. Garbassi, F., Occhiello, E., Polato, F., Brown, A. *J. Mater. Sci.*, 1987, 22:1450.

18. Sutherland, I., Brewis, D.M., Health, R.J., Sheng, E. *Surf. Interface Anal.*, 1991, 17:507.

19. Minto, F., Gleria, M., Bortolus, P., Daolio, S., Facchin, B., Pagura, C., Bolognesi, A. *Eur. Polym. J.*, 1989, 25:49.

20. Chae, K.H., Jang, H.J. *J. Polym. Sci. A*, 2002, 40:1200.

21. Bradley, R. *Radiation Technology Handbook*. Marcel Dekker, New York, 1984.

22. Onyiriuka, E.C., Hersh, L.S., Hertl, W. *Appl. Spectrosc.*, 1990, 44:808.

23. Strobel, M., Lyons, C.S., Mittal, K.L., eds. *Plasma Surface Modification of Polymers: Relevance to Adhesion*. VSP, Utrecht, The Netherlands, 1994.

24. D'Agostino, R., Favia, P., Oehr, C., Wertheimer, M.R., eds. *Plasma Processes and Polymers*. Wiley-VCH, Weinheim, Germany, 2005.

25. Kaplan, M.L., Kelleher, P.G. *Science*, 1970, 169:1206.

26. Renò, F., Sabbatini, M., Cannas, M. *J. Mater. Sci. Mater. Med.*, 2003, 14:241.

27. Stroud, C., Branch, M.C. *Combust. Sci. Technol.*, 2007, 179:2091.

28. Caro, J.C., Lappan, U., Lunkwitz, K. *Surf. Coat. Technol.*, 1999, 116–119:792.

29. Fischer, D., Eysel, H.H. *J. Appl. Polym. Sci.*, 1994, 52:545.

30. Sacristán, J., Reinecke, H., Mijangos, C., Spells, S., Yarwood, J. *Macromol. Chem. Phys.*, 2002, 203:678.

31. Bhole, Y.S., Karadkar, P.B., Kharul, U.K. *Eur. Polym. J.*, 2007, 43:1450.

32. Higuchi, A., Koga, H., Nakagawa, T. *J. Appl. Polym. Sci.*, 1992, 46:449.

33. Xu, F.J., Zhao, J.P., Kang, E.T., Neoh, K.G. *Ind. Eng. Chem. Res.*, 2007, 46:4866.

34. Tsubokawa, N., Abe, N., Seida, Y., Fujiki, K. *Chem. Lett.*, 2000, 29:900.

35. Hu, S., Ren, X., Bachman, M., Sims, C.E., Li, G.P., Allbritton, N. *Anal. Chem.*, 2002, 74:4117.

36. Kato, K., Uchida, E., Kang, E.T., Uyama, Y., Ikada, Y. *Prog. Polym. Sci.*, 2003, 28:209.

37. Spange, S., Meyer, T., Voight, I., Eschner, M., Estel, K., Pleul, D., Simon, F. *Adv. Polym. Sci.*, 2004, 165:43.

38. Chauhan, G.S., Jaswala, S.C., Verma, M. *Carbohydr. Polym.*, 2006, 66:435.

39. Gleria, M., De Jaeger, R. *Top. Curr. Chem.*, 2005, 250:165.

40. Singler, R.E., Bierberich, M.J. In Shubkin, R.L., ed., *Synthetic Lubricants and High-Performance Functional Fluids*. Marcel Dekker, New York, 1993, Chap. 10, p. 215.

41. Singler, R.E., Gomba, F.J. In Rudnick, L.R., Shubkin, R.L., eds., *Synthetic Lubricants and High-Performance Functional Fluids*. Marcel Dekker, New York, 1999, Chap. 13, Vol. 77, p. 297.

42. Tonei, D.M., Bertani, R., De Jaeger, R., Gleria, M. In Gleria, M., De Jaeger, R., eds., *Phosphazenes: A Worldwide Insight*. Nova Science, Hauppauge, NY, 2004, Chap. 28, p. 669.

43. Allcock, H.R. *Phosphorus–Nitrogen Compounds: Cyclic, Linear, and High Polymeric Systems*. Academic Press, New York, 1972.

44. Allcock, H.R. *Chemistry and Applications of Polyphosphazenes*. Wiley, Hoboken, NJ, 2003.

45. Gleria, M., De Jaeger, R., eds. *Phosphazenes: A Worldwide Insight*. Nova Science, Hauppauge, NY, 2004.

46. De Jaeger, R., Gleria, M., eds. *Inorganic Polymers*. Nova Science, Hauppauge, NY, 2007.

47. Allcock, H.R., Kugel, R.L. *J. Am. Chem. Soc.*, 1965, 87:4216.

48. Allcock, H.R., Kugel, R.L. *Inorg. Chem.*, 1966, 5:1716.

49. Allcock, H.R., Kugel, R.L., Valan, K.J. *Inorg. Chem.*, 1966, 5:1709.

50. Allcock, H.R., Kugel, R.L., U.S. patent, 3,370,020, 1968; *Chem. Abstr.*, 68:69555f, 1968; assigned to American Cyanamid Co.

51. Allcock, H.R. *Appl. Organomet. Chem.*, 1998, 12:659.

52. Gleria, M., Bertani, R., De Jaeger, R. *J. Inorg. Organomet. Polym.*, 2004, 14:1.

53. Gleria, M., Bertani, R., De Jaeger, R., Lora, S. *J. Fluor. Chem.*, 2004, 125:329.

54. Allcock, H.R., Steely, L.B., Singh, A. *Polym. Int.*, 2006, 55:621.

55. Singh, A., Steely, L., Allcock, H.R. *Langmuir*, 2005, 21:11604.

56. Gleria, M., Bortolus, P, Minto, F., Flamigni, L. In Laine, R. M., ed., *Inorganic and Organometallic Polymers with Special Properties*. Nato ASI Series E: Applied Sciences, 1992, Vol. 206, p. 375.

57. Bortolus, P., Gleria, M. *J. Inorg. Organomet. Polym.*, 1994, 4:1.

58. De Jaeger, R., Gleria, M. *Prog. Polym. Sci.*, 1998, 23:179.

59. Gleria, M., Bortolus, P., Minto, F. In Gleria, M., De Jaeger, R., eds., *Phosphazenes: A Worldwide Insight*. Nova Science, Hauppauge, NY, 2004, Chap. 17, p. 415.

60. O'Brien, J.P., Ferrar, W.T., Allcock, H.R. *Macromolecules*, 1979, 12:108.

61. Hiraoka, H., Lee, W.Y., Welsh, L.W., Allen, R.W. *Macromolecules*, 1979, 12:753.

62. Reichert, W.M., Filisko, F.E., Baremberg, S.A. *J. Colloid Interface Sci.*, 1984, 101:565.

63. Gleria, M., Minto, F., Galeazzi, A., Scoponi, M. Italian patent, 1,302,510, 2000; assigned to Consiglio Nazionale delle Ricerche.

64. Groenewold, G.S., Cowan, R.L., Ingram, J.C., Appelhans, A.D., Delmore, J.E., Olson, J.E. *Surf. Interface Anal.*, 1996, 24:794.

65. Minto, F., Scoponi, M., Gleria, M., Pradella, F., Bortolus, P. *Eur. Polym. J.*, 1994, 30:375.

66. Allcock, H.R., Nelson, C.J., Coggio, W.D. *Chem. Mater.*, 1994, 6:516.

67. Gleria, M., Minto, F., Bortolus, P., Facchin, G., Bertani, R., Scoponi, M., Pradella, F. *ACS Polym. Prepr.*, 1993, 34(1):270.

68. Bortolus, P., Gleria, M. *J. Inorg. Organomet. Polym.*, 1994, 4:205.

69. Gleria, M., Minto, F., Bortolus, P., Facchin, G., Bertani, R. In Wisian-Neilson, P., Allcock, H.R., Wynne, K.J., eds., *Inorganic and Organometallic Polymers II. Advanced Materials and Intermediates.* ACS Symposium Series. ACS, Washington, DC, 1994, Vol. 572, Chap. 22, p. 279.

70. Minto, F., Gleria, M., Bortolus, P., Scoponi, M., Pradella, F., Fambri, L. In Allen, N.S., Edge, M., Bellobono, I.R., Selli, E., eds., *Current Trends in Polymer Photochemistry.* Prentice Hall, Hemel Hempstead, UK, 1995, Chap. 11, p. 165.

71. Gleria, M., De Jaeger, R. *J. Inorg. Organomet. Polym.*, 2001, 11:1.

72. Allcock, H.R., Kwon, S. *Macromolecules*, 1986, 19:1502.

73. Matsuki, T., Saiki, N., Emi, S. Eur. Pat. Appl. EP, 425,268, 1990; *Chem. Abstr.*, 115:227837v, 1991; assigned to Teijin Ltd.

74. Matsuki, T., Saiki, N., Emi, S. U.S. patent, 5,268,287, 1993; *Chem. Abstr.*, 115:227837, 1991; assigned to Teijin Ltd.

75. Matsuki, T., Saiki, N., Emi, S. U.S. patent, 5,380,658, 1995; *Chem. Abstr.*, 115:227837, 1991; assigned to Teijin Ltd.

76. Allcock, H.R., Fitzpatrick, R.J., Salvati, L. *Chem. Mater.*, 1991, 3:1120.

77. Neenan, T.X., Allcock, H.R. *Biomaterials*, 1982, 3:78.

78. Allcock, H.R., Neenan, T.X. U.S. patent, 4,451,647, 1984; *Chem. Abstr.*, 101:97663h, 1984; assigned to Research Corp.

79. Allcock, H.R., Fitzpatrick, R.J., Salvati, L. *Chem. Mater.*, 1992, 4:769.

80. Allcock, H.R., Morrissey, C.T., Way, W.K., Winograd, N. *Chem. Mater.*, 1996, 8:2730.

81. Ohkawa, K., Matsuki, T., Saiki, N. U.S. patent, 4,959,442, 1990; *Chem. Abstr.*, 114:83683y, 1991; assigned to Teijin Ltd.

82. Lora, S., Palma, G., Bozio, R., Caliceti, P., Pezzin, G. *Biomaterials*, 1993, 14:430.

83. Kolich, C.H., Klobucar, W.D., Books, J.T. U.S. patent, 4,945,139, 1990; *Chem. Abstr.*, 113:154152q, 1990; assigned to Ethyl Corp.

84. Allcock, H.R., Fitzpatrick, R.J., Salvati, L. *Chem. Mater.*, 1991, 3:450.

85. Allcock, H.R., Rutt, J.S., Fitzpatrick, R.J. *Chem. Mater.*, 1991, 3:442.

86. Allcock, H.R., Desorcie, J.L., Riding, G.H. *Polyhedron*, 1987, 6:119.

87. Chandrasekhar, V., Justin Thomas, K.R. *Appl. Organomet. Chem.*, 1993, 7:1.

88. Chandrasekhar, V., Justin Thomas, K.R. In Clarke, M.J., Goodenough, J.B., Ibers, J.A., Jørgensen, C.K., Mingos, D.M.P., Neilands, J.B., Palmer, G.A., Reinen, D., Sadler, P.J., Weiss, R., Williams, R.P.J., eds., *Structure and Bonding: Structures and Biological Effects.* Springer-Verlag, Berlin, 1993, Vol. 81, p. 41.

89. Chandrasekhar, V., Krishnan, V. In Gleria, M., De Jaeger, R., eds., *Phosphazenes: A Worldwide Insight.* Nova Science, Hauppauge, NY, 2004, Chap. 34, p. 827.

90. Pertici, P., Vitulli, G., Gleria, M., Facchin, G., Milani, R., Bertani, R. *Macromol. Symp.*, 2006, 235:98.

91. Gleria, M., Vitulli, G., Pertici, P., Facchin, G., Bertani, R. In De Jaeger, R., Gleria, M., eds., *Inorganic Polymers.* Nova Science, Hauppauge, NY, 2007.

92. Allcock, H.R., Silverberg, E.N., Nelson, C.J., Coggio, W.D. *Chem. Mater.*, 1993, 5:1307.

93. Klabunde, K.J. *Free Atoms, Clusters and Nanoscale Particles*. Academic Press, New York, 1994.

94. Vitulli, G., Pitzalis, E., Aronica, L., Pertici, P., Bertozzi, S., Caporusso, A.M., Salvadori, P., Coluccia, S., Martra, G. In Armelao, L., Barreca, D., Daolio, S., Tondello, E., Vigato, P.A., eds., *Syntheses and Methodologies in Inorganic Chemistry*. Bressanone, Italy, 2000, p. 43.

95. Spitaleri, A., Pertici, P., Scalera, N., Vitulli, G., Hoang, M., Turney, T.W., Gleria, M. *Inorg. Chim. Acta*, 2003, 352:61.

96. Barazzone, L., *Master's thesis*, University of Pisa, Italy, 2004.

97. Pertici, P., Vitulli, G., Salvatori, P., Pitzalis, E., Gleria, M. In Gleria, M., De Jaeger, R., eds., *Phosphazenes: A Worldwide Insight*. Nova Science, Hauppauge, NY, 2004, Chap. 26, p. 621.

98. Allcock, H.R., Coggio, W.D. *Macromolecules*, 1990, 23:1626.

99. Allcock, H.R., Smith, D.E. *Chem. Mater.*, 1995, 7:1469.

100. Exarhos, G.J. *NTIS Spec. Publ.*, 1990, 801:324.

101. Fewell, L.L. *ACS Polym. Prepr.*, 1991, 32(1), 134.

102. Allcock, H.R., Steely, L.B., Kim, S.H., Kim, J.H., Kang, B.K. *Langmuir*, 2007, 23:8103.

103. Adams, B.E., Hansel, R.D., Quinn, E.J. U.S. patent, 4,145,479, 1979; *Chem. Abstr.*, 90:206024c, 1979; assigned to Armstrong Cork Co.

104. Allcock, H.R., Fitzpatrick, R.J., Visscher, K., Salvati, L. *Chem. Mater.*, 1992, 4:775.

105. Blonsky, P.M., Shriver, D.F., Austin, P.E., Allcock, H.R. *J. Am. Chem. Soc.*, 1984, 106:6854.

106. Allcock, H.R., Austin, P.E., Neenan, T.X., Sisko, J.T., Blonsky, P.M., Shriver, D.F. *Macromolecules*, 1986, 19:1508.

107. Allcock, H.R., Cook, W.J., Mark, D.P. *Inorg. Chem.*, 1972, 11:2584.

108. Bennett, J.L., Dembek, A.A., Allcock, H.R., Heyen, B.J., Shriver, D.F. *Chem. Mater.*, 1989, 1:14.

109. Allcock, H.R., Kwon, S., Riding, G.H., Fitzpatrick, R.J., Bennett, J.L. *Biomaterials*, 1988, 9:509.

110. Allcock, H.R., Gebura, M., Kwon, S., Neenan, T.X. *Biomaterials*, 1988, 9:500.

111. Rose, S.H. U.S. patent, 3,515,688, 1970; *Chem. Abstr.*, 73:36299a, 1970; assigned to Horizon Inc.

112. Leong, K.W. U.S. patent 4,311,736, 1982; *Chem. Abstr.*, 96:105952x, 1982; assigned to Kendall Co.

113. Leong, K.W. U.S. patent 4,341,844, 1982, assigned to Kendall Co.

114. Joung, J. U.S. patent 4,318,947, 1982; *Chem. Abstr.*, 95:205234n, 1981; assigned to Kendall Co.

115. Ren, X., Liu, P.Z., Malik, A., Lee, M.L. *J. Microcolumn Sep.*, 1997, 8:535.

116. Grunze, M., Schrenk, M.G. Ger. Offen. DE, 196 13,048, 1996; *Chem. Abstr.*, 125:285036a, 1996; assigned to M. Grunze.

117. Welle, A., Grunze, M. Proceedings of the 19th Annual Meeting of the Adhesives Society, 1996, p. 432; *Chem. Abstr.*, 127:140507 1997.

118. Grunze, M., Welle, A., Tur, D.R. *Annual Technical Conference, Society of Plastics Engineers*, 1998, 56:2713; *Chem. Abstr.*, 130, 86077t, 1999.

119. Welle, A., Grunze, M., Tur, D. *Materials Research Society Symposium Proceedings*, 1998, 489:139; *Chem. Abstr.*, 130:114991g, 1999.

120. Welle, A., Grunze, M., Tur, D. *J. Colloid Interface Sci.*, 1998, 197:263.

121. Grunze, M., Welle, A. PCT Int. Appl. WO, 99 16,477, 1999; *Chem. Abstr.*, 130:272061h, 1999; assigned to Universität Heidelberg.

122. Grunze, M., Welle, A. Ger. Offen. DE, 197 43,373, 1999; *Chem. Abstr.*, 130:272061h, 1999; assigned to Universität Heidelberg.

123. Welle, A., Grunze, M., Tur, D. *J. Appl. Med. Polym.*, 2000, 4:6.

124. Grunze, M. PCT Int. Appl. WO, 01 70,296, 2001; *Chem. Abstr.*, 135:262311, 2001; assigned to Polyzenix GmbH.

125. Grunze, M., Gries, C. PTC Int. Appl. WO, 01 80,919, 2001; *Chem. Abstr.*, 135:348893, 2001; assigned to Universität Heidelberg.

126. Grunze, M. CA, 2,402,949, 2001; *Chem. Abstr.*, 135:262311, 2001; assigned to Polyzenix GmbH.

127. Grunze, M., Gries, C. Ger. Offen. DE, 100 19,982, 2001; *Chem. Abstr.*, 135:322740, 2001; assigned to Universität Heidelberg.

128. Grunze, M., Welle, A. U.S. Pat. Appl. Publ., 2002 0,054,851, 2002; *Chem. Abstr.*, 130:272061h, 1999; assigned to Universität Heilderberg.

129. Grunze, M., Tur, D.R., Pertsin, A. Ger. Offen. DE, 101 13,971, 2002; *Chem. Abstr.*, 137:264178, 2002; assigned to Polyzenix GmbH.

130. Grunze, M., Welle, A. U.S. Pat. Appl. Publ., 1,019,102, 2002; *Chem. Abstr.*, 130:272061h, 1999; assigned to Universität Heidelberg.

131. Grunze, M. PTC Int. Appl. WO, 02 64,666, 2002, assigned to Polyzenix GmbH.

132. Grunze, M. Ger. Offen. DE, 101 00,961, 2002, assigned to Polyzenix GmbH.

133. Grunze, M. CA, 2,434,596, 2002, assigned to Polyzenix GmbH.

134. Grunze, M., Tur, D., Pertsin, A. PCT Int. Appl. WO, 2002 77,073, 2002; *Chem. Abstr.*, 137:264178, 2002; assigned to Polyzenix GmbH.

135. Grunze, M., Gries, C. CA, 2,408,997, 2002; *Chem. Abstr.*, 135:322740, 2001; assigned to Universität Heidelberg.

136. Denk, R., Grunze, M., Schuessler, A. CA, 2,457,018, 2003; *Chem. Abstr.*, 138:193324, 2003; assigned to Polyzenix GmbH, Euroflex Schuessler GmbH.

137. Grunze, M. U.S. Pat. Appl. Publ., 0,099,683, 2003; *Chem. Abstr.*, 135:262311, 2001; assigned to Polyzenix GmbH.

138. Grunze, M. Eur. Pat. Appl. EP, 1,312,635, 2003, assigned to Polyzenix GmbH.

139. Schuessler, A., Grunze, M., Denk, R. PCT Int. Appl. WO, 2003 15,719, 2003; *Chem. Abstr.*, 138:193324, 2003; assigned to Polyzenix GmbH, Euroflex Schuessler GmbH.

140. Schuessler, A., Grunze, M., Denk, R. Ger. Offen. DE, 102 02,467, 2003; *Chem. Abstr.*, 138:193324, 2003; assigned to Polyzenix GmbH, Euroflex Schuessler GmbH.

141. Schuessler, A., Grunze, M., Denk, R. Eur. Pat. Appl. EP, 1,432,380, 2003; *Chem. Abstr.*, 138:193324, 2003; assigned to Polyzenix GmbH, CeloNova Biosciences GmbH.

142. Grunze, M. U.S. Pat. Appl. Publ., 1,265,653, 2004; *Chem. Abstr.* 135:262311, 2001; assigned to Polyzenix GmbH.

143. Grunze, M. U.S. Pat. Appl. Publ., 0,096,969, 2004.

144. Grunze, M. Eur. Pat. Appl. EP, 1,349,582, 2004, assigned to Polyzenix GmbH.

145. Grunze, M. Eur. Pat. Appl. EP, 1,488,817, 2004, assigned to Polyzenix GmbH.

146. Grunze, M., Gries, C. U.S. Pat. Appl. Publ., 0,014,936, 2004; *Chem. Abstr.*, 135:322740, 2001; assigned to Universität Heidelberg.

147. Schuessler, A., Grunze, M., Denk, R. U.S. Pat. Appl. Publ., 0,253,467, 2004; *Chem. Abstr.*, 138:193324, 2003; assigned to Polyzenix GmbH, Euroflex Schuessler GmbH.

148. Grunze, M. Ger. Offen. DE, 601 03,620 T2, 2005; *Chem. Abstr.*, 135:262311, 2001; assigned to Polyzenix GmbH.

149. Grunze, M., Gries, C. U.S. patent, 7,265,199, 2007; assigned to Celonova BioSciences GmbH.

150. Grunze, M., Gries, C. Eur. Pat. Appl. EP, 1,274,471, 2007; *Chem. Abstr.*, 135:322740, 2001; assigned to Polyzenix GmbH.

151. Schuessler, A., Grunze, M., Denk, R. U.S. Pat. Appl. Publ., 0,184,277, 2007; *Chem. Abstr.*, 138:193324, 2003; assigned to CeloNova Biosciences GmbH.

152. Tur, D.R., Korshak, V.V., Vinogradova, S.V., Dobrova, N.B., Novikova, S.B., Il'ina, M.B., Sidorenko, E.S. *Acta Polym.*, 1985, 36:627; *Chem. Abstr.*, 104:39679u, 1986.

153. Nagel, S., Boxberger, M. PCT Int. Appl. WO, 02 13,882, 2002; *Chem. Abstr.*, 136:172834, 2002; assigned to B. Braun Melsungen Ag.

154. Nagel, S., Boxberger, M. Eur. Pat. Appl. EP, 1,179,353, 2002; *Chem. Abstr.*, 136:172834, 2003; assigned to B. Braun Melsungen Ag.

155. Boxberger, M., Nagel, S. CA, 2,424,359, 2003; *Chem. Abstr.*, 136:172834, 2002; assigned to Polyzenix GmbH.

156. Nagel, S., Boxberger, M. U.S. Pat. Appl. Publ., 0,157,142, 2003; *Chem. Abstr.*, 136:172834, 2002.

157. Nitschke, M., Hollander, A., Mehdorn, F., Behnisch, J., Meichsner, J. *J. Appl. Polym. Sci.*, 1996, 59:119.

158. Hegemann, D., Brunner, H., Oehr, C. *Nucl. Instrum. Methods Phys. Res. B*, 2003, 208:281.

159. Guruvenket, S., Mohan Rao, G., Komath, M., Raichur, A.M. *Appl. Surf. Sci.*, 2004, 236:278.

160. Tajima, S., Komvopoulos, K. *J. Phys. Chem. B*, 2005, 109:17623.

161. France, R.M., Short, R.D. *Langmuir*, 1998, 14:4827.

162. Pemberton, L., De Jaeger, R. *Chem. Mater.*, 1996, 8:1391.

163. Pemberton, L., De Jaeger, R., Gengembre, L. *Phosphorus Sulfur Silicon Relat. Elem.*, 1996, 111:667.

164. Dez, I., De Jaeger, R. *Macromolecules*, 1997, 30:8262.

165. Pemberton, L., De Jaeger, R., Gengembre, L. *J. Appl. Polym. Sci.*, 1998, 69:1965.

166. Pemberton, L., De Jaeger, R., Gengembre, L. *Polymer*, 1998, 39:1299.

167. Gleria, M., Minto, F., Fontana, G., Bertani, R., Facchin, G. *Macromolecules*, 1995, 28:4399.

168. Fontana, G., Minto, F., Gleria, M., Facchin, G., Bertani, R., Favero, G. *Eur. Polym. J.*, 1996, 32:1273.

169. Minto, F., Fontana, G., Bertani, R., Facchin, G., Gleria, M. *J. Inorg. Organomet. Polym.*, 1996, 6:367.

170. Milani, R., Gleria, M., Gross, S., De Jaeger, R., Mazzah, A., Gengembre, L., Frere, M., Jama, C., *J. Inorg. Organomet. Polym. Mater.*, 2008, 18:344.

171. Sassi, A., Milani, R., Venzo, A., Gleria, M. *Design. Monomers Polym.*, 2006, 9:627.

172. Silvestrelli, P., Gleria, M., Milani, R., Boscolo-Boscoletto, A. *J. Inorg. Organomet. Polym. Mater.*, 2006, 16:327.

173. Milani, R., Sassi, A., Venzo, A., Bertani, R., Gleria, M. *Design. Monomers Polym.*, 2007, 19:555.

174. Sassi, A., Maggioni, G., Della Mea, G., Milani, R., Gleria, M. *Surf. Coat. Technol.*, 2007, 201:5829.

175. Boscolo-Boscoletto, A., Gleria, M., Milani, R., Meda, L., Bertani, R., *Surf. Interface Anal.* 2009, 41:27.

176. Gimblett, F.G.R. *Plast. Inst. London Trans. J.*, 1960, 28:65.

177. Emsley, J., Udy, P.B. *Polymer*, 1972, 13:593.

178. Djanic, G., Master's Thesis, University of Padova, Italy, 2001.

179. Milani, R., Master's Thesis, University of Padova, Italy, 2003.

180. Dwight, D.W., McGrath, J.E., Beck, A.R., Riffle, J.S. *ACS Polym. Prepr.* 1979, 20(1), 702.

181. Mochel, V.D., Cheng, T.C. *Macromolecules*, 1978, 11:176.

182. Cheng, T.C., Mochel, V.D., Adams, H.E., Longo, T.F. *Macromolecules*, 1980, 13:158.

183. Schneider, A., Kairies, S., Rose, K. *Monatsch. Chem.*, 1999, 130:89.

184. Christova, D., Ivanova, S.D., Velichkova, R.S., Tzvetkova, P., Mihailova, P., Lakov, L., Peshev, O. *Design. Monomers Polym.*, 2001, 4:329.

185. Tsubokawa, N., Tsuchida, H. *J. Macromol. Sci. Chem.*, 1992, A29:311.

186. Rivin, D. *Rubber Chem. Technol.*, 1963, 36:729.

187. Gleria, M., Minto, F., Scoponi, M., Pradella, F., Carassiti, V. *Chem. Mater.*, 1992, 4:1027.

188. Gleria, M, Minto, F, Scoponi, M, Pradella, F, Carassiti, V Italian patent, 1,258,815, 1996, assigned to Consiglio Nazionale delle Ricerche.

PART IV
Drug Delivery Systems

13 Amphiphilic Ionizable Polyphosphazenes for the Preparation of pH-Responsive Liposomes

DAVID GHATTAS

Faculty of Pharmacy, University of Montreal, Montreal, Quebec, Canada

JEAN-CHRISTOPHE LEROUX

Faculty of Pharmacy, University of Montreal, Montreal, Quebec, Canada; Institute of Pharmaceutical Sciences, ETH Zürich, Zürich, Switzerland

INTRODUCTION

A key challenge in the field of drug delivery has been to improve targeting of the active agent to maximize efficacy and reduce toxicity. A promising approach is to provoke site-specific drug release from a vector in response to stimuli that can either be applied externally or be physiologically produced. Several means can thus be exploited for this purpose, such as ultrasound, enzymatic cleavage, temperature, and pH. The latter has peaked the interest of researchers, as variations in acidity are observed in certain pathologies as well as in normal intracellular activity.

Differences in pH that exist between normal vasculature (pH 7.4) and the tissue interstices of tumors, infections, and inflammations (ca. pH 6.5) pushed for the design of a delivery system targeting such extracellular compartments (Schmaljohann, 2006). Yet it has been technically challenging to construct a vector that could respond to such a narrow variation. In contrast, pH-responsive formulations have been shown to improve the cytoplasmic delivery of therapeutic agents rather than simply in the vicinity of the target cells (Drummond et al., 2000; Simoes et al., 2004; Yessine and Leroux, 2004). Upon

Polyphosphazenes for Biomedical Applications, Edited by Alexander K. Andrianov
Copyright © 2009 John Wiley & Sons, Inc.

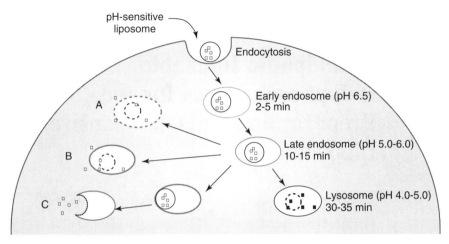

FIGURE 1 Mechanisms of intracellular targeting. Upon endocytosis, the acidification of the endosomal lumen induces one of three possible release mechanisms: destabilization and pore formation of both liposome and endosome (A), destabilization of the liposome and passive diffusion of the active agent (B), or fusion between liposomal and endosomal lamella (C). (Adapted from Simoes et al., 2004, with permission.)

receptor-mediated internalization, the pH gradient established between the endosomal–lysosomal compartments and the cytoplasm is used to induce discharge of the encapsulated material.

Of the vectors explored, pH-sensitive liposomes have received distinctive attention, as controlled release can easily be prompted by destabilization of phospholipid bilayers. Three mechanisms are proposed for the delivery of a liposome-encapsulated agent from the endosomal compartment to the cytoplasm (Karanth and Murthy, 2007) (Fig. 1). The first mechanism presumes that pH-sensitive liposomes can induce pore formation in liposomal and eventually endosomal membranes. The second involves passive diffusion of the drug through the endosomal membrane once liberated from the destabilized vector. This process is limited by the nature of the therapeutic in question. The final pathway suggests fusion between the liposome and endosome for direct release into the cytoplasm. The delivery of the drug is ultimately dependent on the composition of the liposome, the destabilization mechanism, and the interaction of the formulation with the endosomal membrane.

The first generation of pH-sensitive liposomes was prepared by a combination of unsaturated phosphatidylethanolamine (PE) and mildly acidic lipids (Connor et al., 1984), such as oleic acid and cholesterylhemisuccinate. PE alone cannot form liposomes, due to its molecular geometry, and requires the presence of the charged amphiphiles to construct bilayers at neutral pH. Following endocytosis and acidification of the endosomal lumen, the charged lipids are neutralized by protonation, resulting in transition from lamellar to hexagonal (H_{II}) phase, which leads to liposome destabilization and, eventually, fusion with the endosome membrane. Such liposomes have been found to deliver encapsulated agents to the cytosol efficiently when tested in vitro (Drummond et al., 2000). However, moderate stability in the blood and rapid elimination have hampered their efficiency when administered systemically. These problems can be resolved in part by using lipid-conjugated hydrophilic polymers inserted within the bilayer to form a steric barrier, stabilizing the liposomes (Hong et al., 2002; Ishida et al., 2006). In similar fashion, such polymers have been linked to hydrophobic anchors via acid-labile bonds which can be cleaved in the endosome from the surface of the vesicle, to allow fusion after endocytosis (Boomer et al., 2003; Guo and Szoka, 2001).

Peptides and proteins inspired from nature have also been used to improve cytoplasmic delivery of liposomal content. For instance, the pore-forming protein listeriolysin O (LLO) was co-encapsulated with an active agent into PE-based pH-sensitive liposomes (Provoda et al., 2003). Upon release of the liposomal contents, LLO created pores in the endosome membrane, releasing the therapeutic into the cytosol. Similarly, association of derivatives of the influenza virus fusion protein, hemagglutinin, with cationic liposomes has been proven to increase transfection efficiency severalfold (Kamata et al., 1994; Kichler et al., 1997). Many other pH-sensitive fusion peptides have been studied for the destabilization of liposomal and endosomal membranes (Drummond et al., 2000; Li et al., 2004), yet their use poses some challenges. Employing proteins in a drug delivery system incurs the possibility of immunogenicity. Moreover, co-encapsulation of drug and pore-forming elements within the vector may not solve in vivo stability and circulation time issues (Karanth and Murthy, 2007).

An alternative method consists of using synthetic polymers tailored to induce pH-triggered drug release. pH-responsive liposomes have been generated by anchoring polyanions into the lipidic bilayer. Such polymers undergo a coil-to-globule phase transition below a critical pH that elicits destabilization of lipid membranes (Yessine and Leroux, 2004). Table 1 summarizes some of the research employing polyanions for the preparation of pH-sensitive liposomes. It should be noted that copolymers of N-isopropylacrylamide (NIPAM) have been the most investigated so far and that pH-triggered release has been tested predominantly in vitro with fluorescent probes.

NIPAM derivatives have been proposed early for the design of stimuli-responsive liposomes. Original interest was spurred by PNIPAM's sharp lower critical solution temperature (LCST) at 32°C (Heskins and Guillet, 1968; Winnik, 1990). This transition can be tuned to temperatures relevant to physiological applications by introducing a weakly acidic monomer such as methacrylic acid (MAA), which also renders the polymer pH-responsive (Brazel and Peppas, 1996; Chen and Hoffman, 1995). Liposomes formulated with alkylated NIPAM/MAA copolymers rapidly released their contents in an acid environment (Leroux et al., 2001; Meyer et al., 1998; Roux et al., 2002b; Zignani et al., 2000). It was shown that upon collapse, the interaction area between the phospholipids and the copolymers increased (Petriat et al., 2004). The latter introduced a curvature in the bilayer plane, inducing membrane defects (Roux et al., 2003) and release of the entrapped content

TABLE 1 Summary of pH-Sensitive Copolymers Investigated for Liposomes[a]

Polymer	Terminal (T) or Random (R) Anchor	Anchoring Element	Lipids	Marker or Drug Encapsulated	Reference
PEAA	T	DMPE	EPC/DMPE	Calcein	Maeda et al., 1988
PG	T	Decylamine	EPC/PG	Calcein	Kono et al., 1997
P(NIPAM-co-Gly-co-ODA)	R	ODA	POE-SE/Chol or POCP/Chol	HPTS/DPX	Francis et al., 2001
P(NIPAM-co-MAA-co-VP-co-ODA)	R	ODA	EPC/Chol or EPC/Chol/PEG-DSPE	HPTS/DPX	Roux et al., 2002a, 2003
P(NIPAM-co-MAA-co-ODA)	R	ODA	EPC/Chol	HPTS/DPX	Leroux et al., 2001
DODAm-P(NIPAM-co-MAA)	T	DODA	DOPC/Chol	DOX	Leroux et al., 2001
			EPC/Chol	HPTS/DPX	Leroux et al., 2001
			EPC/Chol/PEG-DSPE	HPTS/DPX	Leroux et al., Roux et al., 2001
PPZ (EEE, ABA, $C_{18}(EO)_{10}$)	R	$C_{18}(EO)_{10}$	EPC/Chol	HPTS/DPX	Couffin-Hoarau and Leroux 2004

[a]PEAA, poly(2-ethylacrylic acid); DMPE, dimyristoyl-N-[[4-(maleimidomethyl)cyclohexyl]carbonyl]phosphatidyl ethanolamine; EPC, egg phosphatidylcholine; PG, decylamine-succinylated poly(glycidol); NIPAM, N- isopropylacrylamide; MAA, methacrylic acid; VP, N-vinylpyrrolidone; Gly, glycine acrylamide; ODA, octadecyl acrylate; DODA, dioctadecylamide; HPTS, trisodium 8-hydroxypyrene trisulfonate; DPX, p-xylene-bispyridinium; PEG-DSPE, N-[methoxy(polyethylene glycol) 2000] carbonyl-1,2-distearoyl-sn-glycero-3-phosphoethanolamine; PPZ (EEE, ABA, $C_{18}(EO)_{10}$) carbonyl poly(organophosphazenes); $C_{18}(EO)_{10}$, polyethylene glycol octadecyl ether-grafted poly(organophosphazenes); ethylene oxide diethyl ether-aminobutyric acid-polyethylene glycol octadecyl ether.

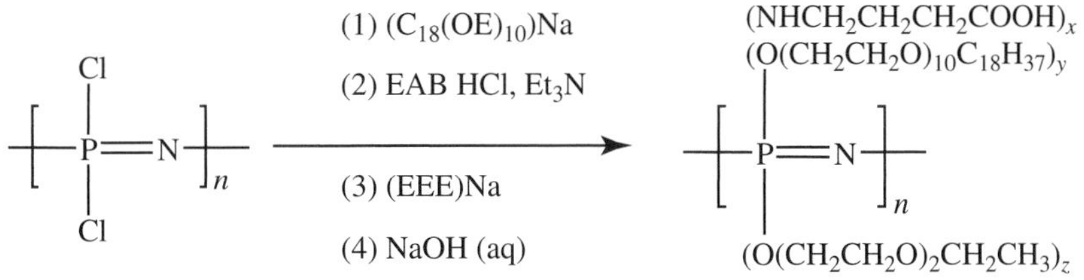

SCHEME 1   Synthesis of trisubstituted amphiphilic, pH-sensitive PPZ.

(Francis et al., 2001). Although no acute toxicity has been observed for NIPAM copolymers (Li et al., 2005; Malonne et al., 2005; Taillefer et al., 2000), their safety following long-term exposure has thus far not been demonstrated, as they are not biodegradable.

Poly(organophosphazene)s (PPZs) have previously been introduced as biodegradable alternatives to NIPAM copolymers (Couffin-Hoarau and Leroux, 2004). It was shown that the properties of PPZs can be tailored by incorporating three critical moieties into the polymer composition: polyethylene glycol octadecyl ether $[C_{18}(EO)_{10}]$, aminobutyric acid (ABA), and ethylene oxide ethyl ether (EEE) (Scheme 1). These units provide for liposome-anchoring capabilities and pH- and temperature-responsiveness, respectively. EEE was selected over other alkoxy side groups since EEE-substituted PPZ possessed an LCST close to the physiological temperature (Allcock and Dudley, 1996). ABA helps modulate the LCST with respect to environmental pH. Furthermore, it can confer biodegradability by mediating intramolecular catalysis of phosphorus–nitrogen bonds (Allcock et al., 1982; Allen et al., 2002). Liposomes prepared with the tri–substituted PPZ displayed pH-dependent release but were unstable under physiological temperature (37°C) at pH 7.4 (Couffin-Hoarau and Leroux, 2004). In the present work, we investigated whether the stability of the formulation at neutral pH could be improved by increasing the content of the ionizable ABA moiety and lowering the molecular weight of the polymer. An advantage of a lower-molecular-weight polymer would be faster excretion after administration. We also examined the degradation of the PPZ under physiological conditions and studied the impact of human serum on the pH sensitivity of the formulations.

MATERIALS AND METHODS

Materials

Cholesteryl 4,4-difluoro-5,7-dimethyl-4-bora-3*a*,4*a*-diaza-*s*-indacene-dodecanoate (Chol-BODIPY), 8-hydroxypyrene-1,3,6-trisulfonic acid (HPTS), and *p*-xylene-bispyridinium bromide (DPX) were obtained from Molecular Probes (Burlington, Ontario, Canada). Egg phosphatidylcholine (EPC) and N-[methoxy(polyethylene glycol) 2000] carbonyl-1,2-distearoyl-*sn*-glycero-3-phosphoethanolamine, sodium salt (PEG_{2000}-DSPE) were purchased from Northern Lipids

(Vancouver, British Columbia, Canada). All other chemicals were obtained from Sigma (Oakville, Ontario, Canada) and used as received, except for the following: Diethyl ether (Et$_2$O), dichloromethane (DCM), and tetrahydrofuran (THF) were run through PureSolv drying columns (Innovative Technologies, Newburyport, Massachessetts); triethylamine (TEA) was distilled over calcium hydride; phosphorus trichloride (PCl$_3$) and sulfuryl chloride (SO$_2$Cl$_2$) were distilled under argon; phosphorus pentachloride (PCl$_5$) was sublimed under vacuum; PEG octadecyl ether [Brij 76, C$_{18}$(EO)$_{10}$] and ethyl 4-aminobutyrate hydrochloride (EAB · HCl) were dried overnight under vacuum over phosphorus pentoxide.

Synthesis and Characterization

Phosphoranimine Synthesis All solid products were weighed in a glove box under an inert argon atmosphere while reactions were performed using standard Schlenk techniques. ^{1}H (400 MHz) and ^{31}P (162 MHz) nuclear magnetic resonance (NMR) spectra were recorded on a Bruker ARX 400 spectrometer (Milton, Ontario, Canada) in deuterated chloroform (CDCl$_3$). Chemical shifts for ^{31}P spectra were recorded with respect to an 85% phosphoric acid standard. Trichloro(trimethylsilyl)phosphoranimine (Cl$_3$P=NSiMe$_3$) was synthesized as reported by Wang et al. (2002). Briefly, lithium bis(trimethylsilyl)amide (10 g, 0.058 mol) was suspended in 200 mL of dry Et$_2$O and cooled to 0°C before the dropwise addition of distilled PCl$_3$ (5.06 mL, 0.058 mol). Completion of the reaction (ca. 1 h) was monitored by ^{31}P NMR from the disappearance of the PCl$_3$ peak (δ = 220 ppm) and appearance of a new species [Cl$_2$PN(SiMe$_3$)$_2$, δ = 186 ppm]. Distilled SO$_2$Cl$_2$ (4.7 mL, 0.058 mol) was then added dropwise at 0°C and allowed to react for 1 h. Complete conversion was evidenced by the appearance of a single peak at δ = −54 ppm in the ^{31}P NMR spectra. The reaction mixture was then filtered through dry celite. Et$_2$O and trimethylsilyl chloride, a reaction side product, were sequentially evaporated at 0°C from the filtrate under reduced atmosphere (200 and 50 mmHg, respectively). Crude Cl$_3$P=NSiMe$_3$, a colorless liquid, was purified by distillation (25°C, 0.1 mmHg of static vacuum) into a liquid nitrogen–cooled trap to collect the final product (10.6 g, 81% yield).

Synthesis of Poly(dichlorophosphazene) Poly(dichlorophosphazene) (PDCP) was obtained by cationic polymerization using PCl$_5$ as the initiator (Allcock et al., 1996). A concentrated solution of Cl$_3$P=NSiMe$_3$ (6.1 g, 0.027 mol) in dry DCM (5 mL) was cannulated to a solution of PCl$_5$ (0.16 g, 7.8 × 10^{-4} mol, Cl$_3$P=NSiMe$_3$/PCl$_5$ molar ratio of 35 : 1) under an inert argon atmosphere to reach a final initiator concentration of 0.035 mol/L. The polymerization reaction was carried out at room temperature and monitored by ^{31}P NMR by following the disappearance of the Cl$_3$P=NSiMe$_3$ peak and the appearance of the PDCP backbone peak (δ = −17 ppm). After 2 h, DCM was evaporated and the crude product was stored under inert conditions at −20°C.

Synthesis of Poly(organophosphazene)s   pH-sensitive poly(organophospha-zene)s (PPZs) were prepared as described earlier (Couffin-Hoarau and Leroux, 2004). Synthesized polymers are termed A_x–P_y, with x and y representing the ratios of the ABA and $C_{18}(OE)_{10}$ moieties, respectively. The following is the typical procedure as performed for the synthesis of PPZ A_7–P_6 (Table 2). Under an inert argon atmosphere, a solution of $C_{18}(EO)_{10}$ (0.72 g, 1.0 mmol) and NaH (0.026 g, 1.0 mmol) stirred overnight was added dropwise to a PDCP solution (obtained from 1.0 mmol of $Cl_3P=NSiMe_3$) dissolved in 10 mL of dry THF. After 6 h at room temperature, a solution of EAB·HCl (0.35 g, 2.0 mmol) treated with 2.8 equivalents of distilled TEA (0.8 mL, 5.7 mmol) was added and the mixture was heated 48 h at 50°C before being cooled to room temperature. Finally, an excess solution of EEE (4.8 mL, 3.5 mmol), treated overnight by NaH (0.88 g, 3.5 mmol), was added dropwise and the reaction was stirred overnight at room temperature. The progression of each substitution reactions was tracked in ^{31}P NMR by the appearance of a peak at $\delta = -8$ ppm corresponding to the substituted phosphazene. After completion of the last reaction, the final solution was filtered from excess salts, concen-trated, and dialyzed against deionized water for 48 h (molecular-weight cutoff 12,000 to 14,000). The resulting aqueous polymer solution was treated by 5 mL of 1 N NaOH for 4 h at room temperature to complete hydrolysis of EAB to ABA. The final PPZ was dialyzed against water for 24 h and lyophilized to obtain 3 g of a yellow oil (75% yield).

Physical Characterization of pH-Responsive Polymers   The degree of substitu-tion was estimated in ^1H NMR by calculating the ratios between the methyl protons of $C_{18}(OE)_{10}$ and EEE ($\delta = 0.9$ and 1.2 ppm, respectively) and a CH_2 of ABA ($\delta = 1.7$ ppm). The percentage of ABA was also confirmed by potentiometric titration using an Accumet AP61 pH meter (Fisher Scientific, Montreal, Quebec, Canada), according to the following procedure: An aqueous solution of the polymer (5 mL, 1 mg/mL) was treated with excess NaOH (3 mL, 0.01 N) to ensure dissolution of the PPZ and complete ionization of the acid functions. Titrations were performed by adding increments of 0.01 N HCl and measuring aqueous pH. During this process, both the amine and carboxylic acid of the ABA molecules were titrated and considered in calculations (Couffin-Hoarau and Leroux, 2004).

The absolute number (M_n) and weight (M_w)-average molecular weights of the polymer samples were determined by size-exclusion chromatography (SEC) using a Breeze system (Waters, Milford, Massachusetts) equipped with a Waters 2410 refractometer and PD2000 light-scattering detector (Precision Detectors, Bellingham, Massachusetts). Measurements were performed in *N,N*-dimethylformamide containing 10 mM lithium bromide at a flow rate of 1 mL/min at 40°C. Molecular-weight separation was achieved using three Waters Styragel columns (HT2, HT3, and HT4) in series and the instrument was calibrated with monodisperse polystyrene standards.

TABLE 2 Characteristics of Synthesized Poly(organophosphazene)s

PPZ	Composition [ABA:C$_{18}$(EO)$_{10}$: EEE] (mol%)		M_w	M_w/M_n	LCST[c] (°C)	ΔH_{LCST}[c] (J/g)	Percent Liposome Fixation[d]
	Theoretical[a]	Experimental[b]					
A$_7$–P$_6$	10:5:85	7:6:87	16,300	1.01	33.5±0.1	17.6±0.8	81.3±7.9[e]
A$_9$–P$_{5.5}$	15:5:80	9:5.5:85.5	19,300	1.01	35.5±0.7	10.1±0.8	97.3±6.18[e]
A$_{9.5}$–P$_{7.5}$	10:7:83	9.5:7.5:83	15,100	1.06	34.0±0.3	12.6±0.4	92.3±16.7[e]
A$_{11}$–P$_{10}$	10:10:80	11:10:79	18,300	1.03	31.7±0.6	6.3±2.5	33.8±18.6[f]
A$_{14}$–P$_{16}$	15:10:75	14:16:70	19,600	1.03	33.0±1.1	4.6±1.3	52.8±14.3[f]

[a]Theoretical values are calculated from the proportions of reagents used for the substitution of polymers.

[b]Experimental values are based on ^{1}H NMR and acid–base titration results.

[c]DSC results for LCST and ΔH$_{\mathrm{LCST}}$ were obtained at pH 5.0 and performed in triplicate.

[d]Efficiency of PPZ fixation to EPC/Chol (3 : 2 mol/mol) liposomes prepared with 1 mol% PPZ as determined by phosphorus content.

[e]PPZ added to lipids before the extrusion process.

[f]PPZ fixed to liposomes after overnight incubation with preformed vesicles at 4°C.

The pH-dependent precipitation of PPZ in aqueous solution was investigated by turbidimetry. PPZs were dissolved in 200 mL of phosphate buffer saline (PBS) (53 mM Na_2HPO_4, 13 mM NaH_2PO_4, 75 mM NaCl) at a concentration of 0.2 mg/mL. The pH of the solution was adjusted to predetermined values and the turbidity of aliquots was measured at 480 nm (37°C) using a Series 2 Aminco Bowman fluorometer (Spectronics Instruments Inc., Rochester, New York) (Roux, et al., 2002b).

The LCST and enthalpy of transition (ΔH_{LCST}) of the PPZs were determined in triplicate on three distinct samples by differential scanning calorimetry on a MicroCal VP-DSC (MicroCal, Northampton, Massachusetts). Polymer samples were dissolved in saline 2-N-(morpholino)ethanesulfonic acid (MES) (100 mM, 110 mM NaCl, pH 5.0) at a concentration of 10 mg/mL. Scans were performed on samples of 0.509 mL at a rate of 20°C/h from 7 to 65°C.

The degradability of the PPZ was tested by incubating 5-mL aliquots of polymer solutions (1.2 mg/mL in 10 mM phosphate buffer, pH 7.4), which were filtered under sterile conditions and then incubated at 37°C for 21 weeks. The samples were lyophilized and changes in M_w were measured by SEC.

Analysis of pH-Sensitive Liposomes

Incorporation of PPZ into Liposomes   The pH-sensitive liposomes were prepared as described before (Leroux et al., 2001; Zignani et al., 2000; Couffin-Hoarau and Leroux, 2004). Briefly, a lipid film was obtained by evaporating chloroform solutions of EPC, cholesterol, and PPZ with a respective molar ratio of 59 : 40 : 1. The polymer/lipid mass ratio was approximately 0.2. In the case of PEGylated liposomes, 5.5 mol% PEG_{2000}-DSPE was included in the lipid bilayer as reported elsewhere (Roux et al., 2004; Yang et al., 2003). The film was then hydrated overnight with HBS (pH 7.4) to obtain a lipid concentration of 40 mM. Finally, the mixture was extruded through 400-, 200-, and 100-nm polycarbonate membranes (Avanti, Alabaster, Alabama) 21 times each. For some formulations, the polymer was post-inserted by incubation with preformed extruded vesicles overnight at 4°C in HBS (PPZ molar ratio of 1%). In both cases, unbound polymer was removed by SEC using a Sepharose 2B column. Liposome size was measured by dynamic light scattering on a Malvern Zetasizer ZS (Malvern, Worcestershire, UK) with a fixed angle of 173° at 25°C. Final vesicle sizes were between 120 and 180 nm, with narrow polydispersity (<0.12).

A procedure adapted from work of Bartlett (1959) was used to measure the total amount of phosphorus in the formulations from which the efficiency of PPZ incorporation was calculated. Chol-BODIPY (0.2 mol% of lipids) was added during the preparation of liposomes as an internal standard to normalize for phospholipid concentrations. The percent PPZ incorporated could then be obtained by subtracting the phosphorus content of bare liposomes from PPZ-liposomes.

In Vitro Release Kinetics   In vitro release kinetics were monitored for EPC/
Chol/PPZ liposomes incorporating the fluorescent markers HPTS (35 mM) and
quencher DPX (50 mM) in HEPES buffer (20 mM) before and after a 1-h
incubation with 50% (v/v) human serum (Han et al., 2006). SEC was
performed to remove nonencapsulated marker/quencher as well as excess
serum components. The release profiles of the various formulations were
measured at different pH values in either HBS pH 7.4 or MES adjusted to
pH 5.0 or 6.0. HPTS release was monitored by fluorescence assay using a Tecan
Safire plate reader (Tecan, Durham, North Carolina) ($\lambda_{ex} = 412$ nm and
$\lambda_{em} = 513$ nm) at 37°C. The percent release at each time point was obtained
from the relative fluorescence intensity with respect to the fluorescence detected
after sample lysis with 0.5% (v/v) Triton x-100.

RESULTS AND DISCUSSION

Synthesis and Characterization of pH-sensitive PPZ

Synthesis   PDCP was synthesized by cationic polymerization of the phosphoranimine monomer as described previously (Allcock et al., 1996; Couffin-
Hoarau and Leroux, 2004). Five different pH-sensitive PPZs (Table 2) were
generated by performing three sequential substitution reactions of P–Cl bonds
in PDCP backbone (Scheme 1). Due to increasing reactivities, $C_{18}(EO)_{10}$ was
added first, followed by EAB and then EEE. Final substitution ratios of 7 to
14% and 5 to 16 mol% were obtained for ABA and $C_{18}(OE)_{10}$, respectively
(Table 2). The slightly lower than theoretical ratios of ABA may result from
metathetical exchange during substitution of some EAB by the stronger
nucleophile, EEE (Allcock, 1977). Moreover, the basic conditions used for
hydrolysis of EAB to ABA might induce cleavage of the aminophosphazene
bond (Allcock et al., 1982), also decreasing the ABA molar ratio. It was
reported previously that PPZ [$M_W = 38,000$, 9% ABA, 5 mol% $C_{18}(EO)_{10}$,
respectively] can provide pH-responsive properties to liposomes (Couffin-
Hoarau and Leroux, 2004), but no further studies were carried out to determine
the relation between structure and properties. The PPZ synthesized here
possessed lower M_w (15,000 to 20,000) and various ABA/$C_{18}(OE)_{10}$ molar
ratios, which allowed examination of the impact of PPZ composition on the
release kinetics.

Physical Characterization   Figure 2 shows the typical pH-dependent phase
transition of three representative PPZs at 37°C. Turbidimetry was used to
detect the polymers' phase separation from the buffered medium under dilute
conditions. With the exception of polymer A_7–P_6, the PPZs were fully soluble
at pH 7.4, while the turbidity markedly increased upon lowering the pH below
6.0. The change in solubility around this pH is what is sought to destabilize the
phospholipid membrane after endocytosis and release the liposomal content.
As shown in Figure 2, sample A_7–P_6, displayed some turbidity near pH 7,

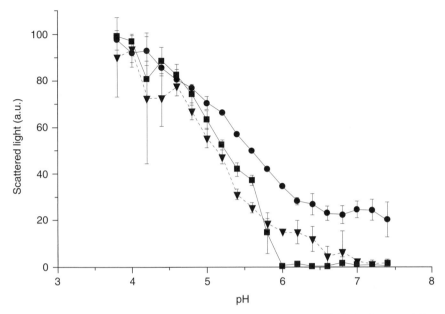

FIGURE 2   pH-dependent phase transition of PPZs A_7–P_6 (circles), A_9–$P_{5.5}$ (triangles), and A_{14}–P_{16} (squares) as determined by turbidimetry in PBS at $37°C$. Mean $\pm$ SD ($n = 3$).

reflecting incomplete dissolution of the polymer. This might be attributed to its lower ABA content, which renders the polymer less hydrophilic. Previously we reported that a PPZ with comparable composition [9 mol% ABA, 5 mol% $C_{18}(EO)_{10}$], but higher molecular weight ($M_w = 38,000$) possessed a LCST of $32.4°C$ at pH 7.4 (Couffin-Hoarau and Leroux, 2004). It has been shown that fully EEE-substituted PPZs have an LCST of $32°C$ at pH 7.4 (Couffin-Hoarau and Leroux, 2004) while the introduction of a sufficient amount of ionizable moiety, such as ABA, can raise the LCST at this pH (Chen and Hoffman, 1995; Hirotsu et al., 1987). Therefore, owing to their better solubility at physiological temperature and neutral pH, PPZs A_{14}–P_{16} and A_9–$P_{5.5}$ are expected to be better candidates than A_7–P_6 for the design of pH-responsive vesicles that would be stable at pH 7.4 and destabilized under mildly acidic conditions.

Differential scanning calorimetric thermograms were recorded for the various PPZs at pH 5.0. LCST values obtained were taken at the maxima of the endotherms and ranged between 32 and $35.5°C$, with transition enthalpies varying from 4.6 to $17.6\,J/g$ (1.1 to $4.2\,cal/g$). As shown in Table 2, all LCSTs were in the same range under acidic conditions. For previously synthesized PPZs, acidification to pH 5.0 decreased the LCST below $30°C$ (Couffin-Hoarau and Leroux, 2004), which is lower than for the PPZs presented here. Feil et al. 1993 have noted that the LCST of NIPAM copolymers was strongly influenced by their overall hydrophilicity and the structuring of water around hydrophobic groups. In the present case, it is difficult to predict the precise variations the substituents impose on the LCST of the PPZ, as three side groups are involved. Moreover, the $C_{18}(EO)_{10}$ side group is by itself amphiphilic, due to the contribution of the $(EO)_{10}$ and C_{18} segments. While the $(EO)_{10}$ chain may raise the LCST as the additional oxygen atoms can increase hydration (Allcock and Dudley, 1996), the alkyl chain may decrease the LCST, depending on whether or not they self-assemble (i.e., exclusion from the solvent) in water.

Interestingly, the changes in enthalpy associated with the phase transition were lower than observed previously for other pH-responsive PPZs (Couffin-Hoarau and Leroux, 2004). It could be hypothesized that the decreased ΔH_{LCST} is a result of the generally higher proportions of $C_{18}(EO)_{10}$ and protonated ABA moieties, which reduce interaction between the polymer and the water molecules and/or increase interactions of the polymer with itself. This tendency was also observed for PPZ A_{14}–P_{16}, which had the highest level of ABA and $C_{18}(EO)_{10}$ while exhibiting the lowest ΔH_{LCST} at pH 5. Indeed, a similar dependence was observed by Laukkanen et al. (2005) for a thermosensitive polymer modified by increasing proportions of an amphiphilic graft.

Biodegradation Study   Poly(aminophosphazene)s have been explored extensively as degradable alternatives to other synthetic polymers (Allcock et al., 1977, 1994; Crommen et al., 1992a, 1992b). The degradation of two PPZs, $A_{9.5}$–$P_{5.5}$ and A_{14}–P_{16}, was compared after a period of 21 weeks at pH 7.4 and 37°C. Only a 20% decrease in M_w was observed for both polymers, showing that the degradation was partial. It is known that the degradation of PPZ involves cleavage of the aminophosphazene bond (Allcock et al., 1982; Lee et al., 1999) catalyzed by the free acid of ABA. However, the extent of degradation is dependent on the nature of the amino acid (Allcock et al., 1982) and its molar ratio (Crommen et al., 1992b; Lemmouchi et al., 1998). It is thus likely that the low ABA content along the PPZ backbone could not promote complete degradation.

Characterization of pH-Responsive Liposomes

Incorporation of PPZ into Liposomes   pH-responsive liposomes were prepared by either of two methods. PPZ with a lower anchor content were incorporated by the inclusion of 1 mol% PPZ in the lipid film. However, for A_{11}–P_{10} or A_{14}–P_{16}, this method failed to produce monodisperse vesicles. For these two polymers, bridging between vesicles may have resulted from the relatively high PPZ/lipid ratio (0.2 w/w) and elevated $C_{18}(OE)_{10}$ content, thus forming a complex network (Meier et al., 1996). The increased viscosity thus could also have made it mechanically difficult to extrude. As a consequence, A_{11}–P_{10} and A_{14}–P_{16} were associated with the lipid membrane by incubating a PPZ solution with preformed extruded vesicles overnight at 4°C. The post-incorporation method involved the addition of PPZ to the vesicle suspension, resulting in a more dilute mixture. Therefore, it permitted the formation of stable liposomes with PPZ inserted solely on the external leaflet of the bilayer.

The extent of polymer incorporation for the various formulations was calculated from the phosphorus content (Table 2). PPZ fixation was significantly higher when included in vesicle preparation, as over 80% PPZ incorporation (0.16 g PPZ/g lipid) was obtained. For PPZ A_{11}–P_{10} and A_{14}–P_{16}, which were incorporated by incubation, anchoring efficiencies of 35 and 50% were achieved, respectively. These findings can be compared to EPC/Chol liposomes prepared with NIPAM/MAA copolymers containing 2% octadecyl acrylate (ODA) for fixation. A two-fold increase in binding efficiency was obtained for P(NIPAM-co-MAA-co-ODA) when included in vesicle preparation rather than post-incorporated (Zignani et al., 2000). This can be explained by the increased surface area available for incorporation on either side of the bilayer and lipid mixing. Furthermore, the post-incorporation of P(NIPAM-co-MAA-co-ODA) yielded a maximum of 0.038 g copolymer/g lipid, which corresponded to a plateau with an efficiency of 30% when prepared with an initial mass ratio of 0.12 g copolymer/g lipid. This is somewhat lower than what was seen with PPZs A_{11}–P_{10} and A_{14}–P_{16} (0.07 and 0.1 g PPZ/g lipid, respectively). Increasing the content of anchoring moiety seems to have improved copolymer fixation. Kono et al. (1999) also observed increased liposome binding for polymers of higher molecular weight while maintaining the proportion of the anchor, suggesting that the binding efficiency improves with an increasing number of anchoring moieties per polymer chain. This general trend is also observed for PPZs with increasing proportions of $C_{18}(OE)_{10}$.

In Vitro Release Kinetics of pH-Responsive Liposomes   pH-sensitive PPZs are required to promote maximal discharge under acidic conditions while permitting complete retention as long as the vector remains in circulation. To test for this character, the release of the encapsulated probe HPTS from pH-responsive liposomes was measured at pH values of 5.0, 6.0, and 7.4 and at a temperature of 37°C. Figure 3 shows the in vitro release kinetics of formulations prepared with PPZs A_7–P_6 (A), $A_{9.5}$–$P_{7.5}$ (B), and A_{14}–P_{16} (C). It can be seen that PPZ induced a marked increase in the release rate of the encapsulated dye as the pH was acidified. Liposomes prepared with PPZ A_7–P_6 released a substantial amount of HPTS at neutral pH (27% within 35 min, Fig. 3a). As discussed above, this polymer is partially dehydrated at pH 7.4 and 37°C and thus can destabilize the lipid membrane. In our previous report, pH-sensitive liposomes prepared with PPZ having an LCST of 32°C at pH 7.4 showed similar profiles under the same experimental conditions (Couffin-Hoarau and Leroux, 2004). In contrast, the other two formulations were significantly more stable with less than 5% dye released after 35 min at neutral pH. $A_{9.5}$–$P_{7.5}$ (Fig. 3b) demonstrated the best triggered-release profile (75 and 47% HPTS released at pH 5.0 and 6.0, respectively). Ideally, as depicted in Figure 1, release should occur within the transit time of the endocytosed material to mature lysosomes (<35 min). A rapid response to the decrease in pH would also improve discharge of the content and delivery to the cytoplasm. PPZ $A_{9.5}$–$P_{7.5}$ exhibited

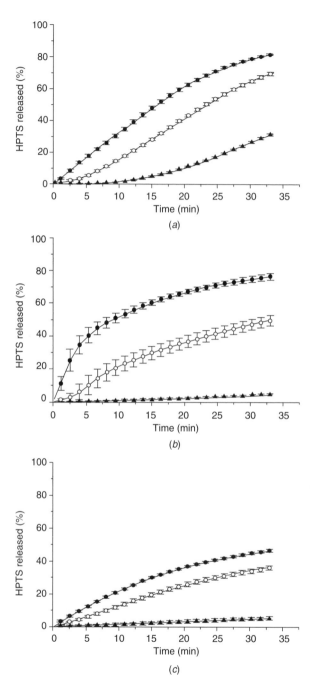

FIGURE 3   Percent HPTS released from EPC/Chol (3 : 2 mol/mol) liposomes (120 to 180 nm) prepared with 1 mol% PPZ A_7–P_6 (a), $A_{9.5}$–$P_{7.5}$ (b), and A_{14}–P_{16} (c) at 37°C and pH 7.4 (solid triangles), 6.0 (open circles), and 5.0 (solid circles). Mean $\pm$ SD (n = 3).

a high marker release over 35 min, yet also showed a triggered discharge within the first 5 min which was not seen for the other PPZ reported here. A_{14}–P_{16} was less efficient in destabilizing the liposomes at acidic pH. After 35 min, about 45% leaked from the vesicles at pH 5.0. The lower performance of A_{14}–P_{16} can be explained by the presence of PPZ only on the outer leaflet of the liposomes, due to the incorporation method. We and others previously reported that pH-responsive liposomes were more readily destabilized when polymers were fixed on either side of the bilayer (Couffin-Hoarau and Leroux, 2004; Hayashi et al., 1999; Roux et al., 2002a; Zignani et al., 2000).

pH-sensitive liposomes, injected intravenously, must circulate for a sufficiently long period to attain target cells. However, EPC/Chol liposomes typically do not survive in the bloodstream, as they are quickly opsonized and eliminated by the mononuclear phagocyte system. Pharmacokinetic studies revealed that their biological half-life ($t_{1/2}$) is less than 35 min in rats after intravenous injection (Roux et al., 2003). PEGylation is well known for providing liposomes with a steric barrier from opsonins and other serum proteins, as well as considerably extending circulation times in the bloodstream (Klibanov et al., 1990; Simoes et al., 2004). pH-sensitive liposomes can additionally be PEGylated to improve their circulation half-life (Roux et al., 2002b, 2003).

The effect of PEG_{2000}-DSPE was therefore evaluated on $A_{9.5}$–$P_{7.5}$ liposomes. This PPZ was chosen, as it showed to the best release kinetics of HPTS. Both polymers were incorporated into the bilayer during vesicle preparation, in the same manner as for the non-PEGylated form. Despite this, only 32% PPZ fixation was achieved, which is a decline of 50% in $A_{9.5}$–$P_{7.5}$ binding efficiency. Steric hindrance caused by the PEG chains may have impaired the anchoring of the PPZ into the bilayer. The HPTS release kinetics of PEGylated pH-sensitive liposomes is reported in Figure 4. In comparison to the unmodified formulation, the amount of dye liberated decreased from 75% to 55% after 35 min at pH 5.0. Also, a lag time was seen for the onset of release. Roux et al. (2003) had shown previously that PEG_{2000}-DSPE contributed to a significant stabilization of pH-sensitive liposomes. The loss in pH responsiveness could therefore be attributed to both the reduced fixation of the PPZ and the stabilizing effect of PEG_{2000}-DSPE on the bilayer.

For a formulation to be clinically viable, it is crucial that it remain stable in the presence of serum. Figure 5 compares the amount of HPTS released after 30 min for $A_{9.5}$–$P_{7.5}$ liposomes with and without PEG, before and after serum incubation. Decreased release at acidic pH was observed when PPZ liposomes were preincubated with 50% (v/v) human serum for 1 h. In other studies, exposure to serum reduced the pH-sensitivity of PEGylated vesicles bearing randomly alkylated P(NIPAM-*co*-MAA) (Roux et al., 2003), whereas no significant desensitization was observed when the anchor was present on the terminus of the polymer chain (Roux et al., 2004). The reduced response may be a result of polymer extraction and/or a shift in transition pH due to protein adsorption (Harvie et al., 1996). Randomly alkylated polymers may affect the

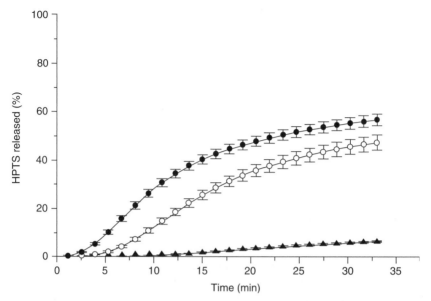

FIGURE 4   Percent HPTS released from EPC/Chol (3:2 mol/mol) liposomes (ca. 120 nm) at 37°C prepared with PPZ $A_{9.5}$–$P_{7.5}$ and 5.5 mol% PEG_{2000}-DSPE. Release performed at pH 7.4 (solid triangles), 6.0 (open circles), and 5.0 (solid circles). Mean ± SD ($n = 3$).

formation of an adequate protective PEG barrier around the liposome, thus allowing protein adsorption. In contrast, terminally alkylated copolymers may facilitate resistance to serum inactivation by allowing uniform polymer distribution on the vesicle surface.

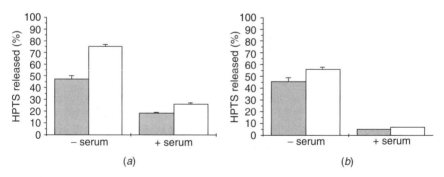

FIGURE 5   Percent HPTS released after 30 min at 37°C from pH-sensitive EPC/Chol (3:2 mol/mol) $A_{9.5}$–$P_{7.5}$ liposomes (ca. 120 nm) prepared without (a) and with (b) 5.5 mol% PEG-DSPE. PH sensitivity was evaluated before and after a 1-h incubation with 50:50 (v/v) human serum at pH 6.0 (solid bars) and 5.0 (open bars). Mean ± SD ($n = 3$).

CONCLUSIONS

Amphiphilic polyelectrolyte PPZs are candidates to regulate the targeted release of liposome-encapsulated agents. The LCST of EEE-substituted PPZ was modified as a function of pH by co-substitution of the acidic moiety ABA. The relatively small proportion of this amino acid grafted seems to have limited the degradability of the PPZ, thus making it preferable at this time to keep the molecular weight low enough to favor renal excretion after administration. Adding $C_{18}(EO)_{10}$ randomly along the backbone permitted efficient anchoring of the pH-responsive PPZ into EPC/Chol liposomes, both during or after preparation of the vesicles. Liposomes formulated with PEG_{2000}-DSPE maintained some pH sensitivity despite a significant reduction of polymer anchoring. However, exposure to serum reduced the pH-responsiveness for both PEGylated and non-PEGylated forms. Additional investigation is thus required to determine the cause of this partial deactivation. In conclusion, the potential of PPZ has been demonstrated further for the development of stimuli-responsive liposomal drug carriers. Steps have been taken to define the parameters required to implement such polymers in an efficient, workable drug delivery system. Consequently, improved systems can possibly be formulated by further fine-tuning of the PPZ structure to allow the preparation of serum-stable pH-sensitive liposomes.

Acknowledgments

This work was supported financially by the CIHR and the Canada Research Chair Program. The authors would like to thank Professor Ian Manners and Keith Huynh sincerely for their advice concerning the synthesis of the monomer.

REFERENCES

Allcock, H.R. 1977. Poly(organophosphazenes): unusual new high polymers. *Angew. Chem. Int. Ed.*, 16:147–156.

Allcock, H.R., Dudley, G.K. 1996. Lower critical solubility temperature study of alkyl ether based polyphosphazenes. *Macromolecules*, 29:1313–1319.

Allcock, H.R., Fuller, T.J., Mack, D.P., Matsumura, K., Smeltz, K.M. 1977. Synthesis of poly[(amino acid alkyl ester)phosphazenes]. *Macromolecules*, 10:824–830.

Allcock, H.R., Fuller, T.J., Matsumura, K. 1982. Hydrolysis pathways for aminophosphazenes. *Inorg. Chem.*, 21:515–521.

Allcock, H.R., Pucher, S.R., Scopelianos, A.G. 1994. Poly[(amino acid ester)phosphazenes]: synthesis, crystallinity, and hydrolytic sensitivity in solution and the solid state. *Macromolecules*, 27:1071–1075.

Allcock, H.R., Crane, C.A., Morrissey, C.T., Nelson, J.M., Reeves, S.D., Honeyman, C.H., Manners, I. 1996. "Living" cationic polymerization of phosphoranimines as an ambient temperature route to polyphosphazenes with controlled molecular weights. *Macromolecules*, 29:7740–7747.

Allen, C., Dos Santos, N., Gallagher, R., Chiu, G.N.C., Shu, Y., Li, W.M., Johnstone, S.A., Janoff, A.S., Mayer, L.D., Webb, M.S., Bally, M.B. 2002. Controlling the physical behavior and biological performance of liposome formulations through use of surface grafted poly(ethylene glycol). *Biosci. Rep.*, 22:225–250.

Bartlett, G.R. 1959. Phosphorus assay in column chromatography. *J. Biol. Chem.*, 234:466–468.

Boomer, J.A., Inerowicz, H.D., Zhang, Z.Y., Bergstrand, N., Edwards, K., Kim, J.M., Thompson, D.H. 2003. Acid-triggered release from sterically stabilized fusogenic liposomes via a hydrolytic dePEGylation strategy. *Langmuir*, 19:6408–6415.

Brazel, C.S., Peppas, N.A. 1996. Pulsatile local delivery of thrombolytic and antithrombotic agents using poly(*N*-isopropylacrylamide-*co*-methacrylic acid) hydrogels. *J. Control. Release*, 39:57–64.

Chen, G., Hoffman, A.S. 1995. Graft copolymers that exhibit temperature-induced phase transitions over a wide range of pH. *Nature*, 373:49–52.

Connor, J., Yatvin, M.B., Huang, L. 1984. pH-sensitive liposomes: acid-induced liposome fusion. *Proc. Natl. Acad. Sci. USA*, 81:1715–1718.

Couffin-Hoarau, A.C., Leroux, J.C. 2004. Report on the use of poly(organophosphazenes) for the design of stimuli-responsive vesicles. *Biomacromolecules*, 5:2082–2087.

Crommen, J.H.L., Schacht, E.H., Mense, E.H.G. 1992a. Biodegradable polymers: I. Synthesis of hydrolysis-sensitive poly[(organo)phosphazenes]. *Biomaterials*, 13:511–520.

Crommen, J.H.L., Schacht, E.H., Mense, E.H.G. 1992b. Biodegradable polymers: II. Degradation characteristics of hydrolysis-sensitive poly[(organo)phosphazenes]. *Biomaterials*, 13:601–611.

Drummond, D.C., Zignani, M., Leroux, J.C. 2000. Current status of pH-sensitive liposomes in drug delivery. *Prog. Lipid. Res.*, 39:409–460.

Feil, H., Bae, Y.H., Feijen, J., Kim, S.W. 1993. Effect of comonomer hydrophilicity and ionization on the lower critical solution temperature of *N*-isopropylacrylamide copolymers. *Macromolecules*, 26:2496–2500.

Francis, M.F., Dhara, G., Winnik, F.M., Leroux, J.C. 2001. In vitro evaluation of pH-sensitive polymer/niosome complexes. *Biomacromolecules*, 2:741–749.

Guo, X., Szoka, F.C. 2001. Steric stabilization of fusogenic liposomes by a low-pH sensitive PEG–diortho ester–lipid conjugate. *Bioconjugate Chem.*, 12:291–300.

Han, H.D., Shin, B.C., Choi, H.S. 2006. Doxorubicin-encapsulated thermosensitive liposomes modified with poly(*N*-isopropylacrylamide-*co*-acrylamide): drug release behavior and stability in the presence of serum. *Eur. J. Pharm. Biopharm.*, 62: 110–116.

Harvie, P., Desormeaux, A., Bergeron, M.C., Tremblay, M., Beauchamp, D., Poulin, L., Bergeron, M.G. 1996. Comparative pharmacokinetics, distributions in tissue, and interactions with blood proteins of conventional and sterically stabilized liposomes containing 2′,3′-dideoxyinosine. *Antimicrob. Agents Chemother.*, 40:225–229.

Hayashi, H., Kono, K., Takagishi, T. 1999. Temperature sensitization of liposomes using copolymers of *N*-isopropylacrylamide. *Bioconjugate Chem.*, 10:412–418.

Heskins, M., Guillet, J.E. 1968. Solution properties of poly(*N*-isopropylacrylamide). *J. Macromol. Sci. Pure Appl. Chem.*, 2:1441–1455.

Hirotsu, S., Hirokawa, Y., Tanaka, T. 1987. Volume-phase transitions of ionized *N*-isopropylacrylamide gels. *J. Chem. Phys.*, 87:1392–1395.

Hong, M.-S., Lim, S.-J., Oh, Y.-K., Kim, C.-K. 2002. pH-sensitive, serum-stable and long-circulating liposomes as a new drug delivery system. *J. Pharm. Pharmacol.*, 54:51–58.

Ishida, T., Okada, Y., Kobayashi, T., Kiwada, H. 2006. Development of pH-sensitive liposomes that efficiently retain encapsulated doxorubicin (DXR) in blood. *Int. J. Pharm.*, 309:94–100.

Kamata, H., Yagisawa, H., Takahashi, S., Hirata, H. 1994. Amphiphilic peptides enhance the efficiency of liposome-mediated DNA transfection. *Nucleic Acids Res.*, 22:536–537.

Karanth, H., Murthy, R.S.R. 2007. pH-Sensitive liposomes: principle and application in cancer therapy. *J. Pharm. Pharmacol.*, 59:469–483.

Kichler, A., Mechtler, K., Behr, J.P., Wagner, E. 1997. Influence of membrane-active peptides on lipospermine/DNA complex mediated gene transfer. *Bioconjugate Chem.*, 8:213–221.

Klibanov, A.L., Maruyama, K., Torchilin, V.P., Huang, L. 1990. Amphipathic poly-ethyleneglycols effectively prolong the circulation time of liposomes. *FEBS Lett.*, 268:235–237.

Kono, K., Igawa, T., Takagishi, T. 1997. Cytoplasmic delivery of calcein mediated by liposomes modified with a pH-sensitive poly(ethylene glycol) derivative. *Biochim. Biophys. Acta.*, 1325:143–154.

Kono, K., Nakai, R., Morimoto, K., Takagishi, T. 1999. Thermosensitive polymer-modified liposomes that release contents around physiological temperature. *Biochim. Biophys. Acta.*, 1416:239–250.

Laukkanen, A., Valtola, L., Winnik, F.M., Tenhu, H. 2005. Thermosensitive graft copolymers of an amphiphilic macromonomer and *N*-vinylcaprolactam: synthesis and solution properties in dilute aqueous solutions below and above the LCST. *Polymer*, 46:7055–7065.

Lee, S.B., Song, S.-C, Jin, J.I., Sohn, Y.S. 1999. A new class of biodegradable thermosensitive polymers: II. Hydrolytic properties and salt effect on the lower critical solution temperature of poly(organophosphazenes) with methoxypoly(ethylene glycol) and amino acid esters as side groups. *Macromolecules*, 32:7820–7827.

Lemmouchi, Y., Schacht, E., Dejardin, S. 1998. Biodegradable poly[(amino acid ester)phosphazenes] for biomedical applications. *J. Bioact. Compat. Polym.*, 13:4–18.

Leroux, J.-C., Roux, E., Le Garrec, D., Hong, K., Drummond, D.C. 2001. *N*-Isopropylacrylamide copolymers for the preparation of pH-sensitive liposomes and polymeric micelles. *J. Control. Release*, 72:71–84.

Li, W., Nicol, F., Szoka, F.C. 2004. GALA: a designed synthetic pH-responsive amphipathic peptide with applications in drug and gene delivery. *Adv. Drug Deliv. Rev.*, 56:967–985.

Li, X., Liu, W., Ye, G., Zhang, B., Zhu, D., Yao, K., Liu, Z., Sheng, X. 2005. Thermosensitive *N*-isopropylacrylamide–*N*-propylacrylamide–vinyl pyrrolidone terpolymers: synthesis, characterization and preliminary application as embolic agents. *Biomaterials*, 26:7002–7011.

Maeda, M., Kumano, A., Tirrell, D.A. 1988. H$^+$-induced release of contents of phosphatidylcholine vesicles bearing surface-bound polyelectrolyte chains. *J. Am. Chem. Soc.*, 110:7455–7459.

Malonne, H., Eeckmann, F., Fontaine, D., Otto, A., De Vos, L., Moës, A., Fontaine, J., Amighi, K. 2005. Preparation of poly(*N*-isopropylacrylamide) copolymers and preliminary assessment of their acute and subacute toxicity in mice. *Eur. J. Pharm. Biopharm.*, 61:188–194.

Meier, W., Hotz, J., Günther-Ausborn, S. 1996. Vesicle and cell networks: interconnecting cells by synthetic polymers. *Langmuir*, 12:5028–5032.

Meyer, O., Papahadjopoulos, D., Leroux, J.C. 1998. Copolymers of *N*-isopropylacrylamide can trigger pH sensitivity to stable liposomes. *FEBS Lett.*, 42:61–64.

Petriat, F., Roux, E., Leroux, J.C., Giasson, S. 2004. Study of molecular interactions between a phospholipidic layer and a pH-sensitive polymer using the Langmuir balance technique. *Langmuir*, 20:1393–1400.

Provoda, C.J., Stier, E.M., Lee, K-D. 2003. Tumor cell killing enabled by listeriolysin O-liposome-mediated delivery of the protein toxin gelonin. *J. Biol. Chem.*, 278:35102–35108.

Roux, E., Francis, M., Winnik, F.M., Leroux, J.C. 2002a. Polymer based pH-sensitive carriers as a means to improve the cytoplasmic delivery of drugs. *Int. J. Pharm.*, 242:25–36.

Roux, E., Stomp, R., Giasson, S., Pézolet, M., Moreau, P., Leroux, J.C. 2002b. Steric stabilization of liposomes by pH-responsive *N*-isopropylacrylamide copolymer. *J. Pharm. Sci.*, 91:1795–1802.

Roux, E., Lafleur, M., Lataste, É., Moreau, P., Leroux, J.C. 2003. On the characterization of pH-sensitive liposome/polymer complexes. *Biomacromolecules*, 4:240–248.

Roux, E., Passirani, C., Scheffold, S., Benoit, J.P., Leroux, J.C. 2004. Serum-stable long-circulating, PEGylated, pH-sensitive liposomes. *J. Control. Release*, 94:447–451.

Schmaljohann, D. 2006. Thermo- and pH-responsive polymers in drug delivery. *Adv. Drug Deliv. Rev.*, 58:1655–1670.

Simoes, S., Moreira, J.N., Fonseca, C., Duzgunes, N., Pedroso de Lima, M.C. 2004. On the formulation of pH-sensitive liposomes with long circulation times. *Adv. Drug Deliv. Rev.*, 56:947–965.

Taillefer, J., Jones, M.C., Brasseur, N., van Lier, J.E., Leroux, J.C. 2000. Preparation and characterization of pH-responsive polymeric micelles for the delivery of photosensitizing anticancer drugs. *J. Pharm. Sci.*, 89:52–62.

Wang, B., Rivard, E., Manners, I. 2002. A new high-yield synthesis of Cl$_3$P$=$NSiMe$_3$, a monomeric precursor for the controlled preparation of high molecular weight polyphosphazenes. *Inorg. Chem.*, 41:1690–1691.

Winnik, F.M. 1990. Fluorescence studies of aqueous solutions of poly(*N*-isopropylacrylamide) below and above their LCST. *Macromolecules*, 23:233–242.

Yang, H., Cheng, R., Wang, Z. 2003. A quantitative analyses of the viscometric data of the coil-to-globule and globule-to-coil transition of poly(N-isopropylacrylamide) in water. *Polymer*, 44:7175–7180.

Yessine, M.A., Leroux, J.C. 2004. Membrane-destabilizing polyanions: interaction with lipid bilayers and endosomal escape of biomacromolecules. *Adv. Drug Deliv. Rev.*, 56:999–1021.

Zignani, M., Drummond, D.C., Meyer, O., Hong, K., Leroux, J.-C. 2000. In vitro characterization of a novel polymeric-based pH-sensitive liposome system. *Biochim. Biophys. Acta.*, 1463:383–394.

14 Poly- and Cyclophosphazenes as Drug Carriers for Anticancer Therapy

YOUN SOO SOHN and YONG JOO JUN

Center for Intelligent Nano-Biomaterials, Ewha Womans University, Seoul, South Korea

INTRODUCTION

The majority of the anticancer drugs currently in clinical use for chemotherapy are low-molecular-weight compounds (less than 1000 Da) that are administered systemically, orally, or locally. Such small-molecular compounds administered systemically or orally are known to have a short half-life (less than 2 h) and a fast clearance rate in the blood circulation system (Hubbard and Jenkins, 1990; Sinko and Kohn, 1993). Furthermore, the small-molecular drugs attack not only tumor cells and tissues but, without selectivity, also normal cells and tissues, which cause severe toxicity and such side effects as nephrotoxicity, neurotoxicity, and cardiotoxicity. Such toxicities are dose-limiting factors in chemotherapy, which prevent effective treatment of cancer. One of the most promising approaches to overcoming such limits associated with small-molecular drugs is to use polymeric drug delivery systems (DDSs), which is called *polymer therapy*. Based on their primary functions, DDSs may be classified into two categories, relating to controlled drug release systems and targeted drug delivery systems.

In controlled drug release systems the role of the polymer carrier is to extend the half-life of drug by chemical or physical control of the rate of drug release. In this case the site of drug release and the site of drug action are not the same. On the other hand, targeted drug delivery systems should carry drugs to the sites of action (Sinko and Kohn, 1993). Generally, controlled release systems may be enough for relatively nontoxic drugs, but targeted delivery systems are urgently necessary to develop severely cytotoxic drugs, such as anticancer drugs. Therefore, targeted drug delivery systems have been under

Polyphosphazenes for Biomedical Applications, Edited by Alexander K. Andrianov
Copyright © 2009 John Wiley & Sons, Inc.

intensive study in recent decades to improve the pharmacological properties of conventional drugs, with focus on their tumor selectivity (Allen and Cullis, 2004; Birgger et al., 2002; Haag and Kratz, 2006; Park et al., 2008; Sinha et al., 2006). Polymeric drug carriers can allow small-molecular drugs to be equipped with tumor selectivity in addition to water solubility and longevity in the plasma by structural modifications of polymer carrier molecules.

In particular, during the last decade, remarkable progress has been made in the technology to improve the tumor selectivity of drugs through diversified and intensive research for tumor targeting by means of polymeric DDSs in various forms of nanoparticles, microspheres, polymeric micelles, hydrogels, and so on. Broadly classifying, there are two different types of tumor targeting: active and passive (Allen and Cullis, 2004; Haag and Kratz, 2006; Park et al., 2008). The basic concept of active targeting using polymers is illustrated in Figure 1.

Active tumor targeting is usually accomplished by chemical attachment of a targeting group, with high affinity to a specific antigen or receptor overexpressed in the tumor cells or tumor tissues to an appropriate polymer backbone or nanoparticles. As a targeting group, antibodies, folates, galacto-amine, or glycoproteins are most frequently employed, but it is very important that the targeting group should not lose its targeting properties after binding to the backbone. To enhance the water solubility of the conjugated drug, a solubilizing group such as poly(ethylene glycol) is employed. Recently, some monoclonal antibody-based therapeutic agents (Brannon-Peppas and Blanch-ette, 2004) were approved by the U.S. Food and Drug Administration (FDA), and many folate receptor–mediated targeting drugs were developed successfully (Hilgenbrink and Low, 2005).

In passive targeting strategy, any specific targeting group is not necessary to incorporate into polymeric DDSs, since the particle size of drug carrier polymers (hydrodynamic volume) or nanoparticles is the major factor in determining the targeting properties. About two decades ago, Matsumura and Maeda (1986) discovered that polymer molecules with appropriate molecular weights are preferentially accumulated in solid tumor tissues. This phenomenon is called the *enhanced permeability and retention* (EPR)

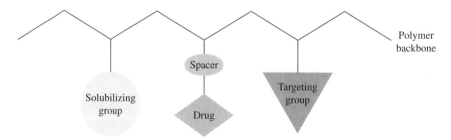

FIGURE 1   Conceptual diagram of tumor-targeting drug delivery systems.

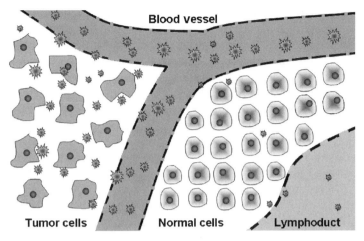

FIGURE 2   Conceptual diagram of the enhanced permeability and retention effect.

effect, which is now a hot issue in DDSs, and a great deal of research has been performed worldwide for development of new polymeric materials showing high tumor selectivity based on the EPR effect (Dreher et al., 2006; Duncan, 1999, 2003; Lundberg and Weinberg, 1999; Maeda et al., 2003; Marecos et al., 1998; Seymour et al., 1995; Torchilin, 2000; Tsuchiya et al., 2000; Vasey et al., 1999). The basic concept of the EPR effect is shown in Figure 2. The EPR effect is known to be based on two factors. First, macromolecules such as polymeric nanoparticles cannot permeate through the blood vessel pores of normal tissues composed of regularly and tightly arrayed normal cells. However, the coarse vasculature and the high vascular pressure of tumor tissues allow polymer particulates of certain sizes to permeate easily through the blood vessels of tumor tissues, as shown in the figure.

Second, there is no lymphatic vessel as a discharge path for polymer particulates in tumor tissues. Therefore, unlike in the normal tissues, it is difficult for polymer particles that have permeated into tumor tissues to be drained off (Maeda et al., 2000), and consequently, polymer particles that have permeated accumulate selectively in the tumor tissues, yielding a high selectivity of polymer particulates to tumor tissues (Maeda et al., 2003). The degree of EPR effect of polymer particles depends on various factors, but their particle size and longevity in the plasma are the most important factors to be considered for the design of effective EPR delivery systems. Although there are many studies to optimize the EPR effect of polymer particles or nano-particles, it seems that the universal size for a maximum EPR effect is difficult to determine, probably because the optimum size may be variable, depending on the nature of both tumor tissues and polymer particulates. However, studies of particulates and liposomes have shown that the tumor vasculature is as large as 200 nm to 2 μm, and the optimum EPR effect was observed for nanoparticles in the range 100 to 200 nm in diameter (Charrois and Allen, 2003; Hashizume et al., 2000; Hobbs et al., 1998; Torchilin, 2001; Yuan et al., 1995). Instead of particle size, the optimum size for an EPR effect of polymeric carriers was frequently represented in terms of molecular weight, which was estimated to be larger than 30 to 50 kDa, corresponding to the renal threshold (Haag and Kratz, 2006; Maeda, 2001). Several nanoparticular polymer–drug conjugates designed based on the EPR effect are in clinical studies (Duncan, 2006; Haag and Kratz, 2006).

During recent decades, researchers have attempted to utilize many different types of organic polymers as drug carriers for active and passive targeting, but it was found that most conventional organic polymers are not suitable as targeting drug carriers, because the drug carrier should satisfy many requirements simultaneously to deliver the right amount of drug to the right site (targeting) at the right time (controlled release), in addition to basic requirements such as water solubility, biocompatibility and biodegradability. Among the various organic polymer carriers attempted for tumor targeting, only a few, for example, N-(2-hydroxypropyl)methacrylamide (HPMA) copolymers, poly(amino acids), and liposomes, have been developed successfully as targeting polymers (Duncan, 2003; Haag and Kratz, 2006). Although HPMA is not biodegradable, it was found that low-molecular-weight HPMA ($<$30 to 50 kDa) is subjected to renal clearance as well as allowing tumor targeting (Duncan, 2003; Maeda, 2001). In fact, HPMA is now most widely used to prepare conjugate prodrugs for many small-molecular anticancer drugs, such as doxorubicin, camptothecin, paclitaxel, methotrexate, and platinum complexes (Duncan, 2003; Haag and Kratz, 2006). However, most other conventional organic polymers are not suitable as targeting drug carriers because of their inherent limitations in molecular structure and physicochemical properties to be tailored to meet the aforementioned requirements. Therefore, a great deal of effort should be devoted to exploiting new types of polymers suitable for tumor targeting.

In this regard, organophosphazenes are excellent resources to use to develop new polymeric drug carriers for tumor targeting, because a wide variety of physicochemical as well as tumor targeting properties can be designed from these relatively new polymers. Soluble linear polyphosphazenes were first synthesized by Allcock and colleagues (Allcock and Kugel, 1965; Mark et al., 2005), who opened a new era of inorganic–organic hybrid polymers. Polyphosphazene is a linear polymer whose backbone consists of alternating phosphorus and nitrogen atoms with two organic or inorganic groups (R) linked to each phosphorus atom as side groups:

$$
\left[\!\!\begin{array}{c} R \\ | \\ N\!=\!\!P \\ | \\ R \end{array}\!\!\right]_n
$$

Depending on the structure of the side groups, a vast range of polymer properties can be designed, including water solubility, biodegradability, amphiphilicity, and stimuli sensitivity, such as thermosensitivity and pH sensitivity. Because of their unique mechanical and physicochemical properties that conventional organic polymers do not have, many commercial products have been made from polyphosphazene elastomers and membranes developed by Allcock and his co-workers (Mark et al., 2005). Also, a great amount of research has focused on biomedical applications of polyphosphazenes, resulting in remarkable progress during recent decades (Allcock, 2006; Gleria and Jaeger, 2005; Mark et al., 2005). In particular, since polymeric drug delivery systems have attracted much attention recently as a major emerging nanobiotechnology for polymer therapy, extensive studies have been directed at new drug delivery systems from polyphosphazenes by many research groups in various fields. For example, polyphosphazene micelles (Allcock et al., 2006; Chang et al., 2002, 2005; Zhang et al., 2005a, b, 2006) and hydrogels (Allcock and Ambrosio, 1996; Allcock and Pucher, 1991; Kang et al., 2006a, b; Lee et al., 2002; Seong et al., 2005) were prepared for sustained release of hydrophobic and small-molecular drugs, biodegradable microspheres, and matrices (Andrianov, 2006; Andrianov and Payne, 1998; Andrianov et al., 2004a, b; Caliceti et al., 2000; Kumbar et al., 2006; Lakshmi et al., 2003; Nair et al., 2004; Veronese et al., 1998) for protein and vaccine delivery, cationic polyphosphazenes for gene delivery (de Wolf et al., 2005, 2007; Luten et al., 2003), and thermosensitive poly and cyclophosphazenes (Jun et al., 2006; Kim, J.I. et al., 2004; Lee et al., 1999a, b; Song et al., 1999; Toti et al., 2007) for controlled release of anticancer drugs. Among the studies on polyphosphazene drug carriers mentioned above, it is worthwhile to note the discovery by Andrianov (2006, 2007) that poly[di(carboxylatophenoxy)phosphazene] (PCPP) is a very potent immunoadjuvant, which was advanced to clinical trials. Details of the progress on PCPP are reviewed in other chapters. Also, extensive studies have been carried out on cancer therapy using polyphosphazene drug carriers, which are reviewed in this chapter.

APPLICATIONS OF POLYPHOSPHAZENES FOR CANCER THERAPY

In general, there are two different methods for the application of drug delivery systems to cancer therapy. One is simply to use a drug carrier substrate for the formulation of known anticancer drugs by encapsulation of a drug using microspheres or micelles, or by homogeneous dispersion of a drug in a biodegradable hydrogel or solid matrix. Drug molecules loaded physically in carriers are released primarily by a diffusion mechanism, but polymer degradation may accelerate the release rate, as shown in Figure 3.

Another method is to conjugate drug molecules directly or by using a biodegradable spacer to the drug carrier molecules by covalent bond, producing a new prodrug. Depending on the final purpose, the molecular structure of

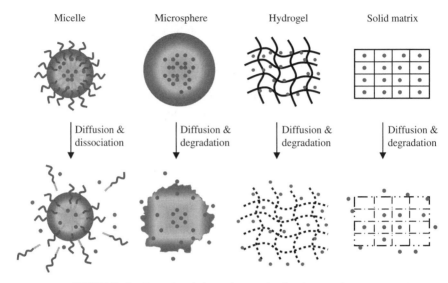

FIGURE 3   Drug-loaded carriers and releasing mechanisms.

drug carriers should be tailored to meet the various requirements, such as water solubility, chemical stability, biodegradability, compatibility with drug, and targeting properties, and consequently, the synthetic method for drug carriers is very important for successful application to cancer therapy. Therefore, application studies of polyphosphazenes for cancer therapy are discussed along with their synthetic methods.

Formulation of Anticancer Drugs Using Polyphosphazenes

The main purposes of anticancer drug formulation is to provide drugs that are release-controlled or solubilized in physiological solution in order to improve their efficacy and toxicity, by physical encapsulation of drugs using polymer micelles or microspheres, or by dispersing drugs in a hydrogel or polymer matrix as described above. Such formulation technologies are relatively simple but depend largely on the inherent properties of carrier polymers. Therefore, despite extensive formulation studies, there are not many commercial products. For well-known anticancer agents in clinical use, such as paclitaxel, doxorubicin, and cisplatin, numerous studies were reported on their formulations using conventional polymeric micelles and hydrogels composed of various amphiphilic diblock copolymers (Hur et al., 2005; Kim et al., 2001; Lee et al., 2003; Liggins and Burt, 2002; Ruel-Gariepy et al., 2004; Soga et al., 2005) and nanoparticles composed of biodegradable polymers such as poly(lactic acid–glycolic acid) (Dong and Feng, 2004; Fonseca et al., 2002; Mu and Feng, 2003). Polyphosphazene micelles and microspheres are relatively new drug delivery systems developed recently but seem to be excellent drug carrier systems for hydrophobic anticancer drugs such as paclitaxel.

In particular, amphiphilic polyphosphazenes can afford not only to solubilize insoluble paclitaxel in aqueous solution but also to change its pharmacokinetic behavior in favorable ways in vivo: for example, long blood circulation for the EPR effect and sustained release for improved efficacy with reduced toxicity (Kataoka et al., 2001; Rösler et al., 2001). However, to the authors' knowledge, there is no study of the formulation of the most widely used anticancer drug, paclitaxel, using polyphosphazenes. Furthermore, there are not many reports on the formulation study for anticancer drugs, despite many efforts to develop polyphosphazene micelles, hydrogels, and microspheres, mentioned above, although there are several reports on formulations for controlled release of protein and anti-inflammatory drugs using polyphosphazene carriers.

Goedemoed and co-workers (1991a, b) carried out studies on controlled release of water-insoluble alkylating agent melphalan using implantable polyphosphazene matrix and injectable microspheres. Matrix devices (tablets) containing the alkylating agent melphalan prepared using a polyphosphazene grafted with glycine ethyl ester as matrix material showed biphasic release profiles with a high initial release of melphalan. However, the use of a more hydrophobic polyphosphazene bearing 50% glutamic acid diethyl ester and 50% glycine ethyl ester reduced the high initial release of melphalan. Devices with these release profiles showed promising therapeutic results in the leukemia L1210 tumor model in mice. In the following study, microspheres containing melphalan and its methyl ester were prepared using a polyphosphazene bearing 50% phenylalanine ethyl ester and 50% glycine ethyl ester. Melphalan methyl ester–loaded microspheres showed gradual and well-sustained release profiles, while melphalan-loaded microspheres revealed very poor release profiles with high initial releases and no sustained delivery. In the lymphatic leukemia L1210 tumor model in DBA2 mice, an intraperitoneal model for metastatic disease promising therapeutic results was observed for the melphalan methyl ester–loaded microspheres.

Schacht et al. (1996) and Lemmouchi et al. (1997) reported studies on the controlled release of anticancer agent mitomycin C from biodegradable polyphosphazene matrices prepared by substitution reactions of poly(dichlorophosphazene) with amino acid ethyl esters. Polyphosphazene grafted only with ethyl glycinate was found to release the drug too fast, but the polymer substituted with ethyl glycinate and hydrophobic ethyl phenyl alanate released the drug slowly. Very recently, Kang et al. (2006a, b) studied the controlled release of doxorubicin from a hydrogel formed from amphiphilic polyphosphazene grafted with hydrophobic isoleucine ethyl ester and hydrophilic α-amino-ω-methoxy-poly(ethylene glycol) as side groups. The release of loaded doxorubicin from the polymer hydrogel was sustained over 20 days, and the releasing rate was effectively controlled by the gel strength. The anticancer efficacy assayed against the mouse lymphoblast of P388D1 was observed to be constant over a prolonged period of time, indicating that the delivery system was an excellent candidate for a locally injectable gel-depot system.

Gene therapy is focused on treating genetic diseases including cancer, and therefore gene delivery is now one of the most important technologies for cancer therapy. Since application of polyphosphazenes as a new nonviral vector for gene delivery does not require covalent bonding to form a polyplex with DNA, it seems reasonable to describe studies on gene delivery using polyphosphazenes in this formulation section. The suitable gene carrier polymers have to show at least a few additional properties in addition to the aforementioned basic requirements as DDSs, such as water solubility, biocompatibility, and biodegradability. First, the carrier polymers should be able to bind to DNA to form condensed nanoparticles called polyplexes, as shown in Figure 4. Therefore, the polymers have to be positively charged in plasma with an appropriate molecular size and shape for effective binding with DNA. The charge density of the polymer is a very important factor. Second, the polyplex formed should have targeting properties to a specific cell. Targeting properties can be afforded by introducing a targeting group such as folate or galactosamine, as mentioned earlier.

Luten et al. (2003) and de Wolf et al., 2005, 2007 have recently performed pioneering work on gene delivery using cationic polyphosphazenes. Polyphosphazenes bearing cationic moieties were synthesized by nucleophilic substitution of poly(dichlorophosphazene) with either 2-dimethylaminoethanol (DMAE) or 2-dimethylaminoethylamine (DMAEA), to obtain cationic polyphosphazene derivatives (pDMAE and pDMAEA). The derivatives were found to bind plasmid DNA, yielding positively charged particles (polyplexes) of size around 80 nm at a polymer/DNA ratio of 3 : 1 (w/w). The polyplexes were able to transfect COS-7 cells in vitro with an efficiency comparable to that of a well-known polymeric transfectant, poly(2-dimethylaminoethyl methacrylate), with lower toxicity. pDMAEA was further subjected to in vivo studies for biodistribution and transfection efficiency of its polyplexes using tumor-bearing mice, and the data were compared with those of polyplexes based on the known nonbiodegradable polyethylenimine (PEI22). Both polyplex systems showed a substantial tumor accumulation of 5% and 8% ID/g for p(DMAEA) and PEI22 polyplexes, respectively, 240 min post-administration. The tumor disposition of p(DMAEA) and PEI22 polyplexes was associated with considerable expression levels of the reporter gene. However, in contrast to PEI22 polyplexes, p(DMAEA) polyplexes did not display substantial gene

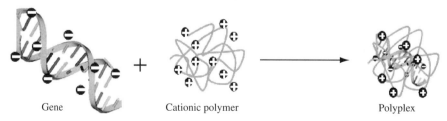

Gene   Cationic polymer   Polyplex

FIGURE 4 Cationic polymer binds to gene to form a polyplex.

expression in the lung or other organs but confined gene expression primarily to tumor tissue. Such preferential tumor gene expression mediated by the p(DMAEA) polyplexes offers a great potential for application of this polymer to deliver therapeutic genes to tumors.

There is another report by Jun et al. (2007) on the synthesis of a new cationic polyphosphazene as a new nonviral vector for gene delivery. In contrast to the homopolymer of p(DMAEA), a dentritic tetralysine, LysLys(LysEt)$_2$, as a cationic moiety, and methoxy-poly(ethylene glycol) with a molecular weight of 350 (MPEG350) as a shielding group, were introduced into the polyphosphazene backbone, yielding a final composition of {NP[MPE-G350]$_{1.55}$[LysLys(Lys(TFA)$_2$)$_2$]$_{0.45}$}$_n$. This polyphosphazene was found to form a polyplex with DNA. The size condensation of the polyplex was observed at 120 nm, and the surface charge was measured to be approximately 27 mV. The cationic polymer exhibited almost no in vitro cytotoxicity against the SK-OV3 cell line probably due to the shielding effect of the MPEG group. However, this polyplex exhibited relatively low transfection yield, probably because of the insufficient number of cationic amine sites of the tetralysine groups grafted to the polymer backbone, and further study to increase the cationic sites of the polymer is necessary.

Polyphosphazene–Anticancer Drug Conjugates

In contrast to the relatively simple formulation technologies mentioned above, versatile and fused high technologies, including chemistry, biology, and polymer science, are involved in the development of new polymer–drug conjugates, which are classified as new prodrugs. Since the beginning of the new millennium, it is generally accepted that polymer therapy using such polymer–conjugate drugs has been one of the most promising tools in cancer therapy, and many conjugate drugs have entered clinical trials (Haag and Kratz, 2006; Sinko and Kohn, 1993).

Among the small-molecular anticancer drugs currently in clinical use, the most widely used are paclitaxel, cisplatin, and doxorubicin, but since paclitaxel was approved relatively late, by the U.S. Food and Drug Administration (FDA) in 1992, the earlier-approved cisplatin and doxorubicin have been employed more frequently for studies on polymer–drug conjugation. In particular, cisplatin [cis-diamminedichloroplatinum(II)] is a simple square-planar coordination complex with two amine (A═NH$_3$) groups as a carrier ligand in the cis position and two chloride ions (X═Cl$^-$) as a leaving group, as shown in Figure 5. It is generally known that the carrier amine ligand is not dissociated from the central platinum(II) atom during metabolism and plays a critical role in anticancer activity as well as in acquired cross-resistance. On the other hand, the anionic leaving group has to be dissociated from the platinum atom in an appropriate stage so that the platinum atom can be chelated by DNA molecules within the cell for anticancer activity, as shown in Figure 5 (Chaney and Vaisman, 2000).

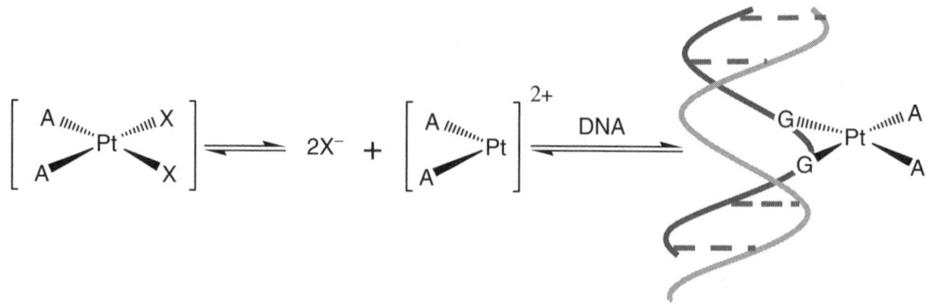

FIGURE 5 DNA binding of cisplatin analogs.

Cisplatin is highly effective against tumors of the ovary, lung (small cell), bladder, head–neck, and testis, but its use is limited, due to its toxic side effects, including nephrotoxicity and neurotoxicity, as well as its acquired drug resistance (O'Dwyer et al., 1999). Moreover, cisplatin has an unfavorable pharmacokinetic profile (Lokich and Anderson, 1998) and a short blood circulation time (Takakura and Hashida, 1996) like most other small-molecular drugs. Therefore, it is believed that one of the most efficient and rational approaches to overcoming such limits of the small-molecular platinum drugs is the design of a polymeric conjugate drug using polymers possessing selective tumor targeting and controlled releasing properties. As mentioned above, many nanosized polymers and liposomes have recently been known to show excellent tumor-targeting properties by the EPR effect (Dreher et al., 2006; Duncan, 1999, 2003; Lundberg and Weinberg, 1999; Maeda et al., 2003; Marecos et al., 1998; Matsumura and Maeda, 1986; Seymour et al., 1995; Torchilin, 2000; Tsuchiya et al., 2000; Vasey et al., 1999). In addition, macromolecular prodrugs exhibit improved body distribution and prolonged blood circulation, due to the dominant pharmacokinetic properties of the macromolecular carrier. Consequently, polyphosphazene–platinum conjugates were studied extensively.

The first phosphazene–platinum(II) conjugates were reported along with crystal structures by Allcock and co-workers in 1977 (Allcock et al., 1977; Allen et al., 1977). Reactions of a polyphosphazene or tetracyclic phosphazenes with K_2PtCl_4 in organic solvents resulted in a polymeric product, $[PtCl_2]_x[NP(NHCH_3)_2]_n$ $(x/n = 1/17)$ or crystalline tetracyclic products, $[PtCl_2][N_4P_4(NHCH_3)_8]$ and $[PtCl_2][N_4P_4(CH_3)_8]$, respectively. The x-ray structure showed a local square-planar structure of platinum coordinated by two skeletal nitrogen atoms of phosphazenes. Therefore, phosphazenes acted as carrier ligands that are usually not dissociated during metabolism. These compounds were subjected to an in vivo survival test against the mouse P388 leukemia cell line and the Ehrlich ascites tumor regression test, but their antitumor activity was not very high.

Water-soluble polyphosphazene–(diamine)platinum conjugate drugs designed rationally based on the structure–activity relationship (McAuliffe et al., 1991) were synthesized by Sohn et al. (1997). Since the antitumor moiety (diamine)Pt^{2+} is dicationic, dicarboxylic amino acids such as aspartic or glutamic diethyl ester as a spacer for platination and methanol or methyl amine as a solubilizing group were introduced into the polyphosphazene backbone by stepwise substitutions of poly(dichlorophosphazene) prepared according to the authors' procedure (Sohn et al., 1995). The resulting polymer conjugate drugs have shown excellent in vivo antitumor activity (ILS > 336%) against the murine leukemia L1210 cell line, but the lyophilized final product was difficult to reconstitute for a drug solution, probably due to the weak solubilizing groups employed. Therefore, the polyphosphazene carrier was subjected to modification studies by changing both the solubilizing group and the spacer group for platination (Jun et al., 2005; Lee et al., 1999b; Song et al., 2005). As a solubilizing group, methoxy-poly(ethylene glycol) (MPEG) was employed instead of the methoxy or methyl amine used in the former study, because FDA-approved PEGs not only have strong hydrophilicity but are also known to suppress protein binding, with long circulation properties in the blood circulation system (Delgado et al., 1992; Lee et al., 1998).

The synthetic route to preparing modified polyphosphazene–platinum conjugate drugs is shown in Scheme 1 (Jun et al., 2005; Song et al., 2005). In the molecular structure of the final product, V, the cationic (dach)Pt(II) (dach = *trans*-($\pm$)-1,2-diaminocyclohexane) moiety was chelated by the gluta-mate anion, forming a thermodynamically unstable eight-membered ring, and consequently, there is a possibility that the (O,O')-chelate formed initially may undergo isomerization to the more stable five-membered (O,N)-chelate isomer

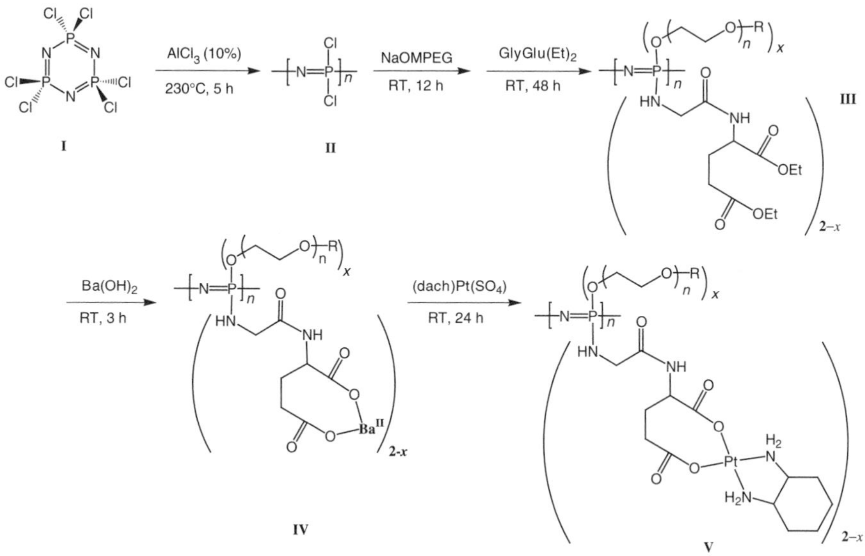

SCHEME 1 Synthetic route to polyphosphazene–Pt conjugate drugs.

with lower antitumor activity. However, it was found from a separate study (Kim, Y.S. et al., 2004) that the N-substituted aminodicarboxylate ligands coordinate to the platinum ion only through the (O,O′)-chelation mode, and their Pt(II) complexes were chemically stable in aqueous solution. The final product, V, was subjected to a biodistribution study (Jun et al., 2005) using male C57 BL/6N mice (8 to 9 weeks old, 25 to 27 g) to examine the EPR effect and long circulation properties of the polymer–Pt conjugate; the platinum distributions in selected organs are shown in Figure 6. We see that the polyphosphazene–Pt conjugate exhibits both long blood circulation and high tumor selectivity compared with the reference carboplatin. Much higher platinum concentrations were observed in the tumor tissue than in other organs at 2 h post-injection of the conjugate drug, probably due to its EPR effect. Furthermore, there is no significant change in platinum concentration even after 24 h post-injection in the polymeric conjugate, probably due to its effective retention in tumor tissue, whereas in the small-molecular carboplatin, the drug concentration decreased remarkably 24 h post-injection.

To examine the molecular size dependency of the EPR effect, polypho-sphazene–Pt conjugates were prepared using polyphosphazenes with molecular weights from 24,000 to 115,000, and the Pt-concentration ratios of tumor to muscle, that is, tumor/tissue ratios (TTRs), were measured. The TTR values listed in Table 1 show clearly that the EPR effect depends on the molecular weight of the polymer–drug conjugates, and the optimum EPR effect was observed for conjugates with a molecular weight in the range 60,000 to 80,000.

Both in vitro and in vivo antitumor activities of the conjugates were assayed against selected human tumor cell lines, but no significant differences were noted in antitumor activity among conjugates with different molecular weights. The overall cytotoxicity of conjugate 1 in the table was comparable to that

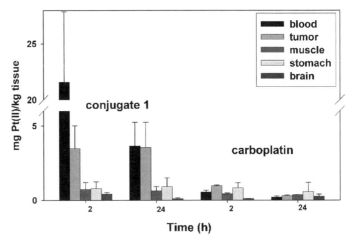

FIGURE 6 Biodistribution profile of the polyphosphazene–platinum(II) conjugate.

TABLE 1   Tumor/Tissue Ratio Based on the Molecular Weight of Pt Conjugates

Conjugate	Tumor/Tissue Ratio		Molecular Weight
	2 h	24 h	
1	4.80	4.60	24,000
2	4.10	5.64	32,600
3	9.40	13.0	62,800
4	5.60	9.40	89,300
5	3.81	7.39	88,300
6	7.70	6.80	115,000
Carboplatin	2.31	0.78	371.25

of cisplatin but much higher than that of carboplatin. In particular, it was surprising that conjugate 1 (IC_{50} = 29.4 µM) was very cytotoxic compared with cisplatin (IC_{50} = 105.5 µM) and carboplatin (IC_{50} = 902.0 µM) against the stomach cancer cell line (YCC-3), which is one of the least responsive to the anticancer drugs currently in clinical use. Therefore, conjugate 1 was advanced to an in vivo xenograft trial on the stomach cancer cell YCC-3 using nude mice, and the results are shown in Figure 7. The tumor growth rate for cisplatin was nearly the same as for the control (saline) and showed no tumor inhibition effect against the stomach cancer cell.

However, conjugate 1 has shown an outstanding effect on tumor inhibition at both high (60 mg/kg) and low dosages (30 mg/kg), and the relative tumor

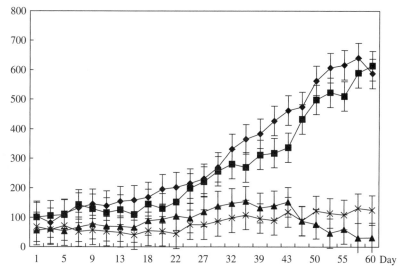

FIGURE 7   Xenograft trials of conjugate 1 and cisplatin against the YCC-3 cell line: conjugate 1 at low dose (30 mg/kg) (▲), and at high dose (60 mg/kg) (×); cisplatin (4 mg/kg) (■); control (saline) (◆).

TABLE 2   Pharmacokinetic Parameters After a Single Intravenous Injection

Parameter[a]	Conjugate 5	Carboplatin
$t_{1/2\alpha}$ (h)	0.257 ± 0.043	0.032 ± 0.003
$t_{1/2\beta}$ (h)	6.188 ± 0.187	0.415 ± 0.042
Cl (L/h·kg)	0.005 ± 0.0003	0.170 ± 0.012
MRT (h)	8.715 ± 0.248	0.562 ± 0.064
V_{dss} (L/kg)	0.044 ± 0.002	0.095 ± 0.009
AUC (nmol·hr/ml)	4019.62 ± 201.00	119.56 ± 7.39

[a] $t_{1/2\alpha}$; half-life in the α phase; $t_{1/2\beta}$; half-life in the β phase; Cl; clearance; MRT; mean residence time; V_{dss}; volume of distribution at steady state; AUC; area under the curve.

growth rate measured throughout the period of trials was less than 5%, which is a very promising result. To confirm quantitatively the long blood-circulating properties and to compare pharmacokinetic behaviors, the polymer conjugate and carboplatin, as reference, were studied pharmacokinetically (Song et al., 2005), and the results are shown in Table 2.

The time-dependent profiles of the plasma concentrations of platinum after injection showed a typical biphasic kinetic pattern like that of cisplatin for both polymer conjugate 1 and carboplatin: A rapid decrease in platinum concentrations was seen at the early stage right after injection (α phase) and the subsequent slow elimination (β phase), but a remarkable difference was noted in the elimination phase. Carboplatin showed a very fast decrease in the plasma concentrations of platinum, and most of the platinum was eliminated in about 5 h post-injection, like other small-molecular drugs, but the polymer conjugate exhibited a very slow platinum elimination rate. Other pharmacokinetic parameters are also listed in the table. Preclinical studies for conjugate 1 were almost completed.

In conclusion, polyphosphazenes bearing a hydrophilic methoxy-poly(ethylene glycol) (MPEG) as a solubilizing group and a glycyl glutamate or glycyl aspartate as a spacer group are excellent carriers for conjugation of the antitumor moiety (diamine)Pt(II) cation. The polyphosphazene–Pt conjgates were prepared by platination of the polymer with the (diamine)Pt(II) cation through (O,O')-chelation of the dipeptide spacer group. These polymer conjugates have exhibited excellent antitumor activity and pharmacokinetic behavior along with outstanding tumor selectivity by the EPR effect in addition to good water solubility and biodegradability.

SYNTHESIS AND APPLICATIONS OF CYCLOTRIPHOSPHAZENES FOR CANCER THERAPY

Thermal reaction of phosphorus pentachloride (PCl_5) and ammonium chloride (NH_4Cl) in an organic solvent such as chlorobenzene yields hexachlorocyclotriphosphazene ($N_3P_3Cl_6$), along with other higher cyclophosphazenes (Cotton and Shaver, 1971; Mark et al., 2005). Among these cyclic phosphazenes, hexachlorocyclotriphosphazene was studied most extensively because of its usefulness as a model compound for polymerization studies and its great potential for applications to a wide variety of new materials, such as flame retardants, luminescent materials, dendrimers, and drug carriers. Furthermore, since (Labarre et al., 1979) discovered the antitumor activity of hexaziridinyl-cyclotriphosphazene, intensive studies have continued to show improved antitumor activity with reduced toxicity. For example (Siwy et al., 2006, 2007) demonstrated recently that water solubility and cytotoxicity could both be improved with reduced toxicity by replacing the aziridinyl groups partially with a crown ether and 2-naphthol or an anthraquinone derivative. However, there are not many studies on anticancer drugs conjugated to cyclotripho-sphazenes other than amphiphilic cyclotriphosphazenes, which are dealt with next.

Synthesis and Applications of Amphiphilic Cyclotriphosphazenes

Many advantages of poly(organophosphazene)s over conventional organic polymers as drug delivery systems were described earlier. However, one of the general problems associated with polymeric drug delivery systems that must be overcome for clinical use is probably how to control exactly and reproducibly the purity and molecular weight of carrier polymers, which applies to poly(organophosphazene)s as well. In this regard, cyclotri(organophospha-zene)s are in a much better position, because the phosphazene trimer backbone is monodisperse and it is much easier to control the purity and molecular weight of the side groups. As a matter of fact, Lee et al. (2000) have succeeded in the synthesis of pure amphiphilic cyclotriphosphazenes by stepwise substitutions of hexachlorocyclotriphosphazene with equimolar hydrophilic PEG and hydrophobic amino acid. A few typical compounds among the products were analyzed to be highly pure ($>99\%$) using MALDI mass spectrometry and high-performance liquid chromatography. The first amphiphilic cyclotripho-sphazenes were prepared according to Scheme 2.

Hexachlorocyclotriphosphazene was reacted initially with 3 mol of sodium salt of a MPEG at low temperature ($< -60°C$) to obtain the intermediate (II) with *cis*-nongeminal conformation, which was further substituted by an amino

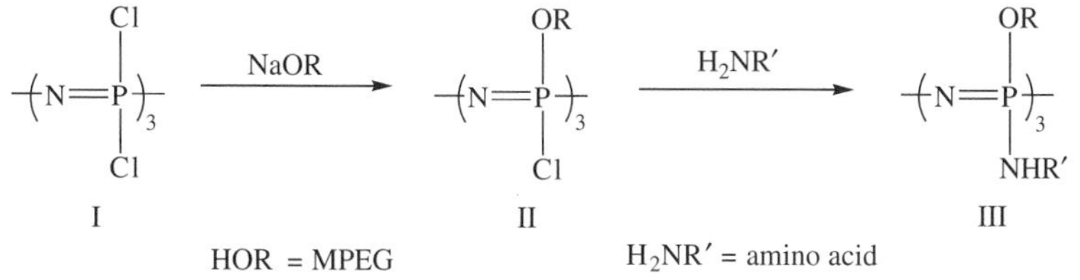

HOR = MPEG H_2NR' = amino acid

SCHEME 2   Synthetic route to amphiphilic cyclotriphosphazenes.

acid. Substitution reactions of cyclotriphosphazene were studied extensively by many research groups, and an excellent review by Allen (1991) is available. In general, when hexachlorocyclotriphosphazene is substituted with 3 mol of a substituent, three isomers (geminal 2,2,4; nongeminal *cis*-2,4,6; nongeminal *trans*-2,4,6) may be formed, but alkoxide groups favor *cis*-nongeminal conformation for an electronic reason. Alkoxy groups are electron donors toward the phosphorus atoms in the trimer ring, which makes the remaining P–Cl bond stronger and less susceptible to further substitution, thus yielding a nongeminal isomer. In addition, the choice of low reaction temperature in this study seems to afford almost exclusively a nongeminal *cis*-2,4,6-trimeric isomer with a peculiar molecular structure of octopus shape, which gave rise to thermosensitive properties. All the cyclotriphosphazene derivatives exhibited thermosensitivity by showing a lower critical solution temperature (LCST) in a wide range from 10.5 to 100°C, depending on their hydrophilic-to-hydrophobic balance. Despite their amphipilicity, no evidence for micelle formation was observed for these cyclotriphosphaznes.

Since these amphiphilic cyclotriphosphazenes exhibited thermosensitivity in a wide temperature range, Song et al. (2003) prepared a series of Pt–cyclotriphosphazene conjugates to see if any thermosensitive conjugate drugs could be obtained. The conjugates were prepared according to Scheme 3. The dicarboxylic esters of the trimer were hydrolyzed with barium hydroxide in methanol and then reacted with (diamine)PtSO$_4$ for platination in aqueous solution. Surprisingly, all the resulting Pt–cyclotriphosphazene conjugates also exhibited an LCST in a temperature range from 12.0 to >100°C. One of the representative conjugates, [NP(PEE)(Asp·Pt(dach))]$_3$ [PEE: 2-(2-proxyethoxy)ethoxy], with an LCST at 15°C, was selected for animal studies on biodistribution and drug-releasing profiles along with in vitro and in vivo assays of antitumor activity. This conjugate drug was administered by intratumoral injection to a CDF1 mouse inoculated with 3LL Lewis lung carcinoma cells, and the time-dependent distributions of platinum in the tumor tissue and other organs were examined along with cisplatin as a reference. The conjugate drug remained mostly in the tumor tissue until 5 h post-injection, but

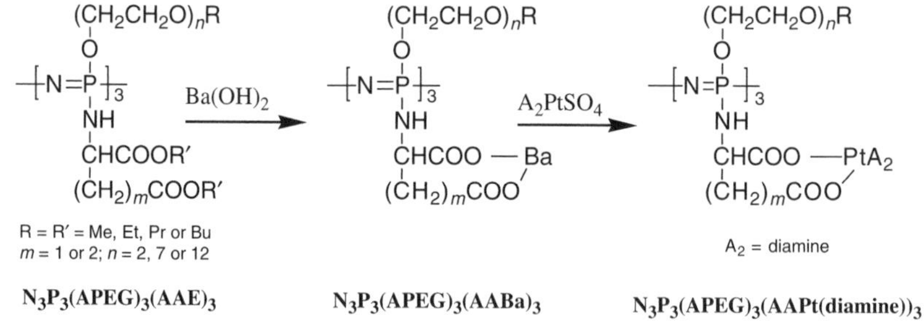

SCHEME 3 Synthetic route to thermosensitive Pt-cyclotriphosphazene conjugates.

only a small fraction of cisplatin remained in the tumor tissue even after 1 h post-injection. Furthermore, this conjugate has shown excellent in vivo antitumor activity against the leukemia L1210 cell line (ILS = >667%). However, this conjugate was found to be hydrolytically degradable, with a half-life of approximately 24 h, which could be a problem of drug stability. According to multinuclear (^{1}H, ^{31}P) NMR study, the hydrolytic instability of the conjugate drug seems to be ascribed to the closely located carboxylate group of amino acids, which may possibly produce carboxylate ions by hydrolysis (Song et al., 1999). Another preclinical candidate compound developed earlier, [NP(L-Glu · Li$_2$)(L-Glu · Pt(dach))]$_3$ (Baek et al., 2000), exposed the same problem. Therefore, further studies were conducted to modify the molecular structure of the cyclotriphosphazene drug carrier, and the problem of drug instability could be overcome by employing oligopeptides instead of amino acids as a spacer group, described in detail in the next section.

Studies on nonplatinum anticancer drugs conjugated to amphiphilic cyclotriphosphazene are rare, but Cho et al. (2005) prepared thermosensitive 5-fluorouracil (FU)–cyclotriphosphazene conjugates by coupling α-substituted glycine derivatives of 5-FU-containing carboxylic groups with cyclotriphosphazenes bearing methoxy-poly(ethylene glycol) (MPEG) or alkoxy ethylene oxide and lysine Et ester (LysOEt). These conjugates exhibited an LCST, and a few of them displayed an LCST below body temperature, which is suitable for local delivery by direct intratumoral injection. The conjugate exhibited gradual degradation at 37°C in both neutral and acidic buffer solutions. All of the conjugates displayed dose-dependent cytotoxicity against the leukemia L1210 cell line and exhibited more pronounced cytotoxic effects than those of free 5-FU.

Synthesis and Applications of Micellar Cyclotriphosphazenes

It is well known that amphiphilic di- or triblock copolymers consisting of both hydrophilic and hydrophobic segments in an appropriate ratio are self-assembled to form micelles in aqueous solution (Allen et al., 1999; Rösler et al., 2001; Zhang et al., 2004). The hydrophobic blocks of the copolymer form the core of the micelle, while the hydrophilic blocks form the corona or outer shell. A variety of drugs, including genes, proteins, and hydrophobic drugs such as paclitaxel, can be incorporated into the hydrophobic core of the micelle. As a hydrophilic block, poly(ethylene glycol) is most widely used, but many low-molecular-weight hydrophobic polymers, such as poly(propylene oxide) and poly(lactic acid), are employed as a hydrophobic block (Lavasanifar et al., 2002; Torchilin, 2001). Nearly all the micelle-forming polymers reported so far are amphiphilic grafted or block copolymers with a linear backbone, but no micellar polymer with a cyclic backbone has yet been reported.

Very recently, Jun et al. (2006) and Toti et al. (2007) have discovered that cyclotriphosphazenes grafted with equimolar amounts of oligopeptide instead of amino acids as a hydrophobic group and poly(ethylene glycol) as a

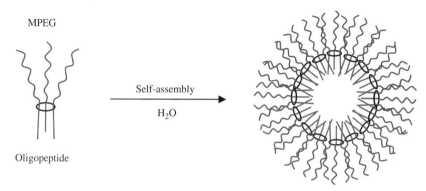

FIGURE 8 Micelle formation from oligopeptide-grafted cyclotriphosphazenes.

hydrophilic group are self-assembled to form micelles in aqueous solution as shown in Figure 8. These micelles were found to be very stable in physiological conditions and exhibited excellent physicochemical properties suitable for cancer therapy as well as protein drug delivery.

These micellar cyclotriphosphazenes were prepared according to the same procedure as that of amphiphilic cyclotriphosphazenes in Scheme 2. The only difference was the use of one of the hydrophobic tri- to hexapeptides instead of a simple amino acid as a hydrophobic group, along with an appropriate MPEG balancing the hydrophobicity of the oligopeptide employed. Typical micellar trimers prepared are listed along with their properties in Table 3.

It was surprising that a simple replacement of amino acid with oligopeptide as a hydrophobic side group made great differences in the physicochemical properties of cyclotriphosphazenes. First, the greatest difference was that oligopeptide-grafted cyclotriphosphazenes could form very stable micelles by

TABLE 3 **Characteristic Properties of Micellar Cyclotriphosphazenes**

| Trimer | Molecular Formula | LCST (°C)^a | | DLS^b (nm) | PDI^c |
		Water	PBS		
1	[NP(MPEG350)(GlyPheLeuAspEt$_2$)]$_3$	29	25	13.9	0.25
2	[NP(MPEG350)(GlyPheLeuGluEt$_2$)]$_3$	26	23	13.3	0.44
3	[NP(MPEG350)(GlyPheLeuGlyPheEt)]$_3$	24	17	13.6	0.24
4	[NP(MPEG350)(GlyPheLeuGlyPheLeuEt)]$_3$	20	13	27.8	0.13
5	[NP(MPEG550)(GlyPheLeuGlyPheLeuEt)]$_3$	48	42	7.4	0.19
6	[NP(MPEG550)(GlyLeuPheGlyLeuPheEt)]$_3$	47	45	8.0	0.35

^{a}The lower critical solution temperatures in water and phosphate-buffered saline (PBS) solutions (0.5%).
^{b}Dynamic light-scattering measurements at 20°C of 0.5% aqueous solutions filtered by a 0.45 μm syringe filter.
^{c}Polydispersity index representing S/d_h, where S is the standard error in hydrodynamic diameter d_h.

self-assembly in aqueous solution, whereas amino acid–grafted cyclotriphosphazenes could not. Second, in contrast to the instability of amino acid–grafted cyclotriphosphazenes due to closely located carboxylate groups, the oligopeptide-grafted cyclotriphosphazenes are chemically very stable, and hydrolytic degradation in aqueous solution was hardly observed, at least for months. These micellar cyclotriphosphazenes as a unimer have a unique molecular structure in which the three hydrophobic oligopeptide side groups are oriented in one direction opposite to the other three hydrophilic PEG groups with respect to the cyclic phosphazene ring shown in Figure 8. Such a steric amphiphile structure is presumed to afford strong intra- and intermolecular hydrophobic interactions to make a highly hydrophobic core structure.

For such a unique structural reason, these micellar cyclotriphosphazenes used as new drug delivery systems offer many advantages over conventional micelles formed from organic di- or triblock linear copolymers. First, these cyclic trimers form very stable micelles with a very low critical micelle concentration (CMC) of 0.1 mg/L, in contrast to the high CMC values (10 to 100 mg/L) of conventional surfactant micelles, which is a critical factor for injectable drug delivery. Second, all these micellar cyclotriphosphazenes exhibit thermosensitivity by showing an LCST, which is a useful property for local drug delivery. Third, the oligopeptide-grafted cyclotriphosphazenes are biodegradable since the oligopeptides employed are all degradable by lysosomal enzymes within cells (Soyez et al., 1996). Fourth, the extended hydophobicity of the amphiphilic cyclotriphosphazenes by employing hydrophobic oligopeptides instead of amino acids allows strong solubilizing power for highly hydrophobic drugs such as paclitaxel. Fifth, the trimer backbone is monodisperse, and consequently, the molecular weight of the carrier polymers is easy to control. Finally, these cyclotriphosphazenes can easily be functionalized by hydrolysis for conjugation with drug molecules. Therefore, these micellar cyclotriphosphazenes can be applied not only as formulating agents for hydrophobic drugs such as paclitaxel and protein drugs but also as drug-conjugated carriers.

Among the application studies for cancer therapy using these micellar cyclotriphosphazenes, Yu et al. (2007) have reported most recently very interesting and promising platinum prodrug candidate compounds that initially form micelles in aqueous solution but aggregate further to larger soft nanoparticles with a diameter of 150 to 200 nm. The micellar Pt–cyclotriphosphazene conjugate compounds were prepared following Scheme 4. The platinum conjugates could be prepared either in an organic solvent (route 1) or in an aqueous solution (route 2), depending on the hydrophobicity of the (diamine)Pt(II) moiety, as shown in the scheme.

It was found further that the hydrophobicity of the drug carrier cyclotriphosphazenes could be extended using the hydrophobic (diamine)Pt(II) moiety, which allowed the micellar conjugates to aggregate to larger nanoparticles. These nanoparticles were very stable in phosphate-buffered saline solution even

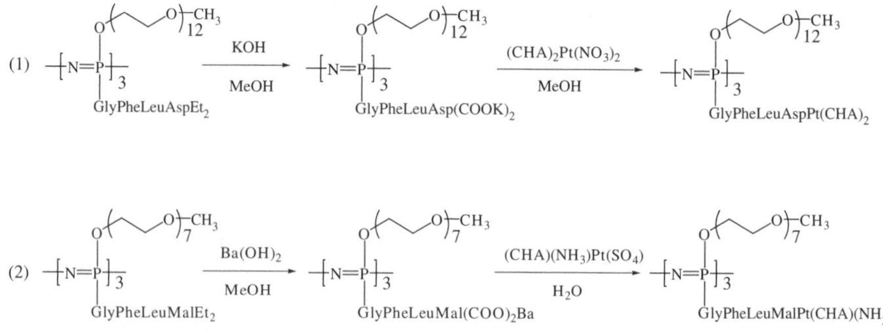

SCHEME 4 Synthetic routes to Pt-cyclotriphosphazene conjugates.

after dilution to $10\,\mu M$ and exhibited an excellent EPR effect by showing a tumor/tissue ratio of 5.2 at 2 h post-injection and 6.5 at 24 h post-injection (results to be published later). The most hydrophobic Pt–cyclotriphosphazene conjugate, $[NP(MPEG550)(GlyPheLeuAsp \cdot Pt(CHA)_2)]$, is now in preclinical studies.

CONCLUSIONS

Remarkable progress has been made in recent decades in the synthesis and application of organophosphazenes as drug carriers for cancer therapy. A variety of novel polyphosphazene drug carriers have recently been developed in various forms, such as micelles, hydrogels, microspheres, and matrices, based on their biodegradability, amphiphilicity, longevity, and other properties, which are useful for both formulation and conjugation with small molecular anticancer drugs to improve their therapeutic values. In particular, the polymeric anticancer drug (diamine)Pt(II)-conjugated polyphosphazene, bearing a hydrophilic methoxy-poly(ethylene glycol) as a solubilizing group and a glycyl glutamate as a spacer group, exhibited excellent antitumor activity and pharmacokinetic behavior along with excellent tumor selectivity by its EPR effect. Most recently it was discovered that cyclotriphosphazenes grafted with equimolar hydrophilic MPEG and hydrophobic oligopeptide were able to form thermodynamically very stable micelles by self-assembly in aqueous solution. These cyclotriphosphazene micelles were found to offer additional advantages as drug carriers, such as strong solubilizing power for highly hydrophobic drugs based on micellar encapsulation and easy control of their molecular weight due to the monodispersity of the trimeric phosphazene backbone. It was also observed that micelles formed initially from very hydrophobic (dicyclo-hexylamine)Pt–cyclotriphosphazene conjugates were subject to further aggregation to large nanoparticles with a mean diameter of 100 to 200 nm, which exhibited outstanding tumor selectivity.

REFERENCES

Allcock, H.R. 2006. A perspective of polyphosphazene research. *J. Inorg. Organomet. Polym. Mater.*, 16:277–294.

Allcock, H.R., Ambrosio, A.M.A. 1996. Synthesis and characterization of pH-sensitive poly(organophosphazene) hydrogels. *Biomaterials*, 17:2295–2302.

Allcock, H.R., Kugel, R.L. 1965. Synthesis of high polymeric alkoxy- and aryloxy-phosphonitriles. *J. Am. Chem. Soc.*, 87:4216–4217.

Allcock, H.R., Pucher, S.R. 1991. Polyphosphazenes with glucosyl and methylamino, trifluoroethoxy, phenoxy, or (methoxyethoxy)ethoxy side groups. *Macromolecules*, 24:23–24.

Allcock, H.R., Allen, R.W., O'Brien, J.P. 1977. Synthesis of platinum derivatives of polymeric and cyclic phosphazenes. *J. Am. Chem. Soc.*, 99:3984–3987.

Allcock, H.R., Cho, S.Y., Steely, L.B. 2006. New amphiphilic poly[bis(2,2,2,-trifluoroethoxy)phosphazene]/poly(propylene glycol) triblock copolymer: synthesis and micellar characteristics. *Macromolecules*, 39:8334–8338.

Allen, C., Maysinger, D., Eisenberg, A. 1999. Nano-engineering block copolymer aggregates for drug delivery. *Colloids Surf. Biointerfaces*, 16:3–27.

Allen, C.W. 1991. Regio- and stereochemical control in substitution reactions of cyclophosphazenes. *Chem. Rev.*, 91:119–135.

Allen, R.W., O'Brien, J.P., Allcock, H.R. 1977. Crystal and molecular structure of a platinum–cyclophosphazene complex: *cis*-dichloro[octa(methylamino)cyclotetraphosphazene-N',N'']platinum(II). *J. Am. Chem. Soc.*, 99:3987–3991.

Allen, T.M., Cullis, P.R. 2004. Drug delivery systems: entering the mainstream. *Science*, 303:1818–1822.

Andrianov, A.K. 2006. Water-soluble polyphosphazenes for biomedical applications. *J. Inorg. Organomet. Polym. Mater.*, 16:397–406.

Andrianov, A.K. 2007. Polyphosphazenes as vaccine adjuvants. *Vaccine Adjuvants Deliv. Syst.*, 355–378.

Andrianov, A.K., Payne, L.G. 1998. Protein release from polyphosphazene matrices. *Adv. Drug Deliv. Rev.*, 31:185–196.

Andrianov, A.K., Chen, J., LeGolvan, M.P. 2004a. Poly(dichlorophosphazenes) as a precursor for biologically active polyphosphazenes: synthesis, characterization, and stabilization. *Macromolecules*, 37:414–420.

Andrianov, A.K., Svirkin, Y.Y., LeGolvan, M.P. 2004b. Synthesis and biologically relevant properties of polyphosphazene polyacids. *Biomacromolecules*, 5:1999–2006.

Baek, H.G., Cho, Y.H., Lee, C.O., Sohn, Y.S. 2000. Synthesis and antitumor activity of cyclotriphosphazene-(diamine)platinum(II) conjugates. *Anti-Cancer Drugs*, 11:715–725.

Birgger, I., Dubernet, C., Couvreur, P. 2002. Nanoparticles in cancer therapy and diagonosis. *Adv. Drug Deliv. Rev.*, 54:631–651.

Brannon-Peppas, L., Blanchette, J.O. 2004. Nanoparticle and targeted systems for cancer therapy. *Adv. Drug Deliv. Rev.*, 56:1649–1659.

Caliceti, P., Veronese, F.M., Lora, S. 2000. Polyphosphazene microspheres for insulin delivery. *Int. J. Pharm.*, 211:57–65.

Chaney, S.G., Vaisman, A. 2000. DNA adduct tolerance bypass. In Kelland, L.R., Farrel, N.P., eds., *Platinum-Based Drugs in Cancer Therapy.* Humana Press, Totowa, NJ, pp. 129–148.

Chang, Y.K., Bender, J.D., Phelps, M.V.B., Allcock, H.R. 2002. Synthesis and self-association behavior of biodegradable amphiphilic poly[bis(ethyl glycinat-*N*-yl)phosphazene]–poly(ethylene oxide) block copolymers. *Biomacromolecules*, 3:1364–1369.

Chang, Y.K., Powell, E.S., Allcock, H.R. 2005. Environmentally responsive micelles from polystyrene–poly[bis(potassium carboxylatophenoxy)phosphazene] block copolymers. *J. Polym. Sci. A*, 43:2912–2920.

Charrois, G.J.R., Allen, T.M. 2003. Rate of biodistribution of STEALTH liposomes to tumor and skin: influence of liposome diameter and implications for toxicity and therapeutic activity. *Biochim. Biophys. Acta*, 1609:102–108.

Cho, Y.W., Lee J-R., Song S-C. 2005. Novel thermosensitive 5-fluorouracil–cyclotriphosphazene conjugates: synthesis, thermosensitivity, degradability, and in vitro antitumor activity. *Bioconjugate Chem*, 16:1529–1535.

Cotton, F.A., Shaver, A. 1971. Some observation on the synthesis of octamethylcyclotetraphosphazene and hexamethylcyclotriphosphazene. *Inorg. Chem.*, 10:2362–2363.

de Wolf, H.K., Luten, J., Snel, C.J., Oussoren, C., Hennink, W.E. 2005. In vivo tumor transfection mediated by polyplexes based on biodegradable poly(DMAEA)–phosphazene. *J. Control. Release*, 109:275–287.

de Wolf, H.K., de Raad, M., Snel, C., van Steenbergen, M.J., Fens, M.H.A.M., Storm, G., Hennink, W.E. 2007. Biodegradable poly(2-dimethylamino ethylamino)phosphazene for in vivo gene delivery to tumor cells: effect of polymer molecular weight. *Pharm. Res.*, 24:1572–1580.

Delgado, C., Francis, G.E., Fisher, D. 1992. The uses and properties of PEG-linked proteins. *Crit. Rev. Ther. Drug Carrier Syst.*, 9:249–304.

Dong, Y., Feng, S.S. 2004. Methoxy poly(ethylene glycol)–poly(lactide) (MPEG-PLA) nanoparticles for controlled delivery of anticancer drugs. *Biomaterials*, 25:2843–2849.

Dreher, M.R., Liu, W., Michelich, C.R., Dewhirst, M.W., Yuan, F., Chilkoti, A. 2006. Tumor vascular permeability, accumulation, and penetration of macromolecular drug carriers. *J. Natl. Cancer Inst.*, 98:335–344.

Duncan, R. 1999. Polymer conjugates for tumour targeting and intracytoplasmic delivery. The EPR effect as a common gateway?. *Pharm. Sci. Technol. Today*, 2:441–449.

Duncan, R. 2003. The dawning era of polymer therapeutics. *Nat. Rev.*, 2:347–360.

Duncan, R. 2006. Polymer conjugates as anticancer nanomedicines. *Nat. Rev.*, 6: 688–701.

Fonseca, C., Simoes, S., Gaspar, R. 2002. Paclitaxel-loaded PLGA nanoparticles: preparation, physicochemical characterization and in vitro anti-tumoral activity. *J. Control. Release*, 83:273–286.

Gleria, M., Jaeger, D.R. 2005. Polyphosphazenes: a review. *Top. Curr. Chem.* 250: 165–251.

Goedemoed, J.H., de Groot, K., Claessen, A.M.E., Scheper, R.J. 1991a. Development of implantable antitumor device based on polyphosphazene, II. *J. Control. Release*, 17:235–244.

Goedemoed, J.H., Mense, E.H.G., de Groot, K., Claessen, A.M.E., Scheper, R.J. 1991b. Development of injectable antitumor microspheres based on polyphosphazene. *J. Control. Release*, 17:245–258.

Haag, R., Kratz, F. 2006. Polymer therapeutics: concepts and applications. *Angew. Chem. Int. Ed. Engl.*, 45:1198–1215.

Hashizume, H., Baluk, P., Morikawa, S., McLean, J.W., Thurston, G., Roberge, S., Jain, R.K., McDonald, D.M. 2000. Openings between defective endothelial cells explain tumor vessel leakiness. *Am. J. Pathol.*, 156:1363–1380.

Hilgenbrink, A.R., Low, P.S. 2005. Folate receptor–mediated drug targeting: from therapeutics to diagnostics. *J. Pharm. Sci.*, 94:2135–2146.

Hobbs, S.K., Monsky, W.L., Yuan, F., Roberts, W.G., Griffith, L., Torchilin, V.P. 1998. Regulation of transport pathways in tumor vessels: role of tumor type and microenvironment. *Proc. Natl. Acad. Sci. USA*, 95:4607–4612.

Hubbard, S.M., Jenkins, J.F. 1990. Chemotherapy administration: practical guidelines. In Chabner, B.A., Collins, J.M., eds., *Cancer Therapy*. J.B. Lippincott, Philadelphia, pp. 450–452.

Hur, K.M., Lee, S.C., Cho, Y.W., Lee, J., Jeong, J.H., Park, K. 2005. Hydrotropic polymer micelle system for delivery of paclitaxel. *J. Control. Release*, 101:59–68.

Jun, Y.J., Kim, J.I., Jun, M.J., Sohn, Y.S. 2005. Selective tumor targeting by enhanced permeability and retention effect: synthesis and antitumor activity of polyphosphazene–platinum(II) conjugates. *J. Inorg. Biochem.*, 99:1593–1601.

Jun, Y.J., Toti, U.S., Kim, H.Y., Yu, J.Y., Jeong, B.M., Jun, M.J., Sohn, Y.S. 2006. Thermoresponsive micelles from oligopeptide-grafted cyclophosphazenes. *Angew. Chem. Int. Ed.*, 45:6173–6176.

Jun, Y.J., Kim, J.H., Choi, S.J., Lee, H.J., Jun, M.J., Sohn, Y.S. 2007. A tetra(L-lysine)-grafted poly(organophosphazene) for gene delivery. *Bioorg. Med. Chem. Lett.*, 17: 2975–2978.

Kang, G.D., Cheon, S.H., Khang, G., Song, S.C. 2006a. Thermosensitive poly(organophosphazene) hydrogels for a controlled drug delivery. *Eur. J. Pharm. Biopharm.*, 63:340–346.

Kang, G.D., Cheon, S.H., Song, S.C. 2006b. Controlled release of doxorubicin from thermosensitive poly(phosphazene) hydrogels. *Int. J. Pharm.*, 319:29–36.

Kataoka, K., Harada, A., Nagasaki, Y. 2001. Block copolymer micelles for drug delivery: design, characterization and biological significance. *Adv. Drug Deliv. Rev.*, 47:113–131.

Kim, J.I., Jun, Y.J., Seong, J.Y., Jun, M.J., Sohn, Y.S. 2004. Synthesis and characterization of nanosized poly(organophosphazenes) with methoxy-poly(ethylene glycol) and dipeptide ethyl esters as side groups. *Polymer*, 45:7983–7089.

Kim, S.C., Kim, D.W., Shim, Y.H., Bang, J.S., Oh, H.S., Kim, S.W., Seo, M.H. 2001. In vivo evaluation of polymeric micellar paclitaxel formulation: toxicity and efficacy. *J. Control. Release*, 72:191–202.

Kim, Y.S., Song, R., Chung, H.C., Jun, M.J., Sohn, Y.S. 2004. Coordination modes vs. antitumor acitivity: synthesis and antitumor activity of novel platinum(II) complexes of N-substituted amino dicarboxylic acids. *J. Inorg. Biochem.*, 98:98–104.

Kumbar, S.G., Bhattacharyya, S., Nukavarapu, S.P., Khan, Y.M., Nair, L.S., Laurencin, C.T. 2006. In vitro and in vivo characterization of biodegradable poly(organophosphazenes) for biomedical applications. *J. Inorg. Organomet. Polym. Mater.*, 16:365–385.

Labarre, J.F., Faucher, J.P., Levy, G., Sournies, F., Cros, S., Francois, G. 1979. Antitumor activity of some cyclophosphazenes. *Eur. J. Cancer*, 15:637–643.

Lakshmi, S., Katti, D.S., Laurencin, C.T. 2003. Biodegradable polyphosphazenes for drug delivery applications. *Adv. Drug Deliv. Rev.*, 55:467–482.

Lavasanifar, A., Samuel, J., Kwon, G.S. 2002. Poly(ethylene oxide)-*block*-poly(L-amino acid) micelles for drug delivery. *Adv. Drug Deliv. Rev.*, 54:169–190.

Lee, B.H., Lee, Y.M., Sohn, Y.S., Song, S.C. 2002. A thermosensitive poly(organo-phosphazene) gel. *Macromolecules*, 35:3876–3879.

Lee, J.H., Kopecek, J., Andrade, J.D. 1998. Protein-resistant surfaces prepared by PEO-containing block copolymer surfactants. *J. Biomed. Mater. Res.*, 23:351–368.

Lee, S.B., Song, S.C., Jin, J.I., Sohn, Y.S. 1999a. A new class of biodegradable thermosensitive polymers: 2. Hydrolytic properties and salt effect on the lower critical solution temperature of poly(organophosphazenes) with methoxypoly(ethylene glycol) and amino acid esters as side groups. *Macromolecules*, 32:7820–7827.

Lee, S.B., Song, S.C., Jin, J.I., Sohn, Y.S. 1999b. Synthesis and antitumor activity of polyphosphazene/methoxy-poly(ethylene glycol)/(diamine)platinum(II) conjugates. *Polymer Journal*, 31:1247–1252.

Lee, S.B., Song, S.C., Jin, J.I., Sohn, Y.S. 2000. Thermosensitive cyclotriphosphazenes. *J. Am. Chem. Soc.*, 122:8315–8316.

Lee, S.C., Kim, C., Kwon, I.C., Chung, H., Jeong, S.Y. 2003. Polymeric micelles of poly(2-ethyl-2-oxazoline)-*block*-poly(epsilon-caprolactone) copolymer as a carrier for paclitaxel. *J. Control. Release*, 89:437–446.

Lemmouchi, Y., Schacht, E., Dejardin, S., Vandorpe, J., Seymour, L. 1997. Biodegradable phosphazenes for drug delivery. *Macromol. Symp.*, 123:103–112.

Liggins, R.T., Burt, H.M. 2002. Polyether–polyester diblock copolymers for the preparation of paclitaxel loaded polymeric micelle formulations. *Adv. Drug Deliv. Rev.*, 54:191–202.

Lokich, J., Anderson, N. 1998. Carboplatin versus cisplatin in solid tumors: an analysis of the literature. *Ann. Oncol.*, 9:13–21.

Lundberg, A.S., Weinberg, R.A. 1999. Control of the cell cycle and apoptosis. *Eur. J. Cancer*, 35:531–539.

Luten, J., van Steenis, J.H., van Sormeren, R., Kemmink, J., Schuurmans-Nieuwenbrek, N.M.E., Koning, G.A., Crommelin, D.J.A., van Norstrum, C.F., Hennink, W.E. 2003. Water-soluble biodegradable cationic polyphosphazenes for gene delivery. *J. Control. Release*, 89:483–497.

Maeda, H. 2001. SMANCS and polymer-conjugated macromolecular drugs: advantages in cancer chemotherapy. *Adv. Drug Deliv. Rev.*, 46:169–185.

Maeda, H., Wu, J., Sawa, T., Matsumura, Y., Hori, K. 2000. Tumor vascular permeability and the EPR effect in macromolecular therapeutics: a review. *J. Control. Release*, 65:271–284.

Maeda, H., Fang, J., Inutsuka, T., Kitamoto, Y. 2003. Vascular permeability enhancement in solid tumor: various factors, mechanisms involved and its implications. *Int. Immun.*, 3:319–328.

Marecos, E., Weissleder, R., Bogdanov, A. 1998. Antibody-mediated versus nontargeted delivery in a human small cell lung carcinoma model. *Bioconjugate Chem.*, 9:184–191.

Mark, J.E., Alcock, H.R., West, R. 2005. *Inorganic Polymers*, 2nd ed. Oxford University Press, New York, pp. 62–153.

Matsumura, Y., Maeda, H. 1986. A new concept for macromolecular therapeutics in cancer chemotherapy: mechanism of tumoritropic accumulation of proteins and the antitumor agent smancs. *Cancer Res.*, 46:6387–6392.

McAuliffe, C.A., Sharma, H.A., Tinker, N.D. 1991. Cancer chemotherapy involving platinum and other platinum group complexes. In Hartly, F.R., ed. *Chemistry of the Platinum Group Metals*, Elsevier, Amsterdam, pp. 546–593.

Mu, L., Feng, S.S. 2003. A novel controlled release formulation for the anticancer drug paclitaxel (Taxol): PLGA nanoparticles containing vitamin E TPGS. *J. Control. Release*, 86:33–48.

Nair, L.S., Bhattacharyya, S., Bender, J.D., Greish, Y.E., Brown, P.W., Allcock, H.R., Laurencin, C.T. 2004. Fabrication and optimization of methylphenoxy substituted polyphosphazene nanofibers for biomedical applications. *Biomacromolecules*, 5:2212–2220.

O'Dwyer, P.J., Stevenson, J.P., Johnson, S.W. 1999. Clinical status of cisplatin, carboplatin, and other platinum-based antitumor drugs. In Lippert, B., ed., *Cisplatin*, Wiley-VCH, Zurich, Switzerland, pp. 1–69.

Park, J.H., Lee, S.K., Kim, J.H., Park, K.G., Kim, K.M., Kwon, I.C. 2008. Polymeric nanomedicine for cancer therapy. *Prog. Polym. Sci.*, 33:113–137.

Rösler, A., Vandermeulen, G.W.M., Klok H-A. 2001. Advanced drug delivery devices via self-assembly of amphiphilic block copolymers. *Adv. Drug Deliv. Rev.*, 53:95–108.

Ruel-Gariepy, G.M., Shive, M., Bichara, A., Berrada, M., Le Garrec, D., Chenite, A., Leroux, J.C. 2004. A thermosensitive chitosan-based hydrogel for the local delivery of paclitaxel. *Eur. J. Pharm. Biopharm.*, 57:53–63.

Schacht, E., Vandorpe, J., Dejardin, S., Lemmouchi, Y., Seymour, L. 1996. Biomedical applications of degradable polyphosphazenes. *Biotechnol. Bioeng.*, 52:102–109.

Seong, J.Y., Jun, Y.J., Jeong, B.M., Sohn, Y.S. 2005. New thermogelling poly(organophosphazenes) with methoxypoly(ethylene glycol) and oligopeptide as side groups. *Polymer*, 46:5075–5081.

Seymour, L.W., Miyamoto, Y., Maeda, H., Brereton, M., Strohalem, J., Ulbrich, K., Duncan, R. 1995. Influence of molecular weight on passive tumour accumulation of a soluble macromolecular drug carrier. *Eur. J. Cancer*, 31A:766–770.

Sinha, R., Kim, G.J., Nie, S., Shin, D.M. 2006. Nanotechnology in cancer therapeutiecs: bioconjugated nanoparticles for drug delivery. *Mol. Cancer Ther.*, 5:1909–1917.

Sinko, P., Kohn, J. 1993. Polymeric drug delivery systems: an overview. In Magda, A., Nokaly, E., Piatt, D.M., Charpentier, B.A., ed., *Polymeric Delivery Systems*, ACS Series 520, ACS Washinton, DC, pp. 18–41.

Siwy, M., Sek, D., Kaczmarczyk, B., Jaroszewicz, I., Nasulewicz, A., Pelczyñska, M., Nevozhay, D., Opolski, A. 2006. Synthesis and in vitro antileukemic activity of some new 1,3-(ocytetraethylenoxy)cyclotriphosphazene derivatives. *J. Med. Chem.*, 49:806–810.

Siwy, M., Sek, D., Kaczmarczyk, B., Joanna, W., Nasulewicz, A., Opolski, A. 2007. Synthesis and in vitro antiproliferative activity of new 1,3-(oxytetraethylenoxy)cyclotriphosphazene derivatives. *Anticancer Res.*, 27:1553–1558.

Soga, O., van Nostrum, C.F., Fens, M., Rijcken, C.J.F., Schiffelers, R.M., Storm, G., Hennink, W.E. 2005. Thermosensitive and biodegradable polymeric micelles for paclitaxel delivery. *J. Control. Release*, 103:341–353.

Sohn, Y.S., Cho, Y.H., Baek, H.G., Jung, O.S. 1995. Synthesis and properties of low molecular weight polyphosphazenes. *Macromolecules*, 28:7566–7568.

Sohn, Y.S., Baek, H.G., Cho, Y.H., Lee, Y.A., Jung, O.S., Lee, C.O., Kim, Y.S. 1997. Synthesis and antitumor activity of novel polyphosphazene–(diamine)platinum(II) conjugates. *Int. J. Pharm.*, 153:79–91.

Song, R., Jun, Y.J., Kim, J.I., Jin, C.B., Sohn, Y.S. 2005. Synthesis, characterization, and tumor selectivity of a polyphosphazene–platinum(II) conjugate. *J. Control. Release*, 105:142–150.

Song, S.C., Lee, S.B., Jin, J.I., Sohn, Y.S. 1999. A new class of biodegradable thermosensitive polymers: I. Synthesis and characterization of poly(organophosphazenes) with methoxy poly(ethylene glycol) and amino acid esters as side groups. *Macromolecules*, 32:2188–2193.

Song, S.C., Lee, S.B., Lee, B.H., Ha, H.W., Lee, K.T., Sohn, Y.S. 2003. Synthesis and antitumor activity of novel thermosensitive platinum(II)–cyclotriphosphazene conjugates. *J. Control. Release*, 90:303–311.

Soyez, H., Schacht, E., Vanderkerten, S. 1996. The crucial role of spacer groups in macromolecular prodrug design. *Adv. Drug Deliv. Rev.*, 21:81–106.

Takakura, Y., Hashida, M. 1996. Macromolecular carrier systems for targeted drug delivery: pharmacokinetic considerations on biodistribution. *Pharm. Res.*, 13:820–831.

Torchilin, V.P. 2000. Drug targeting. *Eur. J. Pharm. Sci.*, 2:S81–S91.

Torchilin, V.P. 2001. Structure and design of polymeric surfactant-based drug delivery systems. *J. Control. Release*, 73:137–172.

Toti, U.S., Moon, S.H., Kim, H.Y., Jun, Y.J., Kim, B.M., Park, Y.M., Jeong, B.M., Sohn, Y.S. 2007. Thermosensitive and biocompatible cyclotriphosphazene micelles. *J. Control. Release*, 119:34–40.

Tsuchiya, K., Uchida, T., Kobayashi, M., Maeda, H., Konno, T., Yamanaka, H. 2000. Tumor-targeted chemotherapy with SMANCS in lipiodol for renal cell carcinoma: longer survival with larger size tumors. *Urology*, 55:495–500.

Vasey, P.A., Kaye, S.B., Morrison, R., Twelves, C., Wilson, P., Duncan, R., Thomson, A.H., Murray, L.S., Hilditch, T.E., Murray, T., Burtles, S., Fraier, D., Frigerio, E., Cassidy, J. 1999. Phase I clinical and pharmacokinetic study of PK1 [*N*-(2-

hydroxypropyl)methacrylamide copolymer doxorubicin]: first member of a new class of chemotherapeutic agents drug–polymer conjugates. *Clin. Cancer Res.*, 5:83–94.

Veronese, F.M., Marsilio, F., Caliceti, P., Filippies, P.D., Giunchedi, P., Loar, S. 1998. Polyorganophosphazene microspheres for drug release: polymer, synthesis, microsphere preparation, in vitro and in vivo naproxen release. *J. Control. Release*, 52: 227–237.

Yu, J.Y., Jun, Y.J., Jang, S.H., Lee, H.J., Sohn, Y.S. 2007. Nanoparticulate platinum(II) anticancer drug: synthesis and characterization of amphiphilic cyclotriphosphazene–plainum(II) conjugates. *J. Inorg. Biochem.*, 101:1931–1936.

Yuan, F., Dellian, M., Fukumura, D., Leunig, M., Berk, D.A., Torchilin, V.P. 1995. Vascular permeability in a human tumor xenograft: molecular size dependence and cutoff size. *Cancer Res.*, 55:3752–3756.

Zhang, J.X., Qiu, L.Y., Zhu, K.J., Jin, Y. 2004. Thermosensitive micelles self-assembled by novel *N*-isopropylacrylamide oligomer grafted polyphosphazene. *Macromol. Rapid Commun.*, 25:1563–1567.

Zhang, J.X., Qiu, L.Y., Jin, Y., Zhu, K.J. 2005a. Physicochemical characterization of polymeric micelles constructed from novel amphiphilic polyphosphazene with poly(*N*-isopropylacrylamide) and ethyl 4-aminobenzoate as side groups. *Colloids Surfaces B*, 16:123–130.

Zhang, J.X., Qiu, L.Y., Jin, Y., Zhu, K.J. 2005b. Thermally responsive polymeric micelles self-assembled by amphiphilic polyphosphazene with poly(*N*-isopropylacrylamide) and ethyl glycinate as side groups: polymer synthesis, characterization, and in vitro drug release study. *J. Biomed. Mater. Res. A*, 76A:773–780.

Zhang, J.X., Li, X.J., Qiu, L.Y., Li, X.H., Yan, M.Q., Jin, Y., Zhu, K.J. 2006. Indomethacin-loaded polymeric nanocarriers based on amphiphilic polyphosphazenes with poly (*N*-isopropylacrylamide) and ethyl tryptophan as side groups: preparation, in vitro and in vivo evaluation. *J. Control. Release*, 116:322–329.

15 Amphiphilic Polyphosphazenes as Drug Carriers

LIYAN QIU

College of Pharmaceutical Sciences, Zhejiang University, Hangzhou, China

CHENG ZHENG

College of Pharmaceutical Sciences and Institute of Polymer Science, Zhejiang University, Hangzhou, China

INTRODUCTION

When the chlorine atoms on a polyphosphazene backbone are substituted by both hydrophilic and hydrophobic substances with appropriate ratios, the product is an amphiphilic polyphosphazene. The substitution style could be designed diversely: hydrophilic polymer chains with hydrophobic small molecules, hydrophilic polymer chains with hydrophobic polymer chains, or amphiphilic polymer chains only (as shown in Fig. 1). These substances are grafted to polyphosphazene backbones in high density; therefore, a complete amphiphilic polyphosphazene polymer chain may look like a polymer brush (Fig. 1b and c).

Early research on amphiphilic polyphosphazenes for drug delivery concerned primarily the controlled drug release property and most of the amphiphilic polyphosphazenes investigated are in hydrogel form, suitable for local drug delivery [1,2]. In addition, Sohn and colleagues [3,4] developed a class of polyphosphazene prodrugs for cancer therapy, which can actually be attributed to amphiphilic polyphosphazenes. Their contributions were discussed in Chapter 14. In this chapter we review recent studies carried out on drug carriers based on amphiphilic polyphosphazenes: more precisely, the self-assembled aggregates of amphiphilic polyphosphazenes with special focus on those containing hydrophobic small molecules.

Nano-sized aggregates of amphiphilic copolymers have attracted much attention recently for their drug delivery potentials [5,6]. Like other amphiphilic

Polyphosphazenes for Biomedical Applications, Edited by Alexander K. Andrianov
Copyright © 2009 John Wiley & Sons, Inc.

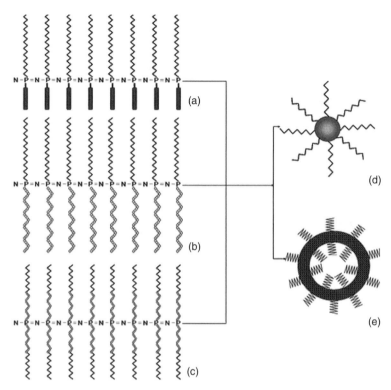

FIGURE 1   Structure of the amphiphilic polyphosphazenes having (a) hydrophilic polymer chains with hydrophobic small molecules; (b) hydrophilic polymer chains with hydrophobic polymer chains; (c) amphiphilic polymer chains only. The chains and small molecules are densely grafted to the polyphosphazene backbones in a "grafting onto" manner. These polymers can be self-assembled into (d) core–shell structured micelles/nanoparticles or (e) polymersomes when they are dispersed in water.

copolymers, amphiphilic polyphosphazenes self-assembled into micelles or nanoparticles in water with a core–shell structure consist of a hydrophobic inner core and a hydrophilic corona (Fig. 1d); Otherwise, amphiphilic polyphosphazenes can also fabricate vesicle-like aggregates called *polymersomes*: hollow spheres with a water core (Fig. 1e). These aggregates could be utilized as vehicles for a drug delivery system (DDS) according to vehicle structure, drug properties, and therapy demands in the clinic. Generally, in core–shell structured spherical aggregates, poorly water-soluble drugs can be encapsulated into the core and redispersed in water for administration, whereas for polymersomes based on amphiphilic polyphosphazenes, both hydrophobic and water-soluble drugs can be encapsulated in the vesicle wall and in the inner aqueous core, respectively.

However, the principles of biomedical applications reflect some special requirements on drug carrier. One of the most important factors is biocompatibility, which means that the polymer materials used for drug delivery should be low or nontoxic and biodegradable or can be eliminated from the body. Hence, virtually all the substitutes grafted to polyphosphazene backbones should be more biocompatible in the context of biomedical applications. Furthermore, drug carrier systems are required to be characterized by other parameters, such as particle size, drug encapsulate efficiency, and drug release profile, which to a great extent are influenced by polymer composition or carrier preparation procedures. This part is discussed later in the chapter.

SYNTHESIS AND CHARACTERIZATION

Synthesis of Amphiphilic Polyphosphazenes

The amphiphilic polyphosphazenes discussed here have been prepared using nucleophilic substitution through a three-step strategy. The synthetic route is shown schematically in Figure 2. First, polyphosphazene backbones are synthesized by ring-opening polymerization of hexachlorocyclotriphosphazene monomer, after which polymer chains and small molecules are grafted to backbones sequentially. For a successful replacement of chlorine atoms, all the substances for grafting should have amino or alkoxy groups. Since a polyphosphazene bearing only a small number of chlorine atoms is liable to cross-link in a moist atmosphere, it is generally necessary to attain almost 100% substitution of polyphosphazene during substitution reaction. Considering that the reaction activity always differs between polymer chains and small molecules, stepwise substitution has been established following the principle that the substance with relatively poorer activity or/and bigger bulk was chosen

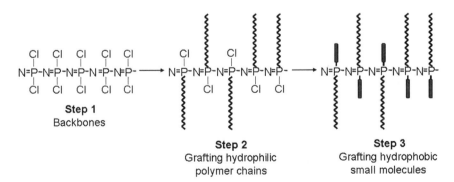

FIGURE 2   Schematic representation of the synthesis of amphiphilic polyphosphazenes. In the first step, polyphosphazene backbones are generated. In the second step, chlorine atoms are partly substituted by hydrophilic polymer chains. In the third step, a sufficient amount (often excess amount) of hydrophobic small molecules is introduced to achieve a complete substitution.

in the first-step substitution reaction. Therefore, complete substitution of chloride atoms on the polymer backbones can be effectively ensured. Moreover, such stepwise substitution favors control of the hydrophobic/hydrophilic ratio in final products and makes the structure of graft polymers more uniform. Synthesis of other amphiphilic polyphosphazenes with different structures, shown in Figure 1b and c, is very similar.

The molecule weight (MW) of amphiphilic polyphosphazenes is determined by the length of polymer backbone (L) and the substances grafted on the backbones. For a defined amphiphilic polyphosphazene chain with fixed branches, the MW lies on the L, which can be tuned by addition of aluminum trichloride during the ring-opening polymerization step [7]. Theoretically, the chlorine atoms on the original backbones can be replaced completely under favorable conditions, and this assumption has been confirmed in experimental work. Thus, the structure of amphiphilic polyphosphazenes can be characterized conveniently by the hydrophilic/hydrophobic ratio f or $x:y$.

Besides, the synthesis of polyphosphazene backbones can also be accomplished through transformation of the phosphoraneimine ($Cl_3P{=}NSiMe_3$) or congeners [8,9]. This solution-state "living" cationic polymerization allows access to polyphosphazene with controlled molecular weight and narrow polydispersity. But the demand of polymerization activity would limit the scope of monomers available. Until now there has been no report about the application in drug delivery of amphiphilic polyphosphazenes synthesized by living polymerization.

Fluorescence Probe Analysis

Although the nuclear magnetic resonance (NMR) spectrum and other analytical means can reveal the structure of amphiphilic polyphosphazenes, nothing is better and more direct evidence of amphiphilic copolymer than concentration-relative self-assembly behavior, which is characterized by critical association concentration (CAC); that is, when the concentration of amphiphilic copolymer in water reaches a critical level, dispersed polymer chains will automatically assemble into aggregates. Fluorescence measurements are frequently performed to study this phenomenon of amphiphilic copolymer in aqueous solutions, and pyrene was most commonly chosen as the probe because of its photophysical and high hydrophobicity [10]. By examining the (0,0) bands in the pyrene excitation spectra and comparing the intensity ratio I_{338}/I_{333}, the CAC can be determined. At low polymer concentrations, this ratio takes the value characteristic of pyrene in water, and at high concentrations it takes the value of pyrene in a hydrophobic environment. The CAC was taken as the intersection of straight-line segments, drawn through the points at the lowest polymer concentrations, which lies on a nearly horizontal line, with that going through the points on the rapidly rising part of the plot.

The strong dependence on concentration of a series of polyphosphazenes with different hydrophilic/hydrophobic ratios is shown in Figure 3 [11]. Sharp

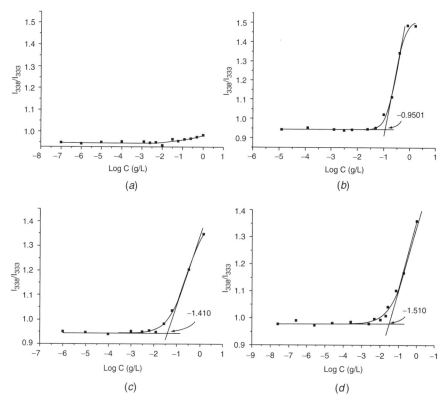

FIGURE 3 Plots of the intensity ratio I_{338}/I_{333} vs. log C for amphiphilic polyphosphazenes PNIPAm/GlyEt-PPPs with various hydrophilic/hydrophobic molar ratios: (a) 2.0/0.0; (b) 1.70/0.30; (c) 1.26/0.74; (d) 0.96/1.04.

transitions are observed in amphiphilic polyphosphazene samples PNIPAm/ GlyEt-PPPs (Fig. 3b–d), which indicate the formation of hydrophobic core induced by the self-assembly process. To the contrary, a fully water-soluble polyphosphazene grafted exclusively with hydrophilic polymer chains does not exhibit obvious changes in I_{338}/I_{333} (Fig. 3a). The values of CAC decreased with the hydrophilic/hydrophobic ratios of amphiphilic polyphosphazenes, consistent with those reported for traditional block copolymers and with the fact that increasing hydrophobic segments would lead to decrease in critical association concentration [12].

SELF-AGGREGATES FROM AMPHIPHILIC POLYPHOSPHAZENES

The self-aggregation behaviors are essential for amphiphilic polyphosphazenes as drug carries, which should be clarified before DDS applications (see the discussion below). There are several features that determine the properties of drug carries: for example, particle morphology and particle size and its distribution. Copolymer composition and solvents selection have a significant effect on these features, the details of which are outlined below.

Temperature-Triggered Self-Aggregation Behaviors of Thermosensitive Amphiphilic Polyphosphazenes

If thermosensitive polymer chains are grafted to polyphosphazene backbones, thermosensitive amphiphilic polyphosphazenes are generated. Thermosensitive poly(N-isopropylacrylamide) (PNIPAm) chains have successfully been grafted to polyphosphazene backbones [13,14], where the grafted PNIPAm segments have one freely mobile end, similar to terminally modified PNIPAm, with hydrophobic segments. Consequently, the thermosensitivity of PNIPAm segments is unaltered, and the amphiphilic polyphosphazenes therefore exhibit thermosensitive behavior with a lower critical solution temperature (LCST), similar to that of PNIPAm at around 32°C.

Self-assembly of the copolymer in aqueous solution was investigated by fluorescence spectroscopy at a temperature below LCST. Figure 4 shows the temperature-dependent changes of I_{338}/I_{333} and I_1/I_3 for a copolymer solution. Obviously, when the temperature increased, I_{338}/I_{333} from the excitation

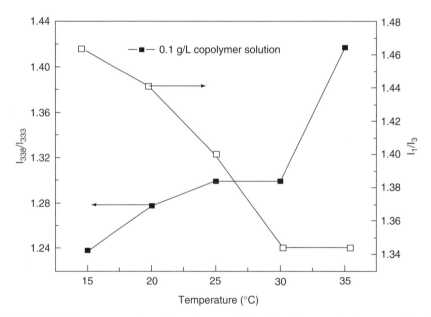

FIGURE 4 Temperature-dependent changes of I_{338}/I_{333} (from excitation spectra) and I_1/I_3 (from emission spectra) for a thermosensitive amphiphilic polyphosphazene of 0.1 g/L (From ref. 14).

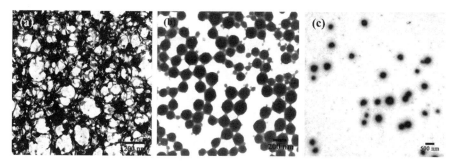

FIGURE 5 TEM images of copolymeric aggregates of PNIPAm-g-PPP-glycine ethyl ester (PNIPAm/GlyEt-PPP) at (a) 15°C, (b) 25°C, and (c) 50°C. (From ref. 14.)

spectra increased, and I_1/I_3 from the emission spectra decreased, suggesting a transfer of probe to a more hydrophobic microenvironment. This result demonstrates that the PNIPAm segments in the copolymer may undergo a phase transition in much the same way that NIPAm homopolymers do.

Further information about the thermosensitive amphiphilic polyphosphazene copolymer assembly in aqueous solution was given by transmission electron microscopic (TEM) measurements. As shown by Figure 5, aggregates with a network structure were observed for the copolymer at 15°C. This morphology was transformed into nanospheres of low dispersity when the temperature was increased from 15°C to 25°C (Fig. 5b). A further increase in temperature led to spheres with a large particle size, as illustrated in Figure 5c. This result corresponded very well with that obtained by fluorescence analysis, indicating a temperature-associated secondary morphology transition related to the PNIPAm segments.

In general, the process of the temperature-dependent self-assembly of thermosensitive amphiphilic polyphosphazene copolymer in aqueous solution may be described as follows: Copolymer chains dissolved into water at lower temperatures self-assemble into network micelles, and then narrowly dispersed nanospheres are formed at higher temperatures, while PNIPAm segments still exhibit an extended conformation at this time, and inter-nanosphere aggregation occurs due to the collapse of PNIPAm surrounding the hydrophobic core only when the temperature is above the LCST of PNIPAm. In addition, as illustrated schematically in Figure 6, this transformation process is reversible.

Control of Amphiphilic Polyphosphazene Self-Assembly: Effects of Solvent and Polymer Composition

Dialysis procedure is widely used to produce aggregates from amphiphilic copolymers [15], where at least one organic solvent is involved and the solvent used for dialysis may have an effect on the self-assembly behavior of amphiphilic copolymers. It is found that amphiphilic polyphosphazene grafted with poly(N-isopropylacrylamide) and ethyl 4-aminobenzoate groups

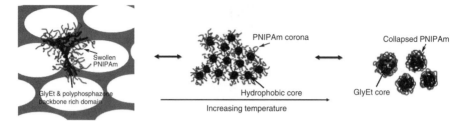

FIGURE 6 Schematic illustration of temperature-triggered self-assembly process of thermosensitive amphiphilic polyphosphazene in aqueous solution.

(PNIPAm/EAB-PPP) exhibits solvent-dependent morphology when it was treated with dialysis method to produce aggregates in water [16].

Figure 7 demonstrates that depending on various solvents employed in dialysis procedure, multimorphological assemblies are observed for the

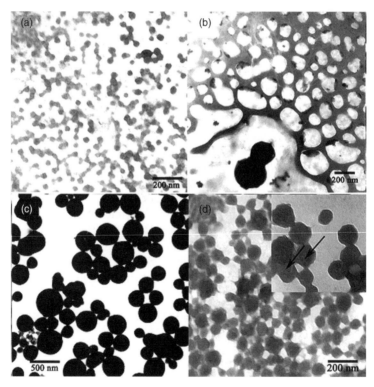

FIGURE 7 TEM images of aggregates made from amphiphilic copolymer PNIPAm$_{(0.68)}$/EAB$_{(1.32)}$-PPP: (a) DMF as solvent; (b) DMAc as solvent; (c) DMSO as solvent; (d) THF as solvent. (From ref. 16.)

copolymer PNIPAm/EAB-PPP containing a relatively high EAB content. When dimethylformamide (DMF) was employed as an organic solvent in dialysis procedure, nanospheres of average diameter 50 nm with a low polydispersity were obtained (Fig. 7a). These micelle-like aggregates of spherical geometry have been identified in many studies on both block and graft copolymers. As dimethylacetamide (DMAc) was employed, however, both spherical nanoparticles and network micelles were observed (Fig. 7b). In the case of dimethylsulfoxide (DMSO), well-defined microspheres were produced (Fig. 7c). In addition, high-genus nanoparticles were obtained as

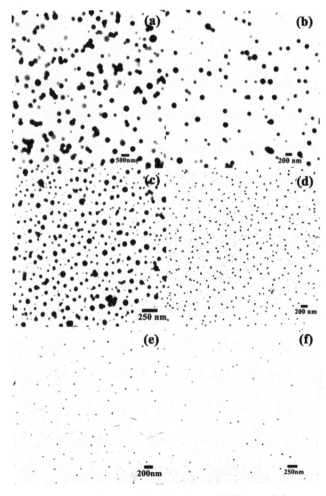

FIGURE 8 TEM images of assemblies based on PEG-containing copolymers: (a) PEG350/EtTrp-PPP, DMF; (b) PEG350/EtTrp-PPP, acetone; (c) PEG350/EtTrp-PPP, DMAc; (d) PEG1100/EtTrp-PPP, DMF; (e) PEG1100/EtTrp-PPP, DMAc; (f) PEG2000/EtTrp-PPP. (From ref. 17.)

tetrahydrofuran (THF) was employed as an organic solvent (Fig. 7d), which seems a new morphology for graft copolymer assemblies. These results provide a novel protocol to architecture supramolecular assemblies with multiple morphologies from graft copolymer. Further investigations are needed to clarify the mechanism of this solvent effect.

For a specific copolymer, particle size and morphology can be modulated by dialysis organic solvent; on the other hand, the size of assemblies can also be controlled by the chain length of the hydrophilic segment. In the case of another class of amphiphilic polyphosphazene graft with ethyl tryptophan and PEG (PEG/EtTrp-PPP) [17], three different PEG chains of molecular weight 350, 1100, and 2000 were employed. Figure 8 shows micelles based on PEG_{350}/EtTrp-PPP and PEG_{1100}/EtTrp-PPP prepared by dialysis using various organic solvents. PEG_{350}/EtTrp-PPP derived polymeric micelles with a mean size of 320, 180, and 60 nm were prepared with DMF, acetone, and DMAc employed as solvent, respectively. As for PEG1100/EtTrp-PPP, assemblies with a mean size of 66 and 45 nm could be prepared using DMF and DMAc as solvent, respectively (Fig. 8d and e). In the case of PEG2000/EtTrp-PPP, assemblies were obtained by dissolving copolymer directly into water, and micelles of 50 nm were thus prepared (Fig. 8f). These results suggested that depending on the type of dialysis solvent or copolymer composition, sphere-shaped nanoparticles ranging from hundreds to tens of nanometers could be obtained from PEG-containing amphiphilic copolymers.

DRUG ENCAPSULATION

There are several ways to encapsulate drugs into self-aggregates of amphiphilic polyphosphazenes, depending on the physicochemical properties of copolymers and drugs. Two drug encapsulation strategies have been employed most (see Fig. 9): solution dilution and emulsion evaporation. It is found that copolymer composition, solvent, and drug-polymer interactions significantly influenced the drug-loading efficiency of polymeric carriers, which will be discussed in the following section. Noteworthy, drug encapsulation may induce topological transformation of polymeric carrier; consequently, DDS formulation should take this into account.

Drug-Loading Methodology

Solution dilution, the dialysis method, is the technique used most widely to produce drug-loaded nanoparticles from amphiphilic copolymers through hydrophobic interaction [18,19]. In solution dilution, copolymers are dissolved in a water-miscible organic solvent (such as DMF, acetone, or tetrahydrofuranthiol) along with the drug and the solution is then diluted by water. As the water content increases, the amphiphilic copolymer self-assembles into nanoparticles with drug encapsulated in the inner core via hydrophobic interaction.

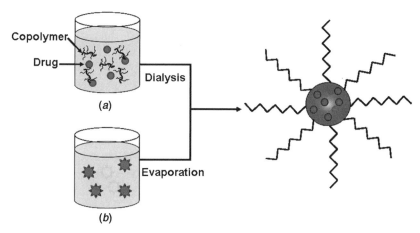

FIGURE 9   Schematic representation of drug encapsulation procedure for a DDS: (a) solution dilution; (b) emulsion evaporation.

This dilution procedure can be achieved either by dialysis via the dialysis membrane or by direct dropwise addition of water to an organic mixed solution of copolymer and drug under stirring.

Drug encapsulation could also be achieved through emulsion evaporation. Here a water immiscible organic solvent is used to dissolve copolymer and drug first, and this solution is then added into a certain amount of water to be emulsified. Afterward organic solvent is evaporated under reduced pressure and a drug carrier solution is obtained.

Drug-Loading Efficiency

It was found that the drug-loading capacity of drug carriers based on amphiphilic polyphosphazenes was determined primarily by copolymer composition and the chemical structure of a drug. In the case of amphiphilic polyphosphazenes grafted with ethyl tryptophan and PNIPAm (PNPAm/ EtTrp-PPP) [20], various drugs are selected to elucidate the effects of copolymer composition and the drug structure on encapsulation. Experimental results revealed that drug-loading capacity and encapsulation efficiency are both increased with an increasing hydrophobic fraction of copolymer; moreover, the corresponding values of drug loading and entrapment efficiency were higher for drugs with a relatively smaller molecular volume than those with a bulky structure.

According to previous reports on the solubilization of hydrophobic compounds in aqueous conventional surfactants and block copolymer micellar solutions, the number of molecules solubilized is closely correlated with the molecular volume of solubilizates, and van der Waals dispersion forces, hydrogen bonding, and dipole–dipole interactions between micellar core and

TABLE 1   Effect of Cosolvents on Drug Loading and Micelle Yield

PNIPAAm$_{(1.8)}$/ EAB$_{(0.2)}$–PPP (mg)	IND (mg)	Cosolvent	Drug Loading (wt%)	Micelle Yield (%)
50	10	DMSO	8.6 ± 1.6	62.1 ± 6.4
		DMF	7.6 ± 1.8	62.2 ± 5.9
		DMAc	9.7 ± 1.7	70.2 ± 6.5
		THF	12.3 ± 2.5	63.5 ± 5.4
		Acetonitrile	7.2 ± 1.2	67.5 ± 5.8

Source: Ref. 23.

solubilizates [21,22]. This might partly explain the lower loading level of bulky drugs in polymeric carriers. In addition, hydrogen-bonding interaction between drug and copolymers could be an important factor governing drug encapsulation into polymeric carriers based on PNIPAm/EtTrp-PPP. The hydrogen-bonding interaction between amide groups in PNIPAm chains on polyphosphazene backbones and carboxyl groups in indomethacin (IND) has been confirmed by infrared measurement, which might contribute to the higher loading of drugs that have a carboxylic group [20].

It must be noted that the cosolvent used for drug encapsulation in the dialysis process also influences the drug-loading efficiency. As shown in Table 1, several organic solvents, including DMSO, DMF, DMAc, THF, and acetonitrile were employed as a cosolvent to incorporate IND into PNIPAAm/EAB-PPP copolymer micelles by the dialysis method. Obviously, when DMAc or THF was employed as a cosolvent, PNIPAAm/EAB-PPP-based micelles with relatively higher IND loading and yield were obtained [23].

Drug Encapsulation–Induced Transformation of a Carrier

Significant morphology transformation was observed during the preparation of polyphosphazene drug carrier through a dialysis procedure [24]. As shown in Figure 10b, a TEM image of amphiphilic polyphosphazene-based aggregates reveals a necklace-like structure for assemblies with an IND content of 6.1%. When the drug content increased to 14.7%, vesicles of irregular shape are formed (Fig. 10c). A further increase in the IND content leads to the formation of spherical vesicles, as shown clearly in Figure 10d. For aggregates prepared from a higher initial IND content of 31.6%, larger spheres of low polydispersity are obtained. The particle size and size distribution of IND-containing aggregates was determined by DLS. As the IND content in the resulting aggregates increased, the particle size of the aggregates decreased, which suggests that a more compact structure was formed with an increase in IND loading.

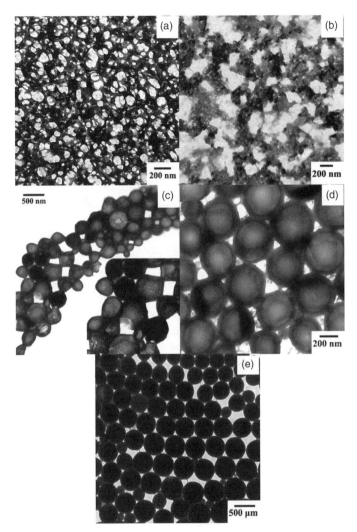

FIGURE 10   TEM images of PNIPAm/EtTrp-PPP-based aggregates (a) without IND; (b) 6.1% IND; (c) 14.7% IND; (d) 24.1% IND; (e) 31.6% IND. (From Ref. 24.)

As discussed previously, the morphology of polymeric aggregates is controlled by the hydrophobicity of amphiphilic copolymers, and drugs with carboxylic groups can interact with the amide group in the PNIPAAm chain, which in turn leads to the formation of a pseudo highly hydrophobic copolymer. Hence, a substantial morphology change can be observed for copolymer aggregates with an increase in IND content. Drug delivery applications often require a fixed morphology and size of drug carriers; therefore, this drug content–related topology must be take into consideration.

DRUG DELIVERY

Drug Release

Prolonged and controlled release are essential for optimal formulations in many drug delivery applications, where an initial release of drug is necessary to bring the drug concentration into the therapeutic window, followed by additional drug release over a longer period of time to maintain a drug concentration within the window throughout the dosage period desired. Drug carriers of amphiphilic polyphosphazenes have a high potential to achieve this goal, and several types of amphiphilic polyphosphazene drug delivery systems have emerged in the literature whose drug release behavior in vitro is determined mainly by the physicochemical characteristics of carriers and drugs. For example, a hydrogel formulation based on amphiphilic polyphosphazene can conduct sustained drug release over a month [1], whereas another drug carrier system based on polyphosphazene micelles released guest drug completely within 24 h [20].

Generally, loaded drugs in most polymer carriers are released by a diffusion pathway. In previous work the data from the release curves of IND from polymeric micelles based on PNPAm/EtTrp-PPP copolymers have been fitted to the Higuchi model [25]. It is found that in the middle stages of the cumulative IND release from 10 to 90%, drug release from micelles provided linear relationships for Higuchi plotting, indicating that Fickian diffusion plays an important role during the entire release period [20]. Similarly, Song et al. [1] observed diffusion-controlled drug release from amphiphilic polyphosphazene drug carriers, where the carrier with higher viscosities provides more sustained release of drugs. Sometimes, temperature plays an important role in the drug release process, especially when the carriers are fabricated by thermally responsive materials or when the drugs have temperature-sensitive solubility in release media [14,26].

Local Delivery of Hydrophobic Drugs

Local delivery is an alternative administration of drug since it not only can improve the therapy effect by increasing the drug concentration but can also avoid gastrointestinal stimulation, associated with oral administration. As usual, it is difficult to manufacture a topical dosage form for a hydrophobic drug; however, self-assemblies based on amphiphilic polyphosphazenes show the potential to be utilized for local delivery of hydrophobic drugs. Furthermore, taking advantage of thermosensitivity of amphiphilic polyphosphazenes containing PNIPAm, sustained local delivery has been achieved. Figure 11 provides the drug concentration in rat plasma after subcutaneous administration of IND solution and IND-loaded micelles of PNIPAAm/EtTrp-PPP, respectively. It is clear that the drug-loaded micelles produce a lower maximum IND concentration than to that of a crude drug solution, while after 3 h a

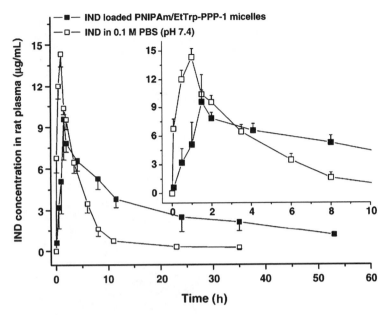

FIGURE 11 Concentration–time curves of IND-loaded PNIPAAm/EtTrp-PPP-1 micelles (34.8% IND) and 0.1 M PBS (pH 7.4) solution containing IND after subcutaneous injection of the same dosage. (From ref. 20.)

higher drug concentration level was observed for micellar formulation throughout the following period. This phenomenon is due to the sustained release profile of IND from micelles in vivo. In vivo pharmacodynamic studies based on both acute and adjuvant arthritis models also indicated the resulting sustained therapeutic efficacy in the case of intraarticular injection of IND-loaded micelles [18].

Cellular Uptake of Amphiphilic Polyphosphazene Carriers

Drug carriers are not only drug containers; in many cases (especially in cancer therapy) they also act as vehicles to transport drugs into object cells. Micelles of amphiphilic polyphosphazene have proved to be candidates for intracellular drug delivery. Figure 12 shows microscopy images of MG-63 cells that were incubated in culture medium with polymeric carriers based on fluorescent amphiphilic polyphosphazene $PEG_{2000}/EtTrp-PPP$. The fluorescent image revealed the internalization of blank carriers (Fig. 12b) and drug-loaded carriers (Fig. 12d). The absence of green fluorescence in the nucleus of MG-63 cells in Figure 12b indicates the localization of carriers in cytoplasm organelles but not in the nucleus [17], which is consistent with that reported by Savic et al. [27].

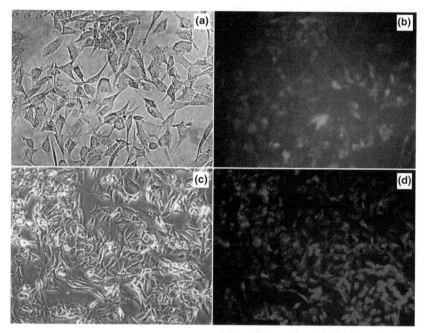

FIGURE 12 Intracellular uptake of PEG2000/EtTrp-PPP micelles by MG63 tumor cells: (a) control micelles with no drug, bright field; (b) control micelles with no drug, excited with blue light; (c) micelles with 5.0 wt% DOX, bright field; (d) micelles with 5.0 wt% DOX, excited with green light.

CYTOTOXICITY AND CANCER THERAPY

The discovery of an enhanced permeability and retention (EPR) effect [28] has aroused great enthusiasm for nanocarriers in cancer therapy. According to widely accepted theory, anticancer drugs encapsulated in nanoparticles have promise of accumulating around solid tumors and exhibiting better therapeutic performance. Studies on amphiphilic polyphosphazene-based drug carriers for cancer therapy are at a very early stage. Very few reports have been published, including the anticancer prodrugs of polyphosphazene–platinum(II) conjugates developed by Sohn's group [3,4]. In recent years our group began to explore this region, and a series of amphiphilic polyphosphazene-based carriers have been investigated. It is found that drug carriers prepared from amphiphilic polyphosphazenes have absolutely enhanced cytotoxicity on various cancer cell lines, generally with lower IC_{50} than that of free drug formulation. This phenomenon was supposed to be attributed to the aforementioned cellular uptake of drug-loaded micelles and drug release in cytoplasma as well. Figure 13 shows the 72 h cytotoxicity of doxorubicin (DOX)-loaded polyphosphazene micelles and blank polymer pon HeLa cells. It is clear that amphiphilic polyphosphazene polymer has no inhibitory effect; therefore, the

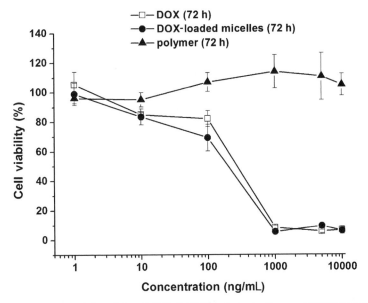

FIGURE 13 Cytotoxicity of free DOX, DOX-loaded micelles, and PEG/EtTrP-PPP-1 polymer against HeLa cells after 72 h incubation.

cytotoxicity of drug-loaded micelles was induced by the DOX released from the micelles exclusively. Despite the progress that has been made, further research is necessary for medical applications of these amphiphilic polyphosphazene drug carriers.

CONCLUSIONS

Designing a successful drug carrier from polyphosphazenes requires a thorough understanding of the physicochemical properties of the polymeric carriers, such as polymer synthesis, self-assembly behavior, drug encapsulation, pharmaco-dynamics, and so on. In this chapter we discussed some of the properties that must be considered in the formulation of a drug delivery system based on amphiphilic polyphosphazenes. Although there are still many hurdles to overcome before the full therapeutic potential of amphiphilic polyphospha-zenes can be reached, there is great promise that amphiphilic polyphosphazene-based drug carriers will greatly benefit drug delivery applications.

REFERENCES

1. Kang, G.D., Cheon, S.H., Song, S.C. Controlled release of doxorubicin from thermosensitive poly(organophosphazene) hydrogels. *Int. J. Pharm.*, 2006, 319:29–36.

2. Seong, J.Y., Jun, Y.J., Kim, B.M., Park, Y.M., Sohn, Y.S. Synthesis and characterization of biocompatible poly(organophosphazenes) aiming for local delivery of protein drugs. *Int. J. Pharm.*, 2006, 314:90–96.

3. Jun, Y.J., Kim, J.I., Jun, M.J., Sohn, Y.S. Selective tumor targeting by enhanced permeability and retention effect: synthesis and antitumor activity of olyphosphazene–platinum (II) conjugates. *J. Inorg. Biochem.*, 2005, 99:1593–1601.

4. Song, R., Jun, Y.J., Kim, J.I., Jin, C., Sohn, Y.S. Synthesis, characterization, and tumor selectivity of a polyphosphazene–platinum(II) conjugate. *J. Control. Release*, 2005, 105:142–150.

5. Nishiyama, N., Kataoka, K. Current state, achievements and future prospects of polymeric micelles as nanocarriers for drug and gene delivery. *Pharmacol. Ther.*, 2006, 112:630–648.

6. Gaucher, G., Dufresne, M.H., Sant, V.P., Kang, N., Maysinger, D., Leroux, J.C. Block copolymer micelles: preparation, characterization and application in drug delivery. *J. Control. Release*, 2005, 109:169–188.

7. Sohn, Y.S., Cho, Y.H., Baek, H., Jung, O.S. Synthesis and properties of low molecular weight polyphosphazenes. *Macromolecules*, 1995, 28:7566–7568.

8. Krogman, N.R., Steely, L., Hindenlang, M.D., Nair, L.S., Laurencin, C.T., Allcock, H.R. Synthesis and characterization of polyphosphazene-*block*-polyester and polyphosphazene-*block*-polycarbonate macromolecules. *Macromolecules*, 2008, 41:1126–1130.

9. Raab, M., Schick, G., Fondermann, R., Dolg, M., Henze, W., Weynand, U., Gschwind, R.M., Fischer, K., Schmidt, M., Niecke, E. A pH-functionalized polyphosphazene: a macromolecule with a highly flexible backbone. *Angew. Chem. Int. Ed.*, 2006, 45:3083–3030.

10. Astafieva, I., Zhong, X.F., Eisenberg, A. Critical micellization phenomena in block polyelectrolyte solutions. *Macromolecules*, 1993, 26:7339.

11. Zhang, J.X., Qiu, L.Y., Jin, Y., Zhu, K.J. Controlled nanoparticles formation by self-assembly of novel amphiphilic polyphosphazenes with poly(*N*-isopropylacrylamide) and ethyl glycinate as side groups. *React. Funct. Polym.*, 2006, 66:1630–1640.

12. Dowling, K.C., Thomas, J.K. A novel micellar synthesis and photophysical characterization of water-soluble acrylamide-styrene block copolymers. *Macromolecules*, 1990, 23:1059.

13. Zhang, J.X., Qiu, L.Y., Zhu, K.J., Jin, Y. Thermosensitive micelles self-assembled by novel *N*-isopropylacrylamide oligomer grafted polyphosphazene. *Macromol. Rapid Commun.*, 2004, 25:1563–1567.

14. Zhang, J.X., Qiu, L.Y., Wu, X.L., Jin, Y., Zhu, K.J. Temperature-triggered nanosphere formation through self-assembly of amphiphilic polyphosphazene. *Macromol. Chem. Phys.*, 2006, 207:1289–1296.

15. Zhang, L., Eisenberg, A. Multiple morphologies and characteristics of "crew-cut" micelle-like aggregates of polystyrene-*b*-poly(acrylic acid) diblock copolymers in aqueous solutions. *J. Am. Chem. Soc.*, 1996, 118:3168–3181.

16. Zhang, J.X., Qiu, L.Y., Jin, Y., Zhu, K.J. Multimorphological self-assemblies of amphiphilic graft polyphosphazenes with oligopoly(*N*-isopropylacrylamide) and ethyl 4-aminobenzoate as side groups. *Macromolecules*, 2006, 39:451–455.

17. Zhang, J.X., Qiu, L.Y., Li, X.D., Jin, Y., Zhu, K.J. Versatile preparation of fluorescent particles based on polyphosphazenes: from micro- to nanoscale. *Small,* 2007, 3:2081–2093.

18. Opanasopit, P., Ngawhirunpat, T., Chaidedgumjorn, A., Rojanarata, T., Apirakaramwong, A., Phongying, S. Incorporation of camptothecin into *N*-phthaloyl chitosan-g-mPEG self-assembly micellar system. *Eur. J. Pharm. Biopharm.,* 2006, 64:269–276.

19. Liu, S.Q., Tong, Y.W., Yang, Y.Y. Thermally sensitive micelles self-assembled from poly(*N*-isopropylacrylamide-*co*-*N*,*N*-dimethylacrylamide)-*b*-poly(D,L-lactide-*co*-glycolide) for controlled delivery of paclitaxel. *Mol. Biosyst.,* 2005, 1:158–165.

20. Zhang, J.X., Li, X.J., Qiu, L.Y., Li, X.H., Yan, M.Q., Jin, Y., Zhu, K.J. Indomethacin-loaded polymeric nanocarriers based on amphiphilic polyphosphazenes with poly(*N*-isopropylacrylamide) and ethyl tryptophan as side groups: preparation, in vitro and in vivo evaluation. *J. Control. Release,* 2006, 116:322–329.

21. Nagarajan, R., Barry, M., Ruckenstein, E. Unusual selectivity in solubilization by block copolymer micelles. *Langmuir,* 1986, 2:210–215.

22. Liu, J.B., Xiao, Y.H., Allen, C. Polymer–drug compatibility: a guide to the development of delivery systems for the anticancer agent. ellipticine, *J. Pharm. Sci.,* 2004, 93:132–143.

23. Zhang, J.X., Yan, M.Q., Li, X.H., Qiu, L.Y., Li, X.D., Li, X.J., Jin, Y., Zhu, K.J. Local delivery of indomethacin to arthritis-bearing rats through polymeric micelles based on amphiphilic polyphosphazenes. *Pharm. Res.,* 2007, 24:1944–1952.

24. Zhang, J.X., Li, X.D., Yan, M.Q., Qiu, L.Y., Jin, Y., Zhu, K.J. Hydrogen bonding-induced transformation of network aggregates into vesicles: a potential method for the preparation of composite vesicles. *Macromol. Rapid Commun.,* 2007, 28:710–717.

25. Higuchi, T. Mechanism of sustained-action medication: theoretical analysis of rate of release of solid drugs dispersed in solid matrices. *J. Pharm. Sci.,* 1963, 52:1145–1149.

26. Zhang, J.X., Qiu, L.Y., Jin, Y., Zhu, K.J. Thermally responsive polymeric micelles self-assembled by amphiphilic polyphosphazene with poly(*N*-isopropylacrylamide) and ethyl glycinate as side groups: polymer synthesis, characterization, and in vitro drug release study. *J. Biomed. Materi. Res. A,* 2006, 76:773–780.

27. Savicć, R., Luo, L.B., Eisenberg, A., Maysinger, D. Micellar nanocontainers distribute to defined cytoplasmic organelles. *Science,* 2003, 300:615–618.

28. Matsumura, Y., Maeda, H. A new concept for macromolecular therapeutics in cancer chemotherapy: mechanism of tumoritropic accumulation of proteins and the antitumor agent SMANCS. *Cancer Res.,* 1986, 46:6387–6392.

16 Synthesis and Characterization of Organometallic Polyphosphazenes and Their Applications in Nanonoscience

C. DÍAZ and M.L. VALENZUELA

Departamento de Química, Facultad de Ciencias, Universidad de Chile, Santiago, Chile

INTRODUCTION

Organometallic compounds as well as polymers are of great interest in materials science [1]. The link between these two types of compounds gives rise to organometallic polymers [2] (Fig. 1). These materials exhibit interesting properties specifically because they combine the electronic properties associated with transition metal compounds and the catalytic characteristic of the organometallic fragment and the processing advantages of organic polymers. Characteristics routinely associated with transition metals, such as multiple oxidation states and magnetism, are achieved only rarely in organic molecules. The incorporation of transition metals into a polymer main chain therefore clearly offers considerable potential for the preparation of processable materials having properties that differ significantly from those of conventional organic or inorganic polymers. Transition-metal-based polymers may also be expected to function as convenient and processable materials for metal-containing ceramics, films, fibers, and coatings with high stability and desirable useful physical properties. As shown in Figure 1, organometallic polymers are also useful precursors to nanomaterials [2a].

Organometallic polymers can conveniently be classified into the following types:

1. Organometallic polymers with transition metals in the main chain
2. Polymers with organometallic moieties in the side-group structure
3. Polymers with the metal–metal bond in the main chain

Polyphosphazenes for Biomedical Applications, Edited by Alexander K. Andrianov
Copyright © 2009 John Wiley & Sons, Inc.

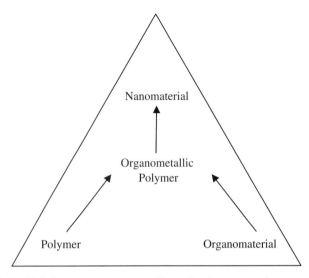

FIGURE 1 Link between organometallic and polymers to give nanomaterials.

These types of polymers are represented schematically in Figure 2. Type I organometallic polymers containing covalent M-spacer bonds have recently been reviewed [2a]. On the other hand, type III organometallic polymers having metal–metal bonds are still scarce [1d,2a] and are not discussed here. Another more general review [2c] of organometallic polymers deals mainly with type I polymers and coordination polymers. Types II and III polymers

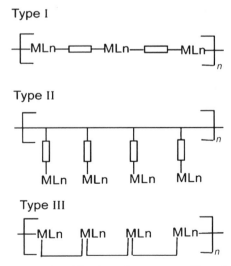

FIGURE 2 Classification of organometallic polymers.

have also been discussed briefly in selected books [1d,1e]. The organometallic function in organometallic polymers can arise either from the MLn fragment (which may contain one or more M–C bonds) or from the –[spacer]–C–M linkage. Polymers of the type where MLn is a coordination fragment are shown as

$$[ML_{n-1} - L' - ML_{n-1} - L' - ML_{n-1} - L']_n$$

The link of [polymer]-MLn units that do not contain an M–C bond are not covered in this chapter. Such polymers, often called *coordination polymers*, have been reviewed elsewhere [1c]. Coordination polymers are also mentioned in references 1c–1d and 2c. Also, polymers such as dendrimers [2a], containing organometallic fragments in the periphery, are not covered. From the standpoint of stability and processability, inorganic polymers are most useful in materials science [3]. Among the inorganic polymers, the polyphosphazenes have been studied most, due to their advantageous properties and broad application [1f,3,4]. We illustrate the synthesis, characterization of organometallic derivatives of polyphosphazene, and their use in nanomaterials science.

GENERAL SYNTHETIC PROCEDURES OF ORGANOMETALLIC POLYPHOSPHAZENE

The synthesis, characterization, and properties of organometallic derivatives of co-polyphosphazenes have been reviewed in detail [5–14]. Two principal methods exist for the synthesis of the organometallic derivatives of copolyphosphazenes: (a) by a direct substitution reaction of the copolyphosphazene ligand with the respective organometallic compound (Scheme 1), and (b) by reaction of $\{[NPR_1R_2]_{1-x}[NPCl_2]_x\}_n$ with the respective phenol ligand containing the organometallic fragment as $HOC_6H_4D \cdot MLn$, in the presence of Cs_2CO_3 and/or K_2CO_3 using tetrahydrofuren (THF) as solvent (Scheme 2).

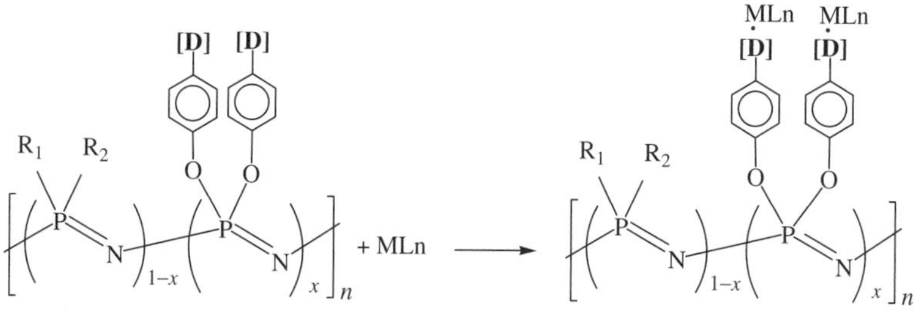

SCHEME 1   General formation of organometallic polymer from a polymer ligand and organometallic fragment.

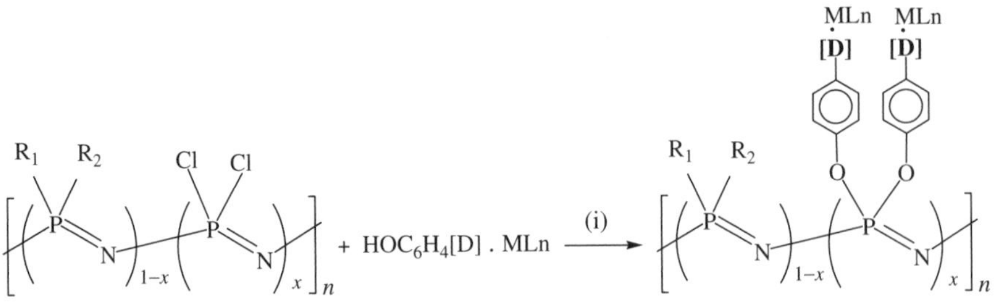

SCHEME 2 General formation of organometallic polymer from $[PCl_2]_n$ and P-functionalized donor phenol. (i) Denotes the reaction conditions: K_2CO_3 and THF as solvent.

The formulas of some organometallic derivatives of polyphosphazenes prepared by these methods are shown in Tables 1 and 2. The organometallic polymers are stable solids whose color depends on the organometallic fragment chromophore: red–brown for 1, 2, 6, and 7; yellow for 3, 4, 8, and 9; green for 5; white for 15; and so on. In general, they are insoluble except for 7, 8, 14, and 15.

GENERAL PROPERTIES OF ORGANOMETALLIC POLYPHOSPHAZENES

Thermal behavior under a nitrogen atmosphere yields pyrolytic residues, in the range 10 to 71%, which convert it in some cases into interesting pre-ceramic materials. As pointed by Allcock [3a], cross-linking of polyphosphazene chains precludes the formation of volatile oligomeric phosphazene species, which in turn yield a low pyrolytic residue. Cross-linking of a polyphosphazene chain can occur in general in physical and chemical ways. Physical methods involve, among others, ultraviolet- and radiation-induced cross-linking [3a,15]. Chemical methods involve cross-linking by metal ions and organometallic fragments. In general, the different ways of cross-linking of polyphosphazene chains by a metal ion are shown in Scheme 3.

A possible mechanism explaining the cross-linking of polyphosphazene chains by the organometallic fragments used in our previous work [5–14] is depicted in Scheme 4. As shown, when the pyrolysis is under air, the final isolated products are oxides and/or metal pyrophosphates. In this case it is crucial that the organometallic fragment have a labile ligand as phosphine (or labile on heating as CO) or other for linking to the donor site of another polymeric chain. The high pyrolytic yield from organometallic polymers shown in Tables 1 and 2 was attributed to the cross-linking of polyphosphazene chains by the organometallic fragment (for the pyrolysis yield of a series of these polymers, see Table 5 of reference 5).

TABLE 1   Formulas for Some Type II Organometallic Derivatives of Polyphosphazene

Polymer	R	MLn	x	y	z	Ref.
1	⟨◯⟩-CH$_2$CN	C$_p$Fe(dppe)	0.80	0.18	0.02	9
2	⟨◯⟩-CH$_2$CN	C$_p$Fe(dppe)	0.55	0.20	0.25	9
3	⟨◯⟩-CH$_2$CN	C$_p$Ru(PPh$_3$)$_2$	0.80	0.18	0.02	9
4	⟨◯⟩-CH$_2$CN	C$_p$Ru(PPh$_3$)$_2$	0.55	0.20	0.25	9
5	⟨◯⟩-CH$_2$CN	Cr(CO)$_5$	0.80	0.18	0.02	6
6	⟨◯⟩-CH$_2$CN	(π-CH$_3$-C$_5$H$_4$)Mn(CO)$_2$	0.55	0.20	0.25	7
7	⟨◯⟩-CH$_2$CN	C$_5$(CH$_3$)$_5$Fe(dppe)	0.80	0.18	0.00	8
8	⟨◯⟩-PPh$_2$	(π-CH$_3$-C$_5$H$_4$)Mn(CO)$_2$	0.60	0.40	0.00	13
9	⟨◯⟩N	W(CO)$_5$	0.65	0.35	0.00	14
10	⟨◯⟩-PPh$_2$	W(CO)$_5$	0.65	0.35	0.00	12

In fact, recent infrared (IR) studies have shown that loss of the carbonyls from transition metal–containing polyphosphazenes by heating produces vacant sites around the metal, which induces the coordination of donor groups of the polymeric chain [11]. In the case of the polyphosphazene, the

TABLE 2 General Formulas for Some Organometallic Derivatives of Polyphosphazenes

Polymer	R_1	R_2	R_3	R_4	MLn	x	y	Ref.
11	$(O_2C_{12}H_8)_{1/2}$	$(O_2C_{12}H_8)_{1/2}$	OC_5H_4N	OC_5H_4N	$\pi\text{-}CH_3C_5H_4Mn(CO)_2$	0.7	0.3	34
12	$(O_2C_{12}H_8)_{1/2}$	$(O_2C_{12}H_8)_{1/2}$	OC_5H_4N	OC_5H_4N	$CpFe(dppe)$	0.7	0.3	34
13	$(O_2C_{12}H_8)_{1/2}$	$(O_2C_{12}H_8)_{1/2}$	OC_5H_4N	OC_5H_4N	$CpRu(PPh_3)_2$	0.7	0.3	34
14	OC_6H_5	OC_6H_5	OC_6H_5	OC_6H_4CN	$CpFe(dppe)$	0.94	0.06	10
15	$(O_2CH_2(C_2B_{10}H_{10}))_{1/2}$	$(O_2CH_2(C_2B_{10}H_{10}))_{1/2}$	$(O_2CH_2(C_2B_{10}H_{10}))_{1/2}$	$(O_2CH_2(C_2B_{10}H_{10}))_{1/2}$	—	0.5	0.5	35

302

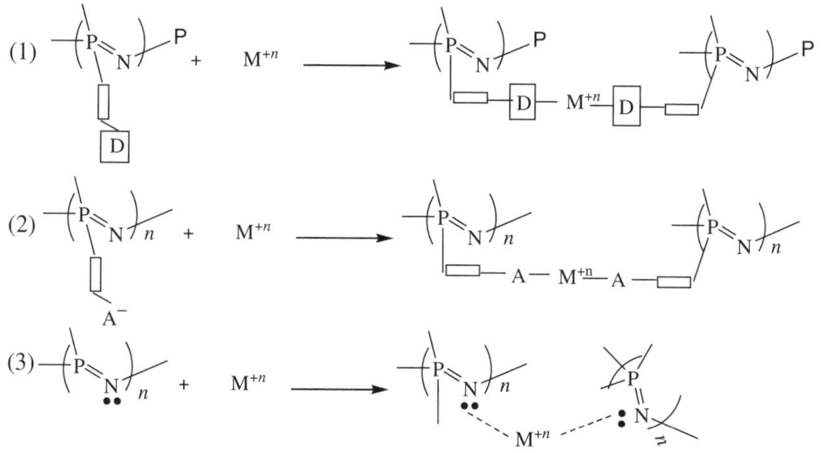

SCHEME 3 Approaches to metal coordination by polyophosphazenes. In some cases M can represent an organometallic fragment.

organometallic fragments are crucial because single metal ions coordinated to the polymer chains as the polymer gel $\{NP[(OC_6H_4C(O)C-OC_6H_5)_2]_x$ $[NP(OC_6H_4C(O)C-OC_6H_5)_2][Cu(BF_4)_2]_y\}_n$ afford low pyrolytic yields, pointing to a low degree of cross-linking upon heating [16].

Glass transition temperature, T_g, values for organometallic polymers have been little discussed. Some recent data have been compiled and discussed [5]. In general, enhancing of T_g in going from the free polymer to the organometallic derivatives has been observed [4,14]. This has been attributed to the very restricted flexibility of the chains, due to the presence of cross-linking capacity in the organometallic moiety. On the other hand, a decrease in T_g has been observed, which can be due to the flexibility of the P=N–P=N chains, caused by a decrease in hindrance of the aromatic ring in the polyspirophosphazenes by the proximity of the organometallic groups.

APPLICATIONS IN NANOSCIENCE

Materials with submicrometer dimensions, such as nanoparticles, represent an exciting new class of materials [17]. As a consequence of their tiny size, nanomaterials often display unique physical and chemical properties that are atypical of bulk materials. Optical, magnetic, and electrical properties, for example, are sensitive to size effects. Furthermore, nanosized particles are also very efficient in the field of catalysis [18], due to their high surface-to-volume ratio, nanosensors, nanoelectronics, and so on. Consequently, numerous processes for nanomaterials synthesis have been investigated with the aim of controlling their size, morphology, structure, and chemical composition. Many studies on the production of nanoparticles have been published [17].

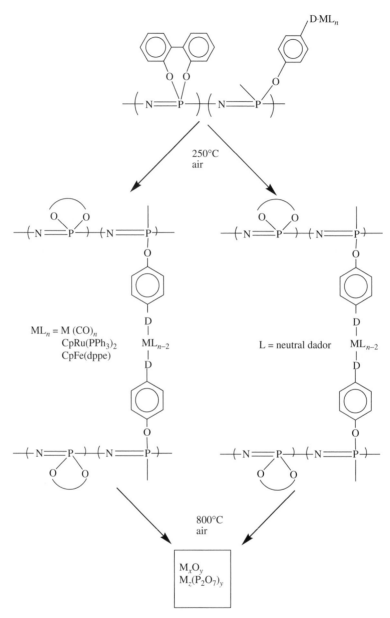

SCHEME 4   Cross-linking of polyphosphazenes by organometallic fragments.

There exist two main routes for their preparation: chemical processes using the aqueous TorKevich method [19], the two-phase Brust method [20], or the sol–gel technique [21], among others, and physical processes using spray pyrolysis or vapor condensation methods.

In this context a solid-phase pyrolytic method is an interesting alternative for preparing nanomaterials, but few methods have been reported. Pyrolysis of poly(ferrocenylsilanes) in air yields a red crystalline material of composition SiO_2/Fe_2O_3 [22]. Solid-state gold nanoparticles have been obtained by vaporizing a toluene solution of the solution generated by gold nanoparticles and pyrolysis of the residue [23]. Another solid-state approach to gold nanoparticles arises from heating solutions of preformed smaller gold particles [24]. Solid-state transition metal nanoparticles included in a SiO_2 matrix have been obtained from a combined sol–gel metal salt inclusion method [25]. Bimetallic Pt/Pd nanoparticles in the solid state have been obtained by pyrolysis of the organometallic precursor *trans*-PtCl(PEt$_3$)$_2$ SnCl$_3$ [26].

A crucial aspect in the formation of nanoparticles in both solution and in the solid state is their stabilization. Stabilization can be electrostatic and steric [27]. Steric stabilization can be achieved through either ligands, polymers, oligomers, or solvents. Polyphosphazene polymers $[N{=}PR_2]_n$ can bind metal ions and organometallic fragments through R-containing donor groups [4–14] (or with the basic nitrogen in some cases). This suggests the possibility of stabilizing metal nanoparticles. In fact, Olshavsky and Allcock [28] reported the preparation in water solution of CdS nanoparticles entrapped in a MEEP [poly(methoxyethoxyethoxyphosphazene)] network. Walker et al. [29] prepared gold nanoparticles by reducing $AuCl_4{}^-$ with $NaBH_4$ in the presence of PMPP [poly(methylphenylphosphazene)] using a biphasic water–toluene system. Recently, Jung et al. [30] prepared gold nanoparticles using the method of Brust–Schiffrin in water–toluene and stabilizing the nanoparticles with thioether groups anchored to a polyphosphazene backbone. In all these methods the nanoparticles are generated in solution by an independent method, and the polyphosphazene acts as a matrix to stabilize them. Here we review our work on a novel method to produce monometallic and bimetallic nanoparticles from the solid-state pyrolysis in air of the organometallic derivatives of the polyphosphazene [31–39]. Most recently, we also show and discuss preliminary results on solid-state pyrolysis of triorganocyclotriphosphazenes affording (similar to using a polyphosphazene precursor) nanostructured metallic materials [39].

PYROLYSIS OF POLYPHOSPHAZENE–CONTAINING ORGANOMETALLIC FRAGMENTS

Pyrolysis in air of organometallic polymers affords nanostructures whose nature depends on the metal, on the organic spacer of the polymer, and on the charge of the copolymer unit containing the organometallic fragment (see Fig. 3). The nature of the pyrolytic products depends strongly on the metal [31–39]: metal oxides for Cr [32], Ru [39], and W [34]; metal pyrophosphate for Mn [33] and Fe [38]; and metal in zero oxidation state for Au [37]. Here we review our last results on the obtention of nanostructured metal containing

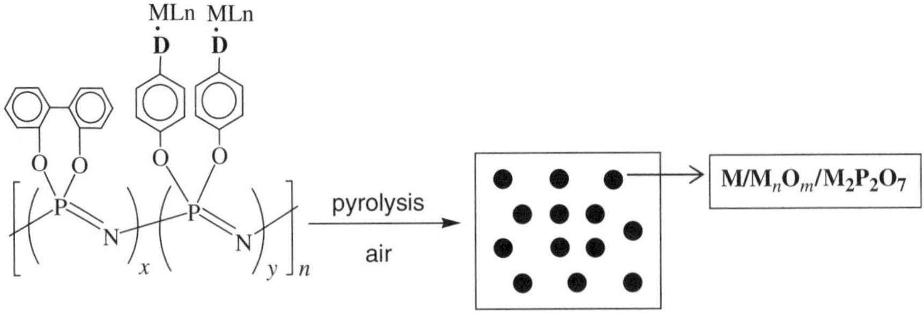

FIGURE 3   Solid-state pyrolysis of the organometallic polyphosphazene (SSPO) method.

materials using the SSPO method. Also, some preliminary results using cyclophosphazenes as the pyrolytic precursor instead of the analog polymer are presented. Results can be shown considering the metal involved or considering the nature of the nanostructured metallic materials. Here we adopt the former option.

COPOLYPHOSPHAZENE CONTAINING METAL NOBLE ORGANOMETALLIC FRAGMENTS

Previously we have studied the pyrolysis of the polymer containing the organometallic fragment AuCl anchored to the polymeric chain, of the formula $\{[NP(O_2C_{12}H_8)]_{0.85}[NP[(OC_6H_4PPh_2 \cdot Au\ Cl)_2]_{0.15}\}_n$ [37]. Nanostructured Au was obtained. Incorporation of silver in the polyphosphazene chain containing pyridine as donor using the organometallic fragment $Ag(OTf)PPh_3$ and method (a), depicted in Scheme 5, gave the polymer $\{[NP(O_2C_{12}H_8)]_{0.7}[NP(OC_5H_4N \cdot Ag(PPh_3))_2]_{0.3}\}_n$. (Scheme 6).

Ag polymer, once precipitated, was insoluble in most common organic solvents, except for 1-methylpyrrolidone (NMP). As observed previously, IR spectra of the polymeric Ag derivative clearly evidenced coordination of the metal fragments $Ag(OTf)PPh_3$ to the pendant pyridines. The strong

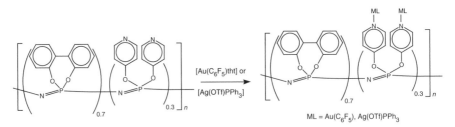

SCHEME 5   Synthesis of Au- and Ag-containing polyphosphazene.

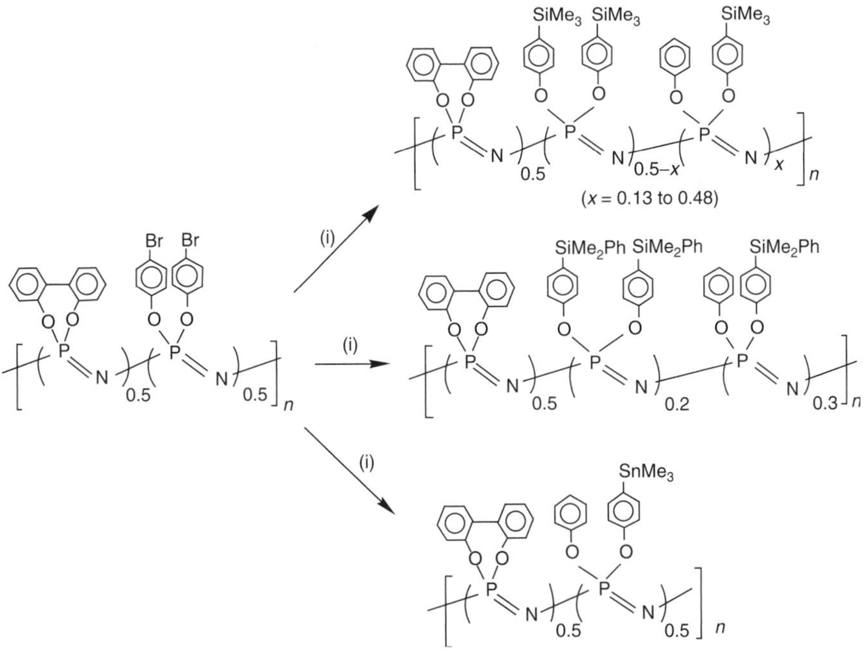

SCHEME 6 (i) *t*-BuLi (3.2 equivalents) THF/−78°C. Followed by SiMe₃Cl or SiMe₂PhCl or SnMₑ₃Cl.

oxypyridine band at 1582 cm^{-1} of the respective starting precursor polymer disappeared after coordination and a new band appeared at 1602 cm^{-1}. The IR spectra also show the absorptions attributable to the trifluoromethanesulfonate unit for the Ag polymer. Pyrolysis of the Ag polymer, in air and at 800°C, affords yellow solid in 12% yield. Figure 4 shows the x-ray diffraction (XRD) pattern recorded for the pyrolytic material generating the lines (111), (200), (220), (311), (222) at $2\theta = 38.23°$, 44.41°, 64.64°, and 77.6°, respectively, depicted as full circles. The position of these lines in XRD is similar to that obtained in silver (JCPDS card 04-0783).

On the other hand, the diffraction peak at 20.9°, 29.78°, 33.37°, 36.69°, 47.9°, 52.8°, and 55.14°, marked as arrows, can be assigned to a Ag_3PO_4 (JCPDS card 70-0702). These fit the main feature of the XRD patterns. However, some small amount of P_4O_7 is seen from the peak at $2\theta = 6.43°$, 17°, and 18.8°, marked as squares, which have been assigned to a P_4O_7 matrix formed during the pyrolysis of the organometallic derivatives of polyphosphazene [31–38]. The scanning electron microscopic (SEM) images in Figure 4b–d exhibit the morphology of the pyrolytic product, which consistent with the XRD analysis, shows two phases, corresponding to Ag and Ag_3PO_4. The Ag product is obtained as a dense monolith with 10 to 20-μm-diameter grains surrounded by bars of Ag_3PO_4 and forming a three-dimensional network. EDAX (energy-dispersive

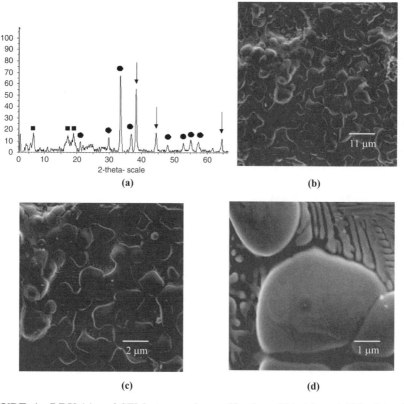

FIGURE 4   RDX (a) and SEM at several magnifications [(b), (c), and (d)] of the Ag polymer.

analysis of x-rays) (see Fig. 5a) shows the presence of silver, phosphorus, and oxygen, which is in agreement with the composition of the two-phase material (Fig. 4a). To our knowledge no similar three-dimensional networks of silver supported by silver phosphate materials have been obtained.

An EDAX linear scanning of the materials (Fig. 5b) shows an increase in silver content together with a decrease in phosphorus content on going from Ag_3PO_4 bars to Ag grains. As seen from SEM images of Figure 4, the morphology is similar to that of metal foams [40–48]. Metal foams are interesting materials with unique combinations of such properties as stiffness, low density, gas permeability, and thermal conductivity [41,42]. Such materials promise to enable new technologies in areas as diverse as catalysis, fuel cells, hydrogen storage, and thermal and acoustic insulation. To date, techniques for making metal foams have been somewhat limited: to dealloying of an Au–Ag alloy [43–45], reduction of metal in aqueous solution using a preformed organic template [46], chemical vapor deposition of metal alloys [47], and self-propagating combustion of complexes [48]. The pyrolysis of metal noble

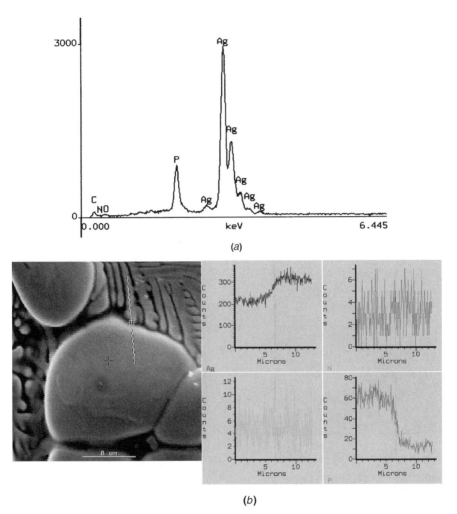

FIGURE 5   EDAX (a) and linear scanning EDAX (b) of the pyrolytic products from an Ag polymer.

organometallic derivatives of polyphosphazenes could be a useful method to prepare metal noble foam materials. Experiments with Pd and Pt organometallic derivatives of polyphosphazenes are currently in process.

COPOLYPHOSPHAZENE CONTAINING Si AND Sn ORGANOMETALLIC FRAGMENTS

The difficulty of preparing polyphosphazene containing anchored silicon groups has been pointed out by Allcock et al. [49–55]. Previous attempts include partial lithiation reaction of the $\{[NP(OCH_2CF_3)]_x[NP(Cl)_2)]_y\}_n$ intermediates, followed by coupling of the chlorosilanes [51], ring-opening addition of the respective phosphazene trimer [50], or by reaction of the intermediate with the respective amines in the presence of triethylamine [53,54]. We have prepared sililated derivatives of polyphosphazenes using a strategy similar to that used by Allcock, of lithiation of the intermediate $\{[NP(O_2C_{12}H_8)]_{0.5}[NP(OC_6H_4Br)_2]_{0.5}\}_n$ and subsequent reaction with $ClSi(CH_3)_3$ or $ClSi(CH_3)_2C_6H_5$ (Scheme 6) [55]. Pyrolysis of a polyphosphazene containing silicon fragments in air at temperatures of 600°C, 800°C, and 1000°C was studied. The morphology and composition of the pyrolytic products depend on the temperature of the pyrolysis, as shown in Figure 6. The product can be formulated as the components of the phase $SiP_2O_7/P_2O_5/SiO_2$ [56–61]. At 800°C some porous morphology was observed, whereas at 600°C a most dense

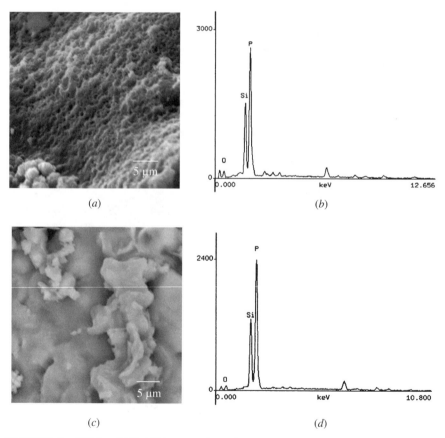

FIGURE 6   SEM and EDAX images of pyrolysis products at several temperatures: (a) and (b) 600°C; (c) and (d) 800°C. The pyrolysis in air of tin-containing polyphosphazene was studied at 800°C.

structure was evident. The EDAX analysis shown at the right side exhibits only the presence of P, O, and Si. The phase $SiP_2O_7/P_2O_5/SiO_2$ is an interesting commercial product with a variety of applications, especially in the area of catalysis [58–62] and microelectronics [63]. However, to date no studies on nanostructured $SiP_2O_7/P_2O_5/SiO_2$ have been reported.

A tin analog derivative was prepared using a route similar to that shown in Scheme 6. High-molecular-weight polyphosphazene containing tin organometallic fragments is scarce [64,65,67–70]. Allcock has reported polyphosphazene containing anchored $SnPh_3$ groups [64], and cyclophosphazenes containing MMe_3 groups (M = Si, Ge, Sn) have also been reported [65]. The Sn polymer has a very regular distribution of the tin organometallic along the chains, as determined by the correlation of T_g with the composition of the polymer [66]. The resulting gray pyrolytic residue was characterized by SEM-EDAX, IR, transmission electron microscopy (TEM), and XRD. By contrast with the Si products, the x-ray powder diffraction pattern of the pyrolysis product [67,70] displayed only the typical peaks corresponding to tin diphosphate SnP_2O_7, and no peaks for SnO_2 were noted Fig. 7.

Figure 7 also shows SEM images and EDAX analysis. The SEM images at two magnification levels display a fibrous three-dimensional network structure. EDAX analysis exhibits the presence of tin, phosphorus, and oxygen atoms, in agreement with formulation of the pyrolytic product. SnP_2O_7 has generated interest as a matrix for lithium batteries of high efficiency [68,69].

MECHANISM OF FORMATION OF NANOPARTICLES

TG studies in air indicate a common and general mechanism of formation of nanoparticles from pyrolysis of the organometallic polymer. In almost all of the TG studies [31–38], the first step involves carbonization of the organic matter (a small weight loss was also seen and attributed to the decomposition of the organometallic fragment and/or loss of solvent molecules trapped inside the polymer). This step produces CO/CO_2 mixtures, forming holes that allow agglomeration of the metallic nanoparticles. The formation of metal oxide nanoparticles can occur in two ways: (1) oxidation of the previously formed metal in oxidation state zero (it can be formed by reduction with CO/CO_2 mixtures of the corresponding M^{+n} ion [25]) or (2) reaction of the M° species with O_2:

$$M^{+n} + CO \rightarrow M^\circ + CO_2 \qquad (1)$$

$$M^\circ + (n/2)O_2 \rightarrow MO_n \qquad (2)$$

The metal pyrophosphate nanostructured salts can be formed by simple reactions of the M^{+n} ions with the "in situ"-formed pyrophosphate anions

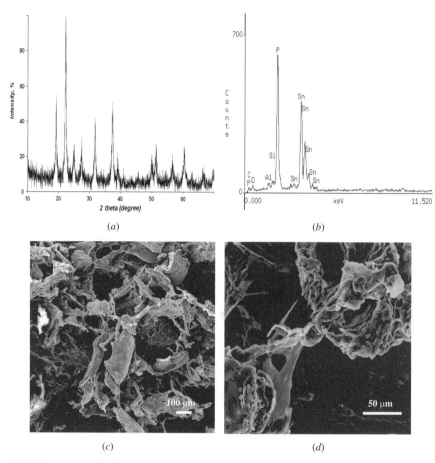

FIGURE 7 (a) RDX pattern, (b) EDAX, and (c) and (d) SEM images at two magnification levels of the pyrolytic product from the polymer.

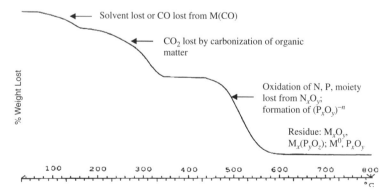

FIGURE 8 Representative TG of an organometallic polyphosphazene.

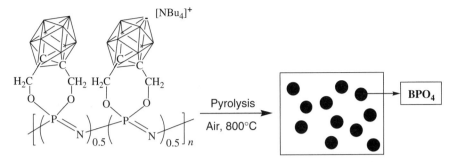

FIGURE 9 Formulation of nanostructured BPO_4 from pyrolysis of carborane-containing polyphosphazene.

from oxidation of the phosphorus phosphazenes atoms:

$$mM^{+n} + zP_2O_7^{2-} \rightarrow M_m(P_2O_7)_z \qquad (3)$$

In all cases, the low metal/P,N ratio produces P_4O_7 molecules in excess, which act as a stabilizing matrix for the nanoparticles:

$$(N{=}P)_n + O_2 \rightarrow N_xO_y + P_xO_y \qquad (4)$$

Figure 8 shows a general, representative thermogravimetric (TG) curve of an organometallic derivative containing polyphosphazene.

On the other hand, and consequent with the mechanism proposed above, the pyrolysis of polyphosphazenes 1–7 (Table 1), without metal affords only P_4O_7 (evidenced by XRD analysis). In agreement with the mechanism proposed, pyrolysis of the carborane polyphosphazene derivative (Fig. 9) yields

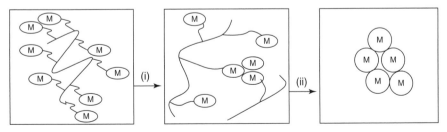

FIGURE 10 Mechanism proposed for the formation of metallic nanoparticles from the pyrolysis of polymer. (i) Metal release from the polymer and partial calcination of the organic matter. (ii) Nucleation and growth of the metal nanostructured material; here M represents the metallic nanoparticle: oxide, phosphate, or metal. The matrix formed by P_4O_7 that surrounds the metallic compound is not indicated.

nanostructured BPO_4 [35], which is usually formed by reaction of B_2O_3 with P_2O_5 [71]. A schematic representation of the general mechanism of the formation of metallic nanoparticles is depicted in Figure 10.

LINK OF ORGANOMETALLIC FRAGMENTS VERSUS CLASSIC COORDINATION COMPOUNDS TOWARD POLYPHOSPHAZENES

With regard to the nature of the metal species linked to polyphosphazene (through the type II mode), no thing has been reported regarding whether the anchored molecule is an organometallic fragment or a coordination classic compound, or of its chemical and physical implications. Allcock [3a] has discussed various ways of cross-linking polyphosphazenes but when the cross-linking agent is a metal, without making a distinction as to whether the metal forms part of an organometallic or a classic coordination ion. To answer such a question we have synthesized and characterized the gels resulting from the interaction of the polymer $[NP(OC_6H_4C(O)C–OC_6H_5)_2]_n$ with $[Cu(BF_4)_2]$ [16] (Scheme 7).

Some thermal implications were mentioned earlier. Reaction of the polymer $[NP(OC_6H_4C(O)C–OC_6H_5)_2]_n$ with $[Cu(BF_4)_2]$ in THF produces light blue gels of general composition $\{[NP(OC_6H_4C(O)C–OC_6H_5)_2]_x[NP(OC_6H_4C(O)C–OC_6H_5)_2]\cdot[Cu(BF_4)_2]_y\}_n$. Increasing Cu(II) content leads to gels of composition $\{[NP(OC_6H_4C(O)C–OC_6H_5)_2]_{0.99}[NP(OC_6H_4C(O)C–OC_6H_5)_2]\cdot[Cu(BF_4)_2]_{0.01}\}_n$, $\{[NP(OC_6H_4C(O)C–OC_6H_5)_2]_{0.82}[NP(OC_6H_4C(O)C–OC_6H_5)_2]\cdot[Cu(BF_4)_2]_{0.14}\}_n$, and $\{[NP(OC_6H_4C(O)C–OC_6H_5)_2]_{0.72}[NP(OC_6H_4C(O)C–OC_6H_5)_2]\cdot[Cu(BF_4)_2]_{0.27}\}_n$. Incorporation of the Cu(II) ions was clearly corroborated by electron paramagnetic resonance (EPR) spectroscopy. As shown in Figure 11, the EPR spectra of the polymer gels at 5.0 K exhibits the expected

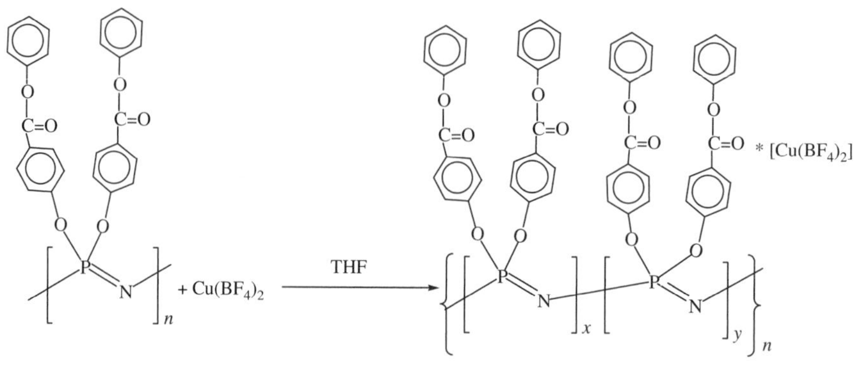

SCHEME 7   Formation of $\{[NP(OC_6H_4C(O)C–OC_6H_5)_2]_x[NP(OC_6H_4C(O)C–OC_6H_5)_2]\cdot[Cu(BF_4)_2]_y\}_n$ gels.

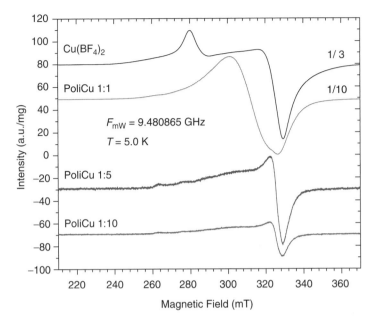

FIGURE 11 EPR spectra of the Cu/polymer gels with ratios 1 : 1; 1 : 5, and 1 : 10 and the Cu precursor salt, $Cu(BF_4)_2$.

behavior corresponding to variations in the Cu(II) ion content. At high Cu(II) concentrations, a wide signal with two g parameters, typical of a symmetry of tetragonal distortion [72], was observed. In contrast, at more diluted Cu(II) ions a single line with axial symmetry but with some hyperfine interactions is observed.

It was found that despite the presence of copper in the polymers, their thermal properties (vitreous transition point, pyrolytic residues) are modified only slightly. It appears that the main changes caused by the incorporation of copper ions inside polyphosphazene are in the porosity of the materials. Porosity as evidenced by SEM images is shown in Figure 12 for polymer gels with several Cu contents.

Thus, the polymeric matrix stabilizes Cu(II) nanostructures as shown in Figure 13. A TEM image of the polymer gel $\{[NP(OC_6H_4C(O)C-OC_6H_5)_2]_{0.72}[NP(OC_6H_4C(O)C-OC_6H_5)_2] \cdot [Cu(BF_4)_2]_{0.27}\}_n$ displays nanostructures of about 200 nm with the typical "raspberry" morphology.

Surprisingly, pyrolysis under air at 800°C affords negligible yields, contrary to the pyrolysis of the organometallic derivatives of polyphosphazenes. This is an important result, an outcome which implies that copper(II) ions do not produce significant cross-linking of the polyphosphazene chains on heating, thus giving rise to some appreciable formation of volatile trimer, which Allcock predicts [3a] will produce a low pyrolytic yield.

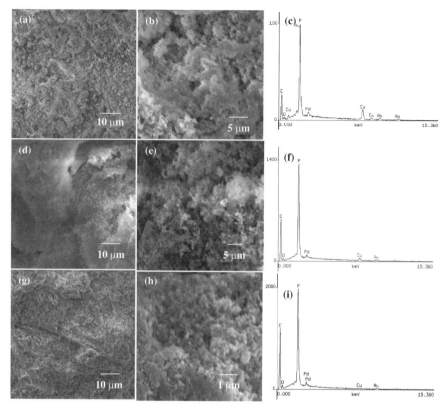

FIGURE 12 SEM images at different magnifications of polymer gels: (a) and (b) polymer gel with Cu/polymer 1 : 1; (d) and (e) polymer 1 : 5; (g) and (h) polymer gel 1 : 10. The respective EDAX are also shown.

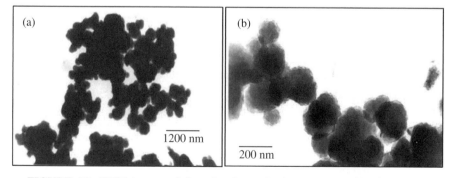

FIGURE 13 TEM images of the gel polymer 1 : 1 at two magnification levels.

PYROLYSIS OF POLYPHOSPHAZENE/METALLIC COMPLEX MIXTURES

With regard to the mechanism of formation of the metal nanostructures from the pyrolysis of organometallic derivatives of polyphosphazene, the problem is whether the metal (as the organometallic fragment) is covalently linked to polymeric chain or rather, if it is located close to the polyphosphazene chain but not necessarily anchored to it covalently. Preliminary results of pyrolysis of mixtures of polyphosphazene with metal complexes such as the $[NP(O_2C_{12}H_8)]_n \| AuCl(PPh_2)$ system indicate that in this case results similar to pyrolysis of the organometallic polymer $\{[NP(O_2C_{12}H_8)]_{0.85}[NP[(OC_6H_4PPh_2 \cdot AuCl)_2]_{0.15}\}_n$ were observed. Figure 14 shows the RDX, EDAX, and the SEM of the products from the pyrolysis at 800°C of the mixture $[NP(O_2C_{12}H_8)]_n \| AuCl(PPh_2)$.

The presence of nanostructured Au was corroborated by x-ray power diffraction analysis (Fig. 14a), showing the typical patterns of cubic Au and by EDAX exhibiting the presence of Au only (23b). The porous two-dimensional shape metal foam was evidenced by its SEM (Figure 14c–e). From these results it appears that often the metal containing the organometallic

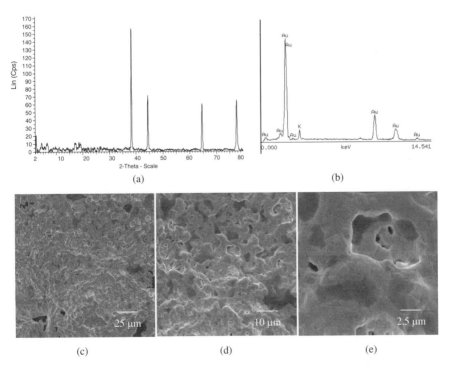

(a)   (b)

(c)   (d)   (e)

FIGURE 14   XRD, EDAX, and SEM images of the pyrolytic product from a mixture of $[NP(O_2C_{12}H_8)]_n \| AuCl(PPh_2)$.

cannot be linked covalently to the polyphosphazene to give rise to nano-structured metal after pyrolysis. However, this partial result is not sufficient to give a conclusive answer, and pyrolysis experiments with several polypho-sphazene–organometallic mixtures are in order.

TRIMER VS. POLYMER PHOSPHAZENE AND THEIR IMPLICATIONS IN THE NATURE OF THE PYROLYTIC PRODUCTS

As pointed out by Allcock et al. [73], small phosphazene molecules are useful models for their most complex counterparts: the polyphosphazenes. In this context, the obvious question is whether the organometallic derivatives of the organocyclotriphosphazenes (see Fig. 15) are useful precursors for the pre-paration of nanostructured metallic materials, such as the SSPO method using the respective organometallic polymers as precursors. Figure 15 shows the representative SSPCO (solid-state pyrolysis of cyclotriphosphazene organo-metallic derivatives) method.

Preliminary experiments with the trimer $[N_3P_3(OC_6H_5)_5(OC_6H_4CH_2CN \cdot CpRu(PPh_2)_3)][PF_6]$ indicate that pyrolysis produces products similar to those obtained starting from the polymer, nanostructured RuO_2 [39]. According to the mechanism of pyrolysis proposed for the polyphosphazenes suggested earlier, the similar results obtained with the trimer could be explained assuming the formation of cyclomatrix (see Fig. 16), which gives rise to a three-dimensional network precluding volatilization of the cyclotriphosphazenes and leading to a significant pyrolytic residue. Although the formation of cyclomatrix is usual in the chemistry of cyclotriphosphazenes [74–76], there is little information about the formation of a cyclomatrix of organometallic derivatives of cyclotriphosphazenes [75]. Formation of the cyclomatrix shown in Figure 16 could be viewed by assuming two subsequent losses of labile

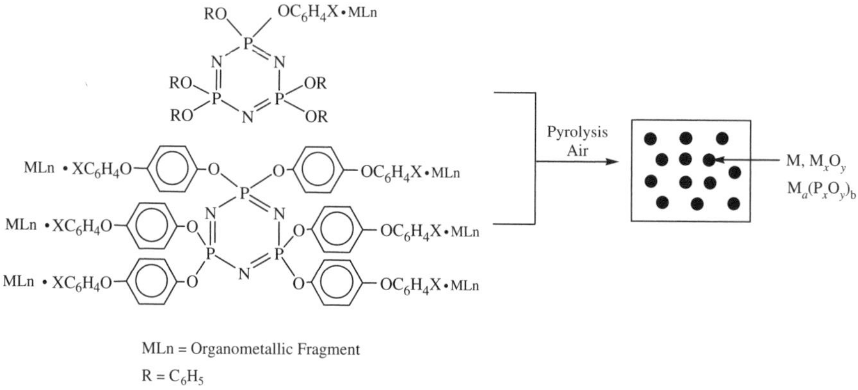

MLn = Organometallic Fragment

R = C₆H₅

FIGURE 15   Solid-state pyrolysis of the organometallic phosphazene trimer.

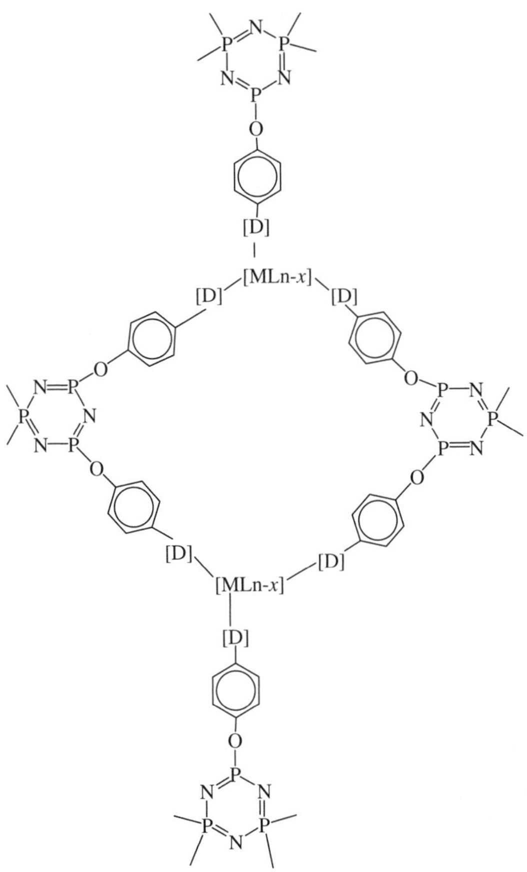

FIGURE 16 Possible formation of cyclomatrix from the pyrolysis of organometallic derivatives of cyclotriphosphazenes. Here the spacer linked to the ring is a OC_6H_4–D unit, with D a group or atom donor.

ligands from the organometallic fragment, followed by the coordination of two phosphazene rings.

CONCLUSIONS

In this chapter, we briefly summarized our recent results on the synthesis and characterization of polyphosphazenes containing anchored Si, Sn, and metal noble organometallic fragments and their conversion to nanostructured metallic material and compared with earlier works. As seen previously, the products depend on the nature of the metal, the nature of the spacer polymer, the charge on the organometallic fragments, and the pyrolysis temperature.

With organometallic fragments containing noble metals such as Au and Ag, interesting metal foams were obtained. On the other hand, by using metals such as Si and Sn, the metallic pyrophospates as well as the metal oxide can be obtained. Results obtained from a single ion such as Cu(II) coordinated to a polyphosphazene polymeric chain suggest that the link of an organometallic fragment, but not a single metal ion, is crucial to obtain a high pyrolytic yield. Preliminary results on the pyrolysis of polyphosphazene–organometallic complex mixtures indicate that in some cases the metal-containing fragment does not need to be linked to the polymeric chain to give nanostructured metal materials. Finally, cyclophosphazenes having coordinated organometallic fragments could also be useful precursors of metallic nanostructures.

Acknowledgments

We acknowledge funding support under Fondecyt project 1030515 and partially, 1085011. We are also grateful to Professor G. A. Carriedo for his valuable collaboration in the work cited herein.

REFERENCES

1. (a) O'Hare, D., Bruce, D.W. *Inorganic Materials*. Wiley, Chichester, UK, 1992.; (b) Coville, N.J., Cheng, L. *J. Organomet. Chem.*, 1998, 571:149; (c) Pickup, P.J. *J. Mater. Chem.*, 1999, 9:1641.; (d) Ciardelli, F., Tsuchida, E., Wohrle, D. *Macromolecule–Metal Complexes*. Springer-Verlag, Berlin, 1996; (e) Pittman, C.U. Jr., Carraher, C.E., Zeldin, M., Sheats, J.E., Culbertson, B.M. *Metal-Containing Polymeric Material*. Plenum, NewYork, 1996; (f) Wisian-Neilson, P., Allcock, H.R., Wynne, K.J. Inorganic and Organometallic Polymers: II. Advanced Material and Intermediates. *ACS Symp. Ser.*, 572. ACS, Washington, DC, 1994; (g) Whole, D., Pomogailo, A.D. *Metal Complexes and Metal in Macròmolecules*. Germany, Wiley-VCH, Weinheim, 2003.
2. (a) Nguyen, P., Gomez-Elipe, P., Manners, I. *Chem. Rev.*, 1990, 99:1515; (b) Manners, I. *Chem. Commun.*, 1999, 857; (c) Kingsborough, R.P., Swager, T.M. *Prog. Inorg. Chem.*, 1990, 48:123.
3. (a) Allcock, H.R. *Adv. Mater.*, 1994, 6:106; (b) Manners, I. *Angew. Chem. Int. Ed. Engl.*, 1996, 35:1602.
4. (a) Gleria, M., De Jaeger, R., eds., *Phosphazenes: A Worldwide Insight*. Nova Science, Hauppauge, NY, 2004; (b) Allcock, H.R.*Chemistry and Applications of Polyphosphazenes*. Wiley, Hoboken, NJ, 2003.
5. Díaz, C., Valenzuela, M.L. In Bregg, R.K., ed., *Horizons in Polymers Deveplop-ments*. 2005 Nova Science, Hauppauge NY.
6. Díaz, C., Castillo, P. *J. Inorg. Organomet. Polym.*, 2001, 11:183.
7. Díaz, C., Castillo, P., Carriedo, G.A., Gómez-Elipe, P., García Alonso, F.G. *Macromol. Chem Phys.*, 2002, 203:1912.
8. Díaz, C., Castillo, P. *Polym. Bull.*, 2003, 50:12.

9. Díaz, C., Valenzuela, M.L., Barbosa, M. *Mater. Res. Bull.*, 2004, 39:9.

10. Carriedo, G.A., García Alonso, F.J., Gómez-Elipe, P., Díaz, C., Yutronic, N. *J. Chilean Chem. Soc.*, 2003, 48:25.

11. Carriedo, G.A., García Alonso, F.J., Díaz, C., Valenzuela, M.L. *Polyhedron*, 2006, 25:105.

12. Carriedo, G.A., García Alonso, F.G., González, P.A., Gómez-Elipe, P. *Polyhedron*, 1999, 18:2853.

13. Carriedo, G.A., García Alonso, F.G., González, P.A., Díaz, C., Yutronic, N. *Polyhedron*, 2002, 21:2579.

14. Carriedo, G.A., García Alonso, F.J., García Alvarez, J.L., Díaz, C., Yutronic, N. *Polyhedron*, 2002, 21:2587, and refs. therein.

15. Allcock, H.R., McDonnell, G., Riding, G.H., Manners, I. *Chem. Mater.*, 1990, 2:425.

16. Díaz, C., Valenzuela, M.L., Magon, C., Rengel, O., Unpublished results.

17. (a) Edelstein, A.S., Cammarata, R.C. *Nanomaterials: Synthesis Properties and Applications.* J.W. Arrowsmith, Bristol, UK, 2000; (b) Klabunde, K.J. *Nanoscale Materials in Chemistry.* Wiley-Interscience, Hoboken, NY, 2001; (c) Rao, C.N., Muller, A., Cheetham, A.K. *The Chemistry of Nanomaterial.* Wiley-VCH, Weinheim, Germany, 2003.

18. (a) Grunes, J., Gabor, J., Somorjai, A. *Chem. Commun.*, 2003, 2257; (b) Bell, A.T. *Science*, 2003, 229:1688.

19. (a) Turkevich, J., Stevenson, P.C., Hillier, J. *Discuss. Faraday Soc.*, 1951, 11:55; (b) Enustun, B.V., Turkevich, J. *J. Am. Chem. Soc.*, 1963, 85:3317.

20. Brust, M., Walker, M., Bethell, D., Schiffrin, D.J., Whyman, R. *J. Chem. Soc. Chem. Commun.*, 1994, 801.

21. Pierre, A., Rutgers, L.K. *Introduction to Sol-Gel Processing.* Kluwer Academic, Dordrecht, The Netherlands, 1998.

22. (a) Ginzburg, M., MacLachlan, M.J., MingYang, S., Coombs, N., Coyle, T.W., Praju, N., Greendan, J.E., Herber, R.H., Ozin, G.A., Manners, I. *J. Am. Chem. Soc.*, 2002, 124:2625; (b) Petersen, R., Foucher, D.A., Tang, B., Lough, A., Praju, N., Greedan, J.E., Herber, R.H., Manners, I. *Chem. Mater.*, 1995, 7:2045.

23. (a) Terenish, T., Hasegawa, S., Shimizu, T., Miyake, M. *Adv. Mater.*, 2001, 13:1699; (b) Shimizu, T., Teranishi, T., Hasegawa, S., Miyake, M. *J. Phys. Chem.*, 2003, 107:2719.

24. Qiu, J., Jiang, X., Zhu, C., Shirai, M., Jiang, N.J. *Angew. Chem. Int. Ed.*, 2004, 43:2230.

25. Leite, E.R., Carreño, N.L.V., Lango, E., Pontes, F.M., Barison, A., Ferreiro, A.G., Maniette, Y., Varela, J.A. *Chem. Mater.*, 2002, 14:3722.

26. Boxall, D.L., Kenik, E.A., Lukehart, C.M. *Chem. Mater.*, 2002, 14:1715.

27. Roucoux, A., Schulz, J., Patin, H. *Chem. Rev.*, 2002, 102:3757.

28. Olshavsky, M.A., Allcock, H.R. *Chem. Mater.*, 1997, 9:1367.

29. Walker, C.H., St. John, J.V., Wisian-Neilson, P. *J. Am. Chem. Soc.*, 2001, 123:3846.

30. Jung, J., Kmecko, T., Claypool, C.L., Zhang, H., Wisian-Neilson, P. *Macromolecules*, 2005, 38:2122.

31. Díaz, C., Valenzuela, M.L. *J. Chilean Chem. Soc.*, 2005, 50:417.

32. Díaz, C., Castillo, P., Valenzuela, M.L. *J. Cluster Sci.*, 2005, 16:515.

33. Díaz, C., Valenzuela, M.L. *J. Inorg. Organomet. Polym. Mater.*, 2006, 16:123.

34. Díaz, C., Valenzuela, M.L. *Macromolecules*, 2006, 39:103.

35. Díaz, C., Valenzuela, M.L. *J. Inorg. Organomet. Polym. Mater.*, 2006, 16:211.

36. Díaz, C., Valenzuela, M.L. *J. Inorg. Organomet. Polym. Mater.*, 2006, 16:419.

37. Díaz, C., Valenzuela, M.L., Carriedo, G.A., García Alonso, F.J., Presa, A. *Polym. Bull.*, 2006, 57:920.

38. Díaz, C., Valenzuela, M.L., Yutronic, N. *J. Inorg. Organomet. Polym. Mater.*, 2007, 17:577.

39. Díaz, C., Valenzuela, M.L., Spodine, E., Moreno, Y., Peña, O. *J. Cluster Sci.*, 2007, 18:831.

40. Banhart, J. *Adv. Eng. Mater.*, 2006, 8:781.

41. Hodge, A.M., Hayes, J.R., Caro, J.A., Biener, J., Hamza, A.V. *Adv. Eng. Mater.*, 2006, 8:853.

42. Biener, J., Hodge, A.M., Hayes, J.R., Volkert, C.A., Zepeda, L.A., Hamza, A.V., Abraham, F. *Nano Lett.*, 2006, 6:2379.

43. Hu, B., Heilderberg, A., Boland, J., Sader, J.E., Sung, X.M., Li, Y. *Nano Lett.*, 2006, 6:468.

44. Ding, Y., Erlebacher, J. *J. Am. Chem. Soc.*, 2003, 125:7772.

45. Ji, C., Searson, P.C. *J. Phys. Chem.*, 2003, 107:4494.

46. Hattori, Y., Konishi, T., Kanoh, H., Kawasaki, S., Kaneko, K. *Adv. Mater.*, 2003, 15:529.

47. Feng, M., Puddephatt, R.J. *Chem. Mater.*, 2003, 15:2696.

48. Tappan, B.C., Huynh, M.H., Hiskey, M.A., Chavez, D.E., Luther, E.P., Mang, J.T., Son, S.F. *J. Am. Chem. Soc.*, 2006, 128:6589.

49. Allcock, H.R., Brennan, D.J. *J. Organomet. Chem.*, 1998, 341:231.

50. Allcock, H.R., Brennan, D.J., Dunn, B.S. *Macromolecules*, 1989, 22:1534.

51. Allcock, H.R., Coggio, W.D., Archibald, R.S., Brennan, D.J. *Macromolecules*, 1989, 22:3571.

52. Allcock, H.R., Coggio, W.D. *Macromolecules*, 1990, 23:1629.

53. Allcock, H.R., Kuhercik, S.E., Nelson, C.J. *Macromolecules*, 1996, 29:3686.

54. (a) Allcock, H.R., Kuhercik, S.E. *J. Inorg. Organomet. Polym.*, 1995, 5:307; (b) Allcock, H.R., Kuhercik, S.E. *J. Inorg. Organomet. Polym.*, 1996, 6:1.

55. Carriedo, G.A., Valenzuela, M.L., Díaz, C., Ushak, S., *Eur. Polym. J.*, 2008, 44:686.

56. Poojaray, D.M., Borade, R.B., Campbell, F.L., Clearfield, A. *J. Solid State Chem.*, 1994, 112:106.

57. Tillmanns, E., Gebert, W. *J. Solid State Chem.*, 1973, 7:69.

58. Krawietz, T.R., Lin, P., Lotterhos, K.E., Torres, P.D., Barich, D.H., Clearfield, A., Haw, J.F. *J. Am. Chem. Soc.*, 1998, 120:8502.

59. Iuliucci, R.J., Meier, B.H. *J. Am. Chem. Soc.*, 1994, 112:106.

60. Shibata, N., Horigudhi, M., Edahiro, T. *J. Non-Cryst. Solids*, 1981, 45:115.

61. Wong, J. *J. Non-Cryst. Solids*, 1973, 20:83.

62. Coetzee, J.H., Mashapa, T.N., Prinsloo, N.M., Rademan, J.D. *Appl. Catal. A*, 2006, 204:3008.

63. Peev, G., Rouseva, M., Zambov, L. *Semicond. Sci. Technol.*, 1994, 9:137.

64. Allcock, H.R., Fuller, T., Evans, T.L. *Macromolecules*, 1980, 13:1325.

65. Calhoun, H.P., Lindstrom, R.H., Oakley, R.T., Paddock, N.L., Todd, S.M. *Chem. Commun.*, 1975, 9:343.

66. Carriedo, G.A., Fidalgo, J.L., García Alonso, F.J., Presa Soto, A., Díaz, C., Valenzuela, M.L. *Macromolecules*, 2004, 37:9431.

67. Gover, R.K.B., Withers, N.D., Allen, S., Withers, R.L., Evans, J.S.O. *J. Solid State Chem.*, 2002, 166:42.

68. Attidekou, P.S., Garcia-Alvarado, F., Connor, P.A., Irvine, J. *J. Electrochem. Soc.*, 2007, 154:217.

69. Attidekou, P.S., Connor, P.A., Wormald, P., Tunstall, D.P., Francis, S.M., Irvine, J. *Solid State Ionics*, 2004, 175:185.

70. Fayon, F., King, I.J., Harris, R.K., Gover, R.K.B., Evans, J.S.O., Massiot, D. *Chem. Mater.*, 2003, 15:2234.

71. Dachille, F., Dent-Glasser, L. *Acta Crytallogr.*, 1959, 12:280.

72. (a) Hathaway, B.J., Billing, D.E. *Coord. Chem. Rev.*, 1970, 5:143; (b) Hanabusa, K., Hashimoto, M., Kimura, M., Koyama, T., Hirofusa, S. *Macromol. Chem. Phys.*, 1996, 197:1853; (c) Hay, R.W., Ali, M.A., Jeragh, A.B. *J. Chem. Soc. Dalton Trans.*, 1988, 2763; (d) Ray, R.K., Kauffman, K.B. *Inorg. Chim. Acta*, 1990, 173:207; (e) Canevali, C., Morazzoni, F., Scotti, R., Cauzzi, D., Moggi, P., Predieri, G. *J. Mater. Chem.*, 1999, 9:507.

73. (a) Allcock, H.R., Al-Shali, S., Ngo, D.C., Visscher, K.B., Parvez, M. *J. Chem. Soc. Dalton Trans.*, 1996, 3549.; (b) Allcock, H.R., Al-Shali, S., Ngo, D.C., Visscher, K.B., Parvez, M. *J. Chem. Soc. Dalton Trans.*, 1996, 3521; (c) Allcock, H.R. *Acc. Chem. Res.*, 1979, 112:351.

74. Gleria, M., De Jaeger, R. *Applicative Aspects of Cyclotriphosphazenes.* Nova Science, Hauppauge, NY, 2004.

75. Chandrasekhar, V. *Inorganic and Organometallic Polymers.* Springer-Verlag, New York, 2004.

76. De Jaeger, R., Gleria, M. *Prog. Polym. Sci.*, 1998, 23:179.

17 Transport Properties of Polyphosphazenes

JOEL R. FRIED

Department of Chemical and Materials Engineering, University of
Cincinnati, Cincinnati, Ohio

INTRODUCTION

A wide range of poly(organophosphazene)s have been synthesized based on the
general structure

where R_1 and R_2 include a wide range of substituent groups, such as alkyl,
alkoxy, aryloxy, and amine, that strongly influence such properties as the glass
transition temperature (T_g), crystallinity, and permeability, as illustrated for
selected polyphosphazenes in Table 1. The T_g can vary from 173 to 428 K [1].
Generally, alkoxy- and aryloxy-disubstituted polyphosphazene are semicrystalline, while amine-substituted polyphosphazenes are amorphous and polyphosphazenes having aromatic amine substituent groups have among the highest T_g
values of any polyphosphazene [2]. Many semicrystalline polyphosphazenes
exhibit both a crystalline melting temperature (T_m) and a sub-T_m transition,
$T(l)$, that represents a transition from a crystalline state to a mesophasic
structure of lower organizational order. As suggested by density functional
calculations of phosphazene trimers by Sun [3], the P–N bonding of the main
chain consists of an ionic σ and a π bond induced primarily by negative

Polyphosphazenes for Biomedical Applications, Edited by Alexander K. Andrianov
Copyright © 2009 John Wiley & Sons, Inc.

TABLE 1 Physical Properties and Gas Permeability[a] of Selected Polyphosphazenes

Polymer (Structure)[b]	T_g (K)	T_1 (K)	T_m (K)	d^{25} (g/cm^3)	P(He)	P(H$_2$)	P(O$_2$)	P(CO$_2$)	P(CH$_4$)	Ref[c]
PnBuP[d] (**11**)	165	—	—	1.047	84.7	170	128	647	196	[41]
PiBuP[e] (**12**)	178	—	—	1.055	106	167	86.6	395	98.7	[41]
PsBuP[f] (**13**)	182	—	—	1.104	67.1	95.0	40.7	177	39.1	[41]
PneoBuP[g] (**14**)	181	—	—	1.032	60.2	65.7	15.4	48.7	7.51	[41]
PDMP[h]	225	—	423	1.25	NA	NA	3.53	29.4	2.52	[33]
PPOP[i] (**15**)	270	—	—	—	—	—	1.61	9.26	—	[37]
MEEP[j] (**1**)	192	—	—	—	17	25.0	7.0	250	11.0	[44]
PDCP[k]	207	—	—	—	—	84	—	543	92	[45]
PMPP[l]	310	—	—	1.2091	—	—	1.8	6.5	0.7	[46]
PTFEP[m] (**5**)	191	339–356	491–518	1.707	94.6	63.2	35.4	196.3	19.0	[40]

[a] Permeability coefficient, units of barrier [1 barrier = 10^{-10} cm^3(STP) · cm/(cm^2 · s · cmHg)].

[b] Reference to the structure in the text.

[c] Reference to permeability data.

[d] PnBuP, poly[bis(*n*-butoxy)phosphazene].

[e] PiBuP, poly[bis(*iso*-butoxy)phosphazene].

[f] PsBuP, poly[bis(*sec*-butoxy)phosphazene].

[g] PneoBuP, poly[bis(*neo*-butoxy)phosphazene].

[h] PDMP, poly(dimethylphosphazene), 70% crystallinity.

[i] PPOP, poly[bis(phenoxy)phosphazene], 20% crystallinity.

[j] MEEP, poly[bis(2-(2-methoxytethoxy)ethoxy)phosphazene].

[k] PDCP, poly(dichlorophosphazene), hydrolytically unstable.

[l] PMPP, polymethylphenylphosphazene, permeability data at 35°C.

[m] PTFEP, poly[bis(2,2,2-trifluoroethoxy)phosphazene], 40% crystallinity.

hyperconjugation. Charges on P and N are close to $+1$ and -1, respectively. The result is a very flexible polymer chain with rotational barriers comparable to that of the C–C bond in ethane (ca. $2\,\text{kcal/mol}$) and an extremely electronegative N site on each repeat unit. In general, main-chain mobility leads to low T_g values and high permeability, similar to properties that characterize polysiloxanes, such as poly(dimethylsiloxane) (PDMS). High permeability makes many polyphosphazenes attractive as membranes for gas, liquid, and vapor separations [4,5]. Other important uses for polyphosphazenes include such diverse applications as controlled drug and protein delivery, microencapsulation [6,7], polymer electrolytes, nonlinear optical and electrooptical materials, and proton-exchange membranes for methanol fuel cells. In this chapter we focus on the transport properties of polyphosphazenes, for which permeant solubility and diffusivity are important criteria. Computational chemistry and molecular simulations provide important tools for understanding how the chemical structure of polyphosphazenes controls transport properties, especially in the areas of gas separations and lithium ion transport in polymer electrolytes for secondary battery applications, as discussed in subsequent sections.

POLYPHOSPHAZENE ELECTROLYTES

Poly(ethylene oxide) (PEO), which has the repeating unit structure

$$-\!\!\left[\!-CH_2-\!\!-CH_2-\!\!-O-\!\right]\!\!-$$

doped with a variety of low-dissociation energy salts such as lithium triflate ($LiSO_3CF_3$) and lithium perchlorate ($LiClO_4$), is a feasible candidate for use as a polymeric electrolyte in lightweight, rechargeable lithium batteries [8]. A polyphosphazene that has been investigated extensively for this purpose is poly[bis(2-(2-methoxyethoxy)ethoxy)phosphazene] (**1**) or MEEP:

$$
\begin{array}{c}
OCH_2CH_2OCH_2CH_2OCH_3 \\
| \\
-\!\!\left[\!-P\!=\!N-\!\right]\!\!- \\
| \\
OCH_2CH_2OCH_2CH_2OCH_3
\end{array}
$$

1

As shown by its repeat unit structure (**1**), the side chain of MEEP has an ethylene oxide composition. Like PEO, MEEP is amorphous with a low T_g (i.e., $189\,\text{K}$).

In 1984, Blonsky and co-workers [9] reported the conductivity measurements of lithium triflate–doped MEEP, $(LiSO_3CF_3)_{0.25}\cdot MEEP$. Between room temperature and 100°C, the conductivity of this complex was reported to be one to three orders of magnitude greater than for PEO, suggesting an important opportunity for MEEP as a polyelectrolyte for thin-film batteries operating at room temperature. In subsequent studies, conductivities up to $10^{-5}\,S/cm$ were reported by Kaskhedikar et al. [10] for poly[(2-methoxyethyl) amino(n-propylamino)phosphazene] (2) with $LiCF_3SO_3$ and 2% Al_2O_3 nanoparticles. Comparable conductivities (up to $7.69 \times 10^{-5}\,S/cm$) have been reported by Allcock and co-workers [11] for MEEP–silicate hybrid networks. Polymer electrolytes of a polyaminophosphazene with oligo(propylene oxide) (3) side chains have been reported by Kaskhedikar et al. [12] to have ionic conductivities of up to $2.8 \times 10^{-4}\,S/cm$ for 20 wt% $LiCF_3SO_3$ at 80°C. Klein and co-workers [13] have shown that ion mobility is about 10 times greater for a poly(methoxyethoxy-ethoxyphenoxyphosphazene) doped with lithium bis(trifluoromethanesulfonyl)imide than for the corresponding ionomer, where the anion is attached covalently to the phenoxy side chain (4) (see Chart 1). Conner and co-workers [14] have reported an ionic conductivity of $1.2 \times 10^{-4}\,S/cm$ for a (60 : 40) mixed-substituent polyphosphazene containing 2-(2-methoxyethoxy) ethoxy and 2-(2-phenoxyethoxy)ethoxy side groups.

Several studies have investigated the potential mechanisms of Li^+ association and transport. Spectroscopic studies of ^{15}N-MEEP [15] suggest that the preferred association for the lithium cation is with the nitrogen atoms along the polyphosphazene backbone over that of the oxygen atoms in the side chains. An illustration of a proposed "pocket" model for lithium association with the N atom of the main chain and the oxygen atoms of the side chain is shown in Figure 1. Using low-level ab initio and density functional calculations and molecular dynamics (Dreiding force field [16]) of a system consisting of a polyphosphazene, lithium triflate, and water. Wang and Balbuena [17] reached the similar conclusion that nitrogen atoms exhibited stronger affinity for Li^+ than the oxygen atoms of the ether side chain and the oxygen atoms of water molecules and would be expected to strongly influence Li^+ migration (see Figs. 2 and 3). The strongest association is expected when Li^+ coordinates with two or three neighboring nitrogen atoms and one methoxy oxygen. Values for the diffusion coefficients, D_α, were obtained from molecular dynamics using the mean-square displacement (MSD) of one gas molecule by means of the Einstein equation in the form [18]

$$D_\alpha = \frac{1}{6N_\alpha} \lim_{t \to \infty} \frac{d}{dt} \left\langle |\mathbf{r}_i(t) - \mathbf{r}_i(0)|^2 \right\rangle \tag{1}$$

where N_α represents the number of diffusing molecules of type α, $\mathbf{r}_i(0)$ and $\mathbf{r}_i(t)$ are the initial and final (at time t) positions of the center of mass of one gas molecule i over the time interval t, and $\left\langle |\mathbf{r}_i(t) - \mathbf{r}_i(0)|^2 \right\rangle$ is the MSD averaged

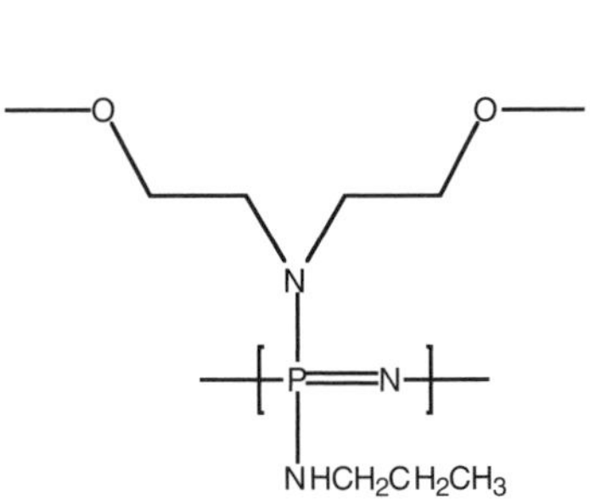

NHCH2CH2CH3

2

OCH3
OCH3

3

OCH2CH2OCH2CH2OCH3
N⁻ Li+
CF3

4

CHART 1

R
R
:O:
Li+
:O:
OR
OR

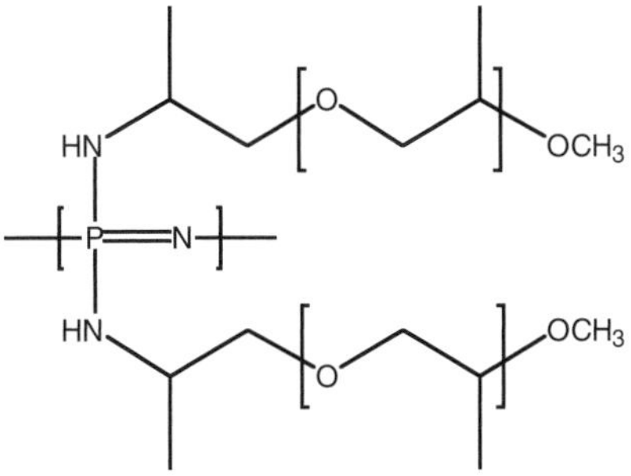

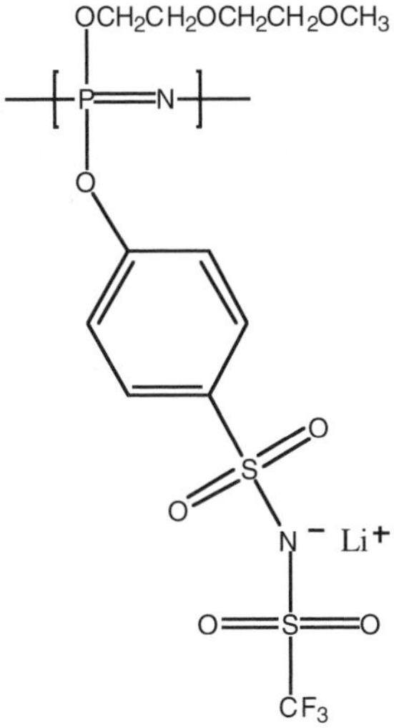

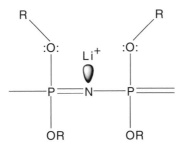

FIGURE 1 Lithium association with nitrogen in polyphosphazenes, showing the chelating effect of ligand oxygens on the lithium cation. (From ref. [15].)

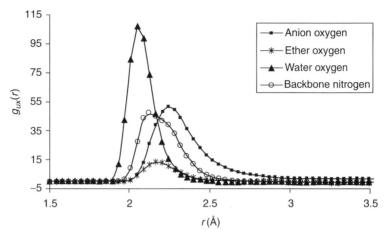

FIGURE 2 Radial distribution function g_{LiX} obtained from molecular dynamics simulations of mixtures of MEEP (1), $LiCF_3SO_3$, and water at 300 K, illustrating distributions of anion oxygen, ether oxygen, water oxygen, and backbone nitrogen atoms around Li^+ in the range 1.5 to 3.5 Å. (From ref. [17].)

over the ensemble. Fried has discussed in detail the use of molecular simulation methods in the study of membrane permeability [19]. Diffusion coefficients obtained by Wang and Balbuena for Li^+, H_2O, and $CF_3SO_3^-$ were 4.4×10^{-6}, 3.1×10^{-4}, and 2.2×10^{-5} cm^2/s, respectively.

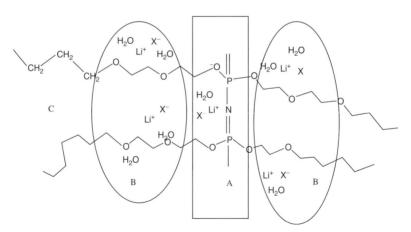

FIGURE 3 (A) High-Li^+ affinity region, including nitrogen of the backbone and ether oxygen of the side chain, (B) low-Li^+ affinity region, consisting of ether oxygen atoms, (Li^+ may jump among the side chains) and (C) hydrophobic region made of a carbon matrix (the high density of the region can prevent the penetration of H_2O). (From ref. [17].)

PROTON-EXCHANGE MEMBRANES

Nafion, a perfluorsulfonate ionomer,

is a widely used proton-exchange membrane (PEM) for hydrogen fuel cells, but its high methanol crossover (i.e., diffusion of methanol from anode to cathode across the PEM) limits its efficient use in direct methanol fuel cells (DMFCs) [20]. Several polyphosphazenes have been proposed for use in DMPCs, due to their low methanol crossover compared to Nafion. These include sulfonated alkyl group–substituted polyphosphazenes based on poly[bis(phenoxy)phosphazenes] (5) [21–23], such as sulfonated poly[(3-methylphenoxy)(phenoxy)phosphazene] (P3MPP) (6) and sulfonated poly[bis(3-methylphenoxy)phosphazene] (7) (Chart 2). Phosphonated polyphosphazenes [24] have also been used as described below. Another advantage is that these polyphosphazenes can easily be cross-linked by a variety of methods, including ultraviolet irradiation to control swelling and improve mechanical performance for fuel cell applications.

In the case of sulfonated P3MPP, Tang and co-workers [21] have reported that diffusion coefficients at 25°C for methanol concentrations between 1.0 and 5.0 M were in the range 8.0×10^{-8} to 4.0×10^{-7} cm^2/s. These are significantly lower than reported for Nafion and therefore offer promise for DMFC applications. In the case of the sulfonated poly[bis(3-methylphenoxy)phosphazene], proton conductivity was unaffected by cross-linking and were reported to be about 30% lower than for Nafion between 25 and 65°C [23]. An advantage for methanol fuel cell applications was that the diffusivity of methanol was only 1.62×10^{-8} cm^2/s, compared to 6.5×10^{-6} cm^2/s for Nafion 117 at 30°C. Membranes were stable up to 173°C and at pressures of 800 kPa.

Allcock and co-workers [25] have looked at a copolymer of a phenyl phosphonic acid functionalized poly[bis(3-methyphenyl)phosphazene] and poly(aryloxyphosphazene)s (8). Proton conductivities were high, between 10^{-2} and 10^{-1} S/cm, with methanol diffusivity about six times lower than for Nafion. Zhou and co-workers [24] have compared the conductivity and methanol permeability of sulfonated and phosphonated poly[(aryloxy)phosphazene]s with Nafion 117 at temperatures up to 20°C. The methanol permeability of a sulfonated membrane was about eight times that of Nafion 117 at room temperature but comparable at 120°C, while the permeability of a

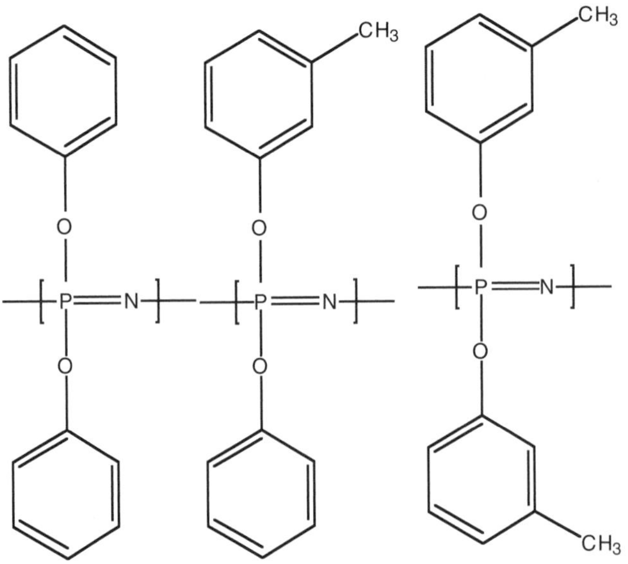

CHART 2

phosphonated polyphosphazene was about 40 times lower at room temperature and about nine times lower at 120°C. Conductivities of the sulfonated and phosphonated polyphosphazenes were about one-half that of Nafion. In a recent study, Dotelli et al. [26] reported extremely high proton conductivity (ca. 10^{-3} S/cm) for a phosphoric acid–doped composite membrane consisting of poly(dipropyl)phosphazene (**9**) with a sulfonated poly[(hydroxy)(propyl) phosphazene] (**10**) (Chart 3).

GAS SEPARATION MEMBRANES

The gas permeability of polyphosphazenes has been reported by several groups. In terms of gas transport properties, the most extensively studied polymers of this polymer group are four poly(butoxyphosphazenes) (**11**, **12**, **13**, **14**), where $R = OC_4H_9$, and poly[bis(2,2,2-trifluoroethoxy)phosphazene] (PTFEP) (**15**), where $R = OCH_2CF_3$ (Chart 4). The poly(butoxyphosphazenes) include poly [bis(*n*-butoxy)phosphazene] (PnBuP) (**11**), poly[bis(*iso*-butoxy)phosphazene] (PiBuP) (**12**), poly[bis(*sec*-butoxy)phosphazene] (**13**), and poly[bis(*neo*-butoxy) phosphazene] (PneoBuP) (**14**). The interest has been primarily for gas separation membranes [4,5], for which several patents have been issued [27–29].

Permeability data are available for arloxy- [30–39], alkoxy- [40,41] [36,42–45], and alkyl-substituted [46] polyphosphazenes. Some limited gas data exist

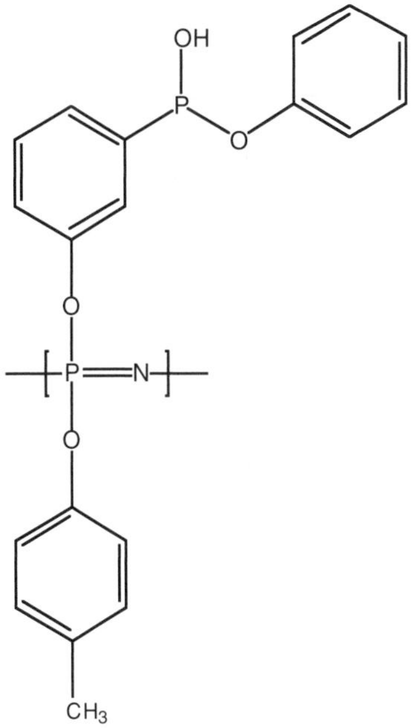

8

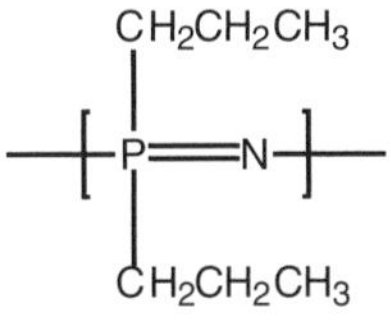

9

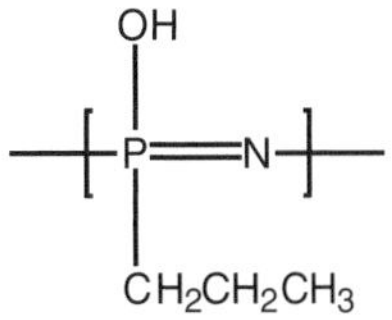

10

CHART 3

for amino-substituted polyphosphazenes [47–52] such as *n*-butylamino and *n*-hexylamino-substituted polyphosphazenes that indicate very high permeability; however, data for these polyphosphazenes are inconsistent [53].

Gas permeability data for representative polyphosphazenes are shown in Figures 4 to 6 in the form of Robeson plots [54,55] for three important commercial gas separations: O_2/N_2, CO_2/CH_4, and He/CH_4, respectively. In these plots, *ideal* permselectivity is plotted against the permeability coefficient. The ideal permselectivity coefficient is defined as

$$\alpha_{\mathrm{A}}^{\mathrm{B}} = \frac{P_{\mathrm{A}}}{P_{\mathrm{B}}} \tag{2}$$

where P_{A} is the permeability coefficient in units of barrers [1 barrer $= 10^{-10}$ cm^3 (STP) $\cdot$ cm/cm$^2 \cdot$ s $\cdot$ cmHg] of the more permeable pure gas A and P_{B} is the permeability coefficient of the less permeable gas B. By comparison to the most permeable polymers, polydimethylsiloxane (PDMS) and poly[1-(trimethylsilyl)-1-propyne] (PTMSP), the polyphosphazenes fall in the region of medium to high permeability. The low-T_{g} (rubbery) polyphosphazenes such as PTFEP, and particularly amorphous poly[bis(*n*-butoxy)phosphazene] (PnBuP) ($T_g = 165$ K), have among the highest gas permeabilities. The upper bounds

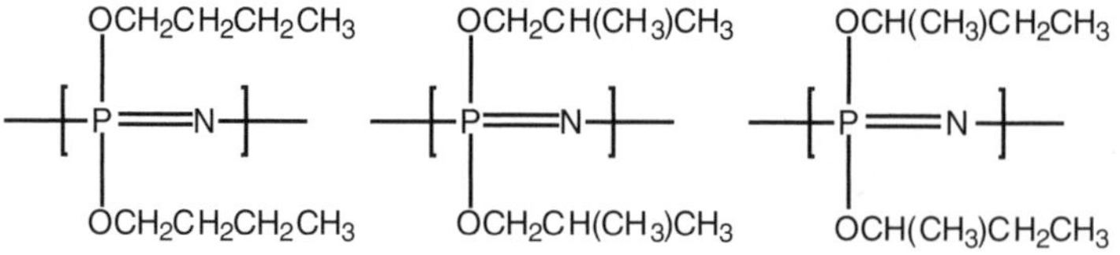

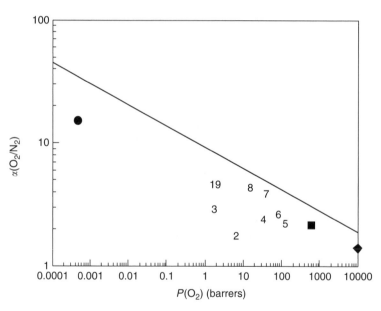

CHART 4

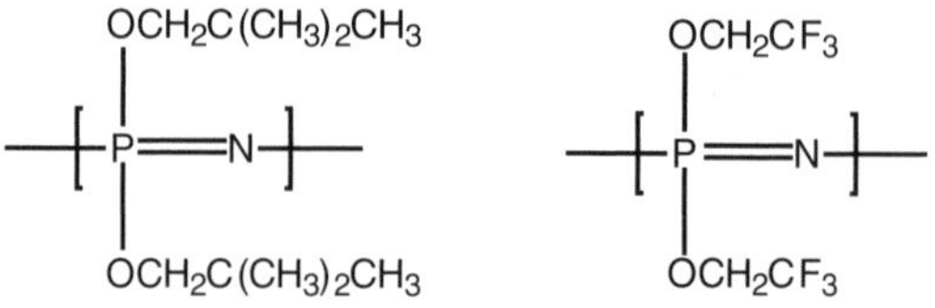

FIGURE 4   Robeson plot [54] of ideal permselectivity (O_2/N_2) versus oxygen permeability at 25°C unless otherwise indicated. (■) Polydimethylsiloxane; (●) Vectra polyester; (♦) poly[1-(trimethylsilyl)-1-propyne] (PTMSP). 1 barrer $= 10^{-10}$ cm^3(STP) cm/(cm$^2 \cdot$ s $\cdot$ cmHg). The solid line represents the upper limit for O_2/N_2 separation. 1, Poly[bis(phenoxy) phosphazene] (PPOP); 2, poly[bis(2-(2-methoxyethoxy)phosphazene] (MEEP); 3, poly-(methylphenylphosphazene) (PMPP); 4, poly(trifluoroethoxyphosphazene) (PTFEP); 5, poly[bis(*n*-butoxy)phosphazene] (PnBuP); 6, poly[bis(*iso*-butoxy)phosphazene] (PiBuP); 7, poly[bis(*sec*-butoxy)phosphazene] (PsBuP); 8, poly[bis(*neo*-butoxy)phosphazene] (Pneo-PeP); 9, poly[bis(4-*t*-butoxyphenoxy)phosphazene] (PTBP).

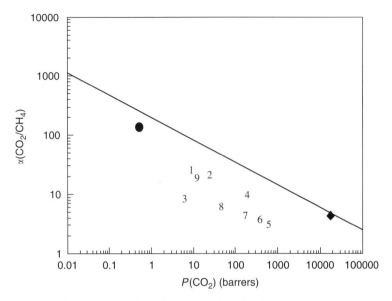

FIGURE 5   Robeson plot [54] of ideal permselectivity (CO_2/CH_4) versus CO_2 permeability at 25°C unless otherwise indicated. (•) Poly(methyl methacrylate) (PMMA) polyester; (♦) poly[1-(trimethylsilyl)-1-propyne] (PTMSP). The solid line represents the upper limit for CO_2/CH_4 separation. For a key to the curves, see Figure 4.

represents those reported by Robeson [54] from over 300 references and are given by

$$P_i = k\alpha_{ij}^n \tag{3}$$

where the values of k and n are given in Table 2.

Molecular Simulation of Gas Transport   Molecular simulation studies of polyphosphazenes have been reported by Fried and co-workers [56–58]. An early version of the COMPASS force field was used in these studies [59]. COMPASS has been fully parameterized and validated for phosphazenes and has been used to simulate the glass transition of four polyphosphazenes using NPT dynamics [60]. The first simulation study of gas transport in polyphosphazenes [57] focused on gas diffusion and solubility of He, Ne, O_2, N_2, CH_4, and CO_2 in two isomeric poly(butoxyphosphazenes) — poly[bis(n-butoxy)phosphazene] (PnBuP, $T_g = 165$ K) and poly[bis(sec-butoxy)phosphazene] (PsBuP, $T_g = 182$ K) — for which extensive gas transport properties have been reported by Hirose and Mizoguchi [41], as given in the preceding section. Self-diffusion coefficients [eq. (1)] were obtained from NVT dynamics using up to 3 ns of simulation time. With the exception of He, good agreement was obtained

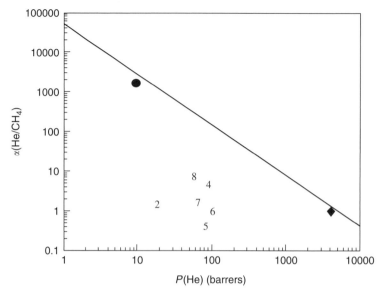

FIGURE 6   Robeson plot [54] of ideal permselectivity (He/CH$_4$) versus He permeability at 25°C unless otherwise indicated. (•) Poly(methyl methacrylate) (PMMA); (♦) poly[1-(trimethylsilyl)-1-propyne] (PTMSP). 1 barrer $= 10^{-10}$ cm^3(STP) cm/(cm^2 · s · cmHg). Solid line represents the upper limit for He/CH$_4$ separation. For a key to the curves, see Figure 4.

between simulation and experimental (time-lag) diffusion coefficients for PnBuP and PsBuP as correlated by the Teplyakov and Meares equation [61]:

$$\log D = K_1 - K_2 d_{\text{eff}}^2 \qquad (4)$$

where K_1 is nearly independent of the polymer, while K_2 increases with increasing cohesive energy density (CED). Diffusion coefficients obtained from simulation for He were significantly larger than experimental values for both polymers. This discrepancy may be attributed to the inadequacy of force parameters for some gases. For example, force-field parameters for He were obtained by fitting condensed-phase data for liquid He at 4 K. The extension of

TABLE 2   Robeson Parameters for Three Representative Separations [eq. (3)]

Gas Pair	k (barrers)	n
O$_2$/N$_2$	389,224	−5.800
CO$_2$/CH$_4$	1,073,700	−2.6264
He/CH$_4$	5,002	−0.7857

Source: Ref. 54.

this parameterization to represent the diffusion of individual He molecules sorbed in a polymer at 298 K may be questionable.

Simulation values for solubility coefficients at 298 K were obtained from grand canonical Monte Carlo (GCMC) simulations of sorption isotherms. The solubility coefficient can be correlated with a frequently used relationship linking the solubility coefficient (S) with the Lennard-Jones potential well depth parameter, ε/k, given by the equation [61,62]

$$\log S = \log S^{\circ} + m(\varepsilon/k) \tag{5}$$

where m has a value of approximately $0.01 \, \mathrm{K}^{-1}$. Values of S° range from 0.005 to about $0.02 \, \mathrm{cm}^3 \mathrm{(STP)}/\mathrm{cm}^3 \cdot \mathrm{atm}$ and depend on the polymer. Comparison between solubility coefficients obtained from permeability measurements and simulation for most gases was good, with the exception that the solubility coefficient for He obtained from GCMC simulations was larger than that predicted by the Lennard-Jones correlation given by eq. (5). In the case of PsBuP, all GCMC-derived solubility coefficients were higher than experimental values but followed the Lennard-Jones correlation. It was suggested that the higher solubilities obtained from GCMC simulation of an amorphous cell of PsBuP may indicate that the experimental sample used by Hirose and Mizoguchi [41] was not completely amorphous. Many polyphosphazenes exhibit two first-order transitions, $T(1)$ and T_m, as discussed earlier. In the case of alkoxy-substituted polyphosphazenes, the possibility for mesophase formation decreases with the length of the substituent group; therefore, PnBuP would more likely be amorphous than PsBuP, for example. Fractional free volumes of PnBuP, PsBuP, and poly[bis(*iso*-butoxy)phosphazene] (PiBuP) calculated from group-contribution methods were approximately equal, in the range 0.084 to 0.097. The distribution of free volumes obtained from transition-state theory (TST) simulation of all three isomeric polybutoxyphosphazenes were each different and explain differences in the dependence of the diffusion coefficients on gas diameter [i.e., the slopes in the Teplyakov–Meares correlation, eq. (4)].

Gas transport properties of PTFEP have been reported by Hirose and coworkers [31,40] and by Starannikova et al. [42]. The most notable feature is the high solubility of CO_2 in this polymer, as shown in the Lennard-Jones plot in Figure 7. For comparison are the solubility plots of two other fluorinated polymers — poly[5,5-difluoro-6,6-bis(trifluoromethyl)norbornene] (PFMNB) and poly(trifluoropropyl methyl siloxane) (PTFPMS) — that have been reported to show elevated CO_2 solubility. Elevated CO_2 solubility has been attributed to interaction between CO_2 and the electron-withdrawing trifluoroethoxy group of PTFEP [40,42]. Interactions between CO_2 and three low-molecular-weight fluoroalkanes (CF_4, CH_3CF_3, and $CH_3CH_2CF_3$) were explored by high-level ab initio calculations (MP2/6-311 + +G**) [56]. Results indicated a weak interaction (up to $-11.5 \, \mathrm{kJ/mol}$) between the quadrupole of CO_2 and the dipole of the fluoroalkyl group. Solubility coefficients obtained

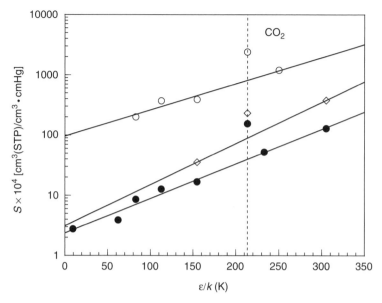

FIGURE 7 Semilogarithmic plot of gas solubility versus the Lennard-Jones potential well-depth parameter, ε/k. Solubility data: PTFEP ($\bullet$), Hirose et al. [72]; PTMPS ($\diamond$), Stern et al. [73]; PFMNB ($\circ$), Yampol'skii et al. [73]. Values of Lennard-Jones potential well-depth parameters were taken from Teplyakov and Meares [61]. Lines are drawn by least-squares fit of experimental data with the exclusion of the CO_2 data point. The dashed vertical line locates solubility data for CO_2. (From ref. 56.)

from GCMC simulation of sorption isotherms for He, N_2, O_2, and Xe in PTFEP followed a dependence on the Lennard-Jones potential well depth given by eq. (5), although values obtained from simulation using an amorphous cell were substantially larger than experimental values, as observed previously for PsBuP. This may be attributed to a mesophasic structure of the experimental samples. Results of pair-correlation analysis indicated a strong correlation of CO_2 with the trifluoromethyl group of PTFEP. These results agree well with the conclusions of the ab initio study of model compounds.

In a subsequent study, Hu and Fried [58] reported results of the simulation of diffusion and solubility of seven gases (He, H_2, O_2, N_2, CH_4, CO_2, and Xe) in an amorphous cell and in an α-orthorhombic crystalline cell of PTFEP. Diffusion coefficients and displacements were similar in both amorphous and crystalline simulations and were comparable to experimental values. This indicates that gas diffusion is unrestricted in the crystalline state of PTFEP as has been reported for poly(4-methyl-1-pentene) (PMP). In the case of PTFEP and PMP, the density of the crystalline phase is very close to that of the amorphous state, due to loose packing of chains in the unit cell. In relation to the solubility of the amorphous cell, it was shown that all solubility data, including that of CO_2, could be correlated by the same line by incorporating a

Flory interaction parameter, χ, in the form

$$\log S = 0.026(\varepsilon/k) - \chi - K \tag{6}$$

where K is a parameter that includes contribution from the partial molar volume of the sorbed gas.

MEMBRANES FOR LIQUID AND VAPOR SEPARATIONS

Polyphosphazenes have been a variety of separation of liquid- and vapor-phase organics as well as the separation of ions and isoptopes. For example, carboxylated PPOP has been used to separate tritiated water [63]. Allen et al. [64] have shown that Cr^{3+} can be separated from Co^{2+} and Mn^{2+} from aqueous solution using a PPOP membrane on the basis of the lower diffusivity of Cr^{3+}, due to its larger radius of hydration. The majority of applications, however, have focused on pervaporation, whereby liquid organics can be separated from aqueous solution by evacuating the downside of the membrane. Organic vapors (VOCs) can been separated from air by a process called *evapomeation*, where VOC-contaminated air is at one side of the membrane while vacuum is applied to the other side. Polyphosphazene membranes have been used for both pervaporation and evapomeation separations, as discussed below.

Vapor-Phase Separations   Sorption isotherms of the vapors of several alcohols, ketones, methyl isobutyl ketone, and aromatic molecules (benzene, toluene, and xylene) in PTFEP and PPOP have been reported by Sun et al. [65] at 35°C. The sorption isotherms were nearly linear — approximately Henry's law behavior. Diffusion coefficients were in the range 10^{-9} to 10^{-8} cm²/s in PPOP and 10^{-8} to 10^{-7} cm²/s in PTFEP. In a more recent publication, Sun et al. [66] reported the solubility and permeability of benzene, cyclohexane, and n-hexane in PTFEP as a function of temperature and vapor activity. The permeability of benzene ranged from several hundred to several thousand barrers. Sun et al. [67] have also investigated the sorption and permeation of water and ethanol vapors in PTFEP. Compared to ethanol, water was selectively permeable. The sorption isotherms for water and ethanol were fitted by Henry's law. Diffusivities for benzene and cyclohexane were determined at different vapor activities. The diffusivity of water and ethanol ranged from 10^{-8} to 10^{-7} cm²/s.

Sun et al. [68] have investigated evaporation and pervaporation of mixtures of benzene and cyclohexane through PPOP. In pervaporation, benzene permeation was selective over that of cyclohexane. Uragami and Morikawa [69] have investigated the pervaporation and evapomeation of aqueous solutions of methanol and ethanol in PTFEP. Suzuki et al. [70] have studied the pervaporation of water and several organic solvents in PTFEP over the temperature range 42 to 80°C. Permeation fluxes increased in the order methanol > ethanol > benzene > water > cyclohexane. An Arrhenius plot of permeation flux changed in slope in the region of the $T(1)$ transition near 75°C. For aqueous methanol pervaporation, methanol permeated more rapidly, while poor permselectivites were observed for aqueous ethanol solutions. The pervaporation of the close-boiling mixture of benzene and cyclohexane showed a separation factor [eq. (2)] of 12 for benzene over cyclohexane.

Orme et al. [71] have studied pervaporation of water–dye, 2-propanol–dye, water–2–propanol, and water–methanol mixtures in a polyphosphazene containing 2-(2-methoxyethoxy)ethanol (MEE), 4-methoxyphenol, and 2-allylphenol pendant groups. The MEE and 4-methoxyphenol groups influence the hydrophilicity of the membrane. Three-dimensional solubility parameters of MEEP were reported. Fluxes of methanol and isopropanol were greater than those for water. For the alcohol–water separations, the alcohol was the favored permeate in all cases, with higher fluxes observed for higher alcohol feed concentrations.

CHALLENGES FOR THE FUTURE

As we discussed, there is an established body of data and some simulation studies available primarily for the diffusion and solubility of fixed gases, due to the significant interest in the application of polyphosphazenes for gas separation membranes. Most of these studies have focused on polyaroxy- and polyalkyloxyphosphazenes. Additional information exists for Li^+ ion conduction in polyphosphazene electrolytes. There is an important need to extend these studies to include the transport of oxygen, carbon dioxide, water, other molecules, and nutrients in polyphosphazenes of special interest to medical and biomedical applications, including controlled drug release and tissue engineering. Molecular simulation methods have a strong opportunity here for molecular design. This will require continued development of class II force fields with parameterization for a wider variety of phosphazene structures.

REFERENCES

1. Allcock, H.R. *Chemistry and Applications of Polyphosphazenes*. Wiley, Hoboken, NJ, 2003.

2. Potin, P., DeJaeger, R. Polyphosphazenes: synthesis, structures, properties, applications. *Eur. Polym. J.*, 1991, 27(4–5):341–348.

3. Sun, H. Molecular structures and conformations of polyphosphazenes: a study based on density functional calculations of oligomers. *J. Am. Chem. Soc.*, 1997, 119:3611–3618.

4. Golemme, G., Drioli, E. Polyphosphazene membrane separations: Review. *J. Inorg. Polym.*, 1996, 6(4):341–365.

5. Stewart, F.F., Harrup, M.K. Membrane separations using linear polyphosphazene polymers. *Polym. News*, 2001, 26:78–85.

6. Cohen, S., et al. Ionically cross-linkable polyphosphazene: a novel polymer for microencapsulation. *J. Am. Chem. Soc.*, 1990, 112:7832–7833.

7. Allcock, H.R. Polyphosphazenes. *J. Inorg. Organomet. Polym. Mater.*, 1992, 2(2):197–210.

8. Shriver, D.F., et al. Structure and ion transport in polymer–salt complexes. *Solid State Ionics*, 1981, 5:83–88.

9. Blonsky, P.M., et al. Polyphosphazene solid electrolytes. *J. Am. Chem. Soc.*, 1984, 106:6854–6855.

10. Kaskhedikar, N., et al. Polyphosphazene based composite polymer electrolytes. *Solid State Ionics*, 2006, 177:2699–2704.

11. Allcock, H.R., Chang, Y., Welna, D.T. Ionic conductivity of covalently interconnected polyphosphazene–silicate hybrid networks. *Solid State Ionics*, 2006, 177:569–572.

12. Kaskhedikar, N., et al. Ionic conductivity of polymer electrolyte membranes based on polyphosphazene with oligo(propylene oxide) side chains. *Solid State Ionics*, 2006, 177:703–707.

13. Klein, R.J., et al. Counterion effects on ion mobility and mobile ion concentration of doped polyphosphazene and polyphosphazene ionomers. *Macromolecules*, 2007, 40:3990–3995.

14. Conner, D.A., et al. Influence of terminal phenyl groups on the side chains of phosphazene polymers: structure–property relationships and polymer electrolyte behavior. *Macromolecules*, 2007, 40:322–328.

15. Luther, T.A., et al. On the mechanism of ion transport through polyphosphazene solid polymer electrolytes: NMR, IR, and Raman spectroscopic studies and computational analysis of ^{15}N-labeled polyphosphazenes. *J. Phys. Chem. B*, 2003, 107:3168–3176.

16. Mayo, S.L., Olafson, B.D., Goddard, W.A., III. DREIDING: a generic force field for molecular simulations. *J. Phys. Chem.*, 1990, 94:8897–8909.

17. Wang, Y., Balbuena, P.B. Combined ab initio quantum mechanics and classical molecular dynamics studies of polyphosphazene polymer electrolytes: competitive solvation of Li^+ and $LiCF_3SO_3$. *J. Phys. Chem. B*, 2004, 108:15694–15702.

18. Trohalaki, S., et al. Estimation of diffusion coefficients for small molecular penetrants in amorphous polyethylene. In Row, R.J., ed., *Computer Simulation of Liquids*. Prentice Hall, Englewood Cliffs, NJ, 1991.

19. Fried, J.R. Molecular simulations of gas and vapor transport in highly permeable polymers. In Yampolskii, Y., Pinnau, I., Freeman, B.D., ed., *Materials Science of Membranes for Gas and Vapor Separation*. John Wiley, Chickester, UK, 2007 p. 93–134.

20. Ramya, K., Dhathathreyan, K.S. Methanol crossover studies in heat-treated Nafion membranes. *J. Membrane Sci.*, 2008, 311:121–127.

21. Tang, H., et al. Polyphosphazene membranes: III. Solid-state characterization and properties of sulfonated poly[bis(3-methoxyphenoxy)phosphazene]. *J. Appl. Polym. Sci.*, 1999, 71:387–399.

22. Wycisk, R., Pintauro, P.N. Sulfonated polyphosphazene ion-exchange membranes. *J. Membrane Sci.*, 1996, 119:155–160.

23. Guo, Q., et al. Sulfonated and crosslinked polyphosphazenes-based proton-exchange membranes. *J. Membrane Sci.*, 1999, 154:175–181.

24. Zhou, X., et al. High temperature transport properties of polyphosphazene membranes for direct methanol fuel cell. *Electrochem. Acta*, 2003, 48:2173–2180.

25. Allcock, H.R., et al. Phenyl phosphonic acid functionalized poly[aryloxyphosphazenes] as proton-conducting membranes for direct methanol fuel cells. *J. Membrane Sci.*, 2002, 201:47–54.

26. Dotelli, G., et al. Proton conductivity of poly(dipropyl)phosphazene-sulfonated poly[(hydroxy)propyl, phenyl]ether–H_3PO_4 composite in dry environment. *Solid State Ionics*, 2005, 176:2819–2827.

27. Kraus, N.A., Murphy, K. Polyphosphazene gas separation membranes. Monsanto Company, Dec. 28, 1983.

28. Murphy, M.K. Multicomponent gas separation membranes having polyphosphazene coatings. Monsanto Company, Mar. 1, 1988.

29. Allen, C.A., et al. Polyphosphazene semipermeable membranes. U.S. Department of Energy, June 7, 1988.

30. McCaffrey, R.R., Cummings, D.G. Gas separation properties of phosphazene membranes. *Sep. Sci. Technol.*, 1988, 23(12–13):1627–1643.

31. Mizoguchi, K., Kamiya, Y., Hirose, T. Gas transport in poly[bis(trifluoroethoxy) phosphazene] above the T(1) transition. *J. Polym. Sci. B*, 1991, 29:695–703.

32. Drioli, E., et al. Gas permeability of polyphosphazene membranes. *Gas Sep. Purifi.*, 1991, 5:252–258.

33. Gallazzi, M.C., et al. Polydimethylphosphazene as a material for new organo-inorganic separation membrane. *J. Mater. Sci. Lett.*, 1993, 12:436–438.

34. Allcock, H.R., et al. Gas permeation and selectivity of poly(organophosphazene). *Macromolecules*, 1993, 26:1493–1502.

35. Peterson, E.S., et al. Mixed-gas separation properties of phosphazene polymer membranes. *Sep. Sci. Technol.*, 1993, 28(1–3):423–440.

36. Peterson, E.S., Stone, M.L. Helium separation properties of phosphazene polymer membranes. *J. Membrane Sci.*, 1994, 86:57–65.

37. Golemme, G., Drioli, E., Lufrano, F. Gas transport properties of high performance polymers. *Poly. Sci.*, 1994, 36(11):1647–1652.

38. Houston, K.S., Weinkauf, D.H., Stewart, F.F. Gas transport characteristics of plasma treated poly(dimethylsiloxane) and phosphazene membrane materials. *J. Membrane Sci.*, 2002, 205:103–112.

39. Orme, C.J., Klaehn, J.R., Stewart, F.F. Gas permeability and ideal selectivity of poly[bis-(phenoxy)phosphazene], poly[bis-(4-tert-butylphenoxy)phosphazene], and poly[bis-(3,5-di-*tert*-butylphenoxy)$_{1.2}$(chloro)$_{0.8}$phosphazene]. *J. Membrane Sci.*, 2004, 238:47–55.

40. Hirose, T., Kamiya, Y., Mizoguchi, K. Gas transport in poly[bis(trifluoroethoxy) phosphazene]. *J. Appl. Polym. Sci.*, 1989, 38:809–820.

41. Hirose, T., Mizoguchi, K. Gas transport in poly(alkoxyphosphazene). *J. Appl. Polym. Sci.*, 1991, 43:891–900.

42. Starannikova, L.E., et al. Gas separation properties of poly[bis(trifluoroethoxy-phosphazene]. *Polym. Sci.*, 1994, 36(11).

43. Nagai, K., et al. Gas permeability of poly(bis-trifluoroethoxyphosphazene) and blends with adamantane amine/trifluoroethoxy (50/50) polyphosphazene. *J. Membrane Sci.*, 2000, 172:167–176.

44. Orme, C.J., et al. Characterization of gas transport in selected rubbery amorphous polyphosphazene membranes. *J. Membrane Sci.*, 2001, 186:249–256.

45. Orme, C.J., et al. Gas permeability in rubbery phosphazene membranes. *J. Membrane Sci.*, 2006, 280:175–184.

46. Wisian-Neilson, P., Xu, G-F Gas permeation studies of silylated derivatives of poly(methylphenylphosphazene). *Macromolecules*, 1996, 29:3457–3461.

47. Kajiwara, M. Gas permeability of poly(organophosphazenes). *J. Mater. Sci.*, 1988, 23:1360–1362.

48. Kajiwara, M. Gas permeability of poly(organophosphazenes). *J. Mater. Sci. Lett.*, 1988, 7:102–104.

49. Kajiwara, M. Oxygen gas permeability of poly(organophosphazene) membranes in water. *Polymer*, 1989, 30:1536–1539.

50. Kajiwara, M. Gas permeability and selectivity of poly(organophosphazene) membranes. *Sep. Sci. Technol.*, 1991, 26(6):841–852.

51. Kajiwara, M. Oxygen gas permeability and the mechanical properties of poly(*n*-butylamino)-(*d-n*-hexylamin)phosphazene membranes. *J. Mater. Sci.*, 1994, 29:6268–6272.

52. Kajiwara, M., Kimura, T. Oxygen gas permeability and the mechanical properties of poly(*n*-butylamino)(*di-n*-hexylamino)phosphazene membranes. ACS, Washington, DC, 1994, pp. 268–278.

53. Stern, S.A. Polymers for gas separations: the next decade. *J. Membrane Sci.*, 1994, 94:1–65.

54. Robeson, L.M. Correlation of separation factor versus permeability for polymeric membranes. *J. Membrane Sci.*, 1991, 62:165–185.

55. Robeson, L.M., et al. High performance polymers for membrane separation. *Polymer*, 1994, 35(23):4970–4978.

56. Fried, J.R., Hu, N. The molecular basis of CO_2 interaction with polymers containing fluorinated groups: computational chemistry of model compounds and molecular simulation of poly[bis(2,2,2-trifluoroethoxy)phosphazene]. *Polymer*, 2003, 44:4363–4372.

57. Fried, J.R., Ren, P. The atomistic simulation of the gas permeability of poly(-organophosphazenes): 1. Poly(dibutoxyphosphazene). *Comput. Theor. Polym. Sci.*, 2000, 10:447–463.

58. Hu, N., Fried, J.R. The atomistic simulation of the gas permeability of poly(organophosphazenes): poly[bis(2,2,2,-trifluoroethoxy)phosphazene]. *Polymer*, 2005, 46:4330–4343.

59. Sun, H., Ren, P., Fried, J.R. The COMPASS force field: parameterization and validation for phosphazenes. *Comput. Theor. Polym. Sci.*, 1998, 8(1–2):229–246.

60. Fried, J.R., Ren, P. Molecular simulation of the glass transition of polyphosphazenes. *Comput. Theor. Polym. Sci.*, 1999, 9:111–116.

61. Teplyakov, V., Meares, P. Correlation aspects of the selective gas permeabilities of polymeric materials and membranes. *Gas Sep. Purifi.*, 1990, 4:66–74.

62. Toi, K., Morel, G., Paul, D.R. Gas sorption and transport in poly(phenylene oxide) and comparisons with other glass polymers. *J. Appl. Polym. Sci.*, 1982, 27:1997–2005.

63. Duncan, J.B., Nelson, D.A. The separation of tritiated water using supported polyphosphazene membranes. *J. Membrane Sci.*, 1990, 157:211–217.

64. Allen, C.A., et al. Separation of Cr ions from Co and Mn ions by poly[bis(phenoxy) phosphazene] membranes. *J. Membrane Sci.*, 1987, 33:181–189.

65. Sun, Y.-M., et al. Sorption and diffusion of organic vapors in poly[bis(trifluoroethoxy)phosphazene] and poly[bis(phenoxy)phosphazene] membranes. *J. Membrane Sci.*, 1997, 134:117–126.

66. Sun, Y.-M., Wu, C.-H., Lin, A. Permeation and sorption properties of benzene, cyclohexane, and *n*-hexane vapors in poly[bis(2,2,2-trifluoroethoxy)phosphazene]. *Polymer*, 2006, 47:602–610.

67. Sun, Y.-M., Wu, C.-H., Lin, A. Sorption and permeation properties of water and ethanol vapors in poly[bis(trifluoroethoxy)phosphazene] (PTFEP) membranes. *J. Polym. Res.*, 1999, 6(2):91–98.

68. Sun, Y.-M., et al. Pervaporation for the mixture of benzene and cyclohexane through PPOP membranes. *AIChE J.*, 1999, 45(3):523–534.

69. Uragami, T., Morikawa, T. Characteristrics of permeation and separation of aqueous alcoholic solutions through a poly[bis(2,2,2-trifluoroethoxy)phosphazene] membrane by evapomeation and pervaporation. *Angew. Makromol. Chem.*, 1996, 234:39–51.

70. Suzuki, F., et al. Pervaporation of organic solvents by poly[bis(2,2,2-trifluoroethoxy) phosphazene] membranes. *J. Appl. Polym. Sci.*, 1987, 34:2197–2204.

71. Orme, C.J., et al. Pervaporation of water–dye, alcohol–dye, and water–alcohol mixtures using a polyphosphazene membranes. *J. Membrane Sci.*, 2002, 197:89–101.

72. Hirose, T., Kamiya, Y., Mizoguchi, K. Gas transport in poly[bis(trifluoroethoxy) phosphazene]. *J. Appl. Polym. Sci.*, 1989, 38:809–820.

73. Stern, S.A., Shah, V.M., Hardy, B.J. Structure-permeability relationships in silicone polymers. *J. Polym. Sci. B*, 1987, 25:1263–1298.

PART V
Biodetection

18 Potentiometric Monitoring of Antibody–Antigen Interactions and Stabilization of Polyaniline Electrodes with *p*-Sulfonated Poly(bisphenoxyphosphazene)

ALEXANDER K. ANDRIANOV

Apogee Technology, Inc., Norwood, Massachusetts

ALOK PRABHU

Department of Chemical and Biological Sciences, Polytechnic Institute of NYU, Brooklyn, New York

VLADIMIR SERGEYEV

Department of Chemistry, Moscow State University, Moscow, Russia

BYEONGYEOL KIM and KALLE LEVON

Department of Chemical and Biological Sciences, Polytechnic Institute of NYU, Brooklyn, New York

INTRODUCTION

Diagnostics of early biomarkers, biological compounds that alert us to a disease at an early pre-symptom stage, is an important task in the development of new tools for the prevention and treatment of serious diseases. As organic electronics are becoming an important player on the health diagnostics market, the use of conductive polymers such as polyaniline (PANI) for monitoring various biological molecules becomes increasingly attractive. However, physicians' point-of-care applications require high stability and reliability of devices, a feature that can be difficult to achieve using existing PANI materials, which contain small molecules as dopants.

Polyphosphazenes for Biomedical Applications, Edited by Alexander K. Andrianov
Copyright © 2009 John Wiley & Sons, Inc.

We develop PANI-based potentiometric ion-sensitive electrodes for the early detection of proteins. Sulfonated polyphosphazenes such as *p*-sulfonated poly(phenoxyphosphazene) (SPPZ) show great promise as macromolecular dopants for PANI materials because of high ionic density, water solubility, molecular weight, and flexibility of backbone. In this chapter we describe studies on the preparation of conductive PANI/SPPZ complexes via polymerization and mixing approaches and its potential importance for potentiometric monitoring of immune reactions.

Electrically Conducting Polymers

Numerous advances have been made during the past decade in our understanding of the properties of solid electroactive materials. Electrically conducting polymers (ECPs) are viscoelastic macromolecules that can easily be converted into materials having metal-like properties. ECPs have been studied intensively because of their relative ease of synthesis by chemical or electrochemical oxidative polymerization of the monomers and of their potential technological importance.

A key discovery in the area of ECPs was made in 1973, when the inorganic polymer polysulfurnitride $(SN)_x$ was found to possess properties of a metal [1]. The conductivity of $(SN)_x$ at room temperature is about 10^3 S/cm and below a critical temperature of about $0.3\,K$, $(SN)_x$ becomes a superconductor [2]. Unfortunately $(SN)_x$ is explosive and commercial application was not possible. In the 1970s, Shirakawa and co-workers showed that a thin film of polyacetylene could be oxidized with iodine vapor, increasing its electrical conductivity a billion times [3,4]. It becomes highly conducting on exposure to oxidizing or reducing agents called *dopants*. Reversible doping of conducting polymers, with associated control of the electrical conductivity over the full range from insulator to metal, can be accomplished by either chemical or electrochemical doping. Concurrent with the doping, the electrochemical potential (the Fermi level) is moved either by a redox reaction or by an acid–base reaction into a region of energy where there is a high density of electronic states; charge neutrality is maintained by the introduction of counter ions. Metallic polymers are, therefore, salts. The electrical conductivity results from the existence of charge carriers (through doping) and the ability of those charge carriers to move along the π-bonded "highway."

Consequently, doped conjugated polymers are good conductors for two reasons:

1. Doping introduces carriers into the electronic structure. Since every repeat unit is a potential redox site, conjugated polymers can be doped *n*-type (reduced) or *p*-type (oxidized) to a relatively high density of charge carriers [5].
2. The attraction of an electron in one repeat unit to the nuclei in neighboring units leads to carrier delocalization along the polymer chain and to charge carrier mobility, which is extended into three dimensions through interchain electron transfer.

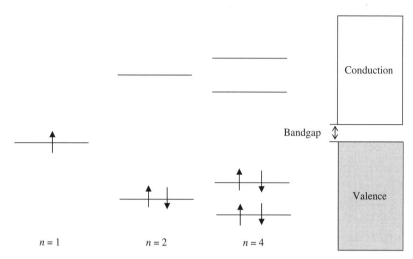

FIGURE 1   Molecular orbitals and bands.

Disorder, however, limits the carrier mobility and limits the electrical conductivity in the metallic state. Indeed, research directed toward conjugated polymers with improved structural order and hence higher mobility is a focus of current activity in the field.

The electrochemical doping of conducting polymers was discovered by Nigrey et al. [6]. The electrochemistry of conducting polymers has developed into a field of its own, with applications that range from polymer batteries and electrochromic windows to light-emitting electrochemical cells [7]. Figure 1 is a schematic diagram of a doping mechanism and related applications and, consequently, can be considered the starting point of the modern area of conducting polymers. ECPs have opened the way to progress in understanding the fundamental chemistry and physics of π-bonded macromolecules and have provided an opportunity to address fundamental issues of importance to condensed-matter physics as well, including, for example, the metal. Finally, ECPs offered the promise of achieving a new generation of polymers [8] consisting of materials that exhibit the electrical and optical properties of metal or semiconductors and that retain the attractive mechanical properties and processing advantages of polymers.

Types of Conducting Polymers

Conducting polymers can be classified into several categories [9]:

1. Conducting polymers can be composite materials comprised of insulating polymers and powdered conductive materials such as metal or carbon black. The electrical conductivity of these materials is due to conductive particles that are in contact with each other; the polymer is only a binding matrix. These materials can be used in antistatic coatings and substituents for soldering metals.

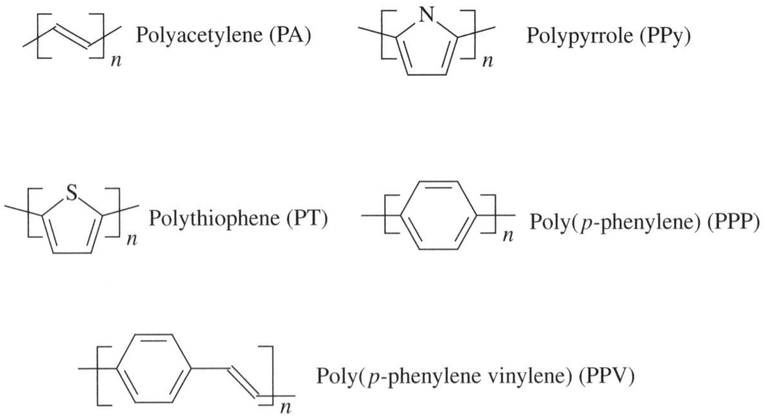

CHART 1 Molecular structures of examples of some conjugated polymers.

2. Ionically conducting polymers, in which electric charge is carried by ions such as ionically doped polyethylene oxide or polyphosphates, can be used in smart window technologies or batteries.

3. Electrically conductive polymers or conjugated polymers contain the electrons that are charge carriers. The electrical conductivity of these polymers is dependent on the conjugated nature of the polymer chain and the mobility of the π-electron. Polyaniline, polythiophene, and polyacetylene are examples of electrically conductive polymers.

The molecular structures of some ECPs, such as polyacetylene, polypyrrole, polythiophene, poly(p-phenylene), and poly(p-phenylene vinylene), are shown in Chart 1. The essential feature of each polymer is its highly conjugated π-system. Conjugation in simple carbon-containing compounds means the presence of alternating single and double carbon–carbon bonds in the molecules. Hence, the simplest conjugated polymer is polyacetylene. In fact, nearly any aromatic or heteroaromatic ring has been used as a monomer unit in conjugated polymers. The π-electron systems can be treated in the most elementary way by using simple free-electron theory. The main postulate in this approximation is that π-electrons are quite mobile, moving freely throughout the π-network. When a positive or negative charge is placed in the π-system of ECPs by a doping process, the resulting conjugation system enables charge movement and provides electrical metallic properties.

Electrical Conduction

The most widely used model for electrical conduction is the one-electron band model. This is based on extending the simple model of a bond between two atoms over an entire crystalline solid. When two identical atoms each having a half-filled orbital are brought together closely enough for their orbital to overlap, the two orbitals interact to produce new orbitals, one of lower energy and one of higher energy. The magnitude of this energy difference is determined by the extent of orbital overlap. The two electrons go into a lower-energy orbital. This process is called *molecular orbital theory*. The filled lower-energy orbital is a bonding orbital, and the empty higher-energy orbital is an antibonding orbital (Fig. 1). In a semiconductor or an insulator there is a *gap* between the highest of the low-energy orbitals and the lowest of the high-energy orbitals. This is caused by the interaction of electrons with the crystalline lattice and increases with decreasing atomic separation. The difference between energy levels within these two sets of orbitals is so small that the *bands* may be regarded as continuous and the electrons can take any energy levels. Since there are N electrons and each of the $N/2$ orbitals within the lower-energy band can accommodate two electrons, the lower-energy band is filled. This band is known as the *bonding* or *valence band*. If electrons are added, they must go into the lowest unoccupied band, known as the *antibonding* or *conduction band*. The nature of the electrical properties of a solid is determined by the extent of occupation of the energy bands and the magnitude of the gap between them. Within each band the electrons are said to be *delocalized*. This means that they are not bound to any one atom but, rather, are spread over several. For any net movement of electrons, there must be orbitals available for the electron to move to. A completely full band, such as the lower one, has no such vacancies, and an empty band, such as the higher one has nothing to move. In a metal, there is no gap, or the number of electrons is such that the highest filled orbital is not at the top of the band. This means that orbitals are freely available for conduction.

Charge Carriers in ECPs

Electrically conducting polymers can be prepared via chemical or electrochemical polymerization. In this reaction a conjugated monomer is polymerized and charge carriers are generated via doping. The doping process is an oxidation or reduction reaction in which electrons are transferred away from or to the polymer chain, respectively. The mechanism of charge transport in ECPs can be described as follows. The general structure of ECPs is an alternating sequence of single and double bonds. In the prototype conjugated polymer, polyacetylene, there is no preferred sense of bond alternation. Most ECPs, however, possess a nondegenerate ground state with a preferred sense of bond alternation. Polypyrrole and polythiophene are two examples of non-degenerate ground-state polymers that possess an aromatic configuration with long bonds between the rings and an aromatic structure within the ring. The other sense of bond alternation, the *quinoid configuration*, is characterized by shortened bonds between the rings and quinoid rings. The quinoid geometry can be considered as an excited-state configuration of the aromatic structure. The degeneracy of the ground state has an important effect on the nature of the charged species that can be obtained via oxidative or reductive doping. Here, oxidative doping will be considered as an example. Reductive doping can be described in a similar way. Oxidation of polyacetylene generates a cation radical. Because there is no preferred sense of bond alternation, the positive charge and the unpaired electron can move independently along the polymer chain, forming domain walls between the two identical parts of bond alternation.

A charge associated with a boundary or domain wall is called a *soliton* [Chart 2 (IV)] in solid-state physics because it has the properties of a solitary wave that can move without deformation and dissipation [10]. From a chemist's point of a view, a soliton is a positive, negative, or neutral radical-like site. In this view the unpaired electron can be considered as a neutral soliton or as an excitation of the system that separates two potential wells of the same energy. It is important to note that neutral and charged solitons are not localized on one carbon atom but are spread over several atoms. The bond alternation changes gradually, giving the soliton a finite width. The positive charge and the unpaired electron of the cation radical formed initially cannot move independently. The structural motif of the chain segment between the positive charge and the unpaired electron is that of a quinoid configuration, which is higher in energy and confines the charge and spin density to a single self-localized structural deformation that is mobile along the chain [Chart 2 (II)]. In condensed-matter physics such a cation radical with an associated lattice deformation is called a *polaron* and carries a spin ($S = \frac{1}{2}$). Two things can happen upon further oxidation of a nondegenerate polymer chain. A second electron can be removed from a different segment of the polymer chain, creating a new polaron, or the unpaired electron of the previously formed polaron is removed [11,12]. The latter produces a spinless dication confined to a single lattice deformation on the chain, which solid-state physicists call a *bipolaron* [Chart 2 (III)]. Bipolarons can also originate from an attractive interaction between two lattice deformations of two polarons in which their unpaired electrons form a bond on a doubly oxidized polymer chain. The latter also results in a single lattice deformation. Conduction by polarons or bipolarons is now generally considered to be the dominant mechanism of intrachain transport. Of course, interchain mechanisms such as hopping are necessary to explain the conductive behavior of bulk materials. Although several theories have been proposed to explain the hopping mechanism, the hopping mechanism of charge carriers between polymer chains is not yet well understood.

Polyaniline

Polyaniline (PANI) was first synthesized in 1862 [13] and described as an octamer existing in four different oxidation states [14]. Its properties as a conducting polymer have been studied extensively [15]. MacDiarmid and co-workers investigated polyaniline as an electrically conducting polymer in 1985,

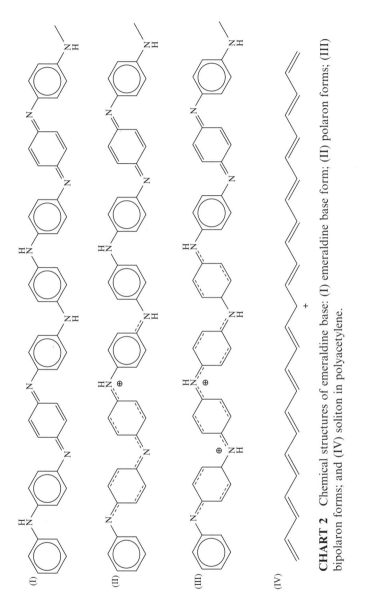

CHART 2 Chemical structures of emeraldine base: (I) emeraldine base form; (II) polaron forms; (III) bipolaron forms; and (IV) soliton in polyacetylene.

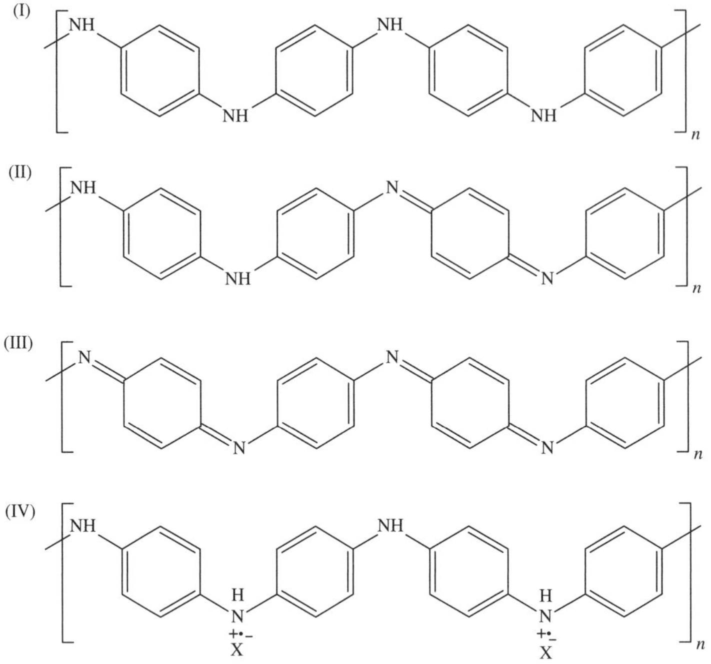

CHART 3 Four forms of PANI: (I) leucoemeraldine base, (II) emeraldine base, (III) pernigraniline base, and (IV) metallic emeraldine salt.

and it has emerged as a promising material. Although it was discovered 134 years ago by Letherby [13], polyaniline has attracted renewed interest from recent researchers, as it is highly conducting and easy to synthesize both chemically and electrochemically. It was also the first commercially available conducting polymer.

Chart 3 summarizes the repeat unit for the various forms: the three insulating forms—fully reduced (I), half-oxidized (II), and fully oxidized (III)—and the metallic emeraldine salt form (IV). The chemical structure of the emeraldine base form of polyaniline can be denoted [(–B–NH–B–NH–) (–B–N=Q=N–)], where B denotes a benzene ring in the benzenoid form, and Q denotes a benzene ring in the quinoid form. Leucoemeraldine has a bandgap of nearly 4 eV and the other two insulators have bandgaps on the order of 2 eV. The scientific interest in PANI originates from the fact that this conjugated polymer can be doped either by charge transfer or by the acid–base (protonation) reaction, the number of π-electrons in the chains remaining constant, which converts the semiconducting emeraldine base form to the conducting emeraldine salt. The excellent environmental and thermal stability in the conducting form and the potential for relatively low cost make PANI an attractive material for applications in batteries, light–emitting diodes, and antistatic packagings and coatings [16–18].

Polymeric Protonating Agents

In general, most substituted polyanilines have a low molecular weight and poor electronic properties. PANI films prepared with small, mobile counter anions, such as chloride or bisulfate, rapidly become electroinactive in neutral solutions. For example, Nyholm and Peter reported that PANI can be electroactive at pH 7 and above but loses electroactivity after only a few potential cycles [19]. McManus et al. [20] explored the influence of protonation on the conductivity of PANI and showed that as the pH changes from 1 to 7, the potential range in which PANI is conducting becomes narrower and shifts toward more negative potentials. To address these problems and to improve the processability of conducting PANI in solution, polymeric protonating agents have been used as polydopants [21]. PANI composite films formed by deposition of the polymer from acidic solutions containing polymeric anions such as poly(styrene sulfonate) (SPS) or poly(vinyl sulfonate) can remain electroactive at pH 7. During electropolymerization in the presence of polymeric anions, the rate of polymer film formation is increased [22,23]. Electron spectroscopy for elemental analysis (ESCA) studies have revealed that there are strong interactions between the SPS anions and the nitrogen atoms of the PANI chains, indicating that the SPS adsorbs selectively and irreversibly on a PANI surface [24]. The immobilization of sulfonate groups within PANI either by formulating it with poly(sulfonic acids) such as SPS, or by sulfonation of the PANI itself, is known to improve the electroactive response of the polymer in solutions with low proton concentrations. Asturias et al. [25] showed that the introduction of SPS anions into PANI films resulted in an extension of the conductivity within the film up to pH 9. Ghosh et al. [26] reported the preparation of a PANI composite with carboxymethyl cellulose and found that the composite exists in the conducting state in a neutral medium.

Sulfonated Polyphosphazenes as Protonating Agents

Polyphosphazene polyacids, polymers with a phosphorus and nitrogen backbone and organic side groups, are an attractive class of compounds which can potentially be used as protonating agents for PANI. In contrast to many other synthetic counterparts, such as SPS, polyphosphazenes contain two side groups per repeating unit, which translates to a higher ionic density, a potentially important feature for a dopant. Recently, a synthesis of *p*-sulfonated poly(bisphenoxyphosphazene), (SPPZ) containing two sulfonic acid groups per repeating unit, was reported (Scheme 1) [27]. In the newly developed approach, sulfonic acid groups are introduced in polyphosphazene at the stage of macromolecular substitution. A macromolecular precursor, poly(dichlorophosphazene) (PDCP), is modified with hydroxybenzenesulfonic acid using quaternary ammonium salts as protective groups. The reaction results in a highly controlled method for the introduction of sulfonic acid groups. Previous methods, based on sulfonation of poly(bisphenoxyphosphazene), were

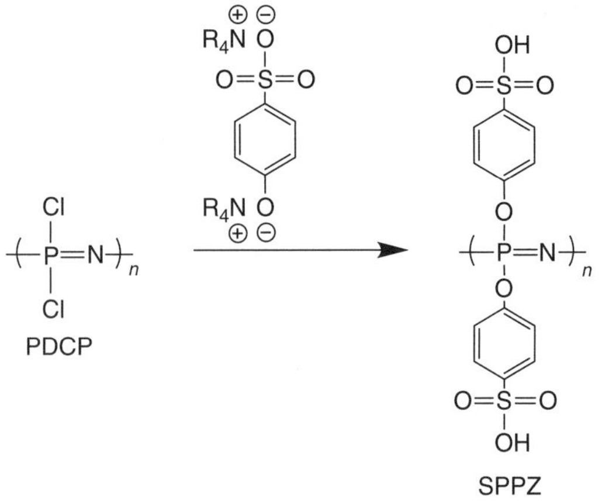

SCHEME 1 Synthesis of SPPZ.

characterized by significant degradation of the polymer and contamination with various reaction by-products.

In addition to high ionic density, polyphosphazenes can offer a number of other advantages as protonating agents. The polyphosphazene backbone is highly flexible, which can result in more efficient complex formation with PANI. Polyphosphazenes are also one of the most structurally diversified classes of polymers. Most polyphosphazenes are synthesized through the reaction of macromolecular substitution of PDCP, so that the variety of homopolymers and copolymers can be synthesized through the choice of an appropriate nucleophile. This allows a tailored synthesis of sulfonated copolymers containing other side groups, which can be critical for conferring their solubility in various solvents.

Template Polymerization

An important approach to the preparation of protonated PANI complexes involves the use of a polyelectrolyte template to align the aniline monomer molecules, which could subsequently be chemically or electrochemically polymerized [28]. The anionic polyelectrolyte template provides charge compensation to the protonated PANI and imparts water solubility to the final molecular PANI–polyelectrolyte complex. Various polyanions, such as SPS, poly(acrylic acid), poly(methacrylic acid), poly(2-acryamido-2-methyl-1-propenesulfonic acid), and poly(methacrylate-*co*-acrylic acid), were used to form a molecular complex, stabilized by noncovalent intermolecular bonds. The advantages of the approach are the following:

1. Polyelectrolyte acts as a template upon which the aniline monomers and/or oligomers preferentially align themselves and form a complex that

leads primarily to *para*-directed synthesis-promoting extended conjugation of the resulting polyaniline chains with limited parasitic branching.

2. This leads to significant improvement and control over the electronic properties of the complex and results in a conducting (redox active) form of polyaniline.

3. Polyelectrolyte actually serves as a large molecular counterion which is integrated and essentially locked to the polyaniline chains.

4. Polyelectrolyte template serves to provide water solubility of the final template–polyaniline complex for facile, inexpensive processing.

5. Anionic dopants, incorporated as part of the molecular complex, are strongly attached to the PANI chain and the conductive state was found to be stable under cycles of heat, solvent, or water.

6. Certain functional groups of the polymeric complex provide strong adhesion to metals and other polymers, an advantageous property for coating applications.

Chart 4 shows examples of template polymers and the schematic structure of template polymerization of PANI in the presence of anionic template polymer. In the first step, the aniline monomers are absorbed onto a polyanion chain dissolved in solution. The resulting adduct, polyanion:(aniline)$_x$, has signatures that can be monitored and verified. In the second step, the attached aniline monomers were oxidatively polymerized to form the polymer complex [29].

Enzymatic Synthesis of PANI

Although utilization of the template polymerization approach has led to significant improvements in the solubility and processability of PANI and PANI complexes, the problem of harsh reaction conditions has remained. Enzymatic reactions have therefore been investigated as an alternatively mild and environmentally compatible approach to the synthesis of conducting polyaniline. Horseradish peroxidase (HRP) has been used as a catalyst for the synthesis of polyphenols and polyanilines in the presence of hydrogen peroxide [30]. The reaction involves an initial two-electron oxidation of the native ferric enzyme to an oxidized intermediate (HRP-I) by hydrogen peroxide. The aniline monomer is oxidized by the HRP-I to produce monomeric radical species, which then undergo coupling to form dimer. Successive oxidation and coupling reactions eventually result in the formation of polymer. Regeneration of the native peroxidase is accomplished by two sequential one-electron reductions through a partially oxidized intermediate (HRP-II). The periodic cycle results in the oxidation of a variety of electron donors, such as phenols and aromatic amines. Scheme 2 shows the schematic representation of template-directed enzymatic synthesis of conducting polyaniline in the presence of SPS as a polyanion. The resulting polyaniline is in its doped state. However, this enzymatic synthesis is limited in that only low-molecular-weight oligomers

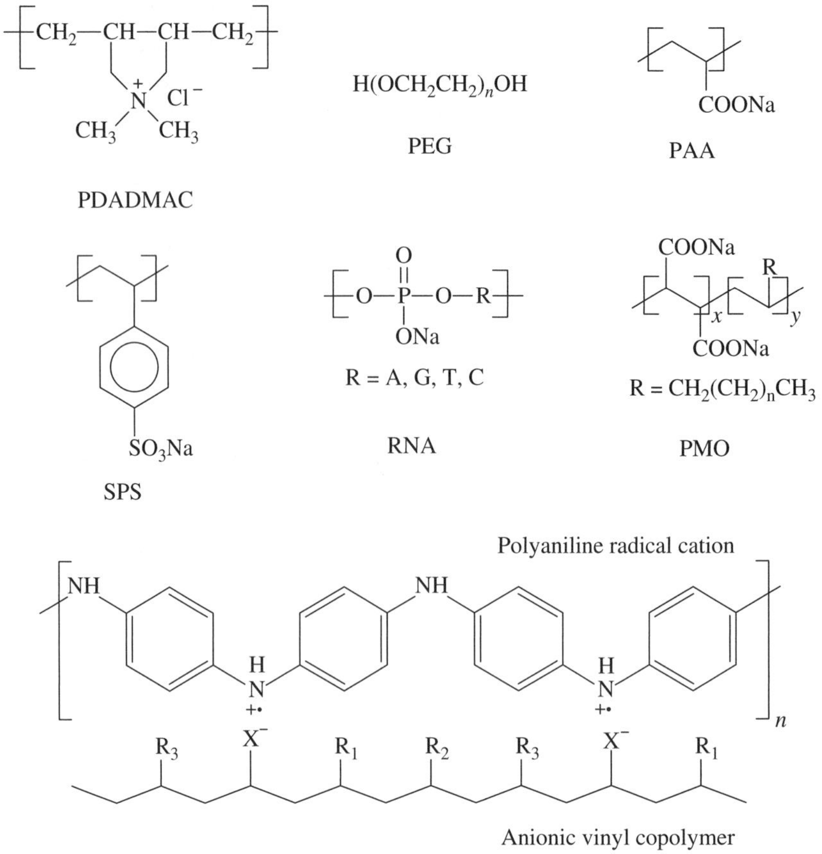

CHART 4 Examples of template polymer and the schematic structure of template polymerization of PANI with the presence of anionic template polymer.

with low electrical properties may be formed under aqueous conditions. The versatility of this biochemical approach to a range of other interesting polyelectrolytes suggests exciting opportunities in the synthesis of new polyanilines. In particular, the extreme mildness of this enzymatic approach is expected to allow for the use of more delicate phosphate-based biological materials as templates.

Organic phosphates/phosphonic acids are present in numerous biologically derived polyelectrolytes. Phospholipids and sterols make up approximately half the mass of biological membranes. Previous work toward the complexation of phosphate-based materials with PANI has included poly(alkylene phosphates) as dopants mixed with PANI in *N*-methylpyrrolidinone (NMP) solution [31]. However, the PANI used in these cases has been chemically polymerized before introduction of the dopant. These systems are not water based and are also prone to spontaneous deprotonation/undoping. Polyaniline has also been

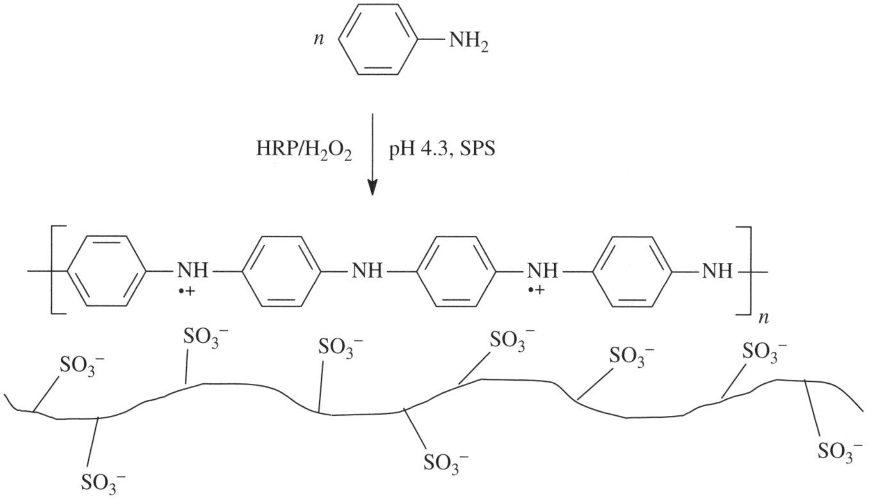

SCHEME 2 Schematic representation of template-directed enzymatic synthesis of conducting polyaniline in the presence of SPS as a polyanion.

synthesized chemically under low-pH conditions in the presence of poly(vinyl phosphonic acid) (PVP) [32]. Although these approaches have been successful toward the polymerization of phosphate-based PANI complexes, the harsh chemical conditions preclude the use of more delicate biological-based phosphate materials as templates. However, the enzymatic approach offers much more biocompatible synthetic conditions. Wei et al. [33] show that there is a direct dependence on the template structure using the various polyelectrolyte templates and the type of polyaniline that is formed. The effective template provides several key "local environmental" conditions to facilitate the reaction. The template provides a lower local pH environment that increases the level of protonation of aniline in the reaction medium. This increase in protonation promotes electrostatic interaction of the aniline monomer to the template. In addition, this allows one to carry out the reaction at a higher pH, to prolong the bioactivity of the enzyme. The template also provides hydrophobic regions that serve to solubilize and orient the monomer molecules prior to reaction. The enzymatic template polymerization method can provide new insight into the selection of appropriate template systems for the synthesis of conducting polyaniline, and the selection of polyelectrolyte as the polyanion for the polymerization of aniline is an important preliminary step toward extending this approach to other biologically derived polyelectrolytes, such as DNA.

π-Conjugated Polymer Electrolytes

The use of polymer electrolytes is well known in electrochemical devices such as sensors, batteries, and fuel cells. Numerous studies of advanced polymer electrolyte materials have been carried out and applied to biomedical devices such as cardiac pacemakers and neurostimulators [34]. The considerable effort has been expanded toward the development of new high-performance π-conjugated polymer materials such as polyaniline (PANI), polypyrrole (PPy), and polythiophene (PT). However, the realization of practical, long-lived π-conjugated polymer electrochemical devices remains an elusive goal because of performance limitations, including poor environmental stability, slow switching speeds, and short lifetimes when cycling electrochemically between oxidation states. These problems derive partly from the electrolytes used in the devices, whether they are based on aqueous, organic, gel, or polymer electrolytes. Improved electrolytes are needed that simultaneously satisfy the requirements of certain applications. To date, most of the research on π-conjugated polymers has been carried out in aqueous electrolytes. However, these systems suffer from narrow electrochemical potential and high volatility. These factors limit the lifetime and performance of a device. For example, electrochemical degradation of polyaniline occurs after only a few cycles in some aqueous electrolytes [35] because of nucleophilic attack in, and hydrolysis of, the polymer [36]. When π-conjugated polymers are cycled electrochemically in the presence of an electrolyte, their color changes; and the absorption wavelength depends on the width of the polymer's bandgap. π-Conjugated polymers are being studied for electrochromic applications because of properties that include ease in the fabrication of large areas, high quality, optically transparent thin films, excellent coloration contrast and matching, rapid coloration rates, and low operational voltages. Recently, research on solid polymeric electrolytes has grown intensively [37]. The interest in these solid-state ionic conductors comes from the possibility of using them to substitute for liquid electrolytes in several electrochemical devices. For example, ionic functional groups substituted conducting polymers such as sulfonated polyaniline (SPANI) or poly(3thiopheneacetic acid) (P3TAA). The major challenge in replacing a liquid or gel electrolyte by a polymeric electrolyte is to remain high operational efficiency, similar to that of electrochemical devices based on liquid junctions. Besides improving the stability of the active interface, allowing long-term durability, a polymer electrolyte eliminates problems concerning evaporation or leakage of solvents. The use of polymer in electrolytes can be divided into three categories: (1) a polymer swollen with a liquid electrolyte, which provides high values of conductivity, but does not eliminate the problems related to the liquid electrolytes; (2) a polymer containing cations or anions attached to the chain, which produces low conductivity values because of the low mobility of charge carriers; and (3) a mixture of a salt in an ion-solvating polymer, which configures a dry solid electrolyte system, can present conductivity values suitable for use in electrochemical devices and eliminates all difficulties relative to the use of liquid or gel electrolytes in commercial applications.

Interpolymer Complexation

Interpolymer complexation is an intermolecular association of two different polymers through secondary binding forces. Interaction between two polymers may lead to the formation of an interpolymer complex (IPC), which essentially possesses properties entirely different from those of the component polymers. A large number of systems that form interpolymer complexes are known and have already found wide applications in technology and medicine [38,39]. The intermolecular association between different polymers in solution is evidenced by a reduction in viscosity arising from the highly compact nature of the associated polymer chains. When the intermolecular interaction between polymers outweighs the polymer–solvent interaction, the two highly associated polymers precipitate out from the solution, and the precipitates are commonly called *interpolymer complexes* (IPCs). Therefore, the formation of precipitates upon mixing two polymer solutions in a common solvent indicates the presence of strong interpolymer interaction. IPCs can be divided into several classes on the basis of the main interaction forces: polyelectrolyte complexes (PECs), which are formed by mixing with oppositely charged polyelectrolytes, due to Coulomb forces [40,41] of polyanions and polycations; hydrogen-bonding complexes, which are stabilized through a hydrogen bond between a polyacid and a polybase [42,43]; charge transfer complexes, which are formed between polymers with electron-donor and electron-acceptor groups [44,45]; stereo-complexes, which are typically formed through van der Waals forces by two polymers with identical chemical structure and complementary stereoisomer-ism [46,47]; and polycomplexes stabilized through hydrophobic interactions in aqueous solution [48]. Hydrophobic force is different from the others because hydrophobic interaction is caused by rearrangement of water molecules rather than direct cohesive force between the molecules. Hydrophobic interaction forces the particles to coil up into compact globules, playing an essential role in stabilization of the polymer complex particles in water.

Interpolyelectrolyte Complexes

Interpolyelectrolyte complexes (IPECs) form a special class of polymeric com-pounds [49–51], consisting of oppositely charged polyions. Depending on the composition, IPEC can be either insoluble but limitedly in swellability, or soluble in water solution. Investigations of IPECs have quite a lengthy history. As early as the 1930s, the attention of colloid chemists was attracted by the fact that upon mixing aqueous solutions of oppositely charged natural polyelectrolytes, char-acterized by a relatively low charge density on the macromolecules (e.g., gelatin) under certain conditions, a phase separation takes place, with the formation of liquid acervates [52]. These investigations were later developed mainly in terms of a description of the conditions of phase separation [53]. Insoluble IPECs have also been known for many years [54,55], at least since the beginning of the 1960s, when a number of synthetic polyelectrolytes with a high linear charge density became available. A certain level of success has been reported for their practical application. In particular, insoluble IPECs show unique efficiency as hydrophilic soil binders, preventing wind and water erosion. They have also been reported for use as biocompatible coatings and other medical items used in contact with blood and other biological fluids [56]. There are also other promising applications, one of which is formation of ultrathin multilayer polymer coatings by complexation of polyelectrolytes at interfaces. Finding soluble IPEC as isolated examples [56] and then establishing the general conditions of soluble IPECs formation valid for most pairs of oppositely charged polyelectrolytes [57,58], marked a serious breakthrough in the field. It brought studies to a qualitatively new level, owing to the application of modern techniques commonly used for studying macromolecules in solution (light scattering, analytical ultracentrifugation, viscometry, gel permeation chromatography, etc.). The fundamental data obtained using these methods allowed a deeper understanding of the structural organization and specific behavior of IPECs not only in solution but also in the concentrated phase. At the same time, these studies opened up an entirely new range of prospective applications of IPECs, related primarily to biomimetics, biotechnology, and medicine [59–61].

Thermodynamics of Interpolymer Complexation

The stability constant (K), degree of linkage (θ), and related thermodynamic parameters of interpolymer complex have been determined using Osada and Sato's method [62]. The thermodynamics of complexes are related to the cooperative hydrogen bonding between a polyacid–polybase pair with stoichiometric composition. For a weak acid with a dissociation constant K_a and concentration C, for $K_a/C \ll 1$,

$$[H^+] = (K_a C)^{0.5} \tag{1}$$

This relation holds for polymeric acid (where C is its normality) in both the presence and absence of complementary polymer. If the apparent dissociation constant of the polyacid is assumed not to vary with complexation, expressing the free acid concentration for these two cases by C and the initial concentration of the polyacid in repeating units by C_0, the degree of conversion (θ), defined as the fraction of bonded polyacid groups, is given by

$$\theta = 1 - \frac{C}{C_0} = 1 - \left(\frac{[H^+]}{[H^+]_0}\right)^2 \tag{2}$$

where $[H^+]$ and $[H^+]_0$ are the hydrogen ion concentrations in the presence and absence of complementary polymers, respectively. The stability constant (K) for the equilibrium between hydrogen-bonded and free sites is given by

$$K = \frac{[\text{complex}]}{[\text{polymeric acid}][\text{polymeric base}]} = \frac{C_0 \theta}{C_0^2 (1-\theta)^2} = \frac{\theta}{C_0 (1-\theta)^2} \tag{3}$$

Therefore, by measuring the pH of the complex solution, it is possible to determine both θ and K. Furthermore, the thermodynamic parameters (e.g., ΔH^0 and ΔS^0) for the interpolymer complexation process can be calculated from the temperature dependence of K:

$$\frac{d(\ln K)}{d(1/T)} = -\frac{\Delta H^0}{R} \tag{4}$$

$$\Delta S^0 = \frac{-(\Delta G^0 - \Delta H^0)}{T} \tag{5}$$

where $\Delta G^0 = -RT \ln K$ is the change in standard free energy for the complexation and R is the molar gas constant. Tsuchida et al. [63] found that ΔS^0 and ΔH^0 depend strongly on the molecular weights of PEO in PMMA–PEO and PAA–PEO complex systems. The thermodynamics for the complexation of PAA with PVPO was recently studied using pH measurements at different temperatures [64], and it was shown that ΔS^0 and ΔH^0 were positive at lower temperatures and decreased continuously with increasing temperature. The positive values of ΔH^0 at lower temperatures were interpreted as being due to hydrophobic interactions and conformational changes during complexation, and the positive values of ΔS^0 were considered as reflecting the release of water during complexation. The kinetics and equilibria of the complexation were studied by Morawetz et al., and it was shown that the complex formation consisted of an initial diffusion-controlled hydrogen-bonding process with low activation energy and an extensive conformational transition of the two polymer chains which induces additional hydrogen bonding, thus stabilizing the complex [65–67].

Complexes of PANI and *p*-Sulfonated Poly(bisphenoxyphosphazene)

Polyaniline (PANI) has attracted considerable scientific interest from recent researchers because of its good electrical and redox properties and easy synthesis both chemically and electrochemically. A large number of applications and potential applications arising from the presence of the various redox states, their interconversions, and their charge transfer interactions have been reported in the patent and scientific literature [68]. The commercial applications are based on the promise of a novel combination of light weight, processability, and electrical conductivity. Polyanilines have been synthesized chemically [69], electrochemically [70], and biochemically [71]. The simplest method for the preparation of powdered samples of PANI is based on the oxidative polymerization of aniline in an aqueous acidic medium with an inorganic oxidant agent. The final PANI structure can exist in various oxidation states, which are characterized by the ratio of amine to imine nitrogen atoms [72]. However, one of the main obstacles to broader commercial development of PANI is its poor solubility in common organic solvents and water.

Multiple approaches to the improvement of PANI solubility and processability have been investigated. One of them includes synthesis of various derivatives of PANI, such as *N*-alkyl or alkyl/methoxy ring–substituted PANIs [73–79]. The nonsynthetic approach involves the use of water-soluble polymeric acid dopants such as poly(styrene sulfonic acid) and poly(acrylic acid), which can improve the water solubility and processability of the resulting complex with PANI due to the remaining ionic groups, which are not involved in complex formation [80,81]. However, the solubility of such "nonstoichiometric" complexes can benefit dramatically from the use of water-soluble polyacids with high ionic density.

In the present work we investigated a new high-ionic-density polyacid sulfonated polyphosphazene (SPPZ), for its ability to form complexes with PANI in NMP–water co-solvent systems. It was anticipated that the use of such a polyelectrolyte dopant may improve the water solubility of the complex and its processability in aqueous solutions, as well as promoting adhesion to metals and polymers, an advantageous property for coating applications. We also attempted the enzymatic polymerization of PANI in the presence of SPPZ as a polyelectrolyte template polymer.

EXPERIMENT

Polyaniline Films Polyaniline films can be obtained by electropolymerization, spin coating, and chemical polymerization over a polymer matrix. We prefer to chemically polymerize the aniline monomer over the nylon 6 films by washing the films with water and acetone twice and drying them by heating to 70°C for 24 h. Then the films are soaked in a solution of aniline 0.2 M with 2 M HCl (37%) for around 24 h at room temperature. The nylon 6 film absorbed with aniline will then be dipped into aqueous 0.25 M ammonium persulfate solutions containing 2 M HCl (persulfate/aniline ratio 1.25). Polymerization takes around 40 to 60 min.

Materials Horseradish peroxidase (HRP) (EC 1.11.1.7) (200 units/mg) was purchased from Sigma Chemicals Co. (St. Louis, Missouri), with RZ > 2.2. Aniline (99.5%) was obtained from the Aldrich Chemical Co., Inc., (Milwaukee, Wisconsin), and used as received. Hexachlorocyclotriphosphazene (Nippon Fine Chemicals, Japan) was used as received. PDCP was synthesized using ring-opening polymerization of hexachlorocyclotriphosphazene in a titanium pressure reactor as described previously. Propyl 4-hydroxybenzoate, 99 + %; benzenesulfonic acid, sodium salt, 98%; and 4-hydroxybenzenesulfonic acid, sodium salt dehydrate, 98% (Aldrich) were dried prior to use in a vacuum oven at 80°C for 8 h. Chlorobenzene, anhydrous; diglyme (methoxyethyl ether), anhydrous; methyl alcohol, 99.9%; methyl sulfoxide, 99.9%; 2-propanol, 99 + % (Aldrich); dimethyldipalmitylammonium bromide, 97 + % (TCI America, Portland, Oregon); *N*,*N*-dimethylacetamide, 99.9% (OmniSolv,

Gibbstown, New Jersey); and tetra-*n*-butylammonium bromide, 98% (Alfa Aesar, Ward Hill, Massachusetts) were used as received.

Analytical Methods   The molecular weight of water-soluble poly[diphenoxyphosphazenedisulfonic acid] was determined by aqueous gel permeation chromatography (GPC) analysis. The polymer was characterized using an Ultrahydrogel linear column (Waters, Milford, Massachusetts) with ultraviolet (UV) (Waters 486 tunable UV/visible absorbance detector) and refractive index detection (Waters 410 RI detector). A mixture of phosphate-buffered saline (PBS; pH 7.4) and methanol (9:1 ratio) was used as a mobile phase. GPC analysis of mixed-substituent copolymers was performed in *N,N*-dimethylacetamide containing 0.1% tetra-*n*-butylammonium bromide using a Waters Styragel HMW 6E column with refractive index detection (Waters 410 RI detector). Molecular weights were calculated using Waters Millennium software and sodium poly(styrene sulfonate) standards (Scientific Polymer Products, Inc., Ontario, New York) for aqueous systems and polystyrene (Polysciences, Inc., Warrington, Pennsylvania) standards for organic systems. ^{31}P, ^{1}H, and ^{13}C nuclear magnetic resonance (NMR) spectra were recorded using a Varian Unity INOVA 400-MHz spectrometer.

Synthesis of the Dimethyldipalmitylammonium Salt of 4-Hydroxybenzenesulfonate (DPSA)   5 g (0.0087 M) of dimethyldipalmitylammonium bromide (DMDPA) was mixed with 5 mL of methanol, and to this suspension was added 300 mL of deionized water. The mixture was then stirred until a clear solution was obtained. A 10-g (0.0042-mol) sample of 4-hydroxybenzenenesulfonic acid sodium salt was dissolved in 100 mL of aqueous solution containing 1.63 g (0.004 mol) of sodium hydroxide. This solution was then added to the solution of DMDPA; the mixture was stirred for 30 min and left at ambient temperature for 120 min. The resulting precipitate of DPSA was filtered and dried under vacuum (yield 4.6 g, 98%).

Synthesis of SPPZ   35 g (0.03 M) of DPSA, prepared as described above, was dissolved in 470 mL of anhydrous monochlorobenzene, placed in a 1-L flask equipped with a stirrer, and kept under nitrogen. Then 1.18 g (0.01 mol) of PDCP in 30 mL of diglyme was added dropwise to the flask at 100°C with constant stirring. The reaction temperature was then increased to 120°C and the reaction continued at this temperature for 5 h. The reaction mixture was then cooled to 70°C and to this mixture 25 mL of ethanol and 32 mL of 12.7 N aqueous potassium hydroxide (0.41 M) were added. The mixture was stirred for an additional 1 h and cooled, and polymer was isolated by precipitation in 1500 mL of methanol. The precipitate was redissolved in 210 mL of 0.6 N aqueous potassium hydroxide (0.13 M) upon stirring at 50°C for 1 h and precipitated with 650 mL of methanol. Polymer precipitation was repeated one more time. Polymer was then converted in the acid form by dissolving in 220 mL of 1 N aqueous hydrochloric acid and precipitating in 1200 mL of methanol. The resulting polymer was dried under vacuum to yield 2.8 g (70%). Polymer structure and purity were determined by ^{31}P, ^{1}H, and ^{13}C NMR, elemental analysis (sulfur), and size-exclusion high-performance liquid chromatography (HPLC) with photodiode array detection (200 to 600 nm). Polymer purity was determined to be in excess of 99%.

Preparation of PANI–SPPZ Complex in NMP–Water Co-solvent   PANI (EB) and SPPZ complex in NMP–water co-solvent was investigated by UV–visible spectroscopy. PANI (based on the approximate EB repeating unit) and SPPZ (based on the approximate –P=N– repeating unit) solutions, both at a concentration of 4.3×10^{-3} M, were prepared in NMP and water, respectively. The PANI–SPPZ complex solutions with different molar ratios were prepared by mixing with appropriate volumes of two solutions under continuous stirring ([PANI]/[SPPZ] ratios were varied from 1 : 9 to 9 : 1). The solvent ratio of NMP–water was 1 : 9 weight ratio in all complex solutions, and both solvents are miscible in each other. The precipitation of PANI–SPPZ complex solutions in 9 : 1 and 8 : 2 molar ratios and PANI were observed after 3 h. All other complex solutions were and soluble in this co-solvent system, and UV–visible spectra were recorded after the precipitates were removed.

PANI–SPPZ Complex Characterization   The UV–visible absorption spectra were recorded by Shimadzu PC-3101 UV–visible/near-infrared spectrophotometer. The NMP–water (1 : 9 weight ratio) co-solvent and distilled water were used as a reference for all UV–visible spectroscopy measurement. The optical absorbance was measured in the wavelength 220 to 1300 nm at a scanning rate of 100 nm/min. Fourier transfor infrared (FT-IR) measurements were carried out on a Nicolet Avatar 360 FT-IR spectrometer. Template-synthesized PANI–SPPZ complex solutions were precipitated with 1 L of methanol and filtered using a Buchner funnel. These solid samples were dried in a vacuum oven at 60°C overnight and mixed with dried KBr powders, pressed into tablets and analyzed. The spectral resolution was 4 cm^{-1} and the number of scans was 32 for the analysis. The conductivity of the PANI–SPPZ complex samples was measured using the four-probe method. A Keithley 220 current source meter and Keithley 197 multimeter were used for conductivity measurements. The conductivity was measured at different sample positions five times and the average value was chosen. The measurement error estimated was 15%. The sample specimens were prepared in pressed pellet form (0.2 mm thickness, 1.0 cm diameter) and dried under dynamic vacuum at 60°C for 1 day to remove any residual moisture during the measurements.

Antibody–Antigen Binding Studies   The PANI–nylon films used were as obtained from Moscow State University. For the investigation of antibody–antigen binding, chrompure rabbit IgG (11.2 mg/mL) and goat anti-rabbit IgG (2.4 mg/mL) were used as the antibody and antigen, respectively, and were obtained from Jackson Immunoresearch, Pennsylvania. The PANI–nylon electrode was allowed to stabilize in pH 7.2 PBS buffer until a stable potential was obtained. All potentials were measured against an Ag/AgCl reference electrode. After stabilization, 10 μL (11.2 μg) of 1 : 10 dilution of the antibody was added to the buffer solution and the potential was recorded. After stabilization, the electrodes were rinsed thoroughly with buffer to remove physically adsorbed antibody, the buffer solution was refreshed, and the electrodes were again allowed to stabilize. Then 10 μL of 1 : 10 (2.4 μg) of antigen was added to the solution and the potential response was recorded. The solutions were stirred gently continuously during the experiment.

RESULTS

PANI-Based Biosensors

A potentiometric protein biosensor (immunosensor) shows promise as an easy-to-operate device which does not require complicated conjugation or labeling chemistry. The system utilizes PANI–nylon as a working electrode and can provide potentiometric monitoring of immunoreactions such as antibody–antigen binding reactions. Figure 2 shows the potentiometric titration results for antibody–antigen immunoreaction in which rabbit IgG antibody was initially deposited on the electrode and then anti–rabbit IgG antigen was added. As shown in the figure, the potential decreases with antibody addition, followed by a positive potential change associated with antigen binding. The method, which allows us to calculate K_d, is a novel approach for easy-to-use and economical point-of-care applications, in which, as was mentioned, no labeling or conjugation is required.

The disadvantage of the system is that PANI protonated with small molecules may suffer from a lack of stability. Thus, the system can be improved through the use of polymeric acids as dopants, which can form a stable interpolymer complex. Polyphosphazene dopant SPPZ was used for the purpose of controlling the viscoelastic properties and improving the stability of the system. The peculiarities of PANI–SPPZ complex formation, potentially useful in biosensors, are discussed below.

UV–Visible Spectroscopy Studies

Formation of interpolymer complexes at various ratios of PANI and SPPZ was studied using UV–visible absorption spectroscopy. The UV profiles of PANI–SPPZ complexes in NMP–water mixtures are shown in Figure 3. The precipitation of PANI–SPPZ complexes at 9 : 1 and 8 : 2 molar ratios and PANI was observed after 3 h, and UV–visible spectra were recorded after removal of the precipitates from the mixture. The insolubility of the complex at 9 : 1 and 8 : 2 ratios suggests that all –SO_3H groups of SPPZ chains were consumed in formation of the complex with PANI, and there were not enough

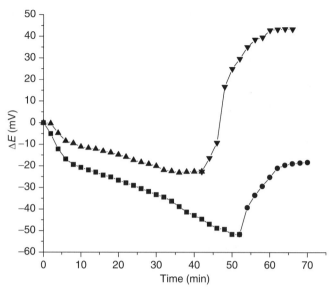

FIGURE 2 Potentiometric responses for antibody binding with a PANI–nylon electrode and antigen binding to the antibody: ▲, antibody binding(11.2 µg) trial 1; ▼, antigen binding trial 1; ■, antibody binding trial 2; ●, antigen binding trial 2.

hydrophilic groups to make the complex soluble in this co-solvent system. Even though the ratios of doping level were close to PANI's maximum doping level (0.5 equivalent proton; one repeat unit of SPPZ has two –SO₃H groups), the amount of sulfonic acid group required to reach the maximum doping level was greater than the theoretical amount due either to conformation hindrance of the SPPZ chain and/or co-solvent effect. The PANI(EB) has two absorption peaks at around 340 and 630 nm which are attributed to π–π^* transition of benzenoid and quinoid rings on the PANI chain, respectively. As the PANI/SPPZ molar ratio decreases, the intensity of the absorption peak at 630 nm diminishes, while new peaks appear at around 440 nm and 830 nm, which are attributed to the characteristic of polaron band transitions [17]. The color of the complex solutions also changed gradually from blue to green due to increased protonation of the imine sites of PANI. This result indicates that the sulfonic acid groups on the SPPZ chains protonate the imine nitrogen atom of PANI in the same way as organic acid dopants. The SPPZ was clearly efficient in solubilizing PANI in a NMP–water mixture.

SPPZ was also investigated as a template for the synthesis of PANI. Figure 4 shows UV–visible absorption spectra of polymerization solutions with molar ratios of aniline (monomer) to SPPZ (repeating unit) at 1 : 0.5 and 1 : 1. The absorption spectrum of the mixture with a 1 : 0.5 molar ratio shows a characteristic band of the exciton transition of polaronic structure at around 440 nm. However, the characteristic band of polaron to bipolaron transition at

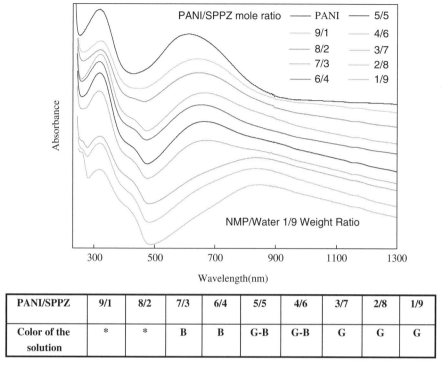

PANI/SPPZ	9/1	8/2	7/3	6/4	5/5	4/6	3/7	2/8	1/9
Color of the solution	*	*	B	B	G-B	G-B	G	G	G

FIGURE 3 UV-visible spectra of PANI–SPPZ complexes with different molar ratios in a 1:9 NMP–water co-solvent and the color changes (B, blue; G, green).

about 830 nm is obviously weak. On the other hand, the absorption spectrum at 1:1 molar ratio exhibits a peak at 440 nm and an intense free carrier tail beginning at 930 nm. This result could be understood from the findings of Gospodinova et al. [82], who demonstrated that PANI synthesis is acid dependent with the linear chain growth of PANI occurring solely in an acidic medium. Thus, a sufficient amount of SPPZ is probably available in the system not only to protonize synthesized PANI, but also to emulsify the aniline monomers prior to polymerization and to afford linear chain growth. However, this contrasts with complexation in the NMP–water mixture, in which the free carrier tail after 830 nm was never observed at all ratios. This result indicates that polarons in the NMP–water system are more localized than in the template polymerization system, perhaps due to the co-solvent effect. This may be attributed to the hydrogen-bonding effect between the carbonyl group of NMP and the sulfonic acid group of SPPZ, and this effect may suppress the protonation of PANI base, and consequently, the protonation of PANI is retarded in an NMP system.

To investigate the stability of PANI–SPPZ complex against deprotonation, the UV–visible spectra of enzymatically synthesized PANI with SPPZ

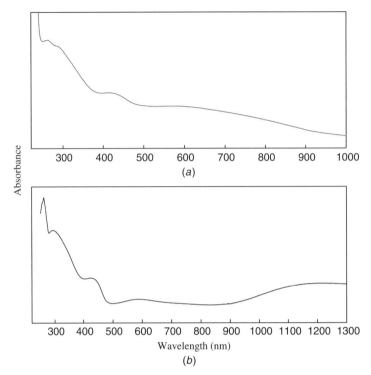

FIGURE 4 UV–visible spectra of PANI–SPPZ complexes with varying molar ratios of aniline to SPPZ in aqueous solution: (a) aniline/SPPZ 1 : 0.5, and (b) aniline/SPPZ 1 : 1 molar ratio.

(1 : 1 molar ratio) at pH ranging from 1.64 to 10.5 were studied (Fig. 5a). All samples were at a concentration of 0.01 wt% in distilled water and the pH was adjusted using 0.01 M aqueous HCl and NaOH. As the pH increases to 8.0, the absorbance ratio of 600 nm to 1200 nm decreases gradually (Fig. 5b), and the intensity of polaron and bipolaron bands transitions at around 440 nm and 830 to 1200 nm remain relatively unchanged. It indicates that doped and dedoped forms of PANI coexist in the same solution at pH below 8.0 and that the PANI–SPPZ complex is stable. The color of the solution changed from green to blue as pH increased from 1.64 to 10.5. PANI becomes dedoped and loses solubility at a pH between 9.0 and 10.5. However, PANI–SPPZ complexes were still soluble in water in these high-pH solutions. These results indicate that the PANI is partially doped by SPPZ, which is also supported by the weak polaron transition bands at around 440 nm.

Infrared Spectroscopy Studies

FT-IR studies of enzymatically synthesized PANI–SPPZ complexes with 1 : 0.5 and 1 : 1 molar ratios confirmed incorporation of the polyphosphazene dopant

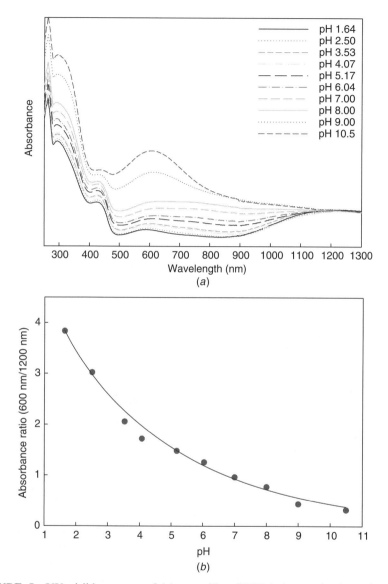

FIGURE 5 UV–visible spectra of (a) an aniline–SPPZ 1:1 complex in various pH aqueous solutions and (b) an absorbance ratio of 600 nm/1200 nm.

in the complex (Fig. 6). The most intense IR bands of PANI–SPPZ complexes are observed at 1130 and 1220 cm^{-1}, which are assigned to the vibrations of the polar P$=$N bond of the phosphazene chain and a strong broad peak appears in the range 2700 to 3400 cm^{-1}, which corresponds to a hydrogen-bonded hydroxyl functional group stretching vibration. Free hydroxyl functional group stretching vibration peaks appear at 3500 cm^{-1}. In addition, the presence of

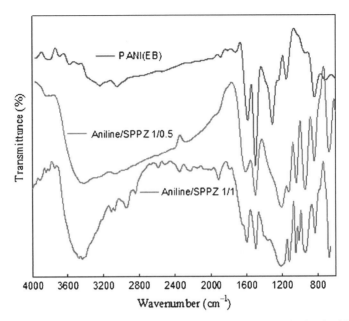

FIGURE 6   FT-IR spectra of PANI(EB) and enzymatically synthesized PANI–SPPZ complexes with aniline/SPPZ molar ratios of 1 : 0.5 and 1 : 1.

asymmetric and symmetric S=O stretching bands at 1000 and 1035 cm^{-1} also confirms the presence of SPPZ in the complex. The peak at 940 cm^{-1} is assigned to a P–O–C bending band. The peak at 1160 cm^{-1} that appears in spectra of PANI–SPPZ complexes can be considered as an indicator of the degree of delocalization in PANI and thus presents an important characteristic of conductive PANI [21]. The stretching band of the –POC– (mainly C–O) group overlapped in the near 1210 to 1270 cm^{-1} region [22,23]. Four major vibration bands—1590, 1495, 1310, and 828 cm^{-1}—are characteristic, of both complexes and PANI(EB). These peaks are attributed to the stretching vibration of the benzene quinoid ring, the stretching vibration of the benzene ring, the stretching vibration of C–N, and the out-of-plane bending vibration of C–H on *para*-disubstituted rings, respectively [20]. The stretching vibration of the benzene quinoid ring at 1495 cm^{-1} and the stretching vibration of the benzene ring at 1590 cm^{-1} were also shown in PANI–SPPZ complexes.

Conductivity Studies

Electrical conductivity of enzymatically template polymerized PANI was measured by the conventional four-point probe method. The sample specimen were prepared using pressed pellets (0.2 mm thickness, 1.0 cm diameter) of sample powder and dried first under dynamic vacuum at 60°C for 1 day to remove any residual moisture for the conductivity measurements. The conductivity of the PANI–SPPZ complex shows 3.6×10^{-6} S/cm and 4.5×10^{-5} S/cm for $1:0.5$ and $1:1$ molar ratios, respectively. As expected, the conductivity increases almost one order of magnitude as the PANI/SPPZ molar ratio increased. This observed increase in conductivity presents evidence that SPPZ promotes the formation of linear and long-chain PANI (delocalized of polarons), resulting in the electrical conductivity of the polymer chain.

CONCLUSIONS

Water-soluble complexes of the conductive polymer PANI were prepared using the sulfonated polyphosphazene dopant SPPZ with various molar ratios of the components. SPPZ was also used as a template for the enzymatic synthesis of a water-soluble conducting PANI. UV–visible and FT-IR spectroscopy studies confirmed formation of the complex. Conductivity studies of enzymatically polymerized PANI demonstrate that SPPZ promotes the formation of linear and long-chain PANI (delocalized of polarons), resulting in electrical conductivity of the polymer chain. SPPZ can present an important dopant for the production of a PANI-based potentiometric ion-sensitive electrode, a potentially important tool for monitoring antibody–antigen interactions.

REFERENCES

1. Walatke, V.V., Labes, M.M., Peristein, J.H. *Phys. Rev. Lett.*, 1973, 31:1139.
2. Greene, R.L., Street, G.B., Suter, L.J. *Phy. Rev. Lett.*, 1975, 34:577.
3. Chiang, C.K., Fincher, C.R., Park, Y.W., Heeger, A.J., Shirakawa, H., Louis, E.J., Gau, S.C., MacDiarmid, A.J. *Phy. Rev. Lett.*, 1977, 39:1098.
4. Shirakawa, H., Ito, T., Ikeda, S. *J. Polym. Sci. Polym. Chem. Ed.*, 1974, 12:11.
5. Chiang, C.K., Gau, S.C., Fincher, J.C.R., Park, Y.W., MacDiarmid, A.G. *Appl. Phys. Lett.*, 1978, 33:18.
6. Nigrey, P.J., MacDiarmid, A.G., Heeger, A.J. *Chem. Commun.*, 1979, 96:594.
7. Pei, Q., Yu, G., Zhang, C., Yang, Y., Heeger, A.J. *Science*, 1995, 269:1086.
8. Ranby, B. In *Conjugated Polymers and Related Materials: The Interconnection of Chemical and Electronic Structures.* Oxford, New York, 1993.
9. Wise, D.L., Wnek, G.E., Trantolo, D.J., Cooper, T.M., Gresser, J.D. *Electrical and Optical Polymer System.* Marcel Dekker, New York, 1998.
10. Rebbi, C. *Sci. Am.*, 1979, 92:240.
11. Pohl, A., Stafstrom, S. *Synth. Met.*, 1999, 101:287.
12. Radhakrishnam, T.P. *Resonance*, 2001, Feb., p. 63.
13. Letherby, H. *J. Chem. Soc.*, 1862, 15:161.
14. Green, G., Woodhead, A.E. *J. Chem. Soc. Trans.*, 1910, 97:2388.
15. Heeger, A.J., Smith, P. *Solution Processing of Conducting Polymers: Opportunities for Science and Technology.* Kluwer, Dordrecht, The Netherlands, 1991.

16. Johnstone, B. *East. Econ. Rev.*, 1988, 17:78.

17. Gustafsson, G., Cao, Y., Treacy, G.M., Klavetter, F., Colaneri, N., Heeger, A.J. *Nature*, 1992, 356:47.

18. DeBerry, D. *J. Electrochem. Soc.*, 1985, 132:1022.

19. Nyholm, L., Peter, L.M. *J. Chem. Soc. Faraday Trans.*, 1994, 90:149.

20. McManus, P.M., Cushman, R.J., Yang, S.C. *J. Phys. Chem.*, 1987, 91:744.

21. Bartlett, P.N., Wang, J.H. *J. Chem. Soc. Faraday Trans.*, 1996, 92:4137.

22. Kuramoto, N., Michaelson, J.C., McEvoy, A.J., Gratzel, M. *J. Chem. Soc. Chem. Commun.*, 1990, 1478.

23. Hyodo, K., Omae, M., Kagami, Y. *Electrochim. Acta*, 1991, 36:357.

24. Hyodo, K., Kobayashi, N., Kagami, Y. *Electrochim. Acta*, 1991, 36:799.

25. Asturias, G.E., Jang, G.W., MacDiarmid, A.G., Doblhofer, K., Zhong, C. *Bunsenges. Phys. Chem.*, 1991, 95:1381.

26. Ghosh, S., Vishalakshi, B., Kalpagam, V. *Synth. Met.*, 1992, 46:349.

27. Andrianov, A.K., Marin, A., Chen, J., Sargent, J., Corbett, N. *Macromolecules*, 2004, 37:4075–4080.

28. Suna, L., Liua, H., Clarka, R., Yang, S.C. *Synth. Met.*, 1997, 84:67.

29. Liu, J.M., Yang, S.C. *J. Chem. Soc. Chem. Commun.*, 1991, 1529.

30. Akkara, J.A., Senecal, K.J., Kaplan, D.L. *Polym. Sci. Polym. Chem.*, 1991, 29:1561.

31. Kulszewicz-Bajer, I., Pretula, J., Pron, A. *J. Chem. Soc. Chem. Commun.*, 1994, 641.

32. Ong, C.H., Goh, S.H., Chan, H. *Polym. Bull.*, 1997, 39:627.

33. Wei, L., Ashok, L.C., Ramaswamy, N., Jayant, K., Sukant, T., Ferdinando, F.B., Lynne, S. *J. Am. Chem. Soc.*, 1999, 121:11345.

34. Holmes, C.F. *Electrochem. Soc. Interface*, 1999, 8:32.

35. Kobayashi, T., Yoneyama, H., Tamura, H. *J. Electroanal. Chem.*, 1984, 161.

36. Ri, Y., Qian, R. *Synth. Met.*, 1993, 53:149.

37. Wright, P.V. *Electrochim. Acta*, 1998, 43:1137.

38. Bekturov, E.A., Bimendina, L.A. *Adv. Polym. Sci.*, 1981, 41:99.

39. Tsuchida, E., Abe, K. *Adv. Polym. Sci.*, 1982, 45:1.

40. Natansohn, A., Eisenberg, A. *Macromolecules*, 1982, 20:323.

41. Desbrieres, J., Rianudo, M. *Eur. Polym. J.*, 1981, 17:1265.

42. Bednar, B., Li, Z., Morwetz, H. *Macromolecules*, 1984, 17:1634.

43. Bednar, B., Li, Z., Morwetz, H., Huang, Y., Chang, L.P. *Macromolecules*, 1985, 18:1829.

44. Rodriguez-Prada, J.M., Percek, V. *Macromolecules*, 1986, 19:55.

45. Panavin, E.F., Svetlova, I.N. *Vysokomol. Soedin. B*, 1973, 15:522.

46. Ikada, Y., Yamashidi, K., Tsuji, H., Hyon, S.H. *Macromolecules*, 1987, 20:904.

47. Spevacek, J., Pierola, I.F. *Makromol. Chem.*, 1987, 188:861.

48. Oyama, H.T., Tang, W.T., Frank, C.W. *Polym. Prepr. Am. Chem. Soc. Div. Polym. Chem.*, 1986, 27:248.

49. Bixler, H.A., Michaels, A.S. *Encyclopedia of Polymer Science and Technology*, Wiley, New York, 1969, Vol. 10, p. 765.

50. Tsuchida, E., Abe, K. *Addv. Polym. Sci.*, 1982, 45:1.

51. Kabanov, V.A., Zezin, A.B. *Macromol. Chem. Suppl.*, 1984, 6:259.

52. de Jong, H.G.B. *Trans. Faraday Soc.*, 1932, 28:27.

53. Michaeli, I., Overbeek, T.G., Voorn, M.J. *J. Polym. Sci.*, 1957, 23:443.

54. Michaels, A.S., Miekka, R.G. *J. Phys. Chem.*, 1961, 65:1765.

55. Michaels, A.S., Falkenstein, G.L., Schneider, N.S. *J. Phys. Chem.*, 1965, 69:1456.

56. Kabanov, V.A. *Usp. Khim.*, 1991, 160:595.

57. Tsuchida, E., Osada, Y., Sanada, K. *J. Polym. Sci. Polym. Chem. Ed.*, 1972, 10:3397.

58. Kabanov, V.A., Zezin, A.B. *Sovi. Sci. Rev. B*, 1982, 4:207.

59. Chen, W., Turro, N.J., Tomalia, D.A. *Langmuir*, 2000, 16:15.

60. Kabanov, V.A., Sergeyev, V.G., Pyshkina, O.A., Zinchenko, A.A., Zezin, A.B., Joosten, J.G., Brackman, J., Yoshikawa, K. *Macromolecules*, 2000, 33:9587.

61. Kabanov, V.A., Kabanov, A.V. *Bioconjugate Chem.*, 1995, 6:7.

62. Osada, Y., Sato, J. *J. Polym Sci. Polym. Lett. Ed.*, 1976, 14:129.

63. Tsuchida, E., Osaka, Y., Ohno, H. *Macromol. Sci. Phys. B*, 1980, 17:683.

64. Gramatges, A.P., Monal, W.A., Covas, C.P. *Polym. Bull.*, 1996, 37:127.

65. Chen, H., Morawetz, H. *Eur. Polym. J.*, 1983, 19:923.

66. Bednar, B., Morawetz, H., Shafer, A. *Macromolecules*, 1984, 17:1634.

67. Chen, H., Morawetz, H. *Macromolecules*, 1982, 15:1445.

68. Salamone, J.C. *The Polymeric Materials Encyclopedia: Synthesis, Properties and Applications*. CRC Press, Boca Raton, FL, 1996.

69. Biersack, J.P., Haggmark, L.G. *Nucl. Instrum. Methods*, 1980, 174:257.

70. Dienes, G.J., Vineyard, G.H. *Radiation Effect in Solids*. Interscience, New York, 1957.

71. Lewis, M.B., Lee, E.H. *Nucl. Instrum. Methods*, 1991, B61:457.

72. Masters, J.G., Sun, Y., MacDiarmid, A.G., Epstein, A.J. *Synth. Met.*, 1991, 41:715.

73. Levon, K., Ho, K.H., Zheng, W.Y., Laakso, J., Karna, T., Taka, E., Osterholm, J. *Polymer*, 1995, 36:2733.

74. Liao, Y.H., Angelopoulos, M., Levon, K. *Proc. ANTEC 95*, 1995, 2:1413.

75. Liao, Y.H., Levon, K., Laakso, J., Osterholm, J.E. *Macromol. Rapid Commun.*, 1995, 16:393.

76. Kim, M.S., Levon, K. *J. Colloid and Interface Sci.*, 1997, 190:17.

77. Liao, Y.H., Angelopoulos, M., Levon, K. *J. Polym. Sci. A*, 1995, 33:2725.

78. Zheng, W.Y., Levon, K., Taka, T., Laakso, J., Osterholm, J.E. *J. Polym. Sci. B*, 1995, 33:15.

79. Zheng, W.Y., Levon, K., Laakso, J., Osterholm, J.E. *Macromolecules*, 1994, 27:7754.

80. Chen, S., Lee, H. *Macromolecules*, 1995, 28:2858.

81. Suna, L., Liua, H., Clarka, R., Yang, S.C. *Synth. Met.*, 1997, 84:67.

82. Gospodinova, N., Mokreva, P., Terlemezyan, L., *Polymer*, 1993, 34:2438.

83. Furugawa, Y., Ueda, F., Hyodo, Y., Harada, I., Nakajima, T., Kawagoe, T. *Macromolecules*, 1988, 21:1297.

84. Salaneck, W.R., Liedberg, B., Inganas, O., Erlandsson, R., Lundstrom, I., MacDiarmid, A.G., Halpern, M., Somasiri, N.L. *Mol. Cryst. Liq. Cryst.*, 1985, 121:191.

85. Young, S.G., Magill, J.H., Lin, F.T., *Polymer*, 1992, 33:3215.

86. Allcock, H.R., Kugel, R.L., Valan, K.J., *Inorg. Chem.*, 1966, 5:1709.

PART VI
Well-Defined Polyphosphazenes: Synthetic Aspects and Novel Molecular Architectures

19 Synthesis and Chemical Regularity in Phosphazene Copolymers

GABINO A. CARRIEDO

Departamento de Química Orgánica e Inorgánica, Universidad de Oviedo, Oviedo, Spain

INTRODUCTION

In phosphazene homopolymers $[NPA_2]_n$ (**1**) (Chart 1) the chemical functions present in the A substituents are not only the same but also have the same averaged chemical environment. In mixed-substituent derivatives of the formula $[NPA_{2-a}B_a]_n$ (**2**), however, the actual situation of the chemical functionalities depends on the relative proportions and sequential distribution of the A and B groups along the chains and on the intermolecular interactions that dictate the conformations in solution and packing in the solid state. Consequently, as for many flexible polymers, not only are the physical properties, such as the glass transition temperature or the formation of mesophases, affected by the chain regularity [1], but also the chemical reactivity, especially the degree of accessibility of the functions to the incoming reagents. The actual extent of the effects of chain regularity on the polymer reactivity will depend on the particular chemical functions and on the reaction conditions, but they would be of particular importance in those reactions more closely related with biochemistry. The situation shows more complexity in trisubstituent derivatives of the formula $[NPA_{2-a-b}B_aC_b]_n$, for which a more intricate relationship between the reactivity and chain constitution can be expected.

The distribution of the units present in a copolymeric chain depends critically on the reaction or sequence of reactions followed for its synthesis. Therefore, the choice of the synthetic methodology may be crucial in designing a polyphosphazene with predetermined chemical properties, particularly in the field connected with the biomedical applications. In this chapter we focus on the possible types of phosphazene copolymers that may be formed and the

Polyphosphazenes for Biomedical Applications, Edited by Alexander K. Andrianov
Copyright © 2009 John Wiley & Sons, Inc.

chain regularity that might be expected using the most common synthetic methods, especially the macromolecular substitution from the appropriate chlorine-containing reactive intermediates, the condensation of phosphorani-mines, and the chemical derivatization of a reactive pendant group in a precursor polyphosphazene. The application of other reactions, such as the ring-opening polymerization of substituted cyclic phosphazene monomers and the decomposition of phosphorus azides for the preparation of phosphazene copolymers, is also discussed.

TYPES OF PHOSPHAZENE COPOLYMERS

Mixed-substituent phosphazene copolymers of the formula $[NPA_{2-a}B_a]_n$ (**2**) may have different monomeric [N=P] units exhibiting a wide range of constitutional or stereoisomers (**3–7**), depending on the composition and sequence of those units along the chains (Chart 1). If the two different units in **3**, **4**, and **6** are randomly distributed along the chains, they are random copolymers, and in the strictly alternating cases, their positions are dictated solely by their relative abundance. In the other extreme case, if the chains of polymer **3** consist of two segments, one of NPA_2 units and the other of NPB_2 units, joined by a single NPA_2–NPB_2 bond, it is a block copolymer. In intermediate cases, the random copolymers may have some blocky and some alternated segments. It should be noted that the phosphazene polymers of formula $[NPAB]_n$ with 50 mol% of each substituent but having only one type of

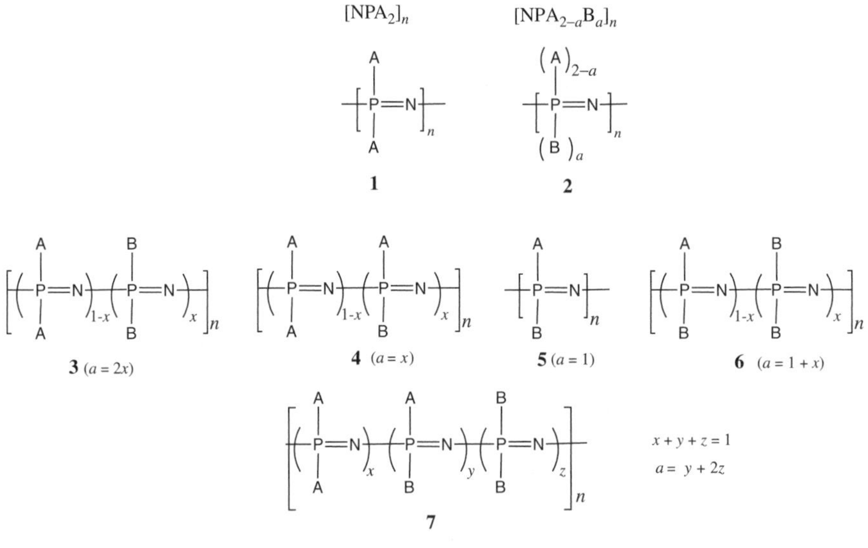

CHART 1

monomeric unit (NPAB units) (**5**) are actually homopolymers. In the latter case, the polymers all have the chemical functions in the same or nearly the same environment, depending only on the tacticity of the chain.

A very common type of two-substituent polyphosphazenes of formula $[NPA_{2-a}B_a]_n$ consists of chains formed by a distribution of three monomeric units; that is, they are random terpolymers that have the general formula $[(NPA_2)_x(NPAB)_y(NPB_2)_z]_n$ (**7**) $(x + y + z = 1; a = y + 2z; \text{mol\% of B} = 50a)$. As can be shown by ^{31}P nuclear magnetic resonance (NMR) spectroscopy, in those polymers, each of the NP units may be surrounded randomly by the others in different ways, including the formation of blocky segments of those units that are more abundant. In A-rich polymers (**7**) (*a* very small) z is negligible and only the units NPA_2 and NPAB are significant, becoming a copolymer like **4** rather than a terpolymer. Similar considerations would easily show the complexity of the possible structures of mixed derivatives of the formula $[NPA_{2-a-b}B_aC_b]_n$.

One special case of polyphosphazenes appears when there are bidentate substituents giving rise to phosphorus heterocycles in the monomeric units (polyspirophosphazenes) (Chart 2). In the polymers $[NP(A-A)]_n$ (**8**) all the cycles are the same, but the combination of units having different cycles gives the copolymers $\{[NP(A-A)]_{1-x}[NP(B-B)]_x\}$ (**9**), and the combination of bidentate and monodentate substituents results in the copolymers of formula (**10**)–(**12**). The structure of (**9**), (**10**), or (**12**) could be strictly alternating or have some alternating and blocky segments (block copolymers of those types are yet unknown). The polymers $\{[NP(A-A)]_{0.5}[NPB_2]_{0.5}\}_n$ (**10**, $x = 0.5$), if strictly alternating, are on average very close to homopolymers with dimeric repeating units $[NP(A-A)NPB_2]_n$ (**11**). Polyspirophosphazenes with one cyclic unit and two different monodentate substituents $\{[NP(A-A)]_{1-x}[NPB_{2-a}C_a]_x\}_n$ having the same mole percentage of B and C are the copolymers $\{[NP(A-A)]_{1-x}[NPBC]_x\}_n$ (**12**). Copolymers (**10**) to (**12**) are interesting because, if

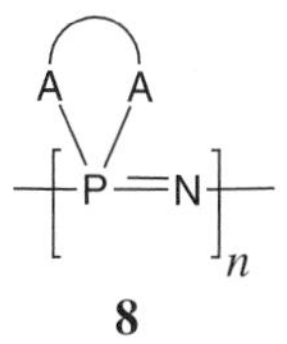

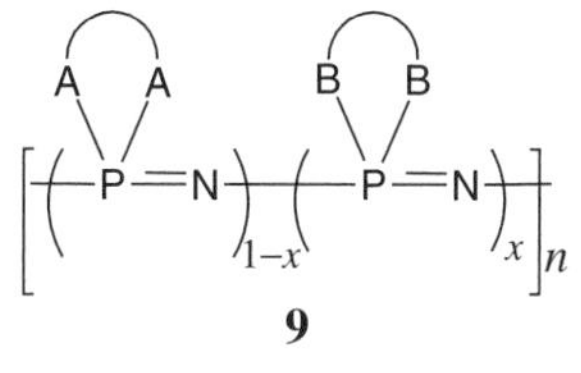

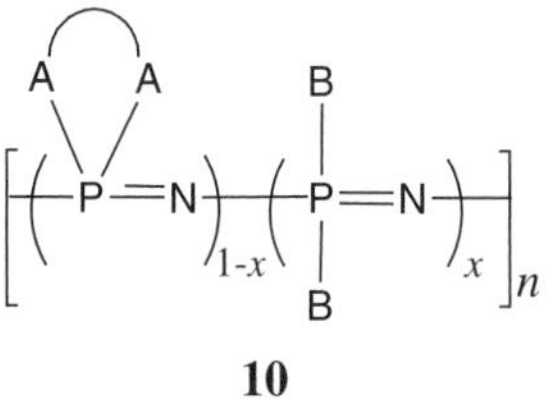

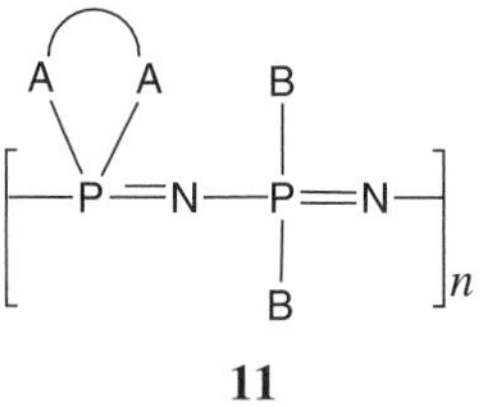

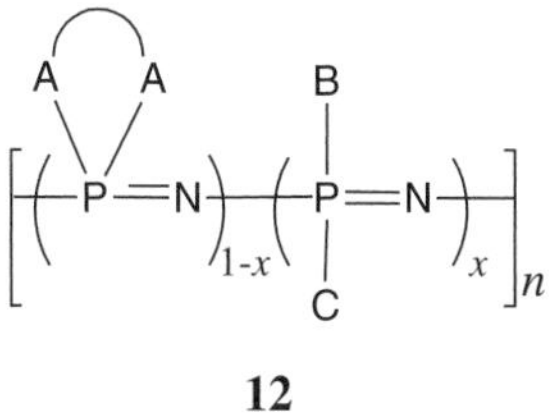

CHART 2

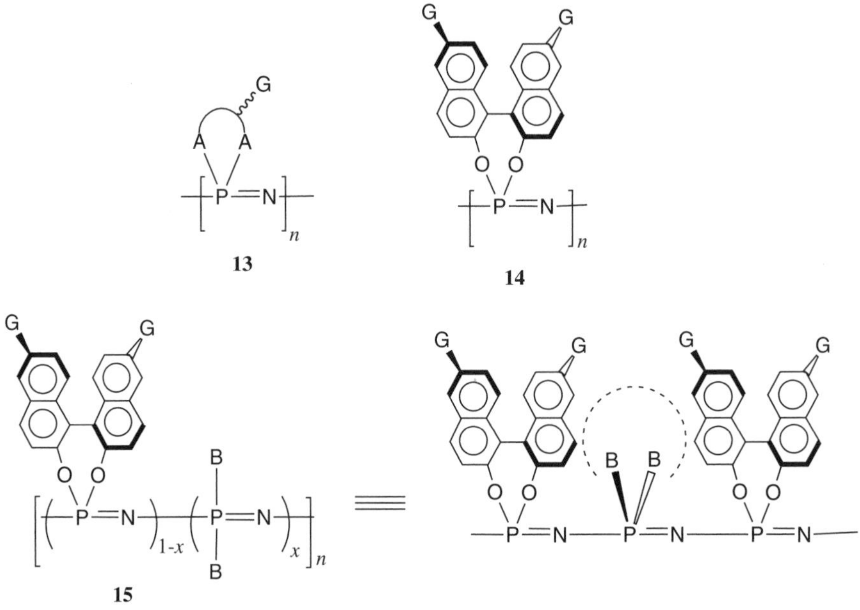

CHART 3

strictly alternating, the chemical function attached to B (or C) appears very regularly distributed along the chains.

Spirophosphazenes may give rise to a completely new type of phosphazene polymer consisting of functionalized cyclic units represented in the general form (**13**) (Chart 3). It is interesting to notice that the tacticity of the chains is dependent on the characteristics of the bidentate A–A substituents. Thus, when there is a bond with restricted rotation, as happens in the C–C' bond of the binaphthoxy phosphazene homopolymers (**14**) (so far, the only polyphosphazenes known of this type), if only one of the enantiomeric biphenoxy groups (R or S) is present, the polymer is chiral and the chain is isotactic. Furthermore, in binaphthoxy phosphazene copolymers of the type **15** with small x, all the monodentated substituents are placed within chiral environments (i.e., inside the *chiral pockets* shown in Chart 3).

SYNTHESIS OF PHOSPHAZENE COPOLYMERS

The general methods for the synthesis of polyphosphazenes have been extensively explained and reviewed [2–4]. This discussion is focused on the general types of copolymers that may be obtained from the different methods and on more recent representative examples.

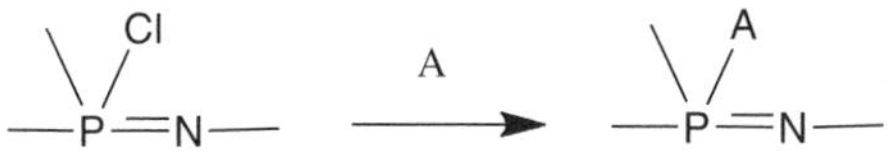

SCHEME 1

Phosphazene Copolymers from Macromolecular Substitution

The macromolecular substitution [5] of chlorines may be carried out using various different types of chlorine-containing precursors to form a variety of phosphazene copolymers (Scheme 1). The typical nucleophiles used in those reactions to introduce the substituent A are sodium alkoxides or aryloxides (A = OR) or the sulfur analogs (A = SR) and the amines NH_2R or NHR_2 (A = NHR or NR_2). However, aryloxy and thioaryloxy phosphazenes can also be obtained by the direct reaction of chorophosphazenes with phenols HOR or thiols using K_2CO_3 or Cs_2CO_3 as proton abstractors, a process that is faster with the latter [6]. With cesium carbonate, it is even possible to use the alcohol $HOCH_2CF_3$, which reacts at room temperature to give the trifluoroethoxy derivatives [7]. The softer reaction conditions and the possibility of using the phenol directly (thus avoiding its previous reaction with Na or NaH to generate the sodium salts) make the use of alkali carbonates advantageous in many cases.

The completeness of chlorine replacement in macromolecular substitution is a crucial factor affecting the nature of the polymers obtained, because even a small proportion of residual NP–Cl bonds in the chains would render the product not only hydrolytically labile but also thermally much less stable [8]. However, the hydrolytical activity of the NP–Cl bonds was noticed early to be advantageous for biomedical applications [9]. It is also known that the presence of a small proportion of NP–Cl defects (ca. 2 mol%) may prevent the formation of mesophases [10].

The macromolecular substitution may present a problem when using bifunctional nucleophiles, because of the possibility of cross-linking (pathway b in Scheme 2). However, the reaction of $[NPCl_2]_n$ with biphenols and chiral binaphthols using K_2CO_3 or Cs_2CO_3 as proton abstractor allowed the preparation of un-cross-linked lineal polyphosphazenes with phosphorus cycles in the units (pathway a in Scheme 2) [11,12]. It is likely that those reactions defying the high probability of cross-linking are facilitated by the generation of

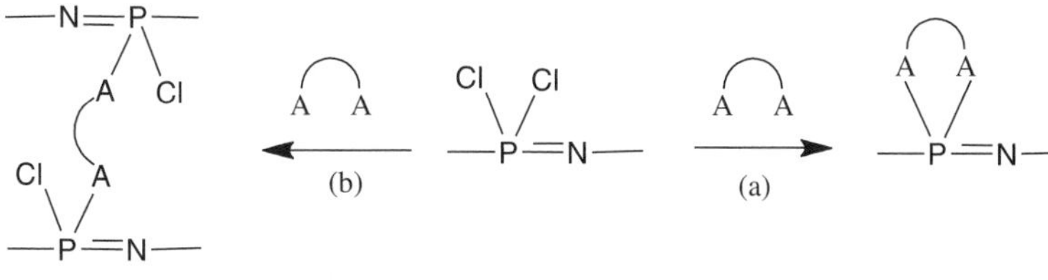

SCHEME 2

a ring, as occurs with the chelate effect in metal–ligand complexes. However, the difficulty of obtaining similar products using the corresponding disodium salts suggests that the cyclization mechanism may also be favored by the action of the alkali carbonate.

Furthermore, it has also been found that heterobifunctional nucleophiles such as the phenol amine $OH-C_6H_4-NH_2$ may react selectively with $[NPCl_2]_n$ in the presence of K_2CO_3. At room temperature only the NH_2 groups are activated to give aminophosphazenes (see also later), but under sterically more demanding conditions and at the refluxing temperature, the reaction goes through the HO groups to give aryloxy derivatives [13]. In the first case, however, the presence of potassium carbonate may not be decisive because it has been reported that the bifunctional L-tyrosine ethyl ester $[NH_2CH(C_6H_4-OH)CH_2COOEt]$, having amino and phenolic groups, also reacts selectively with $[NPCl_2]_n$ by the NH_2 [14].

An important factor affecting the nature of the polymers obtained by the macromolecular substitution is the origin of the chlorophosphazene precursors. Apart from the chemical composition and structure (see the discussion below), those precursors are usually obtained from the parent $[NPCl_2]_n$, and this can be formed by a variety of methods [15]. The polymers derived from $[NPCl_2]_n$ prepared by the ring-opening polymerization of $[N_3P_3Cl_6]$ in either the melt [15b] or solution [15b,7] have higher molecular weight and higher polidispersity index (PDI) than those derived from a $[NPCl_2]_n$ prepared by condensation polymerization.

On the other hand, it have been suggested that some of the polymers obtained by macromolecular substitution from a $[NPCl_2]_n$ prepared by the ring-opening polymerization of $[N_3P_3Cl_6]$ exhibit anomalous behavior in solution, suggesting the formation of intermolecular aggregations and raising the question of the presence of chain imperfections, mainly as branching points [16], a fact that could affect their thermal stability [17].

Copolymers with Monodentated Susbtituents   When obtained by macromolecular substitution, mixed-substituent copolymers of formula $[NPA_{2-a}B_a]_n$ (**2**) are always random copolymers and, in principle, can have the compositions **3** to **7** (Scheme 3). Block copolymers are not possible by this method except in the case of replacing the chlorines in a block $NPCl_2$-containing precursor available by other routes (see the section "Phosphazene Copolymer from Phosphoranimines"), such as **16** (Scheme 4). The chain regularity achieved (i.e., the actual type of the copolymer formed) in the macromolecular substitution (Scheme 3) is strongly dependent on the nucleophile and on the reaction conditions. Starting from $[NPCl_2]_n$ and first using the nucleophile B in substoichiometric amounts, a strictly geminal mechanism would give intermediate polymers I^1 that could subsequently be reacted with A (sequential substitution) to give copolymers **3**. If, however, the mechanism of the reactions with B is strictly nongeminal, the copolymers **4** and **6** and the mixed-substituent homopolymer **5** may be formed through the intermediates I^2 to I^4.

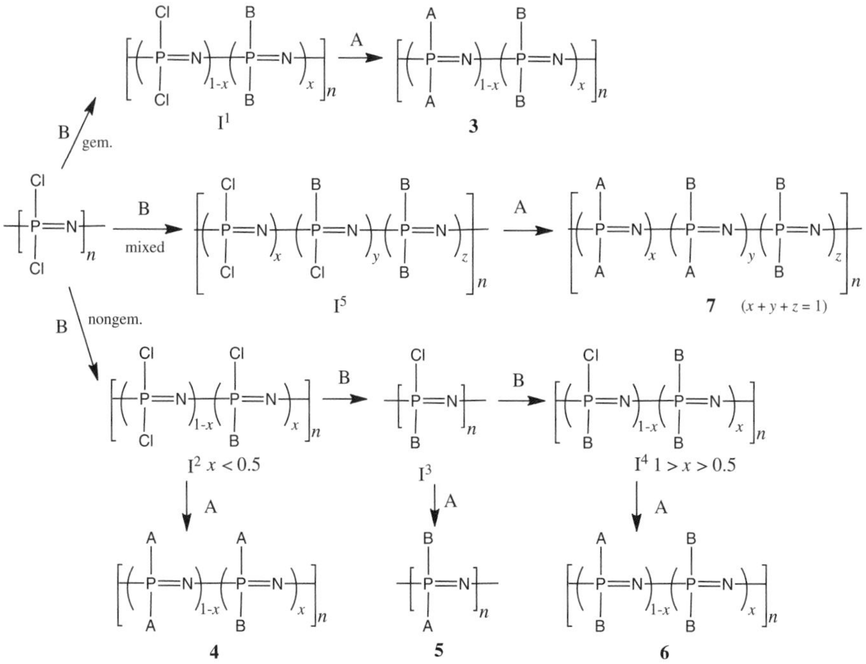

SCHEME 3

Chlorine-containing species I^2 to I^5 are usually too hydrolytically unstable to be isolated, and they are generated in solution only as intermediates for the second substitution. However, some intermediates with high chlorine contents and with unusual hydrolytic stability have been isolated [18].

In practice, perfectly geminal or nongeminal mechanisms are rather unusual, and as a result, the very regular copolymers **3** and homopolymers **5** are almost inaccessible by the macromolecular substitution (for **5** there are other methods, as discussed later). This is because, even in favorable cases, the formation of [NPA$_2$] units becomes more likely as the composition of the substituted products approaches the 50% chlorine substitution, leaving fewer [NPCl$_2$] units available in the intermediates (I^2 with x close to 0.5). In fact, the sequential macromolecular substitution most frequently gives the terpolymers **7** through

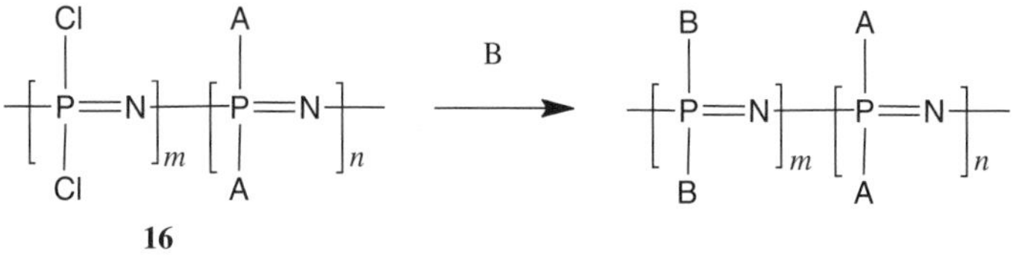

SCHEME 4

the less regular intermediate I^5. The formation of terpolymers 7 is even more likely if the two nucleophiles A and B are placed together in the appropriate ratio from the beginning of the reaction. It is, however, noteworthy to mention that almost all the data available about geminal or nongeminal substitution with alkoxy or aryloxy groups applies to reactions with amines and sodium alkoxy or aryloxy reagents. Little is known on the mechanism of the macro-molecular substitution with phenols promoted by alkali carbonates M_2CO_3.

Although nucleophiles reacting in a perfect geminal way, leading to copolymers 3, are rare, an example is known. Thus, the polymer 17 [19] was obtained by reacting $[NPCl_2]_n$ first with HSC_6H_4Br (geminal) and Cs_2CO_3 in tetrahydrofuran (THF) followed by the addition of HOC_6H_4Br. It is known that the fluoroalcoxy groups [20a], especially the very frequently used OCH_2CF_3, exhibit a certain tendency to react by geminal mechanisms by activating the nearby P–Cl bonds, and also tend to replace other groups [20b], especially aryloxides, by methatetical substitutions giving frequently irregular terpolymers of the type 7, depending on the reactions conditions [20c].

17

In general, the nongeminal mechanism is favored by the steric effects [5], and therefore the copolymers 4 (see Scheme 3) can be achieved in practice using very bulky aryloxy groups B and, more favorably for $x \ll 0.5$, but always only as good approximations. Closer to the nongeminal substitution pathways are the amines [21], with which it might be possible to obtain copolymers such as 4 or 6 in Scheme 3 (A or B = NHR). Thus, the products of the chemoselective reaction of $NH_2C_6H_4OH$ with $[NPCl_2]_n$ and K_2CO_3 at room temperature and, correspond-ingly, their derivatives formed in their subsequent reactions with NH_2Bu, were formulated as the copolymers 18 ($x = 0.1$ to 0.6) and 19, respectively [13]. However, the products of the reactions of poly(dichlorophosphazene) with amines are formulated very frequently as terpolymers (e.g., 20), obtained by the selective direct reaction of $[NPCl_2]_n$ with L-tyrosine ethyl ester and glycine [14].

18   19

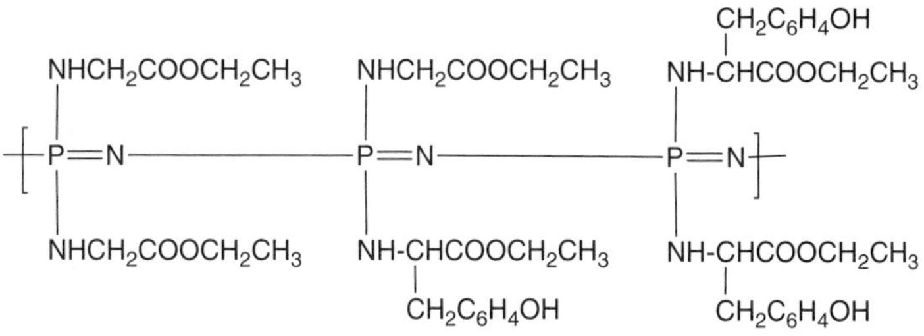
CH2C6H4OH
NHCH2COOCH2CH3
NHCH2COOCH2CH3
NH-CHCOOCH2CH3
P=N
P=N
P=N
NHCH2COOCH2CH3
NH-CHCOOCH2CH3
NH-CHCOOCH2CH3
CH2C6H4OH
CH2C6H4OH

20

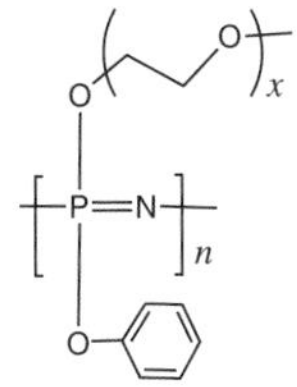

21 (x = 2, 3, 7.3)

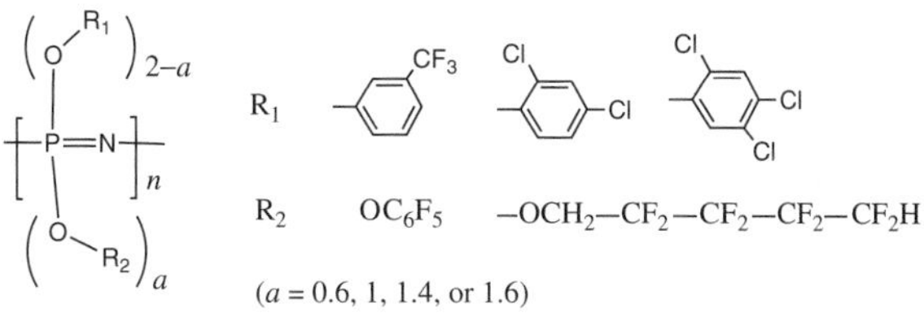
R1
CF3
Cl
Cl
Cl
Cl
Cl
R2
OC6F5
–OCH2–CF2–CF2–CF2–CF2H

(a = 0.6, 1, 1.4, or 1.6)

22

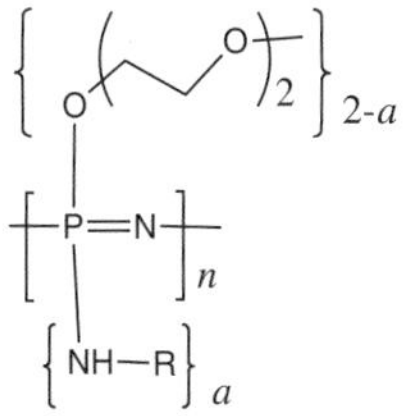

23

(R = Pr, Bu, Pent, Hex, a = 1, 1.2, 1.4, or 1.6)

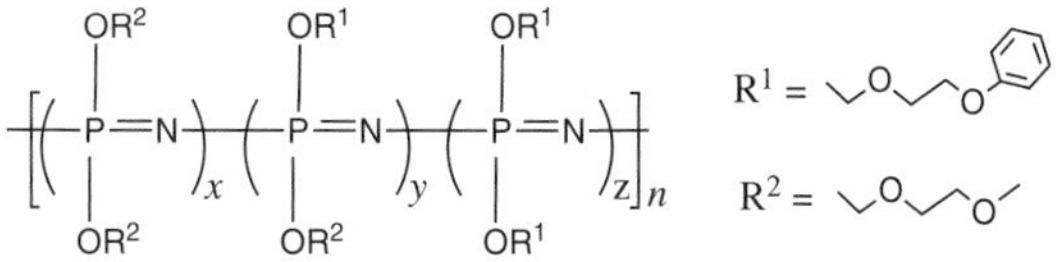
OR2
OR1
OR1
OR2
OR2
OR1
R1 =
R2 =

24

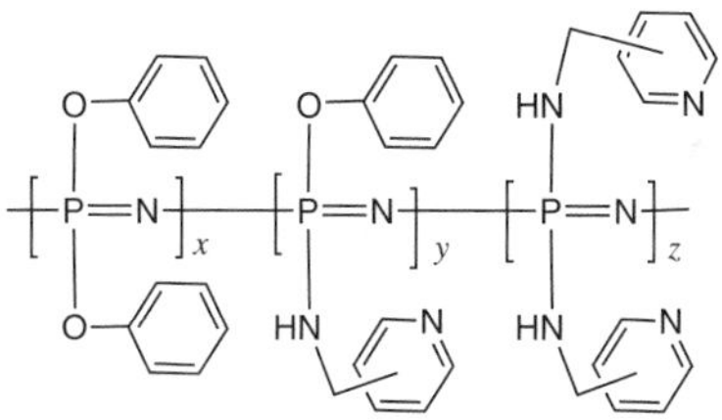

25

CHART 4

As a general conclusion, it can be assumed that except in the case of very B-rich **6** and very A-rich **4** with bulky B, the random terpolymeric **7** is the most probable structure of the mixed-substituent poplyphosphazenes $[NPA_{2-a}B_a]_n$ obtained by the sequential macromolecular substitution, as very

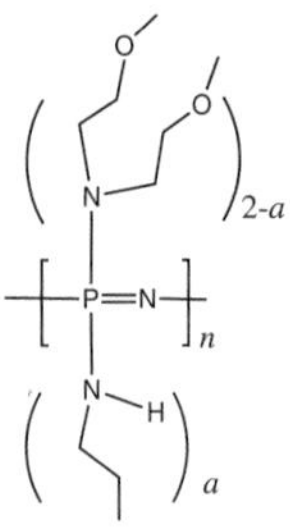

26

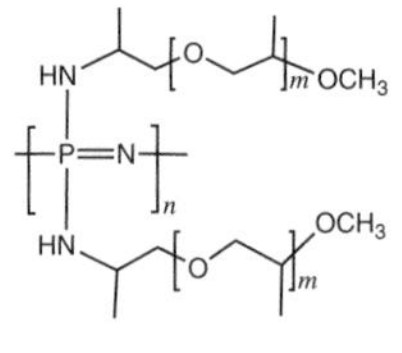

27

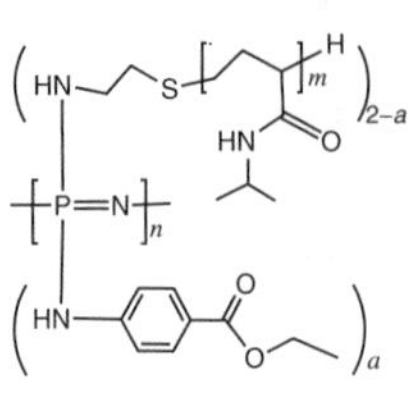

28 (*a* = 0.2, 1, 1.32)

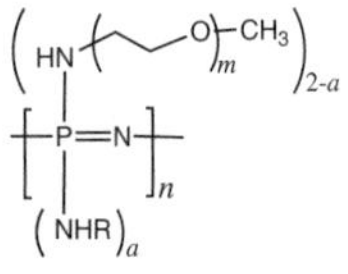

29 (*m* = 11, 16)

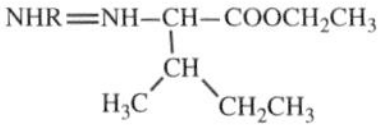

(IleOEt)

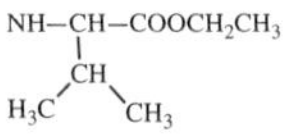

(ValOEt)

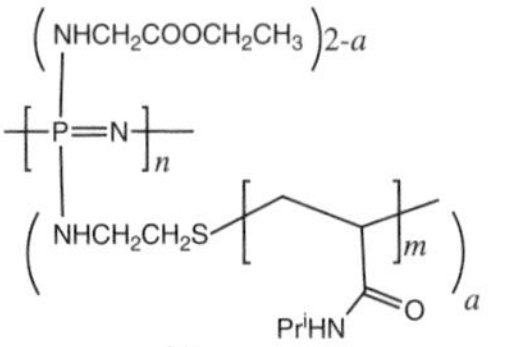

30

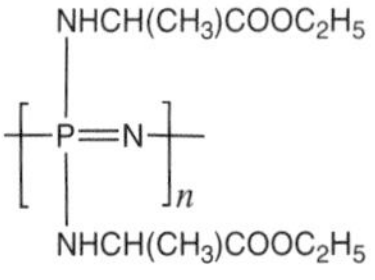

31

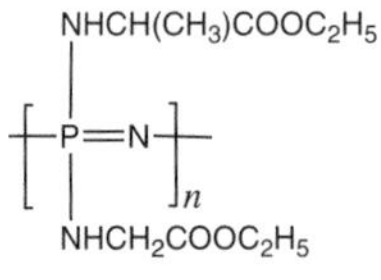

32

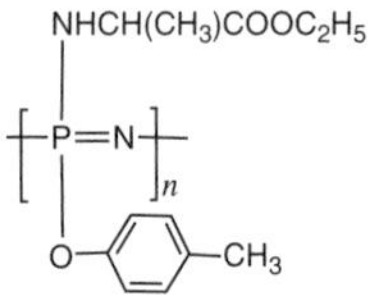

33

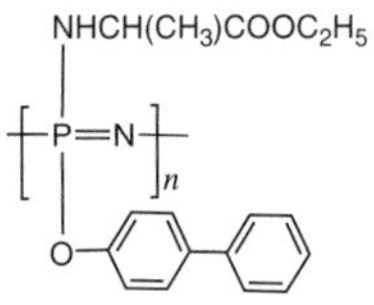

34

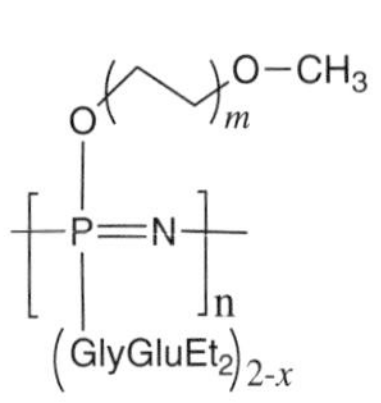

m = 3 or 4; *x* = 0.5 – 1.5; *n* = 180–220

35

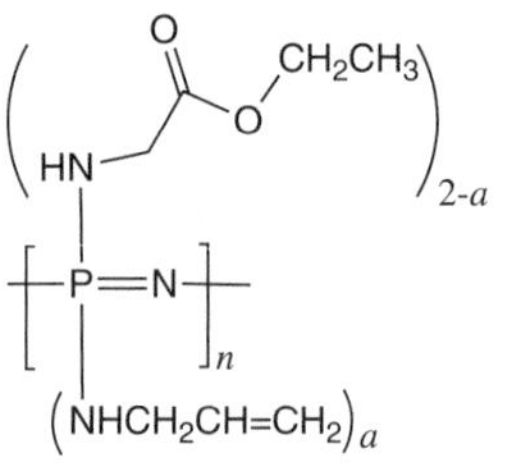

36 (*a* = 2, 1.33, 0.67)

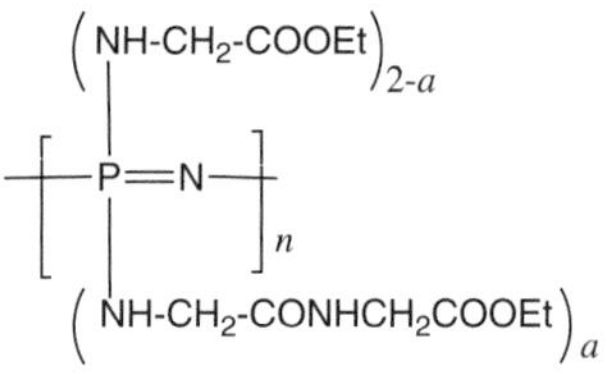

37 (*a* = 0, 0.5, 1, 1.5)

CHART 5

frequently demonstrated by ^{31}P NMR spectroscopy. Examples include the polymers **21** (Chart 4), with 50% of phenoxy and polyether substituents that have 60% [NPAB] units and 20% each of [NPA$_2$] and [NPB$_2$] units [22]; the polymers **22** [23] and **23** [24]; the series **24** [25], which includes one member to which the formula $[NP(OR^1)(OR^2)]_n$ (i.e., $x = 0$, $y = 1$, $z = 0$) was attributed only on the basis of the ratio of OR^1 and OR^2 used (ca. $1:1$); and the terpolymers **25** [26], which have 25, 50, 75, and 100% amine groups (i.e., $y + 2z = 0.5, 1, 1.5,$ or 2).

Numerous mixed-substituent copolymers with NHR in their composition have been prepared by macromolecular substitution [2,3]. Recent examples include **26** (Chart 5) [27], where a varies between 0.8 and 1; **27** [28], **28** [29], and the amphiphilic polymers **29** [30] (x in the range 0.89 to 1.43), which carry hydrophilic α-amino-ω-methoxy-poly(ethylene glycol) and hydrophobic amino acid ester side groups, the self-assembling nanoparticle-forming phosphazene **30** [31], which may be useful as injectable drug carriers for the delivery of hydrophobic compounds; the biodegradable high-molecular-weight film-forming polyphosphazenes **31** to **34**, containing 50% ethyl alanato side groups together with 50% ethyl glycinato, p-methylphenoxy, or p-phenylphenoxy side groups [32]; and the new biocompatible and thermosensitive poly(organophosphazene)s **35**, with short tri- or tetraethylene glycol chains as a hydrophilic group and the dipeptide glycyl-l-glutamic diethyl ester (GlyGluEt$_2$), which had a lower critical solution temperature (LCST) below body temperature and interest for local delivery of hydrophobic drugs [33].

Other examples are the polyphosphazenes **36**, with different ratios of glycino Et ester and allylamine substituents, the ^{31}P NMR spectra of which showed only two distinct signals, suggesting the presence of only two different units [34]; and the new bioerodible dipeptide polyphosphazenes **37**, the blends of which with poly(lactide-co-glycolide) are beneficial for tissue engineering matrices, drug delivery, and device design [35].

Tri-substitituent phosphazene copolymers $[NPA_{2-a-b}B_aC_b]_n$ obtained by three-sequence macromolecular substitution are more difficult to characterize structurally in terms of the real monomeric units actually present in the chains. Examples are materials of composition $[NP(OR^1)_x(OR^2)_y(OR^3)_z]_n$ ($OR^1 =$ 2-allylphenoxy, $OR^2 = $ 4-isobutylphenoxy, $OR^3 = $ 4-methoxyphenoxy) **(38)** [36], which are useful for gas transport membranes [37], and the amino acid esters–containing polymer $[NP(NHR^1)_{0.8}(NHR^2)_{0.8}(NHR^3)_{0.4}]_n$ ($R^1 = C_6H_5CH_2CHCO_2Et$, $R^2 = CH_3CHCO_2Et$, $NHR^3 = CH_2CO_2Et$) **(39)**, which has been tested for in vitro culture of rat endothelial cells [38].

Many copolymers of type $[NPA_{2-a-b}B_aC_b]_n$ with amino acid ester side groups have deserved interest for biomedical applications, as, for example, the thermosensitive poly(amino)phosphazenes **(40)**, useful as a model for drugs-controlled release, which bears hydrophobic l-isoleucine ethyl ester side groups, hydrophilic α-amino-ω-methoxy-poly(ethylene glycol) groups, and a fraction of ethyl-2-(O-glycyl)lactate groups to increase their biodegradability [39]. Other examples are the biodegradable phosphazene copolymers **(41a** to **c**, Chart 6), with about 50 to 70% PheOEt and 30 to 40% GlyOEt bearing 0.1 to 5.8 mol% of the galactose substituents, which were synthesized aiming to develop new scaffold materials with controlled surface properties for tissue engineering [40].

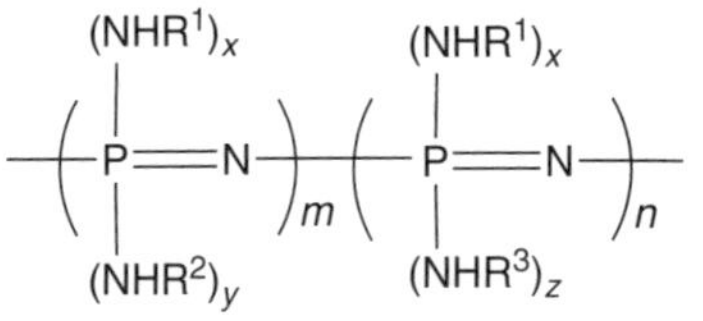
(NHR1)x
(NHR1)x
P
N
P
N
m
n
(NHR2)y
(NHR3)z

40

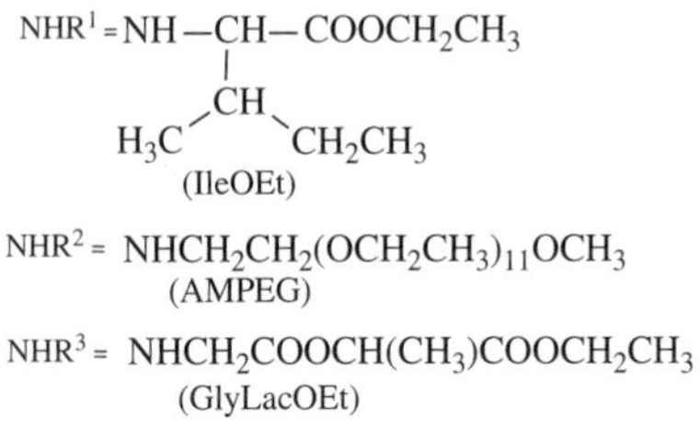
NHR1 = NH—CH—COOCH2CH3
CH
H3C
CH2CH3
(IleOEt)
NHR2 = NHCH2CH2(OCH2CH3)11OCH3
(AMPEG)
NHR3 = NHCH2COOCH(CH3)COOCH2CH3
(GlyLacOEt)

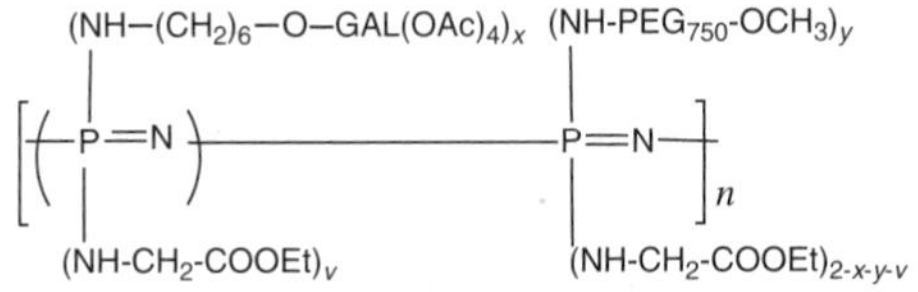
(NH—(CH2)6—O—GAL(OAc)4)x
(NH-PEG750-OCH3)y
P
N
P
N
n
(NH-CH2-COOEt)v
(NH-CH2-COOEt)2-x-y-v

41a

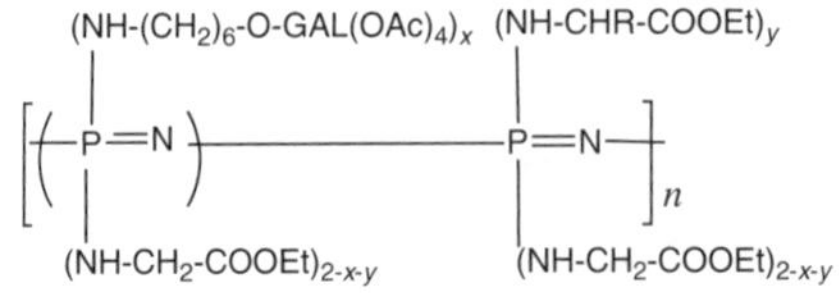
(NH-(CH2)6-O-GAL(OAc)4)x
(NH-CHR-COOEt)y
P
N
P
N
n
(NH-CH2-COOEt)2-x-y
(NH-CH2-COOEt)2-x-y

41b

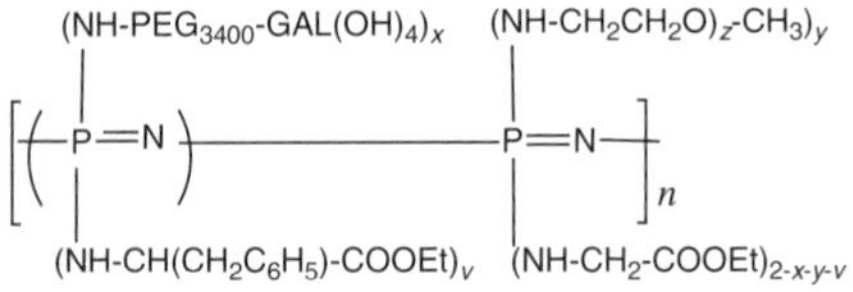
(NH-PEG3400-GAL(OH)4)x
(NH-CH2CH2O)z-CH3)y
P
N
P
N
n
(NH-CH(CH2C6H5)-COOEt)v
(NH-CH2-COOEt)2-x-y-v

41c

CHART 6

Four-substituent polyphosphazene random copolymers such as (**42**) [41], which are less frequent, are even more difficult to characterize in terms of actual composition of the N=P units along the chains, but little, if any, regularity is to be expected.

$$[NP(OR1)_{0.11}(OR2)_{0.07}(OR3)_{0.58}(OEt)_{0.24}]_n$$

42

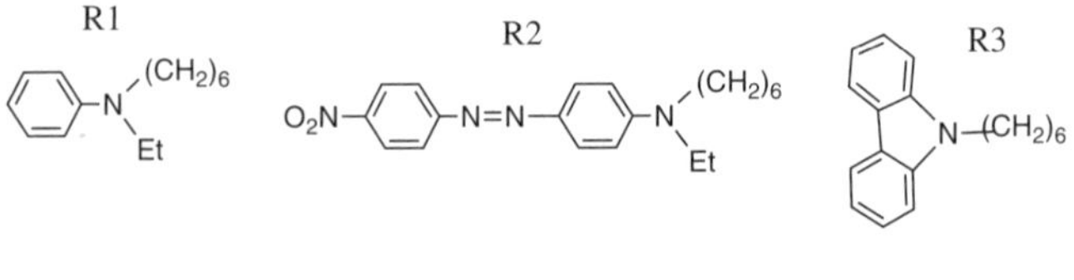
R1
(CH2)6
N
Et
R2
O2N
N=N
N
(CH2)6
Et
R3
N
(CH2)6

Polymers with Phosphorus-Cycles in the Units (Polyspirophosphazenes) So far, the only chlorine-containing precursors used to obtain spiro-phosphazene copolymers that have been described are the intermediates **43** and **44**, (G = H) (Scheme 5), first obtained by the reaction of $[NPCl_2]_n$ with substoichiometric amounts of 2,2′-dihydroxy-1,1′-biphenyl [42] or 2,2′-dihydroxy-1-1′-binaphtyl [43], respectively, in the presence of K_2CO_3 or Cs_2CO_3. Other intermediates, such as **44** (G = Br), have similarly been formed to be used for the preparation of derivatives of the type **47**, (G = Br) [44]. Those Cl-containing intermediates are normally generated in situ from $[NPCl_2]_n$, but recently it was found that when x is not higher than 0.5, the biphenoxy derivatives (**43**) can be isolated as stable white solids and stored for months, to be used in further substitution reactions. Theoretical calculations on closely related models showed that the chlorine atoms are well hindered by the biphenoxy groups, avoiding the presence of water molecules in their vicinity unless in THF solution [45].

The large number and variety of polymers of the types **45** to **47**, carrying various functionalities, including H, Br, NH_2, COOR, COOH, pyridine, phosphines, metal carbonyl, and organometallic complexes, and even pendant polyamide chains in the A or B groups that have been prepared from **43** and **44**, have been already reviewed [12]. Due to the steric effects of the biphenoxy groups during the formation of **43**, the substitution is favored in those $NPCl_2$ units that are not bonded directly to an already substituted $NP(O_2C_{12}H_8)$ unit. Therefore, the reaction sequence leads to a random distribution of the two units

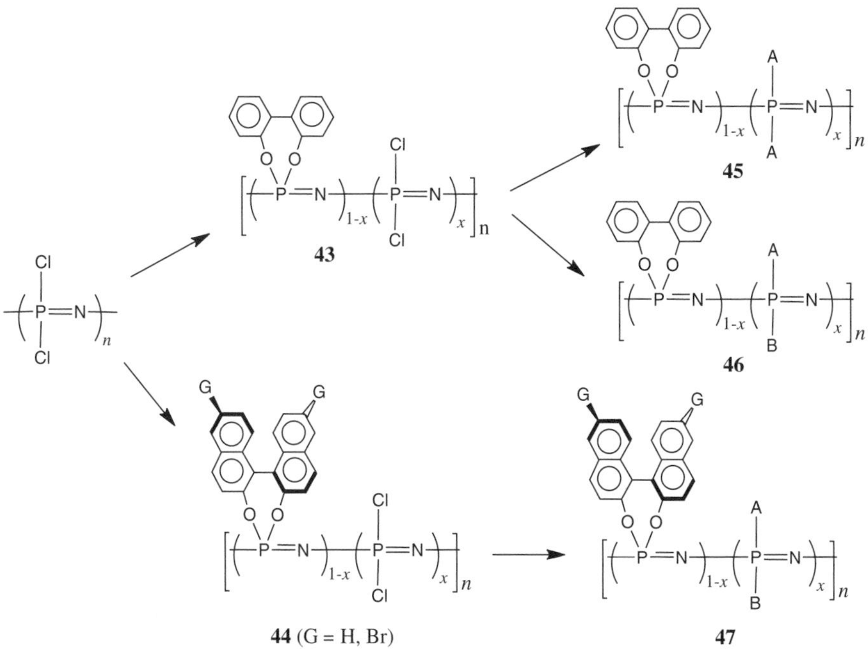

SCHEME 5

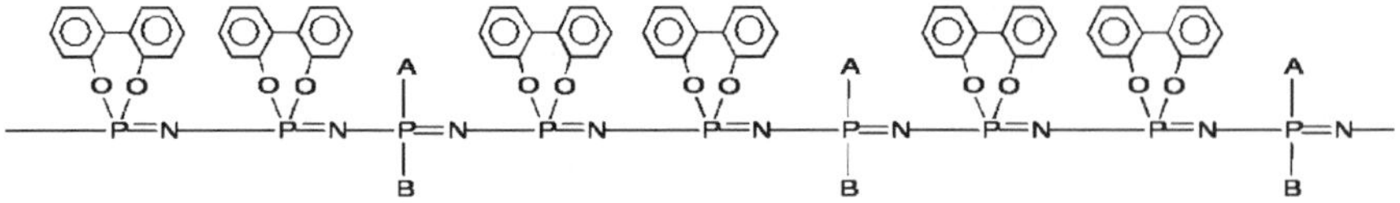

SCHEME 6

which on average should be very regular. In fact, experimental evidence based on the variation of the glass transition of the series (**45**) as a function of x indicate that those polymers behave like strictly alternating random copolymers [46]. As a consequence, the chemical functions attached to A or B along the chains in copolymers (**45**) and (**46**) are on average approaching the regular disposition shown in Scheme 6 for $x = 0.33$.

This may substantially affect the physical properties of the spirophosphazenes because of the higher probability of strong π–π or π–H interactions between the aromatic rings in the solid state and in solution, as demonstrated experimentally with closely related cyclic models [47].

Very similar considerations can be made concerning the binaphthoxy phosphazene derivatives and, in fact, the variation of the T_g values in series **47** (G = H) were also consistent with the strictly alternating chains [46]. As a consequence, in those copolymers all of the NPAB groups are surrounded by two binaphthoxy units, especially in the case of small values of x, and therefore, all the chemical functions pendant from the A or B groups are placed inside chiral pockets or C_2 symmetry (see Chart 3).

Phosphazene Copolymers from Phosphoranimines

Phosphoranimines may give polyphosphazenes by two main well-known processes: One is the PCl_5-catalyzed polymerization and the other is noncatalyzed thermal polycondensation. In both cases, the resulting polyphosphazenes have lower (10^3 to 10^4) and narrower (IPD = 1.05 to 1.5) molecular-weight distributions than those obtained by macromolecular substitution from $[NPCl_2]_n$.

Catalyzed Condensation of Phosphoranimines   The PCl_5-catalyzed condensation of phosphoranimines [48] is a room-temperature living cationic polymerization that may be used to synthesize a variety of mixed-substituent homopolymers (perfectly regular) with controlled molecular-weight (Scheme 7). Thus, the polymer **48** could be the poly(dichlorophosphazene) (A = B = Cl), a monocloro-organophosphazene (A = Cl, B = alkyl or aryl) or a diorganophosphazene (A and B are alkyl or aryl groups). It is interesting to notice that unless terminated by deactivating the terminal groups, **48** is a living polymer. Thus, in the case of the dichlorophosphazene, the product is $([Cl_3P=N-(PCl_2=N)_n-PCl_3][PCl_6]$ and therefore the addition of a second

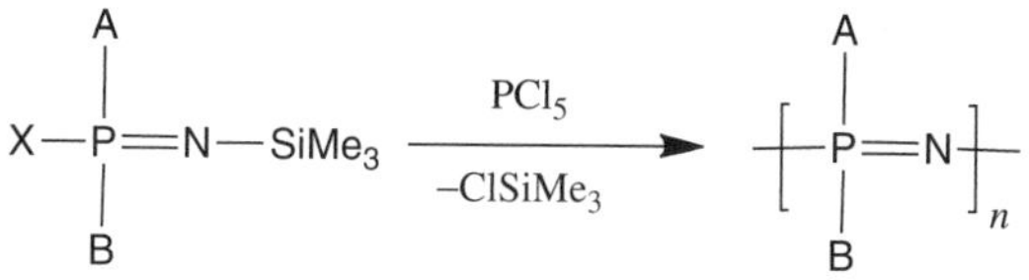

SCHEME 7

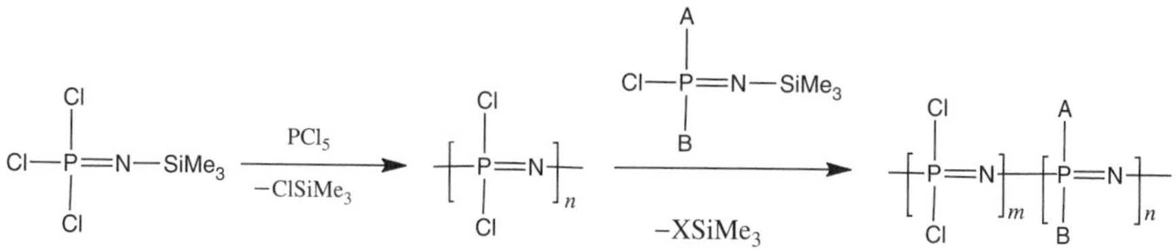

SCHEME 8

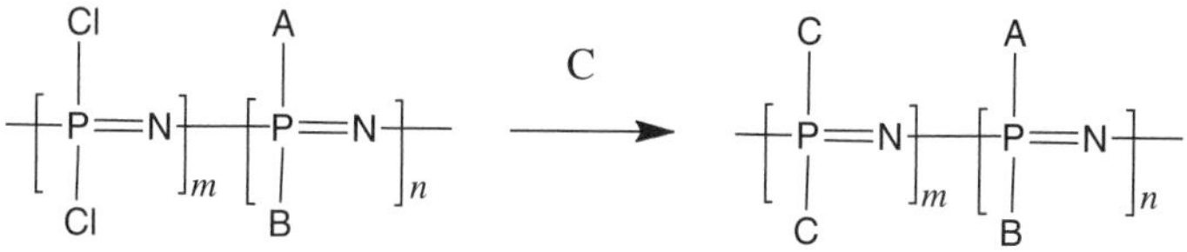

SCHEME 9

phosphoranimine gives the block copolymers $[NPCl_2]_m[NPAB]_n$ (**49**) (Scheme 8). It appears that formation of the related random copolymers using a mixture of phosphoranimines is very hampered by the strong tendency for one of them to polymerize first, finally giving block copolymers.

It has already been noted (Scheme 4) that the copolymers **49** can be useful to obtain block copolymers by macromolecular substitution, as, for example, **50** (Scheme 9). Using two substituents, C and D, in a sequential manner, the reaction of Scheme 9 would give block copolymers in which one of the branches (NPAB) is perfectly regular and the other (NPCD) would have a random distribution of the pendant groups that would depend on the mechanism (geminal or nongeminal) of the macromolecular substitution path as discussed earlier for a single phosphazene chain.

The cationic polymerization of phosphoranimines is also a very convenient method to use to prepare a variety of hybrid block copolymers of the general type **51**:

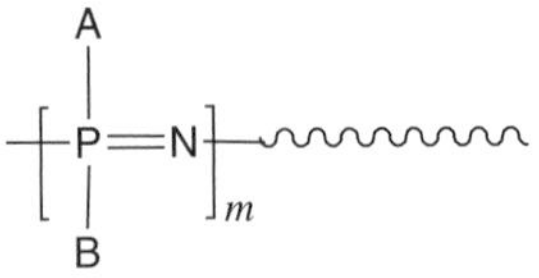

$$\underset{B}{\overset{A}{\left[\!\!\begin{array}{c} | \\ P=N \\ | \end{array}\!\!\right]_m}}\!\!\sim\!\!\sim\!\!\sim\!\!\sim$$

51

Thus, in a recent example, starting with the phosphoranimine **52** as the initiator to form the living polydiclorophosphazene **53**, and adding the poly(styrene)–phosphoranimine **54** as the terminator, resulted in the formation of **55**, which was converted in the stable micelle, forming an amphiphilic polystyrene–phosphazene block copolymer (**56**) by a final macromolecular substitution step (Scheme 10) [49].

Thermal Polycondensation of Phosphoranimines This reaction [50] is a high-temperature process that can be used to prepare two substituent phosphazene homopolymers $[NPAB]_n$ (**48**) or, by means of a co-polycondensation process of two different phosphoranimines, a variety of random copolymers of the type $\{[NPAB]_{1-x}[NPCD]\}_x\}_n$ (**57**, Scheme 11). In all those processes the reactivity of the phosphoranimine decreases depending on the leaving group in the order $X > OAr > OCH_2CF_3 > OR > NR_2$.

In principle, it could also be possible to obtain spirophosphazenes **58** starting from cyclic phosphoranimines (Scheme 12). However, so far the only known example is a poly(phospholene) of formula $[NP(CH_2CMe = CMeCH_2)]_n$ [51]. In practice, the polycondensation of phosphoranimines

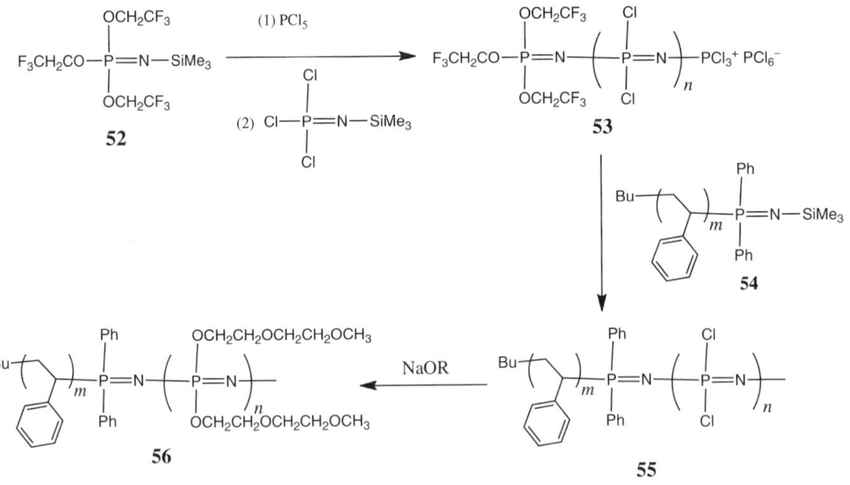

SCHEME 10

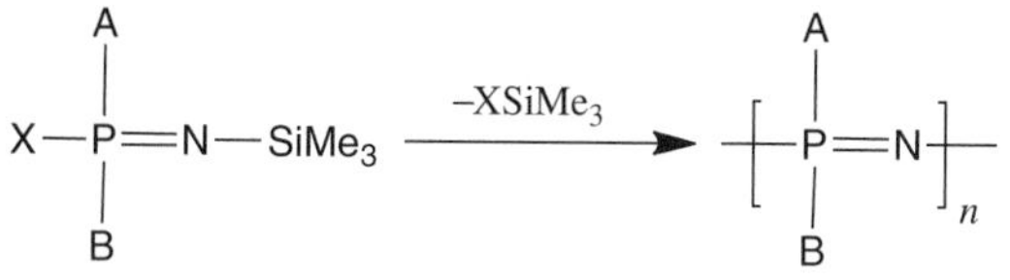

48

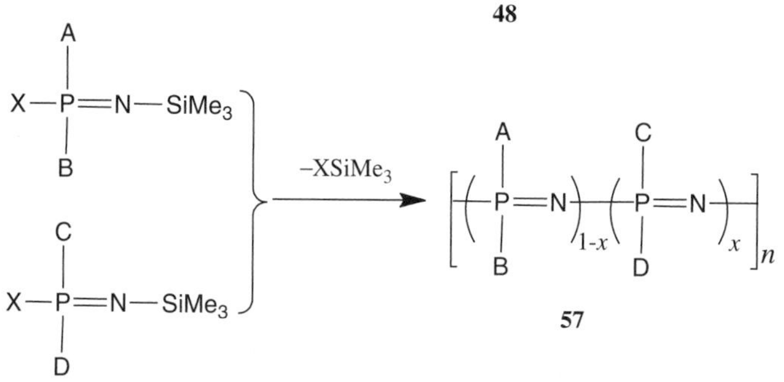

57

SCHEME 11

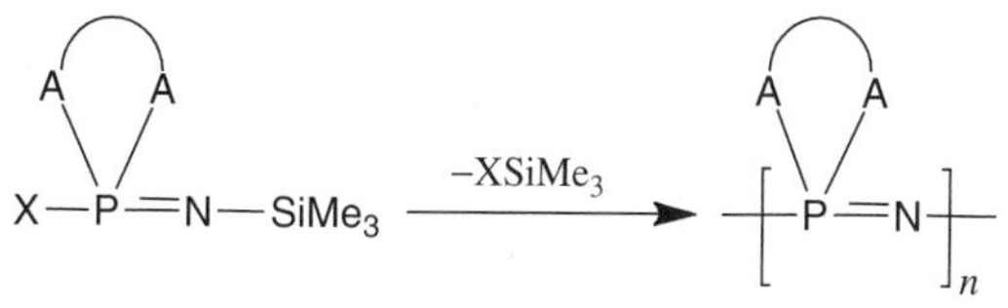

58

SCHEME 12

is used primarily for the preparation of mixed-substituent homoploymers $[NPAB]_n$ (**48**). Recent examples are the polyphosphazene **59**, obtained from $(PhO)_2(Pr^n)P{=}NSi(CH_3)_3$ under dynamic vacuum conditions [52], and (**60**) [53].

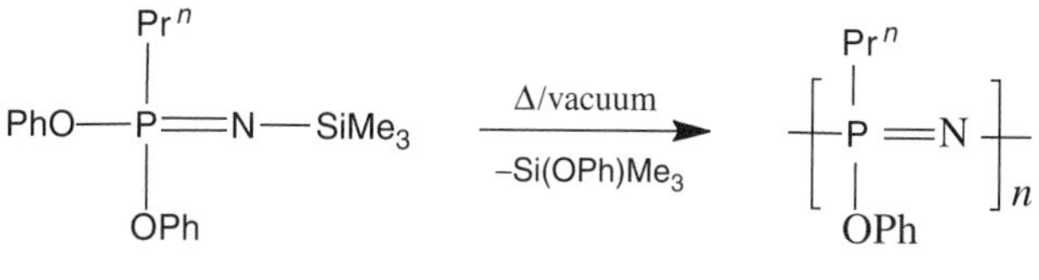

59

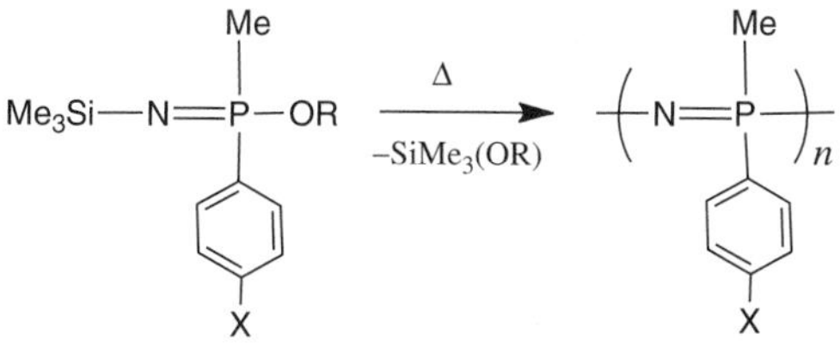

60

R = OCH_2CF_3, C_6H_5
X = Cl, Br, OMe, CF_3

The thermal polycondensation of a phosphoranimines with alcoxide side groups can be catalyzed by Lewis bases: especially, by the fluoride anion. This anionic-initiated polymerization [54] allows a substantial reduction in reaction temperature.

Other interesting variations of the formation of polyphosphazenes from phosphoranimines include the reactions of $(CF_3CH_2O)_2RP=NSiMe_3$ with trifluoroethanol which, except for $R = Pr^n$, Bu^n, or Pr^i, give the corresponding cyclotriphosphazenes $[NP(OCH_2CF_3)(R)]_3$. In the case of $R = Ph$ the reaction results in the formation of the homopolymer $[NP(OCH_2CF_3)Ph]_n$ (**61**) [52]:

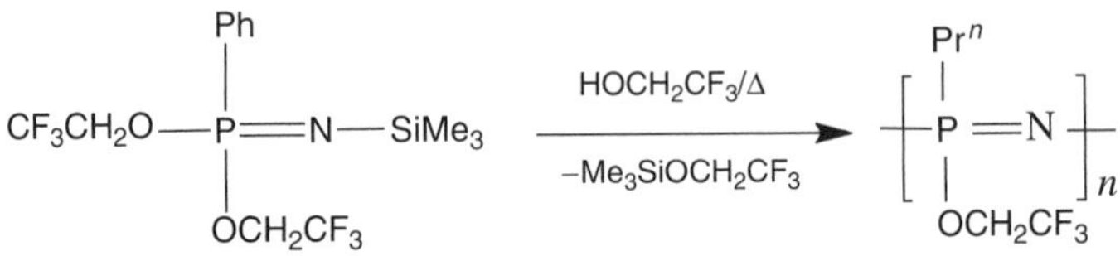

61

A recently discovered route with significant potential advantages over the usual thermal polycondensation is the reaction of the *N*-silylphosphoranimines with $P(OMe)_3$ to give quantitatively the two substituent homopolymers $[NPR^1R^2]_n$ $(R^1 = R^2 = Me; R^1 = Me, R^2 = Ph)$ (**62**) [55].

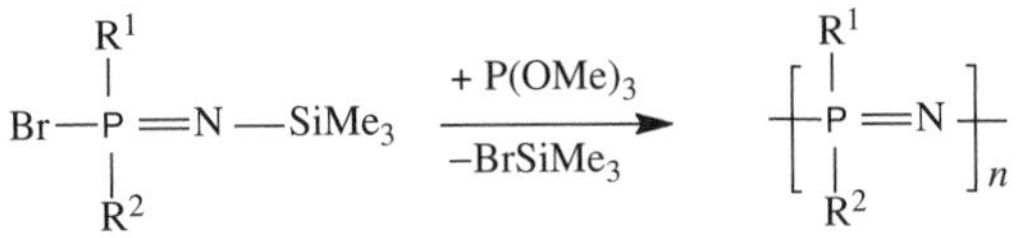

$R^1 = R^2 = Me; R^1 = Me, R^2 = Ph)$

62

In practice, the preparation of polyphosphazenes $[NPAB]_n$ from an appropriate phosphoranimine requires not only successful synthesis of the latter in the pure state, but also the resistance of both A and B substituents to the reaction conditions, which in many cases represents a difficult problem. In fact, the macromolecular substitution route is used more frequently for polyphosphazenes that carry very complex substituents (such as those frequently found in polymers of biological interest) that are liable to undergo collateral chemical transformations.

Macromolecular Chemical Derivatization of Polyphosphazenes

The chemical derivatization of a polyphosphazene can be achieved by a reaction or a sequence of reactions carried out in a precursor polymer to transform pendant organic groups without affecting the main PN chain.

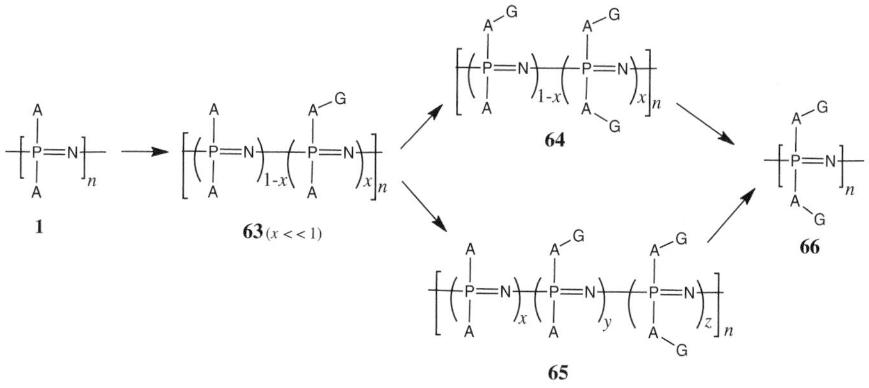

SCHEME 13

Derivatization of Polyphosphazenes on Monodentated Substituents   Al shown in Scheme 13, starting with the homopolymer **1** having the reactive group A during the first stages of the macromolecular derivatization (i.e., when the degree of conversion of A into A–G is small), the derivatized product is very likely a random copolymer of the type **63**, and if the conversion reaches 100%, the homopolymer **66** can be obtained. The progress of the reaction from **63**, however, could give the intermediate copolymers **64** or terpolymers **65**, depending on electronic and steric effects. This is because chemical modification of one of the A groups on a particular $N{=}PA_2$ unit could have a small or a large effect on the reactivity of the other A group, making the next reaction more or less favorable on the same $N{=}PA_2$ than in another unit. Thus, the progress of the derivatization depends on the nature of the A–P–A electronic connections as well as on the steric effects. If the A groups are electronically independent, only the steric effects will be relevant. This might be the case for the phenoxy derivatives, in which only weak electronic interactions between the X groups through the $X{-}C_6H_4{-}O{-}P{-}O{-}C_6H_4{-}X$ bond system have been noted by attaching organic radicals in related cyclic models [56]. Thus, in the functionalization of phenoxy phosphazene polymers, when the steric effects around the already modified A–G groups are sufficiently large, the A groups of the untouched $N{=}PA_2$ units are more accessible than those of the $N{=}PA(A{-}G)$ units, and the reaction proceeds through the copolymers **64**.

On the other hand, to reach the totally transformed homopolymer **66**, the reaction must be very efficient. This means that the total derivatization will be possible only for those reactions that would be almost quantitative when carried out in models with similar nonsupported A groups. Although some of the most established and best known reactions on phenyl groups may approach this limit (sometime using a very large excess of reagents), most frequently the degree of conversion of A into A–G groups is in the range 80 to 90%, and as a result, the polymers **66** would carry a fraction of nonmodified A groups distributed randomly along the chains as chemical imperfections. On the other hand, reactions that are always accompanied by other secondary processes will give even more irregular derivatives. Frequently, however, the target is a polymer carrying only a small fraction of A–G groups (i.e., polymers like **63**, with x not larger than 0.3). In those cases, distribution of the G groups along the chain can be expected to be randomized and globally regular except for those sites that failed to react or underwent a collateral reaction.

Although, in principle, countless reactions are possible, in practice the most studied are those based on the classical reactivity of methyl and phenyl groups, the latter either as simple $-N{=}P{-}C_6H_5$ or $-N{=}P{-}OC_6H_5$ substituents or as their most commonly reactive derivatives $-C_6H_4{-}X$ with X = halogen, NO_2, NH_2, CH_3, OCH_3, CN, CHO, COOR, and other [57]. Although in most of those cases, the transformation of the X group into other derivatives is highly efficient, on some occasions it is difficult to attain complete transformation. A well-known example is the sililation reactions in aryloxyphosphazene derivatives (Scheme 14), which lead to partially sililated derivatives, especially when the R substituents are bulky [58]. Usually, the fraction of lithiated sites that failed to be converted into SiR_3 groups is completed by hydrogen atoms in the isolated materials after the workup, affecting only the regularity of the chains. In most of the reactions used for the chemical modification of phosphazenes, the main chain remains unaffected. Only when using acidic reaction media can the chain be degraded to lower-molecular-weight distributions [59].

The chemical transformation of phosphazenes carrying carboxylate groups into the corresponding carboxylic acid derivatives is very commonly used to obtain a variety of water-soluble phosphazenes with biochemical applications [60]. Other examples of direct derivatization of polyphosphazenes include the formation of polymers **67** [61] and **68** [62] (Chart 7) by the regioselective azo-coupling of a precursor carbazole homopolymer with 4-nitrophenyldiazonium salts and the formation of the randomly branched phosphazene copolymers **69** [63] through a mulitistep derivatization sequence.

Although synthesis of the hybrid poly(methylmethacrylate)-*graft*-poly(phosphazene) (**70**) [64] is also a chemical derivatization, it occurs in a terminal reactive group of the telechelic polymer and not in the lateral groups along the chains. Many other examples of this type of polymer modification are known with mono- and ditelechelic polyphosphazenes that produce di- and triblock

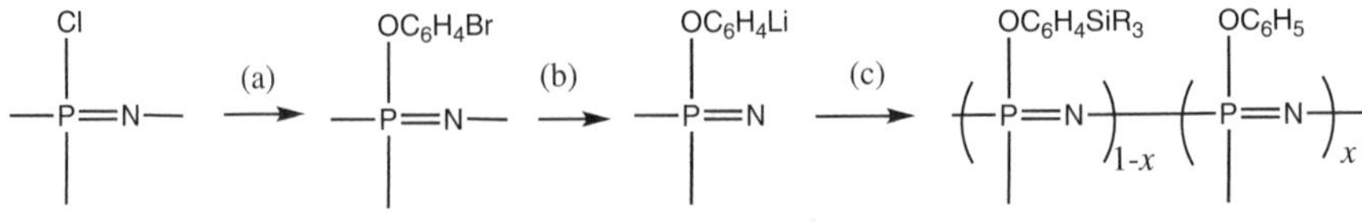

SCHEME 14   (a) OC_6H_4Br; (b) LiBu; (c) SiR_3Cl followed by workup.

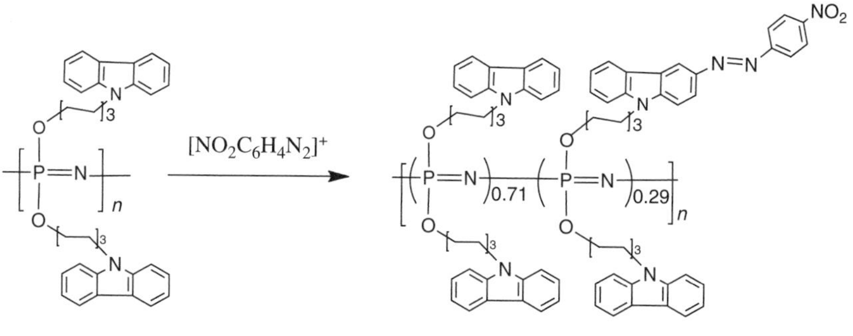

67

68

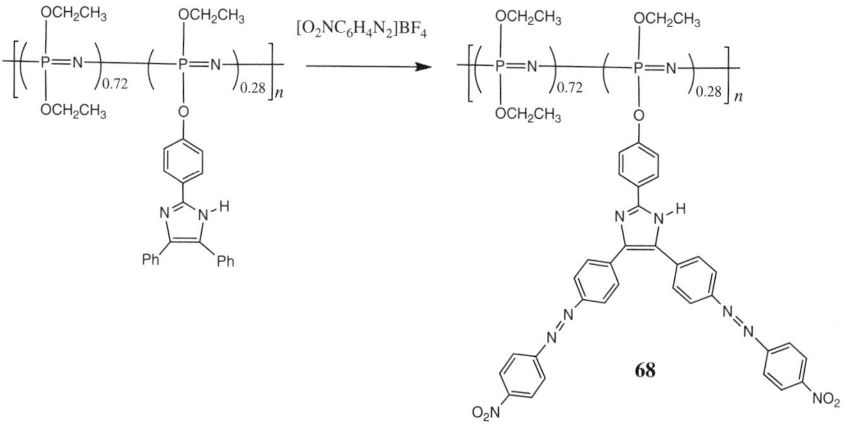

69

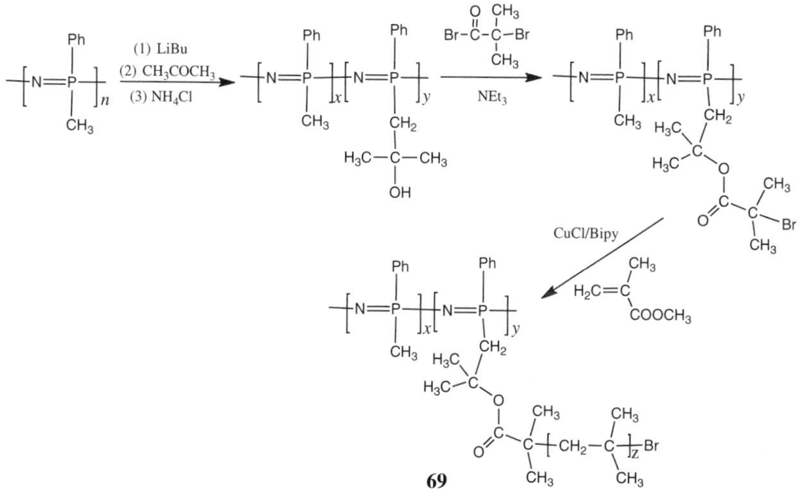

CHART 7

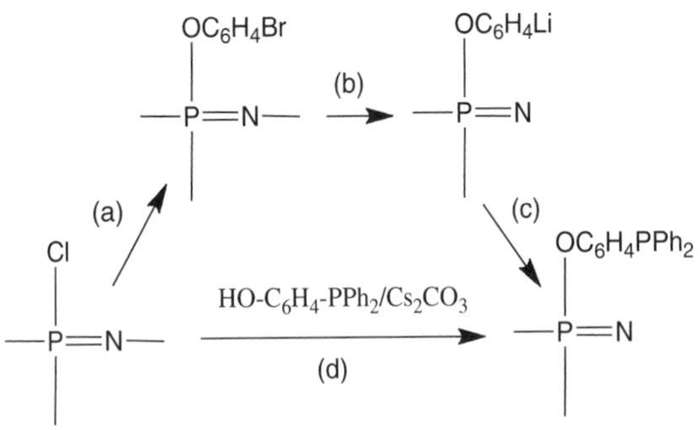

SCHEME 15   (a) OC_6H_4Br; (b) LiBu; (c) PPh_2Cl.

hybrid copolymers [65]. In those cases where direct chemical derivatization of a polyphosphazene leads easily to the formation of products with irregularities or even with totally uncontrollable composition, macromolecular substitution using the already chemically modified reagent (A–G) as the incoming nucleophile may be a useful alternative to obtaining polymers of the types **63** to **66** in Scheme 13. For example, the introduction of a PPh_2 group on aryloxy polyphosphazene can be carried out starting from a chlorophosphazene in three steps, two of them based on the chemical modification of a precursor (Scheme 15) or in only a single step in which the phosphine is introduced directly as a phenol in the presence of Cs_2CO_3 [66].

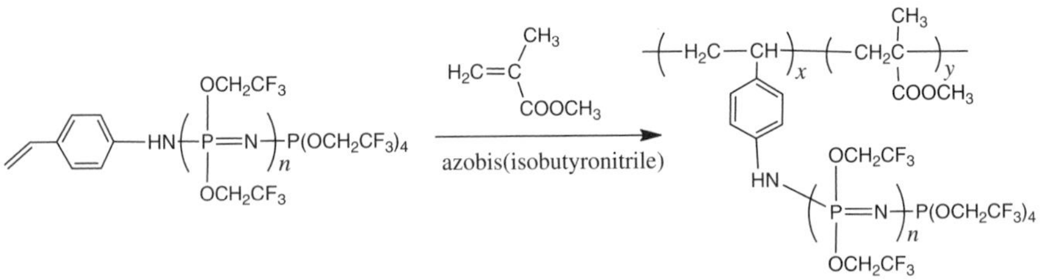

70

Another case is the introduction of terminal $–OC_6H_4–NH_2$ groups (Scheme 16) in which direct substitution may give aryloxyphosphazenes with terminal $–NH_2$ groups in one step (d), avoiding the alternative four-step route, consisting first of a macromolecular substitution with $NaOC_6H_5$, followed by nitration and reduction of the $–NO_2$ group to NH_2. (The nitrophenoxy group can be introduced easily and directly using $HOC_6H_4NO_2$ and K_2CO_3) [6a]. Thus, direct reaction of the intermediate **71** with $OH–C_6H_4–NH_2$ in the presence of Cs_2CO_3 in refluxing THF gives the copolymer **72** [13].

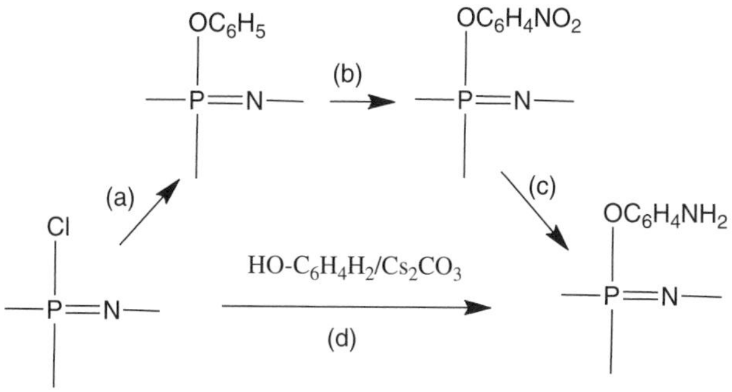

SCHEME 16   (a) $NaOC_6H_5$; (b) HNO_3/H_2SO_4; (c) PtO_2/H_2 or $NaBHS_3$ or Zn/HCl or $Na_2S_2O_4$.

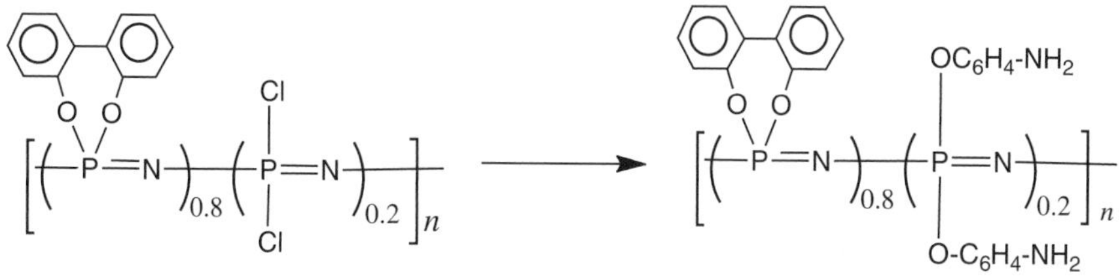

71   **72**

A recent very interesting example is the sulfonation of a $NP–O–C_6H_5$ substituent, which usually results in the formation of irregular polymeric structures and allows little or no control over the position and degree of $–OSO_3$ groups introduced. The newly developed alternative consists of the use of the dimethyl-dipalmithyl diammonium salt of the 4-hydroxybenzene sulfonate ($R_2^1R_2^2N–OC_6H_4–SO_3–NR_2^2R_2^1$) (Scheme 17) [67].

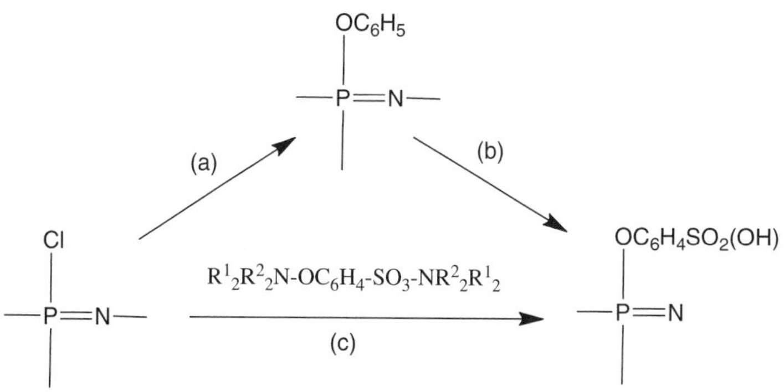

SCHEME 17   (a) OC_6H_5; (b) H_2SO_4 or SO_3.

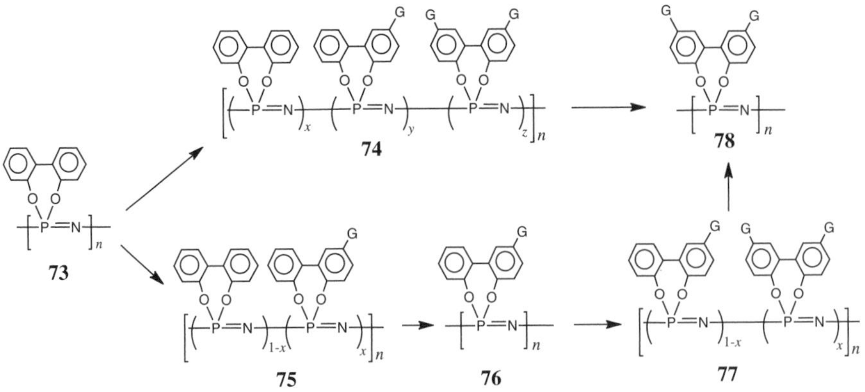

SCHEME 18

Derivatization of Polyphosphazenes on Bidentate Substituents   The derivatization of polyspirophosphazenes is, in principle, different from that of polymers with monodentated substituents, because the chemical connectivities between the possible reaction sites are more intense. For example, a chemo-selective reaction in the 2,2′-dioxybiphenyl phosphazene units of the homopolymer **73** (Scheme 18) would be, at the beginning, equally probable at the C5 and C5′ carbon sites, but after modification of the C5 sites, the reaction could have a different rate at C5′. However, only if the reaction rate at the C5′ position becomes drastically diminished by the substitution at position C5 would the path going through the intermediates **75** to **76** be favored. If the effect on the C5′ site is not sufficiently forceful, the most likely reaction path of the macromolecular modification of **73** would proceed through the terpolymeric intermediates **74**, which might be more or less richer in units with monofunctionalized biphenoxy rings. The formation of **74** would be even more likely under strong reaction conditions (i.e., at elevated temperatures and using high concentrations of reagents). As a consequence, the ratio $x/y/z$ in **74** would be very dependent on the chemical process involved and on the experimental procedure.

So far, the direct derivatization of **73** has been almost entirely unexplored. Thus, it has been observed, for example, that its reaction with H_2SO_4/HNO_3 or with $[IPy_2]BF_4$ gives **74** with $G = NO_2$ or I, respectively [68]. Similar considerations may be applied to the binaphthoxy derivative **79**, but in this case, the reactivity is normally centered on the 6,6′ atoms, which are only weakly chemically connected. Therefore, the formation of randomly functionalized derivatives in both the 6 and 6′ positions are the most likely reaction outcomes. So far, it has been observed that although the free binaphthol can be readily brominated, the direct reaction of **79** with Br_2 is highly inefficient and, in fact, the homopolymer **80** or copolymers **81** had to be obtained by macromolecular chlorine substitution in $[NPCl_2]_n$ with the already brominated binaphthols [44,69].

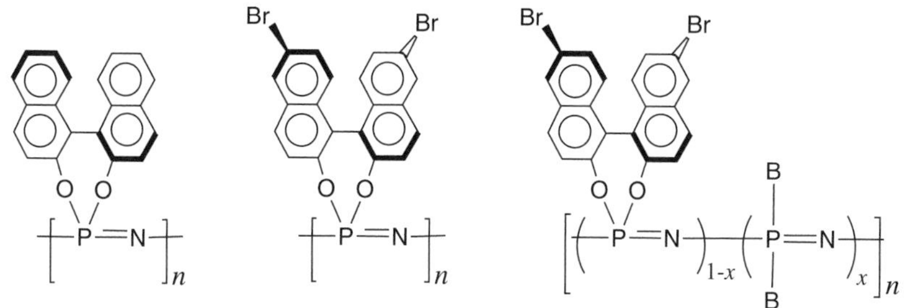

SCHEME 19

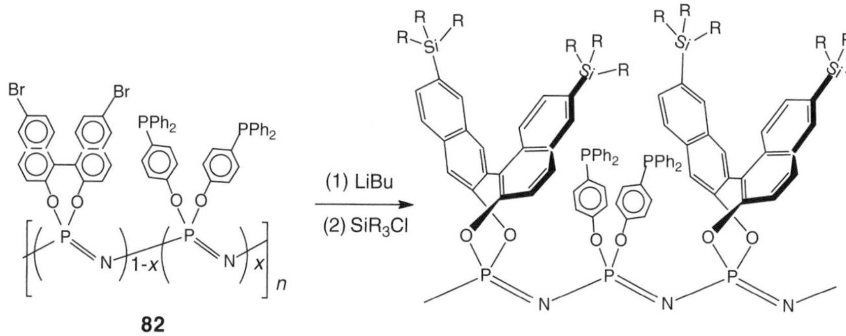

The same occurs with the trimethylsilyl acetylene–substituted binaphthols [69].

The bromobinaphthoxy polymers **81** are useful intermediates in the design of macromolecules with special characteristics. Thus, the direct derivatization of **82** to **83** allowed amplification of the chiral pockets where the phosphine side groups are located (Scheme 19) [70]. In those reactions, the –PPh$_2$ sites may easily become oxidized during the workup, but the resulting –P(O)Ph$_2$ groups may be turned back to free phosphines by treatment with SiHCl$_3$/PPh$_3$ in toluene.

Other Less Common Synthetic Methods for Phosphazene Copolymers

Ring-Opening Polymerization of Cyclic Phosphazenes   This reaction has been thoroughly discussed and reviewed [71]. In the theoretically more general case (Scheme 20), depending on the opening-ring bond breaking, the products could be perfectly regular homopolymers with trimeric units [NPAB–NPCD–NPEF] (or the corresponding homopolymers if the broken bond is 2 or 3), or random copolymers with distributions of trimeric units (if all bonds are liable to be broken). Various homopolymers [NPX$_2$]$_n$ (A = B = C = D = E = G = Cl, F,

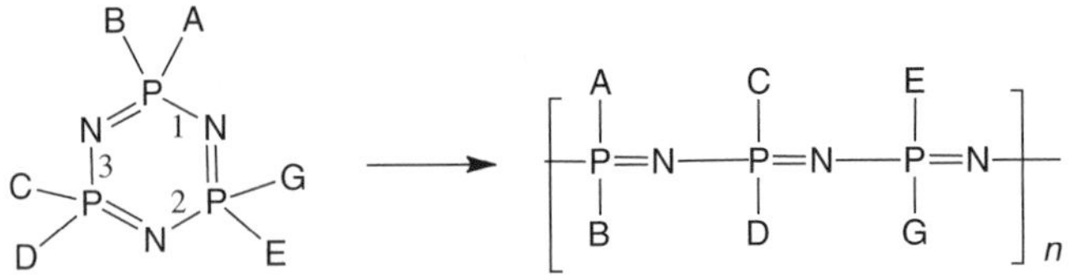

SCHEME 20

or SCN) that are very important as starting materials for other polymers are obtained by this route. However, although in principle any type of mixed-substituent homopolymers or copolymers could be possible, in practice those reactions require high temperature, especially when groups A to G are neither Cl or F, and the thermal stability of those groups may represent important limitations. Also, for the synthesis of a particular designed polymer, the preparation of the cyclic precursor that has the required substitution pattern in a very pure form may not always be straightforward. Therefore, despite the variety of chemical composition that is possible and the high chain regularity that could be achieved by this method, in practice it is applied primarily (but not exclusively) to the preparation of highly chlorine-containing precursors to be used in a subsequent macromolecular substitution leading to phosphazene copolymers.

Polyphosphazenes from Phosphorus Azides The monoazidophosphanes $PR_2(N_3)$ (λ^3-phosphorus azides) are thermally unstable compounds (frequently undergoing detonations) that, among other processes, may decompose, forming cyclic oligomeric phosphazenes $(NPR_2)_n$ and, in some cases, low-molecular-weight polymers [72]. Thus, it has long been known that the halogenoazides $PX_2(N_3)$ formed in reaction of PX_3 with NaN_3 or LiN_3 decompose very easily to give $[NPX_2]_n$ cyclic oligomers [73], and analogous results have been obtained by reacting fluorophosphines PR_2F with $SiMe_3(N_3)$ (trimethyl silyl azide) to generate active $PR_2(N_3)$ intermediates [74]. Therefore, the azidophosphanes have been considered as attractive starting materials for the preparation of phosphazene compounds [75].

Despite the formation of cyclic oligomers and the possibility of explosions, phosphorus azides of the type $PAB(N_3)$ are potentially interesting intermediates for the synthesis of mixed-substituent phosphazene homopolymers $[NPAB]_n$, starting from the corresponding phosphorus compounds PABCl or PAB(OR), as shown in Scheme 21. Thus, the two-substituent phosphazene homopolymer $[PPh(C_6H_5-Me-2)]_n$ was obtained together with variable amounts of cyclic trimer and tetramer in the reaction of the phosphazane $PPh(C_6H_5-Me-2)(OCH_2CF_3)$ with $SiMe_3(N_3)$, which gives first the azide $PPh(C_6H_5-Me-2)(N_3)$ [76] and low-molecular-weight polyphosphazenes have

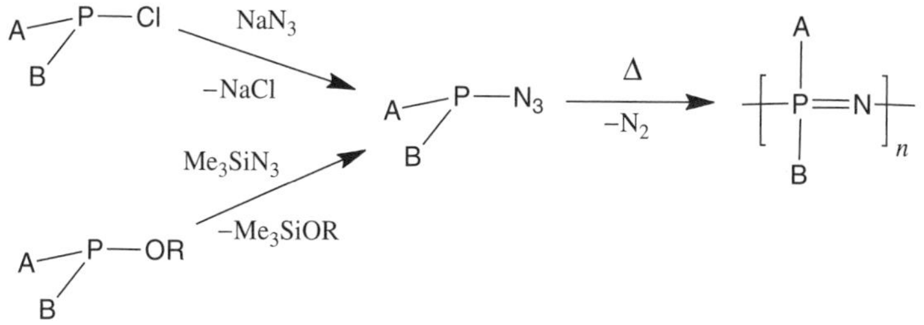

SCHEME 21

been obtained reacting $P(OPh)_{3-n}(CF_3CH_2O)_n$ ($n = 0$, 1, 2) or $P(OPh)_2$ $(CH_3OCH_2CH_2O)$ with azidotrimethylsilane [77].

Studying the reactions with $SiMe_3(N_3)$ of trialkyl or triaryl phosphites $P(OR)_3$ or the mixed derivatives $PR(OR)_2$ and $PR_2(OR)$, it was observed that the resulting polyphosphazenes $[NP(OR)_2]_n$, $[NPR(OR)]_n$, or $[NPR_2]_n$ could be formed by two different pathways. One begins by the elimination of $SiMe_3(OR)$ to give the phosphorus azide, which subsequently loses N_2 to give the phosphazene. In the second route, N_2 is eliminated first to give a trimethyl silyl phosphoranimine of the type $(OR)_{3-n}R_nP=N–SiMe_3$ that can be condensed into the final phosphazene. The trimethyl silyl phosphoranimine route is more favorable in the order $P(OR)_3 > PR(OR)_2 > PR_2(OR)$, and therefore the azide route was considered to be a very interesting method for the synthesis of poly(diarylphosphazene)s [78]. It has also been observed that the polymer $[NP(OCH_2CF_3)Ph]_n$ was formed together with the expected phosphoranimine in the reaction between $P(OCH_2CF_3)_2Ph$ and $SiMe_3(N_3)$, probably through the azide intermediate [79].

In principle, poly(spirophosphazene)s could also be obtained using the corresponding cyclic phosphorus azides (Scheme 22). For example, heating the cyclic phosphite-azide $P[CH_2(6-t-Bu_4-Me-C_6H_2O)_2](N)_3$ gave the corresponding tetraspirocyclophosphazene together with a polymeric fraction that could not be isolated in the pure state [80]. Similarly, the azide **85** could be transformed into soluble low-molecular-weight poly(2,2'-dioxi-1,1'-biphenyl phosphazene) (**73**) (Scheme 23) and an insoluble product consisting of a polymeric matrix of the same cross-linked polymer together with entrapped

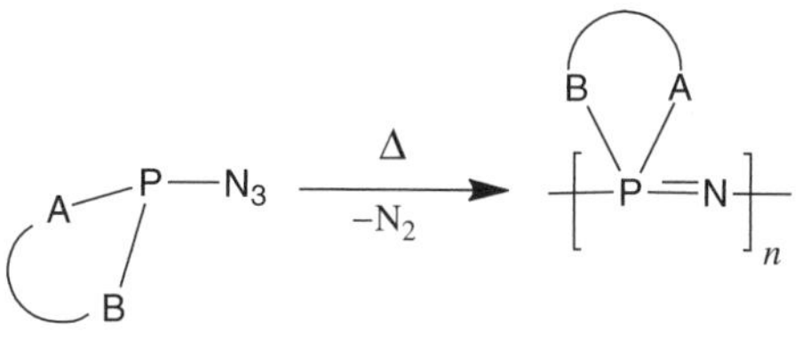

SCHEME 22

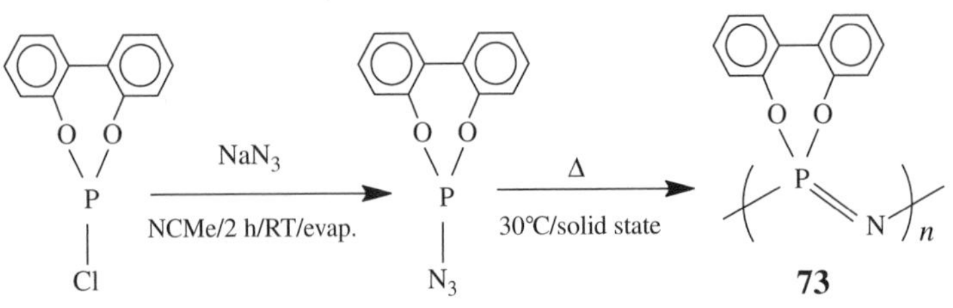

SCHEME 23

and probably interlooped large cyclic oligomers [81]. So far, no attempts have been reported to obtain random copolymers by decomposing mixtures of azide intermediates.

SEQUENTIAL MULTISTEP SYNTHESIS OF POLYPHOSPHAZENES

The knowledge accumulated on the various reactions leading to phosphazene copolymers described in preceding sections allows the design of multistep synthetic strategies that can be used successfully to obtain polyphosphazenes carrying one or various differing substituents on the side groups more or less regularly distributed along the chains. In most cases, the synthesis begins with a macromolecular substitution starting from $[NPCl_2]_n$, which is followed by one or various functionalization steps or further substitution reactions. One example is the synthesis of the amphiphilic block copolymer **84** (Scheme 24) [82].

Although the multistep synthesis may allow the preparation of materials with well-defined chemical composition, it is not always possible to form

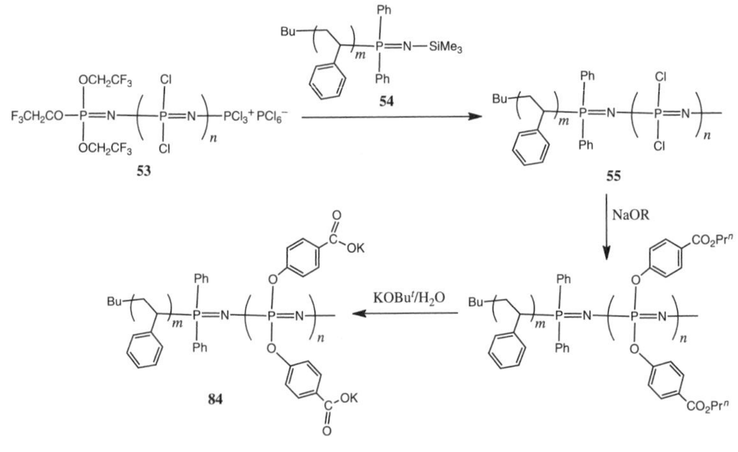

SCHEME 24

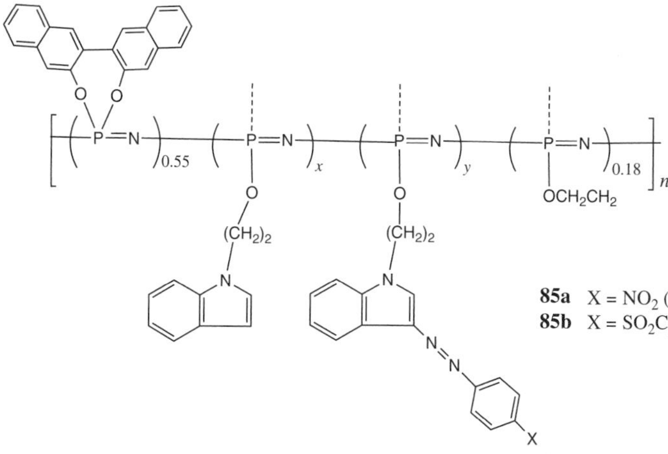

85a X = NO_2 (x = 0.08, y = 0.19)
85b X = $SO_2C_2H_5$ (x = 0.08, y = 0.19)

SCHEME 25

polymers with a very regular structure. This might be the case for the polymers **85**, which were prepared by reacting $[NPCl_2]_n$ first with a binaphthol and the corresponding indole in the presence of triethylamine, followed by the addition of sodium ethanoate to give a fully substituted polymer that was susbsequently reacted (post azo coupling) with 4-nitrophenyl diazonium tetrafluoroborate (for **85a**, Scheme 25) or with 4-ethylsulfonylbenzenediazonum tetrafluoroborate (for **85b**) [83].

REFERENCES

1. Allegra, G., Valdo Meille, S. *Macromolecules*, 2004, 37:3487–3496.
2. Allcock, H.R. *Chemistry and Applications of Polyphosphazenes*. Wiley, Hoboken, NJ, 2003.
3. De Jaeger, R., Gleria, M., eds. *Phosphazenes: A Worldwide Insight*. Nova Science, Hauppauge, NY, 2004, Vol. I, Part 1.
4. Gleria, M., De Jaeger, R. *Top. Curr. Chem.*, 2005, 250:165–251.
5. For a general discussion, see ref. 2, Chap. 7, p. 218.
6. (a) Carriedo, G.A., Fernandez Catuxo, L., García Alonso, F.J., Gómez Elipe, P., González, P.A., Sánchez, G. *J. Appl. Polym. Sci.*, 1996, 59:1879–1885. (b) Carriedo, G.A., García Alonso, F.J., González, P.A. *Macromol. Rapid Commun.*, 1997, 18:371–377.
7. Carriedo, G.A., García Alonso, F.J., Gómez Elipe, P., Fidalgo, J.I., García Alvarez, J.L., Presa Soto, A. *Chem. Eur. J.*, 2003, 9:3833–3836.
8. Maynard, S.J., Sharp, T.R., Haw, J.F. *Macromolecules*, 1991, 24:2794–2799.
9. Mark, J.E., Allcock, H.R., West, R. *Inorganic Polymers*. Prentice-Hall, Englewood Cliffs, NJ, 1962, p. 90.

10. Tur, D.R., Provotorova, N.P., Vinogradova, S.V., Bakhmutov, V.I., Galakhov, M.V., Zhukov, V.P., Dubovik, I.I., Tsvankin, D.Y., Papkov, V.S. *Makromol. Chem.*, 1991, 192:1905–1919.

11. Carriedo, G.A., García Alonso, F.J., Ref. 3, Chap. 8, p. 171.

12. Carriedo, G.A. *J. Chilean Chem. Soc.*, 2007, 52:1190–1195.

13. Carriedo, G.A., Fidalgo Martínez, J.I., García Alonso, F.J., Rodicio González, E., Presa Soto, A. *Eur. J. Inorg. Chem.*, 2002, 1502–1510.

14. Allcock, H.R., Singh, A., Ambrosio, A.M.A., Laredo, W.R. *Biomacromolecules*, 2003, 4:1646–1653.

15. (a) Ref. 2, Chap. 13, p. 421. (b) Ref. 2, Chap. 5, p. 137. (c) Ref. 3, Chap. 2, p. 25.

16. Tarazona, M.P. *Polymer*, 1994, 35:819–829.

17. Allcock, H.R., Moore, G.Y., Cook, W.J. *Macromolecules*, 1974, 7:571–575.

18. Orme, C.J., Klaehn, J.R., Stewart, F.F. *J. Membrane Sci.*, 2004, 238:47–55.

19. Carriedo, G.A., Jiménez, J., Gómez Elipe, P., García Alonso, F.J. *Macromol. Rapid Commun.*, 2001, 22:444–447.

20. (a) Ref. 2, p. 239. (b) Ref. 2, p. 231. (c) Ref. 2, p. 244.

21. (a) Ganapathiappan, S., Krishnamurthy, S.S. *J. Chem. Soc. Dalton*, 1987, 585–590. (b) Allen, C.W. *Chem. Rev.*, 1991, 91:119–135. (c) Ref. 2, p. 246.

22. Allcock, H.R., Clay Kellam, E., III. *Solid State Ionics*, 2003, 156:401–414.

23. Allcock, H.R., Bender, J.D., Chang, Y. *Chem. Mater.*, 2003, 15:473–477.

24. Chen-Yang, Y.W., Hwang, J.J., Huang, A.Y. *Macromolecules*, 2000, 33:1237–1244.

25. Conner, D.A., Welna, D.T., Chang, Y., Allcock, H.R. *Macromolecules*, 2007, 40:322–328.

26. Diefenbach, U., Stromburg, B. *J. Inorg. Organomet. Polym. Mater.*, 2006, 16:295.

27. Paulsdorf, J., Kaskhedikar, N., Burjanadze, M., Obeidi, S., Stolwijk, N.A., Wilmer, D., Wiemhöfer, H.-D. *Chem. Mater.*, 2006, 18:1281–1288.

28. Kaskhedikar, N., Paulsdorf, J., Burjanadze, M., Karatas, Y., Wilmer, D., Roling, B., Wiemhöfer, H.-D. *Solid State Ionics*, 2006, 177:703–707.

29. Zhang, J.X., Qiu, L.Y., Jin, Y., Zhu, K.J. *Macromolecules*, 2006, 39:451–455.

30. Cho, Y.W., An, S.W., Song, S.-C. *Macromol. Chem. Phys.*, 2006, 207:412.

31. Zhang, J.X., Qiu, L.Y., Jin, Y., Zhu, K.J. *React. Funct. Polym.*, 2006, 66:1630–1640.

32. Singh, A., Krogman, N.R., Sethuraman, S., Nair, L.S., Sturgeon, J.L., Brown, P.W., Laurencin, C.T., Allcock, H.R. *Biomacromolecules*, 2006, 7:914–918.

33. Seong, J.-Y., Jun, Y.J., Kim, B.M., Park, Y.M., Sohn, Y.S. *Int. J. Pharm.*, 2006, 314:90–96.

34. Yin, L., Huang, X., Tang, X. *Polym. Degrad. Stabil.*, 2007, 92:795–801.

35. Krogman, N.R., Singh, A., Nair, L.S., Laurencin, C.T., Allcock, H.R. *Biomacromolecules*, 2007, 8:1306–1312.

36. Orme, C.J., Stewart, F.F. *J. Membrane Sci.*, 2005, 253:243–249.

37. Bac, A., Damas, C., Guizard, C. *J. Membrane Sci.*, 2006, 281:548–559.

38. Conconi, M.T., Lora, S., Baiguera, S., Boscolo, E., Folin, M., Scienza, R., Rebuffat, P., Parnigotto, P.P., Nussdorfer, G.G. *J. Biomed. Mater. Res. A*, 2004, 71A:669–674.

39. Kang, G.D., Cheon, S.H., Khang, G., Song, S.-C. *Eur. J. Pharm. Biopharm.*, 2006, 63:340–346.

40. Heyde, M., Moens, M., Van Vaeck, L., Shakesheff, K.M., Davies, M.C., Schacht, E.H. *Biomacromolecules*, 2007, 8:1436–1445.

41. Li, Z., Quin, J.-G., Li, S.-J., Ye, C. *Chin. J. Chem.*, 2003, 21:1395.

42. Carriedo, G.A., Fernández Catuxo, L., García Alonso, F.J., Gómez Elipe, P., González, P.A. *Macromolecules*, 1996, 29:5320–5325.

43. Carriedo, G.A., García Alonso, F.J., Gómez Elipe, P., García Alvarez, J.L., Tarazona, M.P., Rodriguez, M.T., Sáiz, E., Vázquez, J.T., Padrón, J. *Macromolecules*, 2000, 31:3671–3679.

44. Carriedo, G.A., García Alonso, F.J., Presa Soto, A. *Eur. J. Inorg. Chem.*, 2003, 4341–4346.

45. Carriedo, G.A., Presa-Soto, A., Valenzuela, M.L., Tarazona, M.P., Saíz, E., *Macromolecules*, 2008, 41:1881–1885.

46. Carriedo, G.A., Fidalgo, J.I., García Alonso, F.J., Presa Soto, A., Diaz Valenzuela, C., Valenzuela, M.L. *Macromolecules*, 2004, 37:9431–9437.

47. Ainscough, E.W., Brodie, A.M., Chaplin, A.B., Derwahl, A., Harrison, J.A., Otter, C.A. *Inorg. Chem.*, 2007, 46:2575–2583.

48. Ref. 2, Chap. 6, p. 188, and Allcock, H.R., in Ref. 3, Chap. 3, p. 49.

49. Allcock, H.R., Powell, E.S., Chang, Y., Kim, C. *Macromolecules*, 2004, 37:7163–7167.

50. Ref. 2, Chap. 6, p. 206, and Wisian-Neilson, P., in ref. 3, Chap. 5, p. 109.

51. Gruneich, J.A., Wisian-Neilson, P. *Macromolecules*, 1996, 29:5511.

52. Neilson, R.H., Klaehn, J.R. *J. Inorg. Organomet. Chem. Mater.*, 2006, 16:319–326.

53. Neilson, R.H., Wang, B. *J. Inorg. Organomet. Polym. Mater.*, 2007, 17:407–412.

54. Ref. 5, p. 211.

55. Huynh, K., Lough, A.J., Manners, I. *J. Am. Chem. Soc.*, 2006, 128:14002. See also: Huynh, K., Manners, *I. J. Organomet. Chem.*, 2007, 692 2649–2653.

56. Carriedo, G.A., García Alonso, F.J., Gómez Elipe, P., Brillas, E., Labarta, A., Juliá, L. *J. Org. Chem.*, 2004, 69:99–104.

57. Ref. 2, Chap. 8, p. 262, and Wisian-Neilson, P., in ref. 3, Chap. 5, p. 112.

58. (a) Allcock, H.R., Nelson, C.J., Coggio, W.D., Manners, I., Koros, W.J., Walker, D.R.B., Pessan, L.A. *Macromolecules*, 1993, 26:1493–1502. (b) Carriedo, G. A., Valenzuela, M. L., Diaz Valenzuela, C., Ushak, S. *Eur. Polym. J.*, 2008, 44:686–693.

59. Carriedo, G.A., García Alonso, F.J., García Alvarez, J.L., Presa Soto, A., Tarazona, M.P., Laguna, M.T., Marcelo, G., Mendicuti, F., *Macromolecules*, 2008, 41:8483–8490.

60. Andrianov, A.K. In Singh, M., ed., *Vaccine Adjuvants and Delivery Systems*. Hoboken, NJ, Wiley, 2007, Chap. 15, p. 355.

61. Zhang, L., Huang, M., Jiang, Z., Yang, Z., Chen, Z., Gong, Q., Cao, S. *React. Funct. Polym.*, 2006, 66:1404–1410.

62. Pan, Y., Tang, X., Zhu, L., Huang, Y. *Eur. Polym. J.*, 2007, 43:1091–1095.

63. Cambre, J.N., Wisian-Neilson, P. *J. Inorg. Organomet. Polym. Mater.*, 2006, 16:311–318.

64. Allcock, H.R., Powell, E.S., Maher, A.E., Berda, E.B. *Macromolecules*, 2004, 37:5824–5829.
65. Ref. 2, p. 201.
66. Carriedo, G.A., García Alonso, F.J., González, P.A., Gómez Elipe, P. *Polyhedron*, 1999, 18:2853.
67. Andrianov, A.K., Marin, A., Chen, J., Sargent, J., Corbett, N. *Macromolecules*, 2004, 37:4075–4080.
68. Carriedo, G.A., Presa Soto, A., Valenzuela, M.L. *Macromolecules*, 2008, 41:6972–6976.
69. Carriedo, G.A., García Alonso, F.J., García Alvarez, J.L., Soto, A.P. *Inorg. Chim. Acta*, 2005, 358:1850–1856.
70. Carriedo, G.A., García Alonso, F.J., Presa Soto, A. *Macromolecules*, 2006, 39:4704–4709.
71. Ref. 2, Chap. 5, p. 137.
72. Ref. 2, p. 214. See also Böske, J., Niecke, E., Ocando-Marvarez, E., Majoral, J.-P., Bertrand, G. *Inorg. Chem.* 1986, 25:2695–2698.
73. Dillon, K.B., Platt, A.W.G., Waddintong, T.C. *J. Chem. Soc. Dalton*, 1980, 1036–1041, and references therein. Seger, J., Kouril, M., Alberti, M. *Z. Chem.* 1990, 30:215–216.
74. Riesel, L., Friebe, R., Sturm, D. *Z. Anorg. Allgem. Chem.*, 1993, 619:1685–1688.
75. Riesel, L., Friebe, R., Bergemann, A., Sturn, D. *Heteroatom. Chem.*, 1991, 2: 469–472.
76. Franz, U., Nuyken, O., Matyjaszewski, K. *Macromolecules*, 1993, 26:3723–3725.
77. White, M.L., Matyjaszewski, K. *J. Polym. Sci. A*, 1996, 34:277–289.
78. Matyjaszewski, K., Franz, U., Montague, R.A., White, M.L. *Polymer*, 1994, 23:5005–5011.
79. Matyjaszewski, K., Montague, R., Dauth, J., Nuyken, O. *J. Polym. Sci. A.*, 1992, 30:813–818.
80. Kommana, P., Kumaraswamy, S., Kumara Swamy, K.C. *Inorg. Chem. Commun.*, 2003, 6:394–397.
81. Carriedo, G.A., Presa Soto, A., Valenzuela, M.L. *Eur. Polym. J.*, 2008, 44:1577–1582.
82. Chang, Y., Powell, E.S., Allcock, H.R. *J. Polym. Sci. A.*, 2005, 43:2912–2920.
83. Li, Z., Qin, J., Yang, Z., Ye, C. *J. Appl. Polym. Sci.*, 2007, 104:365–371.

20 Supramolecular Structures of Cyclotriphosphazenes

ALEXANDER STEINER

Department of Chemistry, University of Liverpool, Liverpool, United Kingdom

INTRODUCTION

The supramolecular structure is controlled by the molecular shape and the spatial arrangements of functional groups that are able to interact intermolecularly via noncovalent interactions. These include hydrogen bonding, metal coordination, van der Waals forces, π–π interactions, and electrostatic effects [1]. Over recent decades the now well-established concepts of supramolecular self-assembly, molecular recognition, host–guest chemistry, and crystal engineering have emerged and a plethora of molecular building blocks have found applications as sensors, complexing agents, catalysts, gas storage devices, and liquid crystals, to name but a few.

Cyclotriphosphazenes **A** are prominent inorganic heterocycles that consist of repeat units of tetrahedral phosphorus centers carrying two substituents and divalent nitrogen centers featuring lone pairs of electrons. The great stability of the six-membered ring and the ease of introducing a wide variety of substituents at the phosphorus centers facilitates the preparation of numerous derivatives [2]. The most important precursors are chloro- and fluorophosphazenes, which react smoothly with many nucleophiles to form both *homo*- (e.g., $[(RNH)_2PN]_3$, $[(RN_2)_2PN]_3$, $[(RO)_2PN]_3$, $[R_2PN]_3$) and *hetero*-substituted derivatives [3]. The spatial arrangement of side groups X is determined largely by the conformation of the ring (Chart 1). Cyclotriphosphazenes **A** feature a planar ring structure that provides a rigid, D_{3h} symmetrical support for the six P-bonded substituents X, three anchored at either side of the ring plane. On the contrary, higher homologs, such as cyclotetraphosphazenes **B**, display puckered rings of variable conformation. It is the unique combination of a planar ring system

Polyphosphazenes for Biomedical Applications, Edited by Alexander K. Andrianov
Copyright © 2009 John Wiley & Sons, Inc.

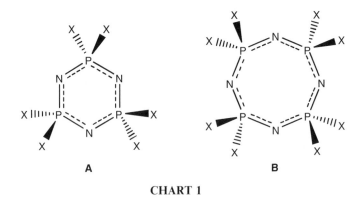

CHART 1

with tetrahedral centers that makes cyclotriphosphazenes interesting building units for supramolecular assemblies.

The ring nitrogen centers of cyclophosphazenes feature lone pairs that are able to engage in intermolecular interactions via hydrogen bonds and metal coordination. The basicity of the nitrogen atom depends largely on the electronic properties of the substituents X at adjacent phosphorus centers. Strong electron donors at P, such as amino groups, enhance the basicity of the ring nitrogen atoms, thus providing strong donor sites for noncovalent interactions. On the contrary, the presence of alkoxy groups reduces the basicity of the ring N centers, often turning them into innocent bystanders without direct involvement in strong and directional intermolecular contacts.

In addition, cyclophosphazenes serve as valuable molecular models for the corresponding polymers [4]. In particular, single-crystal x-ray analyses of cyclophosphazenes provide accurate structure determinations that are not available for polymeric systems. From a supramolecular standpoint, this gives important information regarding the spatial description of noncovalent interactions, substrate binding, and host–guest assemblies. In the following two chapters we discuss how both *form* and *function* of cyclophosphazenes determine their supramolecular structures. *Form* is controlled by the unique geometrical arrangement of substituents around the phosphazene ring, and *function* describes the direct participation of ring nitrogen centres in noncovalent intermolecular bonds.

FORM

As mentioned earlier, cyclotriphosphazenes contain three tetrahedral centers in a planar six-membered ring, which offers a unique array of three binding sites for substituents at either ring face. This arrangement supports distinctly shaped molecular building blocks which, in turn, generate unique supramolecular structures. In this chapter we discuss systems in which the phosphazene ring serves purely as a molecular support, with no significant involvement in supramolecular interactions. Cyclophosphazenes equipped with a rigid sphere of substituents form a series of clathrates, whereas those carrying flexible ligands show dendritic structures that assemble into distinct supramolecular architectures such as columnar and calamatic mesophases.

Clathrates

Clathrates are molecular inclusion compounds in which guest molecules are confined within the lattice of host molecules. The guests can be either trapped in closed cavities or accommodated in channels or between layers [5]. Cyclotriphosphazenes generate a wide variety of inclusion compounds (Chart 2). The most prominent examples are spirocyclic systems of type **C**, which assemble in the presence of suitable guests to form open-channel structures that accommodate a variety of guest molecules. Other examples include basket-shaped molecules **D**, as well as a range of other systems that facilitate clathrate formation.

The first trispirocyclic phosphazenes of type **C** were reported by Allcock in the 1960s, and it was soon realized that these compounds form inclusion compounds in the presence of solvent molecules [6]. In recent years

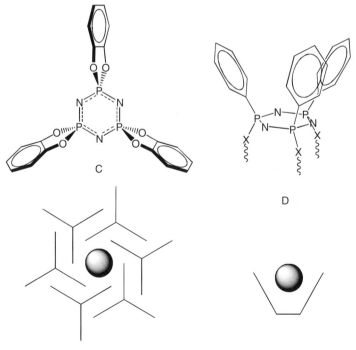

CHART 2

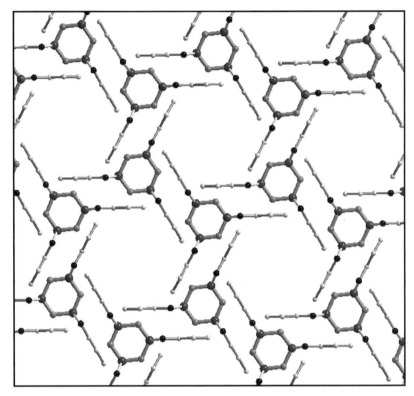

FIGURE 1   Open-channel structure of **1** viewed along the crystallographic z-axis.

these compounds have regained attention, due mainly to the current quest for porous materials that find applications in gas-storage devices and fuel cells [7]. The paddlewheel-shaped tris(o-phenylenedioxy)cyclotriphosphazene **1** exhibits a hexagonal crystal structure in the presence of guest molecules. It consists of alternating layers of interlocking paddlewheels which generate channels along the crystallographic z-axis exhibiting diameters of 4.5 to 5.0 Å (Fig. 1) [8]. A monoclinic guest-free structure of **1** is obtained either by vacuum sublimation or by heating the clathrate above 170°C, which leads to the removal of volatile guest molecules and an exothermic collapse of the hexagonal open-channel structure. The resulting monoclinic phase consists of pure **1** and does not exhibit any channels or larger voids. The effective packing of guest-free **1** is achieved by a distortion of the paddlewheel structure from threefold symmetry. Two phenylene paddles are bent backward, enabling two molecules to pair up via their opened sides (see also Fig. 2) [9]. However, upon exposure to guest molecules the monoclinic crystal disintegrates and the hexagonal clathrate structure is regained. The guest-free porous hexagonal structure can be obtained by removing benzene (or other volatile guests) below the exothermic transition at 75°C under reduced pressure. The channels of the

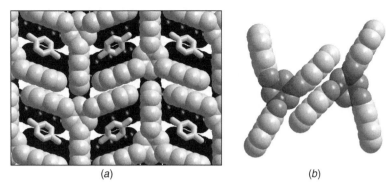

FIGURE 2   (a) Orthorhombic clathrate structure of **2** hosting *p*-xylene molecules; (b) the paired-up assembly of two molecules in the guest-free monoclinic form of **2**.

porous solid were explored using hyperpolarized ^{129}Xe nuclear magnetic resonance (NMR) spectroscopy. This showed that the gas atoms diffuse along one-dimensional channels and interact with the π-electron-rich environment of the aromatic channel walls [10].

Several inclusion compounds of **1** have been prepared that feature the hexagonal host structure and accommodate a variety of guests, ranging from gaseous compounds to linear polymers. Gas-sorption studies revealed that **1** exhibits high storage capacities for CO_2 and CH_4, but poorly absorbs N_2, O_2, and H_2 [11]. This makes these clathrates interesting candidates for gas-storage and purification devices. A combination of x-ray diffraction and solid-state NMR techniques revealed that methane forms significant CH $\cdots$ π interactions with the phenylene groups of **1**. Iodine readily diffuses into the channels along the crystallographic *z*-axis. The diffusion process is visible, since the initially colorless rodlike crystals slowly turn purple from both ends to the center. The resulting iodine-saturated crystals showed high thermal stability, with no significant weight loss of I_2 in vacuum at room temperature. The chains of iodine molecules accommodated in the channels show a conductivity similar to that of pure iodine, which is a two-dimensional semiconductor and one of the best characterized n-type molecular donors for the formation of $n \rightarrow \sigma^*$ charge-transfer complexes [12]. Linear polymers such as polyethylene oxide and polyethylene have also been included in the channels of **1** [13]. The resulting composite materials have very high melting points that greatly exceed that of pure **1** as well as that of the neat polymer. The inclusion of long-chain polymers is favored over that of monomers and short oligomers, thus promising interesting applications for the purification and separation of polymers [14]. In addition, clathrates containing polyconjugated molecules have been reported [15]. It is also possible to carry out polymerization reactions within the channels. The polymerization of vinylic and acrylic monomers is initiated by irradiation with γ-rays [16]. In the case of 1,3-dienes, the clathrate system provides an environment that facilitates the exclusive formation of 1,4-*trans*-polydienes [17].

In addition to the tris(*o*-phenylenedioxy) derivative **1**, a number of other clathrates of trispirocyclic cyclotriphosphazenes have been described. Considerably wider channels are present in the corresponding tris(2,3-naphthalenedioxy) derivative **2**, which crystallizes as a hexagonal clathrate in the presence of suitable guests, such as benzene. Here the channel diameter is 9 to 10 Å, owing to the extended reach of the naphthalene groups, which interact via π-stacking interactions with neighboring molecules [18]. In the presence of *p*-xylene and *p*-chlorotoluene, **2** forms an orthorhombic clathrate that features enclosed cavities rather than open channels (Fig. 2) [19]. Due to the defined shape of the cavities, clathrates of **2** can very effectively separate *para*-isomers from a mixture of disubstituted benzenes. The molecular structure of **2** in the orthorhombic clathrate exhibits a noticeable deviation from the threefold symmetry. The molecules of the guest-free monoclinic structure of **2** show an even greater distortion, which enables them to pair up along their open sides and allows effective crystal packing (Fig. 2) [20].

The aromatic paddles provide much scope for modification. For example, the tris(3,6-dimethyl phenylenedioxy) derivative, which is equipped with additional methyl groups in axial positions, forms cagelike clathrates. However, these are considerably less stable than those formed by **1** [21] (Chart 3). The tris(9,10-phenanthrenedioxy)cyclotriphosphazene **3** forms inclusion adducts with a range of cyclic guest molecules. In contrast to **1**, the guest greatly controls the structure of the host lattice of **3** [22]. Other trispirocyclic systems

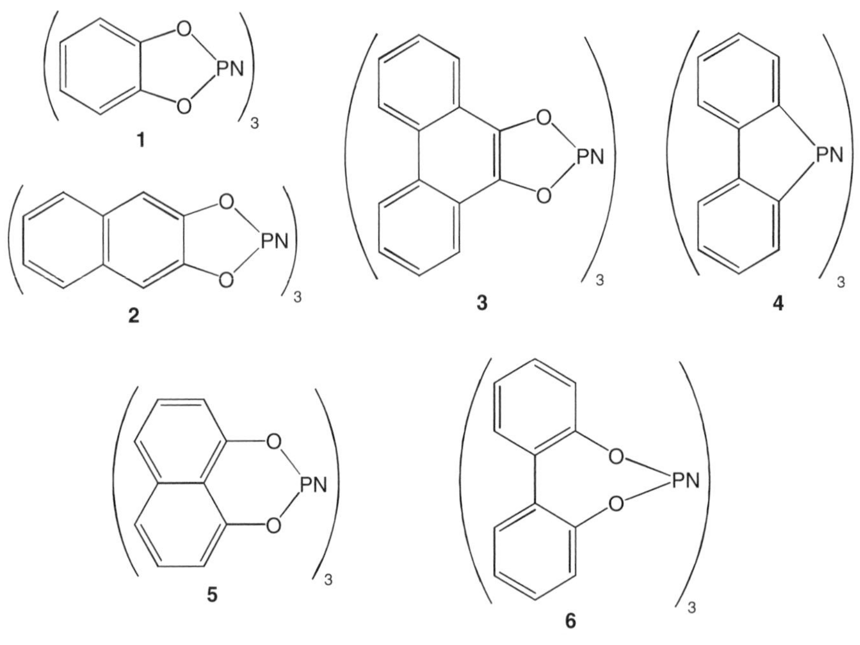

CHART 3

with highly rigid five-membered exo-rings that furnish a threefold molecular symmetry include the tris(2,2'-biphenyl) (**4**) and related (3,3'-bithienyl) derivatives [23]. Spirocyclic systems with larger exo-rings include the tris(1,8-naphthalenedioxy) and the tris(2,2'-dioxy-1,1'-biphenyl) compounds **5** and **6**, respectively [24]. Although **5** forms a clathrate with *p*-xylene, the nonplanar conformation of six-membered PO_2C_3 rings does not support the D_{3h} symmetrical paddlewheel arrangement of **1** [25]. The 2,2'-dioxybiphenyl derivative **6** does not form inclusion compounds. It also lacks the familiar paddlewheel arrangement, but adopts a D_3 propeller shape, since its seven-membered PO_2C_4 rings are highly twisted [26]. Moreover, the twisted conformation of the biphenyl moieties enables effective crystal packing involving $CH \cdots \pi$ and $\pi \cdots \pi$ interactions [27]. Correspondingly, the related but enantiomerically stable tris-(2,2'-dioxy-1,1'-binaphthyl) derivative does not form clathrates [28].

Arrays of three aryl groups that are bonded directly to P centers at one face of the cyclotriphosphazene ring furnish a basket-type arrangement. These systems resemble calixarenes [29] the way they feature a cavity surrounded by π-electron-rich walls. These cavities have the potential to host a range of guest molecules. However, in contrast to calixarenes, the aromatic groups show some conformational flexibility due to the free rotation of the P–C(aryl) bonds. When hexaphenyl cyclotriphosphazene **7** is crystallized from tetrahydrofuran (THF), it forms a 1:1 complex. The three phenyl groups that accommodate the thf molecule are well aligned to form a basket-type arrangement, while the phenyl groups at the other side of the ring rotate into a position that allows them to interlock with the corresponding set of phenyl groups of the neighboring molecule (Fig. 3) [30]. It should be noted that solvent-free crystals are obtained in the absence of suitable guest molecules [31].

Recently, Wisian-Neilson and co-workers developed a route toward cyclophosphazenes (**8**) that feature a set of three phenyl groups at one side of the phosphazene ring plane, while the other side carries alkyl or functionalized alkyl groups [32] (Chart 4). Although the supramolecular chemistry of these compounds has not been studied in detail, they offer interesting applications for self-assembly, bioactive receptors, and chemical sensors.

Other examples of cyclophosphazene clathrates include the hexa(1-aziridinyl) derivative **9** (Chart 5), which forms clathrates with benzene and tetrachloromethane, $9 \cdot 1/2\ C_6H_6$ and $9 \cdot 3\ CCl_4$, respectively. Both compounds crystallize in rhombohedral space groups; the phosphazene rings are located on threefold rotation axes (Fig. 4). The solvent molecules interact with the markedly pyramidal N centers of the azirinidyl substituents. The benzene molecule in the layered structure of $9 \cdot 1/2\ C_6H_6$ is encircled by six molecules of **9** via $CH \cdots N$(azirinidyl) interactions. In addition, there are weak $CH \cdots N$ contacts between azirinidyl groups and the nitrogen atoms of the phosphazene rings [33]. In contrast, $9 \cdot 3\ CCl_4$ forms a three-dimensional network in which the phosphazenes are arranged in stacks along the *z*-axis and surrounded by CCl_4 molecules. Each phosphazene molecule interacts with six CCl_4 molecules

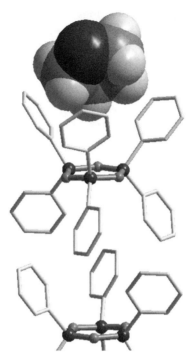

FIGURE 3   Crystal structure of **7** THF, emphasizing the host–guest interaction with THF and the effective interlocking of phenyl groups with those of a neighboring molecule.

via rather short N(azirinidyl)···Cl contacts of 3.14 Å [34]. Solvent-free modifications of **9** were crystallized from *m*-xylene and carbon disulfide, respectively. In the presence of *m*-xylene, an orthorhombic polymorph, and from CS$_2$ solution, a monoclinic polymorph, were obtained [35].

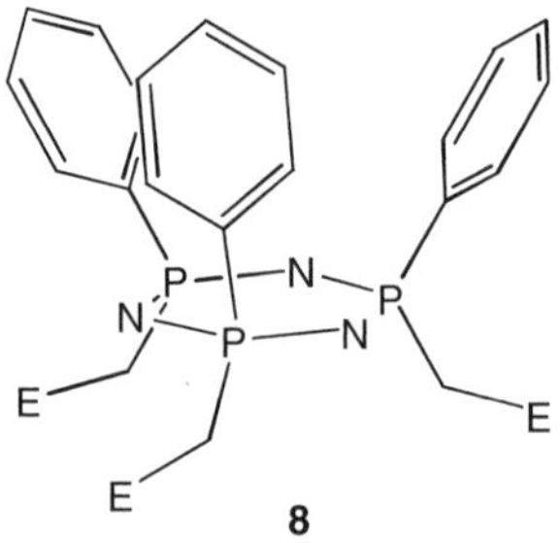

8

E = Me, SR, SO$_2$R, Cl, Br, I

CHART 4

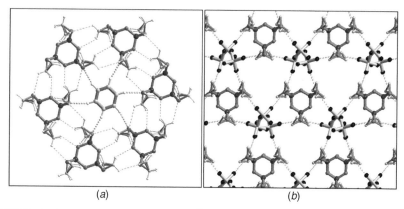

FIGURE 4   Crystal structures of (a) $9 \cdot \frac{1}{2}°C_6H_6$ and (b) $9 \cdot 3CCl_4$ viewed along the crystallographic z-axes.

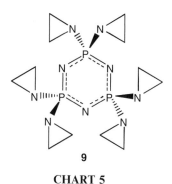

9

CHART 5

Another interesting clathrate structure is formed by the salt $[P_3N_3(DMAP)_6]$ Cl_6 (**10**) (Chart 6), which contains the hexacation $[P_3N_3(DMAP)_6]^{6+}$ {DMAP = 4-(dimethylamino)pyridine} and hosts 19 molecules of chloroform per formula unit (Fig. 5) [36]. Two chloride ions are accommodated by the hexacation in basket-type arrangements of three DMAP ligands at either side of the phosphazene ring, while the *ortho*-H atoms of the DMAP ligands provide three tetradentate binding sites for chloride ions at equatorial positions around the ring. The hexacations are linked by the sixth chloride ion via the methyl groups of two DMAP ligands, resulting in a networked structure. The solvent-accessible volume of **1** amounts to 71%. It is occupied by 19 molecules of chloroform per formula unit, 17 of which coordinate to chloride ions.

Dendrites

The possibility of anchoring six substituents around the cyclotriphosphazene ring, three at either side of the ring plane, provides a unique platform for dendritic molecules, promising interesting applications in supramolecular

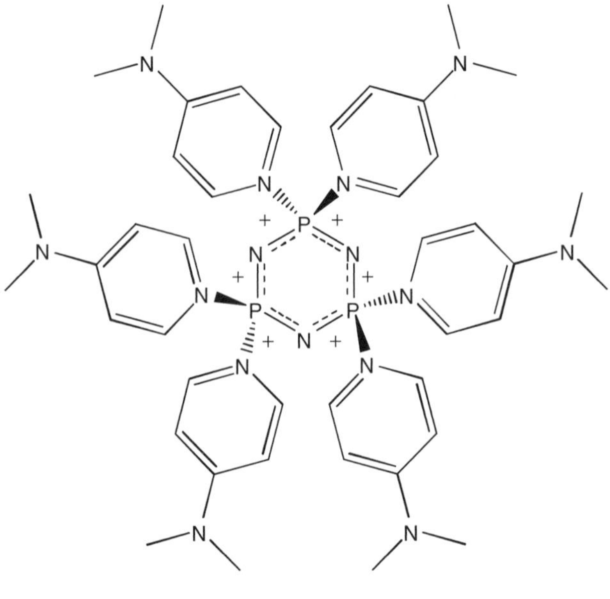

$[P_3N_3(DMAP)_6]^{6+}$

CHART 6

assemblies. A variety of systems have been investigated in view of liquid-crystal behavior, hydrogels, and self-assembled nanostructures. Liquid crystals play a major role in display technology [37]. As a consequence, there is growing interest in specifically functionalized mesogens [38]. Phosphazene-based liquid crystals are promising candidates, owing to their synthetic versatility, chemical

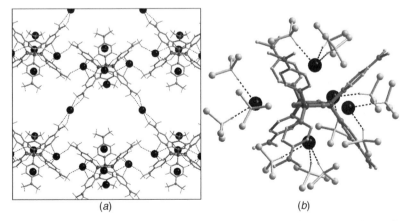

(a)   *(b)*

FIGURE 5   (a) Networked structure of **10** · 19 CHCl$_3$ (chloroform molecules omitted); (b) effective complexation of $\{[P_3N_3(DMAP)_6]Cl_5\}^+$ units by 13 chloroform molecules.

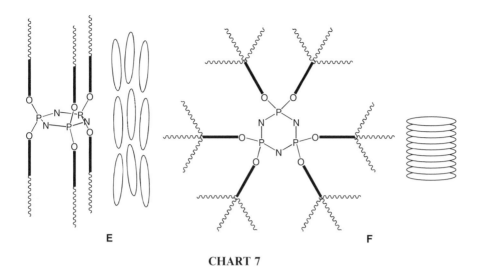

CHART 7

stability, low flammability, and optical transparency from near-infrared to about 210 to 190 nm in the ultraviolet spectrum [39]. Substituents consisting of linear mesogens tend to bundle together at either side of the phosphazene ring, resulting in rodlike (calamitic) arrangements of type **E** (Chart 7). Smectic phases have been obtained using linear Schiff's base and biphenoxy-based substituents [40]. In contrast, phosphazenes carrying mesogenic substituents that branch out adopt disklike (discotic) conformations **F**, which form columnar stacks. The resulting columnar mesophases have been shown to be stable at subambient temperatures [41].

Cylindrical phosphazenes that carry intermolecular binding sites at the terminal positions of substituents are able to self-assemble into rods in the presence of complementary linkers [42]. This can be achieved using the approaches depicted in Scheme 1. In **G**, phosphazenes co-crystallize with compatible difunctional molecules. In one example the phosphazene carries 4-pyridyl end groups that interact with linear dicarboxylic acids via hydrogen bonding [43]. Other examples include phosphazenes equipped with terminal 4-iodo phenyl groups that make N···I contacts to 4,4′-bipyridyl- or, vice versa, pyridyl-substituted phosphazenes that are bridged by 1,4-diiodotetrafluoro-benzene molecules [44]. An alternative route is the co-crystallization of two phosphazene derivatives with compatible functions X and Y (**H**). This has been accomplished with X = 4-pyridyl and Y = COOH groups [45]. Another way to link terminal sites uses metal ions, which results in the formation of coordination polymers **J**. Here silver ions have proven to be versatile linkers that coordinate to phosphazene-bound 4-pyridyl groups. In the presence of nitrate ions the supramolecular rods form a densely packed hexagonal structure, while alkylsufonates (C_{12}, C_{14}, C_{16}, C_{18}) furnish a lamellar structure in which the alkyl chains of neighboring rods interdigitate [46].

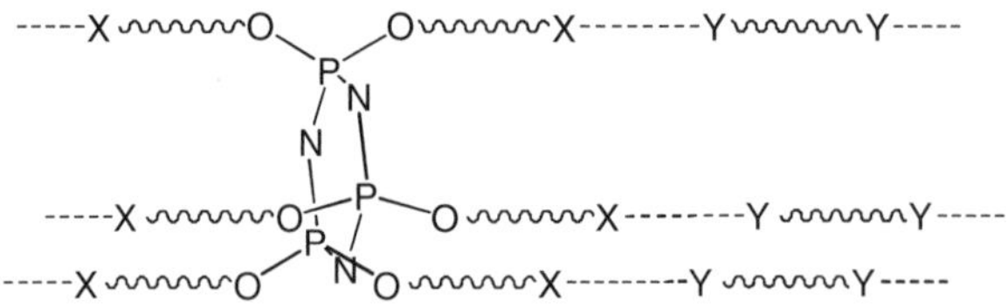

G

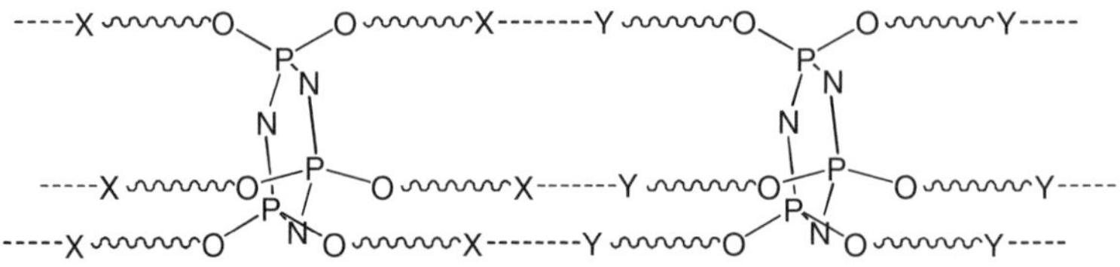

H

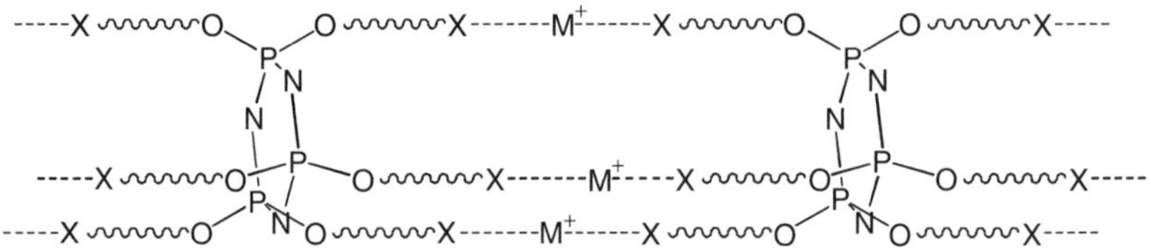

J

SCHEME 1 Rodlike assemblies of terminally functionalized phosphazenes with compatible linkers.

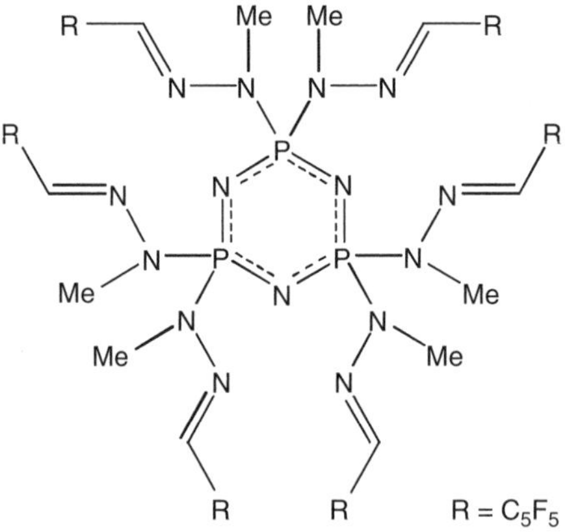

11

CHART 8

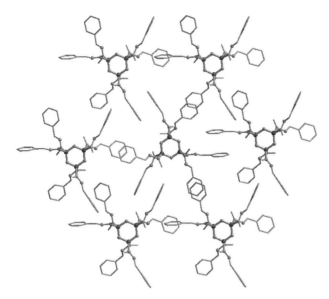

FIGURE 6   Crystal structure of **11**.

It should be noted that the choice of substituent and the nature of inter-molecular interaction have a profound influence on the supramolecular structure. For example, the hydrazone derivative **11** (Chart 8), which features six terminal pentafluorophenyl groups, does not assemble into a rodlike structure but aggregates into a three-dimensional network via π-stacking interactions of aromatic rings (Fig. 6) [47].

A series of cyclophosphazenes have been prepared that carry anionic groups such as carboxlyates [48], sulfonates [49], and phosphonates [50] at the end of their side chains (typically, grafted to the *para* position of phenoxy and anilino substituents). These compounds promise interesting applications for hydrogels, membranes, proton conductors, and composite materials. Cyclophosphazenes were also equipped with terminal quaternary pyridinium and ammonium centers in order to generate polycations. In the presence of weakly binding anions such as $(CF_3SO_3)_2N^-$, these materials exist as viscous ionic liquids [51]. In addition, cyclophosphazenes have been used as core units for hyperbranched dendrimers. A range of macromolecular systems have been prepared that contain specific functionalities at the periphery, encapsulate various substrates, and form well-defined arrays with nanoparticles [52].

FUNCTION

In this chapter we discuss supramolecular structures in which the phosphazene ring participates directly in intermolecular contacts. The nitrogen centers carry

TABLE 1 pK'_a Values of Cyclophosphazenes $X_6P_3N_3$ in Nitrobenzene at 25°C

X	pK'_{a1}	pK'_{a2}
Cl	< -6.0	
OEt	0.2	
Ph	1.5	
Et	6.4	
NHEt	8.2	-1.3

lone pairs; thus, they are potential binding sites for Lewis acids such as metal ions and hydrogen-bonding donor molecules. The basicity of the ring nitrogen atoms is strongly correlated to the electronic nature of the substituents at the adjacent phosphorus centers. Table 1 lists pK'_a values for a series of cyclophosphazenes [53]. The chloro compound $Cl_6P_3N_3$ is a very weak base. It interacts only with strong Lewis acids or with metal ions in the presence of very weakly coordinating anions [54]. In contrasts, aminophosphazenes such as $(RNH)_6P_3N_3$ engage readily in hydrogen bonding and metal coordination with various substrates over a wide pH range.

Hydrogen Bonding

Hydrogen bonds between electronegative centers constitute strong intermolecular interactions, which often have a profound effect on the supramolecular assembly [55]. The concept of assembling solids supramolecularly via hydrogen bonds has been used extensively in the field of crystal engineering. Polycarboxylic acids and ureas, as well as nitrogen heterocycles, are one of the most frequently used building blocks. Their molecular rigidity enables the formation of highly directional interactions, which allows the rational design of supramolecular architectures [56]. Hydrogen bonding in cyclophosphazenes has been studied to a lesser extent, although the unique arrangement of potential hydrogen-bonding sites around the phosphazene ring promises interesting applications. Much of the research has focused on polyaminophosphazenes, in particular $(RNH)_6P_3N_3$, which are very robust compounds that resist extreme conditions. This includes exposure to strongly basic and acidic media as well as hydrothermal treatment. From a supramolecular point of view, they have plenty to offer. The ring N centers can act as hydrogen-bond acceptor sites and the exocyclic NH groups as hydrogen-bond donor sites. It should be noted that polyamine-based compounds are of great interest in many areas, such as anion detection and separation, as well as supramolecular assembled nanosized materials [57].

Cyclophosphazenes carrying amino substituents NH_2 and NHR, respectively, are able to form intermolecular NH$\cdots$N interactions between NH groups and ring nitrogen atoms. The crystal structures of $(RNH)_6P_3N_3$ (**K**) show an unprecedented variety of supramolecular network topologies, which is caused by only subtle alterations in the size or shape of R [58]. The amphiphilic

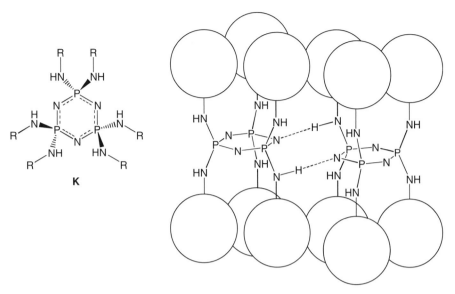

SCHEME 2   Lipid-bilayer arrangement of hydrogen-bonded phosphazenes $(RNH)_6P_3N_3$.

molecules feature an equatorial belt of three nitrogen atoms and six NH groups sandwiched between hemispheres of lipophilic substituents R. Hence, phosphazene molecules tend to interact in a "side-on" fashion rather than "face to face" resulting in some type of lipid-bilayer formation, as illustrated in Scheme 2. Free rotation of the exocyclic P–N bond enables variable directionalities of NH bonds, which results in a variety of H-bridging modes. Ten principal types of intermolecular H-bridges were observed, including single (**a**), double (**b, c, d**), triple (**e, f ,g, h**), quadruple (**j**), and sextuple (**k**) H-bridges (Scheme 3).

The crystal structures of a selection of phosphazenes $(RNH)_6P_3N_3$ are displayed in Fig. 7. The *tert*-butyl derivative **12** does not exhibit intermolecular NH$\cdots$N interactions in the crystal structure ($Z' = 2$), due to the steric bulk of the substituents. However, the cyclohexyl compound **13** forms a centrosymmetric dimer linked by a **c**′-bridge. It is interesting to note that this arrangement crystallizes in two different polymorphic forms, a triclinic and a monoclinic modification, which differ only in the way the dimers are packed in the crystal. The *iso*-propyl derivative **14** forms a discrete hexameric ring structure of S_6 symmetry, which is held together by six **e**-bridges. The benzyl derivative **15** forms zigzag chains via **c** bridges, which exist as triclinic ($Z' = 4$) and monoclinic ($Z' = 2$) polymorphs. The *iso*-butyl derivative **16** displays a chain structure that consists of repeat units containing five crystallographically unique molecules ($Z' = 5$) that are connected by an alternating **g–j–f–h–g** bridge pattern. The *n*-propyl compound **17** exhibits a double-chain type of arrangement containing two crystallographically independent phosphazene molecules ($Z' = 2$): One interacts with three and the other with two neighboring molecules via **g**-bridges. The allyl

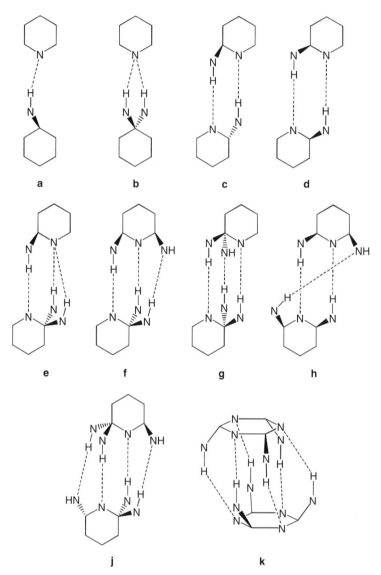

SCHEME 3   Hydrogen-bonding modes observed in $(RNH)_6P_3N_3$ and $(H_2N)_6P_3N_3$.

derivative **18** forms honeycomb sheets ($Z' = 2$), in which molecules are linked via **g**- and **c**-bridges. The supramolecular structure of the ethyl compound **19** can be described as a rectangular grid, in which each molecule interacts with four neighbors via **a**- and **c**-bridges. Finally, the structure of the methyl derivative **20** consists of hexagonal close-packed sheets. Each phosphazene molecule interacts with its six neighbors via bifurcate **b**-bridges. Each of the NH groups acts as an H-donor, while all N(ring) sites operate as H-acceptors. It is interesting to note

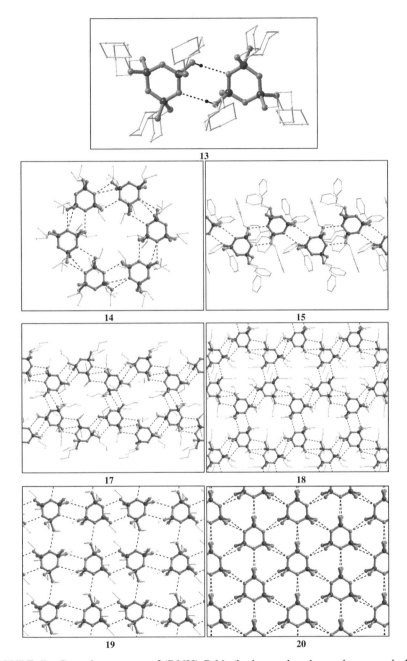

FIGURE 7   Crystal structures of $(RNH)_6P_3N_3$ (hydrogen bonds are drawn as dashed lines).

that the majority of crystal structures of **K** exhibit low-symmetry space groups and often contain more than one crystallographically unique molecule ($Z' > 1$).

As stated above, the variation of hydrogen-bonding modes is furnished by the conformational freedom of the exocyclic P–N bonds. Both the R group and the H atom are hinged on the freely rotating nitrogen center. Thus, the steric interactions between R groups and the directionality of NH bonds are strongly correlated. This subtle interplay enables a smooth transition between different H-bonding modes. A scatterplot of N–H$\cdots$N angles vs N$\cdots$N distances shows that interactions up to 3.4 Å are fairly linear (Fig. 8). The majority of interactions range between 2.9 and 3.3 Å, which indicates moderately strong H-bonds.

Aryl amino derivatives show additional NH$\cdots\pi$ interactions in the solid state. The phenyl derivative $(PhNH)_6P_3N_3$ (**21**) crystallizes with molecules lined up in a head-to-tail fashion along the twofold axis connected by symmetrical **b**-bridges. The resulting one-dimensional chain interacts with neighboring chains via NH$\cdots$aryl interactions to form two-dimensional sheets (Fig. 9). The direction of the head-to-tail assembly is maintained in neighboring sheets, which results in a polar crystal structure. In the structure of the *p*-tolyl derivative, molecules are linked by unsymmetrical **b**-bridges showing one short and one long NH$\cdots$N contact to form a linear chain. In addition, intermolecular NH$\cdots\pi$ interactions occur within the supramolecular chain. In contrast to $(PhNH)_6P_3N_3$, the crystal structure of $(p\text{-tolNH})_6P_3N_3$ (**22**) lacks intermolecular NH$\cdots\pi$ interactions between neighboring chains, presumably due to the steric interference of the *para*-positioned methyl groups. In the neighboring chains the direction of the head-to-tail assembly is reversed, yielding a centro symmetric crystal structure.

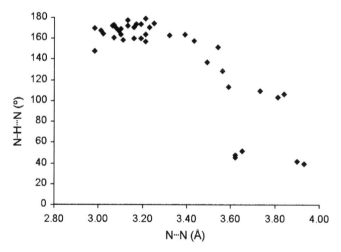

FIGURE 8 Scatterplot of N–H$\cdots$N angles vs. N$\cdots$N distances in **K**.

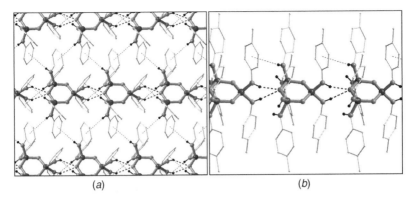

FIGURE 9   Crystal structures of (a) **21** and (b) **22**.

In contrast to **K**, its parent homolog $(H_2N)_6P_3N_3$ lacks a lipophilic periphery and features 12 H-donor sites. It forms a complex three-dimensional H-bonded network in the solid state in which each molecule interacts with eight neighbors involving all N and H atoms [59]. The ammonia solvate $(H_2N)_6P_3N_3 \cdot 1/2NH_3$ also exhibits a three-dimensional networked structure. Here, however, two phosphazene molecules are paired via six NH bonds to form a **k**-type dimer [60]. The hydrogen-bonding modes listed above have also been observed in several other cyclotri- [61] and cyclotetraphosphazenes [62] carrying amino substituents NHR and NH_2, respectively. Notably, the centrosymmetric **c**-bridge is the most frequently occurring H-bonding motif.

As a result of their large numbers of potential H-donor and acceptor sites, phosphazenes **K** form a range of solvate structures. Figure 10 shows two examples where **K** coordinates thf molecules. The 2-phenylethyl derivative produces the solvate **23** · thf, which consists of **f**-bonded chains. Each phosphazene binds one thf molecule, which is accommodated in a cavity formed by

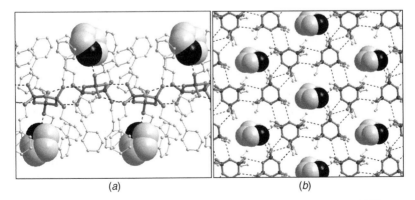

FIGURE 10   Crystal structures of (a) **23** · THF and (b) **19** · 1/2THF.

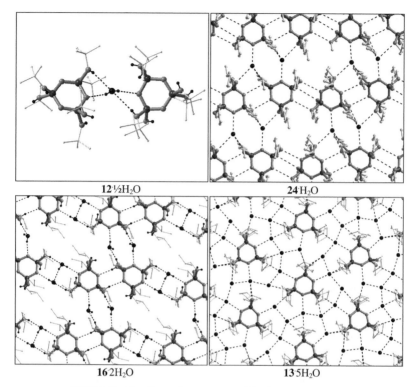

$12\cdot\tfrac{1}{2}H_2O$ $24\cdot H_2O$

$16\cdot 2H_2O$ $13\cdot 5H_2O$

FIGURE 11   Crystal structures of hydrates $\mathbf{K}\cdot n H_2O$.

four 2-phenylethyl groups. In contrast, the ethyl derivative forms a honeycomb sheet in the presence of thf. The thf molecules in the resulting solvate $\mathbf{19}\cdot 1/2\mathrm{thf}$ are located in the hexagonal cavities of the host structure. In both cases the phosphazenes bind to the solvent via $\mathrm{NH}\cdots\mathrm{O}$ interactions.

Water molecules can bind to $\mathbf{K}$ via both $\mathrm{NH}\cdots\mathrm{O}$ and $\mathrm{OH}\cdots\mathrm{N}$ bonds. A number of distinct hydrates $\mathbf{K}\cdot n H_2O$ are readily formed. These display a variety of aggregation patterns that again are controlled by the steric demand of R (Fig. 11) [63]. The *tert*-butyl derivative **12** is hygroscopic and crystallizes from a solution of hexane that is open to the air in the form of the hydrate $\mathbf{12}\cdot 1/2H_2O$. The water molecule is encapsulated effectively by the two sterically demanding phosphazene molecules. The hydrate of the *n*-butyl derivative $\mathbf{24}\cdot H_2O$ consists of H-bonded sheets: The phosphazene molecules are lined up via $\mathbf{c}$- and $\mathbf{g}$-bonds to form one-dimensional chains, which are interconnected by water molecules. Similar structures are found in the hydrates of **17** and **19**, which contain an extra site between the phosphazene chains that is partially occupied by water. These nonstoichiometric hydrates contain between 1 and 1.5 water molecules per formula unit. The hydrate of the *iso*-butyl derivative $\mathbf{16}\cdot 2H_2O$ contains $\mathbf{c}$-bridged phosphazene dimers that are networked via water molecules, resulting in a two-dimensional structure.

The cyclohexyl derivative crystallizes from wet thf as the pentahydrate **13** · 5H$_2$O. The phosphazene molecules are surrounded by 12 water molecules, resulting in a complex two-dimensional network.

It should be noted that the water molecules are effectively shielded by a hydrophobic layer of R groups. Indeed, the two-dimensional structures of **24** · H$_2$O, **16** · 2H$_2$O, and **13** · 5H$_2$O as well as **19** · 1/2thf display the lipid bilayer arrangement that is observed in the crystal structures of solvent-free **K**. It appears as if the lipid-bilayer formation is a general feature of **K**. Remarkably, some derivatives of **K**, including **14**, **18**, and **20**, have so far resisted the formation of hydrates or other solvates. The lack of hygroscopic behavior could be attributed to more effective crystal packing: All three exhibit hexagonal structures (pseudohexagonal in the case of **18**). This shows that **K** can be tailored to either absorb or repel certain substrates solely by slight modification of their lipophilic periphery.

Phosphazenes **K** are protonated at ring nitrogen centres in the presence of Brönsted acids HX, yielding phosphazenium ions (Scheme 4). Weak acids, including carboxylic acids, protonate one nitrogen center, while strong acids, such as HCl, protonate two nitrogen sites, generating mono- and dications **KH**$^+$ and **KH**$_2^{2+}$, respectively. The resulting salts are often highly soluble in methanol or chloroform and readily crystallize upon evaporation of the solvent. The crystal structures exhibit ion pairing via strong directional hydrogen bonding through both exocyclic and ring NH functions. X-ray structure analysis revealed that the P–N bonds adjacent to the protonated nitrogen center are elongated substantially. This information can be important to assign ring NH sites, in particular when the X-ray data are poor. We have investigated the solid-state structures of **KH**$^+$ and **KH**$_2^{2+}$ in the presence of a variety of anions. Although the bulk of this research has not yet been published, some of our current findings are highlighted here, since it illustrates the rich potential of cyclophosphazenes in crystal engineering and self-assembly.

The methyl derivative [**20**H]Cl, which is obtained by gas-phase diffusion of HCl into a solution of **20** in CHCl$_3$, forms direct NH···N bonds between two nitrogen ring sites. This unique feature is facilitated by the low steric demand of

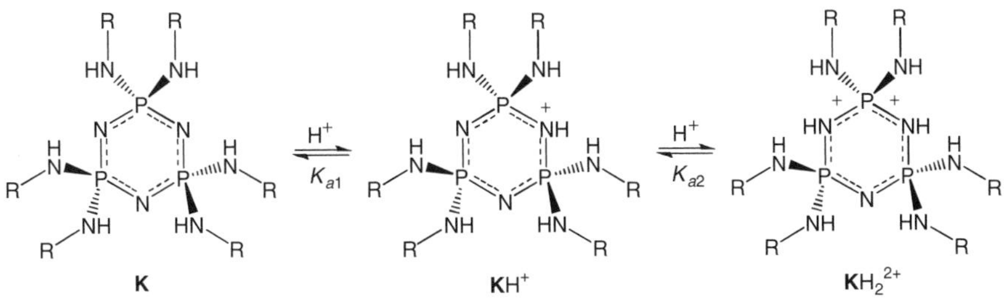

SCHEME 4   Protonation equilibria of **K**.

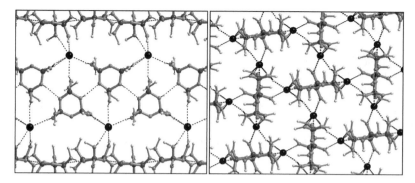

FIGURE 12   Two views of the crystal structure of [**20**H]Cl.

the methyl groups. However, due to the small size of the methyl group, the familiar lipid-bilayer arrangement cannot be maintained. Phosphazenium and chloride ions form ribbons, which in turn bind to neighboring ribbons in orthogonal fashion. The result is a three-dimensional hydrogen-bonded network (Fig. 12).

Salts containing phosphazenium ions **KH**$^+$ and **KH**$_2$$^{2+}$, respectively, are often hygroscopic and exist as hydrates in the solid state. For example, **13** crystallizes in the presence of aqueous HCl in the form of the hydrate [**13**H$_2$]Cl$_2$ 3H$_2$O. It forms a lipid-bilayer arrangement similar to that of the pentahydrate **13** · 5H$_2$O. Figure 13 illustrates how the central ionic, highly polar structure is well shielded by the two hydrophobic layers of cyclohexyl substituents.

Highly networked structures are obtained when **K** is treated with half an equivalent of a dicarboxylic acid. The resulting salts are of the general formula [**KH**]$_2$[OOC-*Sp*-COO], where *Sp* represents a spacer unit between the two carboxylate groups. Again, the lipid-bilayer arrangement is a common feature

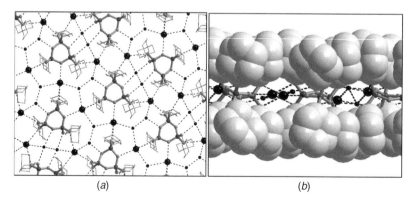

(a)   (b)

FIGURE 13   (a) Crystal structure of [**13**H$_2$]Cl$_2$ · 3H$_2$O; (b) view across its lipid-bilayer arrangement.

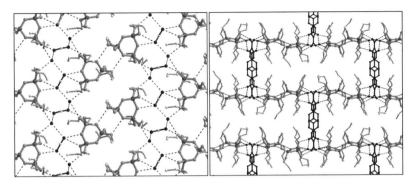

FIGURE 14 Two views of the crystal structure of [**17**H]$_2$[OOC–(C$_6$H$_4$)$_2$–COO]·2H$_2$O.

in these compounds. Bulky R groups and short spacers give two-dimensional structures, while less bulky R groups in combination with a long spacer can give rise to three-dimensional networks. One such example is depicted below. When **17** is crystallized in the presence of 4,4'-biphenyldicarboxylic acid, [**17**H]$_2$ [OOC–(C$_6$H$_4$)$_2$–COO]·2H$_2$O is obtained. Here, the dicarboxylate ions form pillars that link the lipid bilayers of phosphazenium ions (Fig. 14). The phosphazenium ions pair up via direct NH⋯N interaction (**c**-bridges). It should be noted that direct hydrogen bonding between organic cations is rarely observed.

Once more, slight variances of R groups, as well as the choice of solvent, play an important role in controlling the network topology of structures containing **K**H$^+$ ions. Protic solvent molecules, such as water, methanol, and ethylene glycol, are often incorporated into the crystal lattice and contribute to the hydrogen-bonded network. The ease of single-crystal growth is often accomplished by the subtle interplay of adaptable hydrogen-bonding modes and the packing efficiency of the lipophilic substituents. As a result, rather complex systems can be crystallized. Large and multiply charged anions are embedded into a well-ordered array of phosphazenium cations. Thus, the incorporation of anionic metal complexes furnishes crystals containing distantly spaced metal centers, such as in the 1,1'-ferrocenedicarboxylate [**16**H]$_2$[Fc(COO)$_2$]·2H$_2$O (Fig. 15). Some nano-sized anions, such as polyoxymetallates, form single crsytals in combination with phosphazenium ions. Tetraanionic Keggin clusters are effectively complexed via hydrogen bonds by four phosphazenium ions. In the structure of [**16**H]$_4$[SiW$_{12}$O$_{40}$], each **16**H$^+$ provides five hydrogen-bond donor sites, including one ring NH and four exocyclic NH groups (Fig. 15) [64].

Another route to phosphazenium ions is the alkylation of ring nitrogen centers. Alkylated N centers cannot take part in hydrogen bonding, and the nucleophilicity of the vacant ring N sites is reduced on successive alkylation. As a result, the free ring N site of the dialkylated ion [**K**R'$_2$]$^{2+}$ is mainly inactive, while

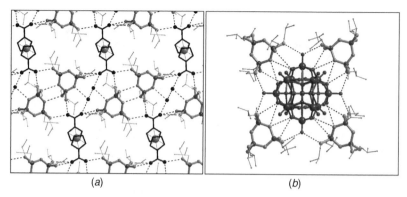

(a)   (b)

FIGURE 15 Crystal structure of (a) $[\mathbf{16H}]_2[Fc(COO)_2] \cdot 2H_2O$ and (b) $[\mathbf{16H}]_4[SiW_{12}O_{40}]$.

both available ring N sites of $[\mathbf{KR'}]^+$ are able to participate in hydrogen bonding (Chart 9). The type of R' groups that have been attached to the ring include methyl, ethyl, allyl, benzyl, and cinnamyl. The crystal structures of a series of iodide salts $[\mathbf{KMe}]I$ illustrates that subtle changes of R affect the topology of the network and control the degree of direct intercationic binding (Fig. 16) [65]. In the n-propyl derivative $[\mathbf{17Me}]I$, each phosphazenium ion forms **c**-bridges to two neighbors, resulting in cationic one-dimensional chains. These chains extend into a two-dimensional sheet structure via iodide ions. The *iso*-butyl $[\mathbf{16Me}]I$ derivative consists of **c**-bridged dimers. These are linked via iodide ions to generate one-dimensional chains. Bulkier R groups do not support intercationic NH$\cdots$N bonds. Thus, the *iso*-propyl $[\mathbf{14Me}]I$ derivative exhibits only NH$\cdots$I interactions, yielding a two-dimensional sheet. The bulky *tert*-butyl derivative $[\mathbf{12Me}]I$ does not form an extended structure. Two phosphazenium ions are connected via two iodide ions in a discrete molecular arrangement.

When two phosphazenes are linked with a covalent spacer Sp' via quaternized N sites, the dications $[\mathbf{K}_2Sp']^{2+}$ are obtained [66] (Chart 10). These are relatively

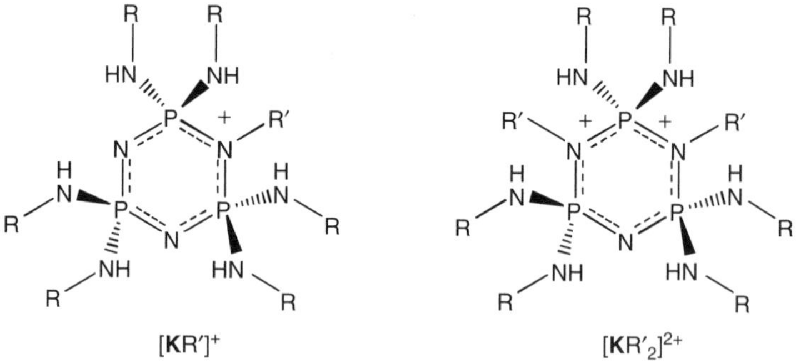

$[\mathbf{KR'}]^+$ $[\mathbf{KR'}_2]^{2+}$

CHART 9

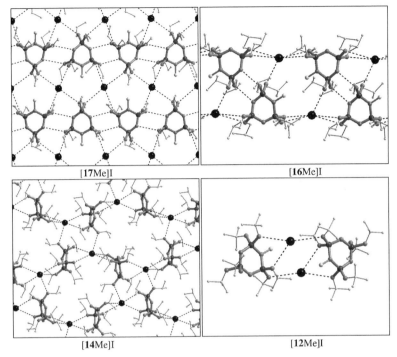

[**17**Me]I   [**16**Me]I

[**14**Me]I   [**12**Me]I

FIGURE 16   Crystal structures of [**K**Me]I.

large molecular entities that contain 12 protential H-donors, in the form of exocyclic NH groups. In addition, four ring N atoms can act as H-acceptors. These ions also exhibit intercationic NH$\cdots$N bonds in the presence of less bulky substituents such as *iso*-butyl in [16$_2$*o*-xyl]Br$_2$ (Fig. 17). This gives the option to link phosphazene rings via both covalent and hydrogen bonds [67].

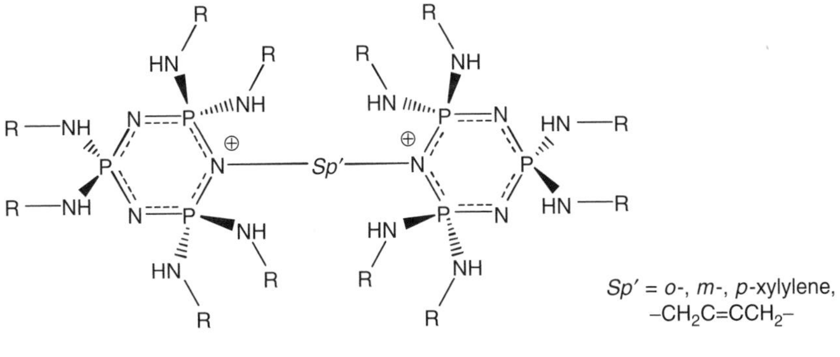

$Sp' = o$-, m-, p-xylylene,
   $-CH_2C=CCH_2-$

[**K**$_2$Sp]$^{2+}$

CHART 10

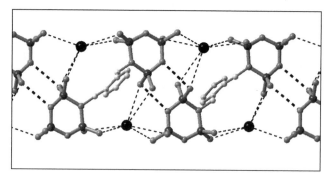

FIGURE 17   Crystal structure of $[16_2o\text{-xyl}]Br_2$.

The polyaminophosphazenium ions presented here display various hydrogen-bonding modes in the presence of anions (Scheme 5). Besides simple monodenate $NH\cdots X^-$ interactions, the bidentate interactions **m** and **n** frequently occur in halide structures. The tridentate mode **p** is featured in halides of N,N'-dialkyl phosphazenium ions $[KR'_2]^{2+}$. The monoprotonated $[KH]^+$ contains a ring NH site that binds anions in either monodentate or, supported by an exocyclic NH group, in bidentate fashion **q**. This coordination mode is highly compatible with carboxylate ions (**q'**). Furthermore, the carboxylate group is able to bridge two phosphazenium ions sideways via $NH\cdots O$ bonds.

The substituents R in aminophosphazenes $(RNH)_6P_3N_3$ can be equipped with hydrogen-bond donor and acceptor groups. Amino acids are particularly interesting side chains, since they can mimic the familiar hydrogen-bonding interactions of peptides. The crystal structure of **25**, which carries six glycin ethylester groups, exhibits a dense network of both $NH\cdots N$ and $NH\cdots O$ bonds (Fig. 18). The result is a double-chain structure in which all NH and ring N functions are involved in hydrogen bonding [68]. Scheme 6 highlights the

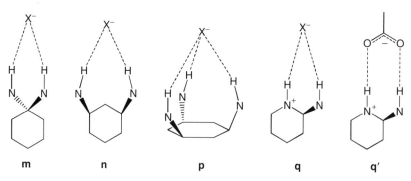

SCHEME 5   Hydrogen-bonding modes of phosphazenium ions toward anions.

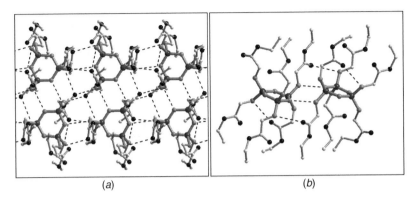

FIGURE 18 Crystal structure of **25**: (a) double-chain arrangement; (b) quadruple hydrogenbond.

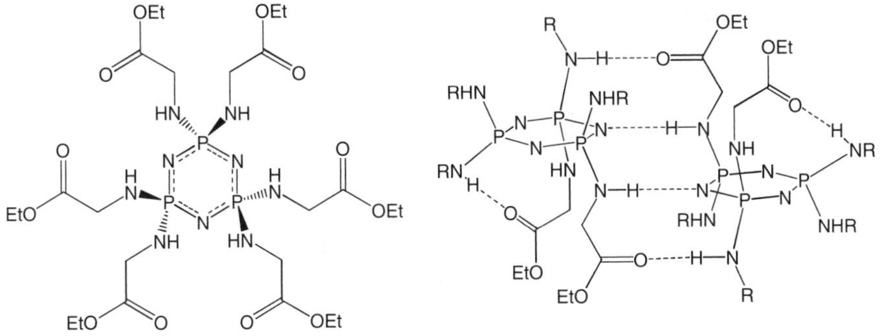

SCHEME 6 Schematic representation of quadruple H-bond interaction in the solid-state structure of **25**.

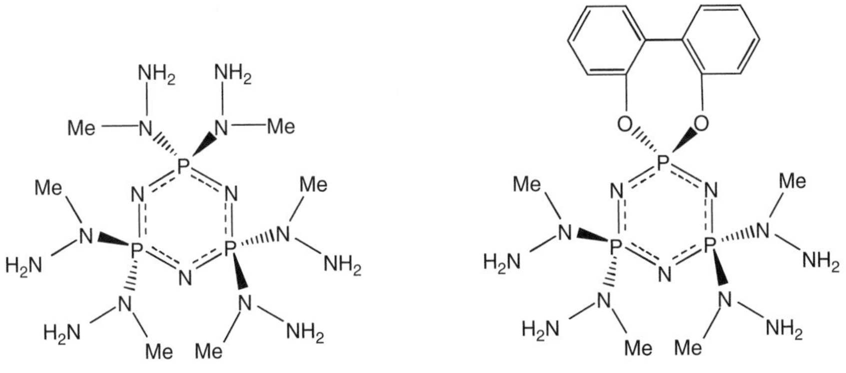

26

27

CHART 11

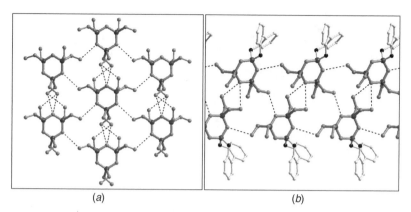

(a)   (b)

FIGURE 19   Crystal structures of (a) **26** and (b) **27**.

interaction between two molecules, which are connected via four hydrogen bonds, including two central NH $\cdots$ N and two peripheral NH $\cdots$ O bonds. This bonding pattern is reminiscent of the β-sheet arrangement of peptides. In addition, there are intramolecular NH $\cdots$ O bonds between substituents that are grafted onto the same ring.

Apart from amino groups, hydrazino substituents provide both H-bond donor and acceptor sites and also render the ring nitrogens sufficiently basic to act as H-acceptors. The hydrazino derivatives **24** and **25** are sterically less restricted than amino derivatives **K**, owing to the peripheral positions of the NH$_2$ groups. In addition, the freedom of rotation around both exocyclic P–N and N–N bonds offers great flexibility. Compound **26**, which carries six hydrazino substituents, assembles in a close-packed sheet in which each molecule interacts with six nearest neighbors (Chart 11). All three ring nitrogen centers and the six NH$_2$ groups are connected via hydrogen bonding. The crystal structure of **27**, which is equipped with four hydrazino substituents, consists of a hydrogen-bonded double-chain arrangement (Fig. 19) [69].

Metal Coordination

One important area of supramolecular chemistry is the generation of coordination polymers in which the metal centers are linked via multitopic ligands, and vice versa. The topology of the system is controlled by the coordination geometry of both the metal center and the ligand [70]. Organonitrogen heterocycles have been widely employed as bridging ligands in coordination polymers. The stereorigid arrangement of N-donor sites provides great control over the topology of the supramolecular network [71]. Cyclotriphosphazenes feature a stable six-membered ring system with a trigonal planar array of nitrogen centers (**L**, Chart 12). Furthermore, the tetrahedral environment of the

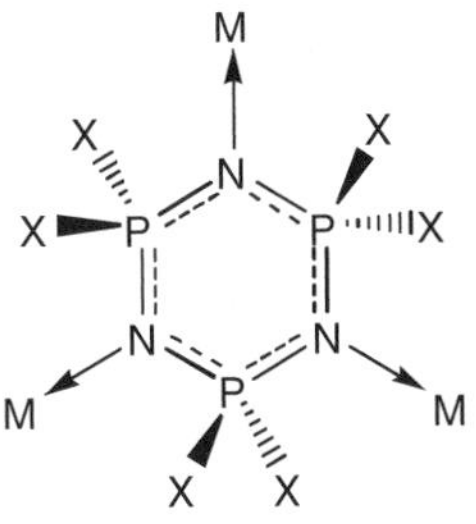

L

CHART 12

ring phosphorus atoms places the exocyclic substituents above and below the phosphazene ring plane. This provides ample space for metal coordination to nitrogen(ring) sites and facilitates the introduction of additional donor functions into the side chains.

Cyclotriphosphazenes carrying organoamino groups $(RNH)_6P_3N_3$ produce coordination polymers in conjunction with silver salts via linear N–Ag–N connections [72]. The donor strength of the anion and the steric demand of the exocyclic R group control the topology of the coordination network. The n-propyl derivative $(\mathbf{17})_2(AgClO_4)_3$ exhibits a graphite-type (6,3) network structure. All three nitrogen(ring) atoms of the phosphazene ligand coordinate to silver ions, which, in return, form linear bridges between two phosphazene ligands (Fig. 20). In contrast, $\mathbf{17}(AgNO_3)_2$ exists of zigzag chains featuring one bridging silver ion and one terminally coordinated silver ion per ligand molecule. The terminally located Ag(I) ions are connected to silver ions of neighboring chains via nitrate ions resulting in a two-dimensional network. The sterically more demanding cyclohexyl derivative generates in the coordination polymer $\mathbf{13}(AgClO_4)$. The phosphazene ring binds two silver ions silver ions, which leaves one N(ring) site vacant and gives one-dimensional zigzag chain arrangements (Fig. 20). The corresponding coordination compounds of the benzyl derivative, $\mathbf{15}(AgClO_4)_2$ and $\mathbf{15}(AgNO_3)_2$, resemble that of $\mathbf{13}(AgNO_3)_2$, but show additional Ag–π(aryl) interactions between the terminally arranged silver ions and two adjacent benzyl groups (Fig. 20). In addition to metal coordination, the phosphazene ligands undergo hydrogen bonding to anions via NH groups.

The cyclotriphosphazene **18**, which is equipped with six pendant allylamino groups, operates as a multitopic, hemilabile ligand in the presence of Ag(I) [73]. The flexibility of the donor sidearms enables the smooth switch between intra- and intermolecular coordination modes, which provides highly adaptable linkers. The combination of **18** and $AgClO_4$ in methanol leads to the immediate formation of a precipitate, which undergoes in situ crystallization. The monoclinic crystals consist of the coordination polymer $\mathbf{18} \cdot 3AgClO_4 \cdot 4H_2O$,

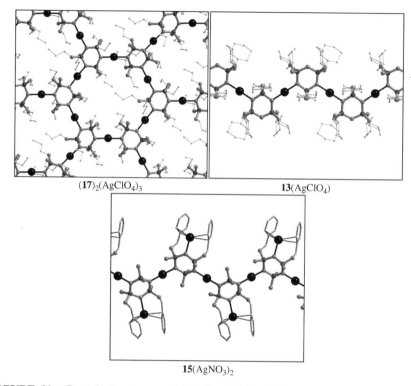

$(\mathbf{17})_2(\text{AgClO}_4)_3$ $\mathbf{13}(\text{AgClO}_4)$

$\mathbf{15}(\text{AgNO}_3)_2$

FIGURE 20   Crystal structures of $(\mathbf{17})_2(\text{AgClO}_4)_3$, $\mathbf{13}(\text{AgClO}_4)$, and $\mathbf{15}(\text{AgNO}_3)_2$ (anions and noncoordinating R groups are omitted).

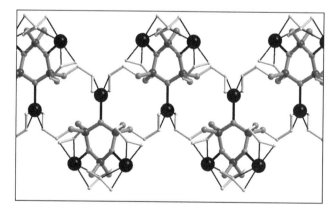

FIGURE 21   Crystal structure of $\mathbf{18}(\text{AgClO}_4)_3 \cdot 4\text{H}_2\text{O}$ (anions and solvent molecules are omitted).

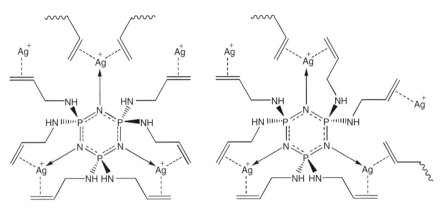

SCHEME 7   Connectivity of $[18Ag_3]^{3+}$ units in the monoclinic (left) and orthorhombic isomer (right).

comprising a one-dimensional chain assembled from $[18Ag_3]^{3+}$ units. Two Ag(I) ions are coordinated by two allyl groups in an intramolecular fashion, while the third Ag(I) ion coordinates intermolecularly to two neighboring $[18Ag_3]^{3+}$ units via Ag–olefin bonds (Fig. 21). When the crystals were kept in methanol for several weeks, they gradually decomposed, giving a black material alongside an orthorhombic modification. It consists of an isomeric three-dimensional coordination network $18 \cdot 3AgClO_4$. Again, this isomer is also assembled from $[18Ag_3]^{3+}$ units interacting via silver ions that are coordinated by two allyl groups (Scheme 7). However, the connectivity pattern of intra- and intermolecular Ag–olefin interactions in the one-dimensional monoclinic and three-dimensional orthorhombic structure are different. One silver ion of the $[18Ag_3]^{3+}$ unit is coordinated by two intramolecular allyl groups, while the other two Ag(I) ions form both one intra- and one intermolecular olefin contact. This distinct connectivity pattern allows each $[18Ag_3]^{3+}$ unit to make contacts with four neighbors via Ag–olefin interactions, resulting in a diamandoid network topology.

The spontaneous recrystallization of the primary precipitate into single crystals is controlled by diffusion of $[18Ag_3]^{3+}$ units through the overlaying methanol solution. The flexible coordination behavior of the $[18Ag_3]^{3+}$ units, which probably involves a rapid exchange between intra- and intermolecular coordination modes, facilitates this process. In addition to silver(I) coordination, the allylamino groups engage in hydrogen bonding to perchlorate ions and solvent molecules. Since both NH and allyl groups are hinged to the same P–N bond, metal coordination and hydrogen bonding are correlated to some extent.

Numerous metal complexes of cyclophosphazene ligands have been prepared, and many of these feature nitrogen atoms as ligand donor sites. Particular emphasis has been placed on the development of multisite ligands [74]. A range of donor sidearms have been grafted onto the

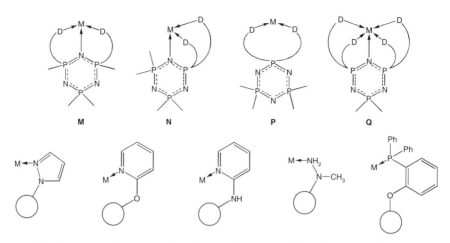

SCHEME 8   Various coordination modes of multisite phosphazene ligands.

phosphazene ring that mainly contain N or P functions and display various coordination modes (Scheme 8). Substituents that carry donor sites in close proximity to the phosphazene ring, such as 2-pyrazolyl and hydrazino groups, support the tridentate coordination modes **M** (nongeminal) and **N** (geminal), respectively [75]. A geminal pair of donor side arms can also give rise to the bidentate mode **P**, which lacks participation of a ring nitrogen atom. The pentadentate mode **Q** is maintained by a set of four 2-pyridiloxy groups grafted onto the P-centers adjacent to the coordinating ring N site [76]. Cyclophosphazenes carrying 2-pyridiloxy groups also form coordination polymers in conjunction with silver salts. The Ag(I) ions are coordinated by two ligands; one binds the metal via a ring N and a pyridiloxy group, while the other binds in **N** fashion via two 2-pyridyloxy groups and a long-range Ag–N interaction [77].

Various types of multidentate substituents have been attached to phosphazene rings. The bis(pyridylpyrazolyl) derivative **S** forms a monomeric complex with lanthanum(III) nitrate. The La^{3+} center is chelated by the tetradentate ligand and further coordinated by three nitrate ligands [78]. A series of copper complexes of the 1,10 phenanthroline derivative **S** have been reported (Scheme 9) [79]. The x-ray structure of the dimeric complex $[(CuS)_2](PF_6)_2$ exhibits a helical arrangement. Each copper(I) ion is coordinated in a distorted tetrahedral environment by phenantroline groups of both ligands. The structure of [CuSCl]Cl shows the copper(II) ion to be in a distorted square-base pyramidal environment. It is coordinated to the four nitrogen atoms of the two phenanthroline groups and a chloride ion.

PNP crowns such as **28** (Chart 13) are obtained when a polyether chain is tethered to two phosphorus atoms of a cyclophosphazene. As a result, the phosphazene ring is part of the macrocyclic arrangement. Alkali metal ions control the regioselectivity of the PNP-crown formation as well as further

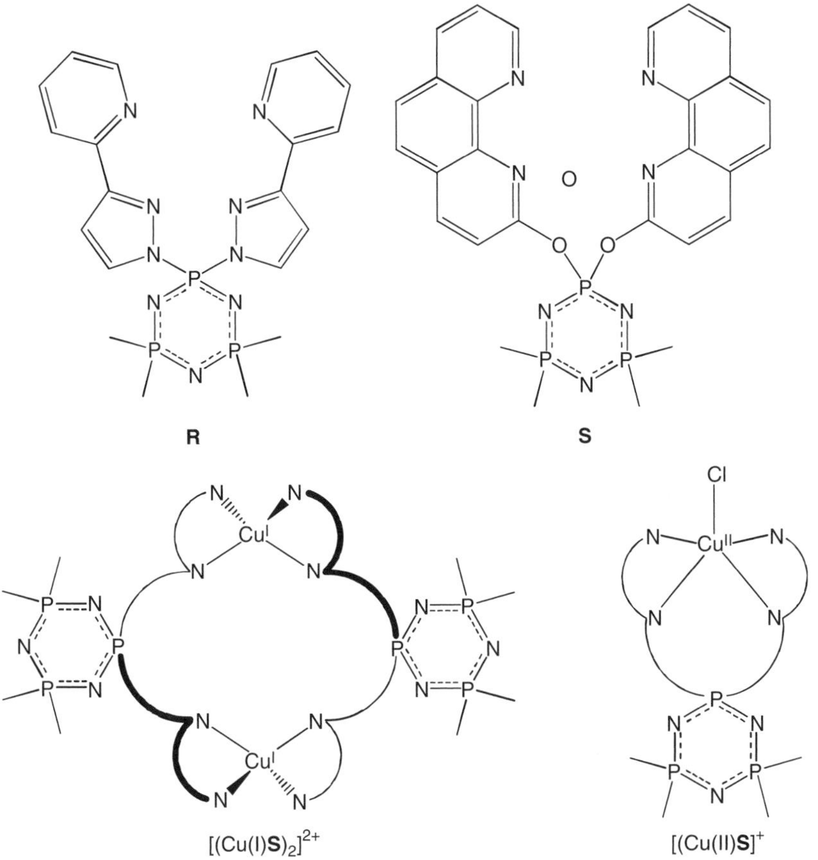

SCHEME 9 Phosphazene ligands carrying multidentate substituents (top) and the connectivity of copper complexes of **S** (bottom).

substitution reactions at P centers [80]. Crystal structures of alkali metal complexes revealed that the ring nitrogen center is not involved in metal coordination since the phosphazene ring is tilted with respect to the mean plane of the crown ether [81]. In addition, macrocyclic systems have been synthesized in which two cyclotriphosphazene units are linked via two polyether chains, as in **29** [82]. However, structurally characterized metal complexes of these have not yet been reported. These systems bear great potential for supramolecular applications: for example, as macrocyclic components in catenanes and rotaxanes.

Polyphosphazenes carrying polyether groups form stable films and exhibit high ion conductivities for lithium salts, making them promising electrolyte materials for secondary lithium batteries [83]. Cyclophosphazenes, which feature amino groups with ether functions (30), are able to accommodate lithium chloride via the following in situ method [84]. Treatment of **30** with HCl gives the salt [**30**H$_2$]Cl$_2$, which is then deprotonated by *n*-butyllithium to yield

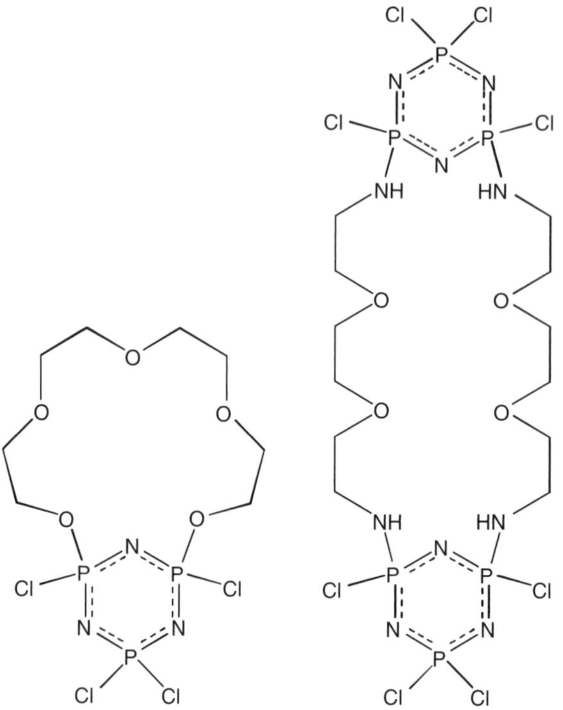

28   **29**

CHART 13

the lithium chloride complex **30**(LiCl)$_2$ (Scheme 10). Although **30** exists as an oil at room temperature, **30**(LiCl)$_2$ is a colorless hygroscopic solid which melts at 65 °C. The x-ray crystal structure revealed that the supramolecular structure consists of a one dimensional coordination polymer. Dimeric kite-shaped

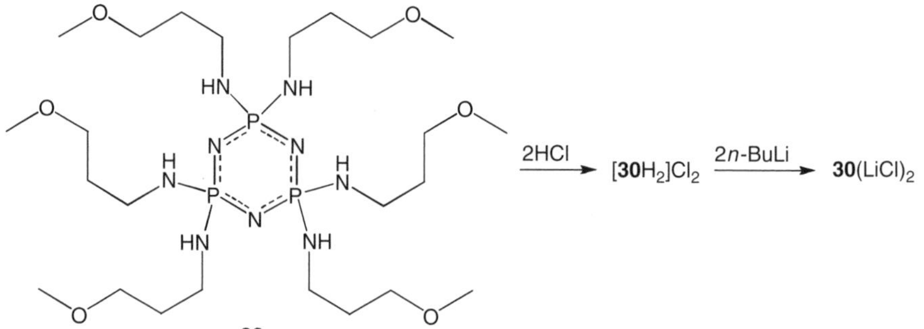

30

SCHEME 10   In-situ route to **30**(LiCl)$_2$.

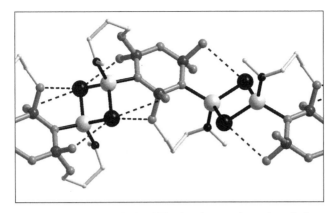

FIGURE 22   Crystal structure of **30**(LiCl)$_2$ (noninteracting ether chains are omitted).

(LiCl)$_2$ units interact with phosphazene molecules via Li–N(ring), Li–O, and NH$\cdots$Cl interactions (Fig. 22). The structure of **30**(LiCl)$_2$ illustrates that cyclophosphazenes can offer extensive arrays of potential binding sites for both cations and anions.

Multianionic phosphazenate ligands are able to accommodate large numbers of metal ions in their coordination spheres [85]. The hexa-anionic ligand **T**$^{6-}$ is obtained by deprotonation of all NH functions of (RNH)$_6$P$_3$N$_3$ (**K**) with strong organometallic bases [86]. The deprotonation pathway selectively yields the trianionic phosphazenate **TH**$_3$$^{3-}$ after treatment with 3 equivalents of base. Exclusively, amino groups at one side of the ring plane are deprotonated. The P$_3$N$_3$ ring in **TH**$_3$$^{3-}$ adopts a chair conformation; the deprotonated N centers are spread into equatorially positions, which enables effective charge distribution. This gives a unique hexadentate ligand surface consisting of three equatorial and three ring nitrogen centers (Scheme 11).

These multianionic ligand systems can act as templates to support well-defined cation–anion arrays. One example is the reaction of the anisidyl derivative **31**H$_6$ with 12 equivalents of butyllithium in tetrahydrofuran

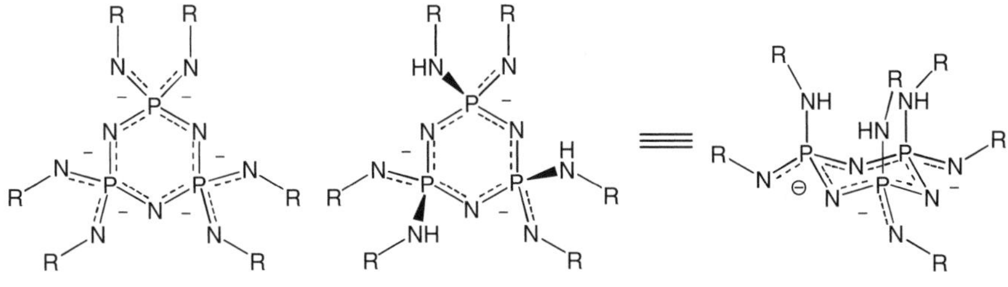

SCHEME 11   Multianionic phosphazenate ligands **T**$^{6-}$ and **TH**$_3$$^{3-}$.

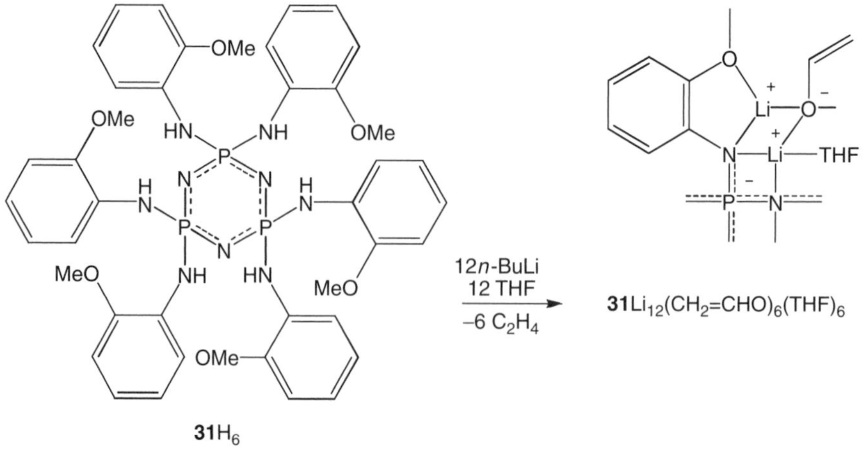

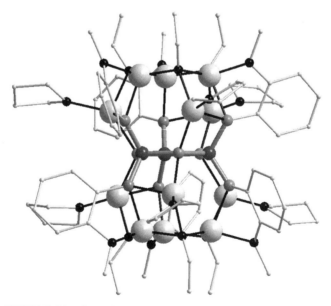

SCHEME 12   Formation of $30Li_{12}(CH_2\!=\!CHO)_6(THF)_6$.

(THF) to give the complex $31Li_{12}(CH_2\!=\!CHO)_6(thf)_6$ (Scheme 12) [87]. The resulting hexaanion 31^{6-} acts as a pentadecadentate ligand, which hosts 12 lithium ions in bidentate chelation sites (Fig. 23). Six lithium ions are coordinated at an inner rim consisting of N(exo)–P–N(ring) sites, while the other six lithium ions are located at the outer rim in N(exo)–C$_6$H$_4$–O sites. Six enolate anions cap the complex from either side. The enolate ions are the

FIGURE 23   Crystal structure of $31Li_{12}(CH_2\!=\!CHO)_6(THF)_6$.

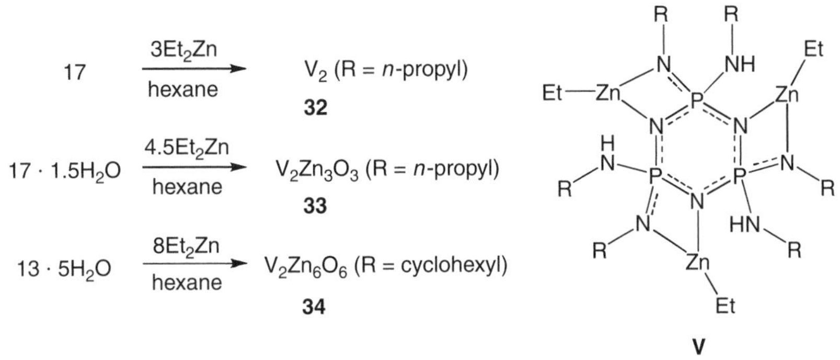

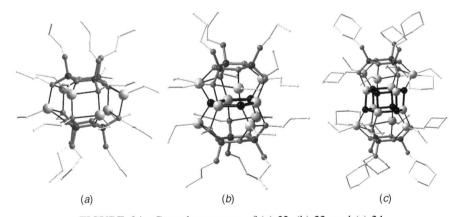

17 $\xrightarrow[\text{hexane}]{3Et_2Zn}$ V$_2$ (R = *n*-propyl)
32

17 · 1.5H$_2$O $\xrightarrow[\text{hexane}]{4.5Et_2Zn}$ V$_2$Zn$_3$O$_3$ (R = *n*-propyl)
33

13 · 5H$_2$O $\xrightarrow[\text{hexane}]{8Et_2Zn}$ V$_2$Zn$_6$O$_6$ (R = cyclohexyl)
34

SCHEME 13 Preparation of templated zinc oxide clusters.

product of THF ring cleavage. Usually, this is a very slow reaction; however, the instantaneous formation of **30**Li$_{12}$(CH$_2$=CHO)$_6$(thf)$_6$ suggests a concerted mechanism aided by the multianionic ligand and the lithium ions.

Multianionic phosphazeante ligands can also template zinc oxide clusters. The *n*-propyl derivative **17** reacts with 3 equivalents of diethylzinc to generate the dimeric complex **32**, in which six ethylzinc units are sandwiched by two trianionic phosphazenate ligands. Alternatively, this arrangement can be viewed as a dimer of two tris(ethylzinc) phosphazenate segments **V**. These feature a bowl-shaped coordination surface of three Lewis acidic Zn centers and three Lewis basic N(exo)-sites, which provides a cast for planar {(ZnO)$_3$} rings and hexagonal {(ZnO)$_6$} prisms. When appropriate amounts of diethylzinc are added to the phosphazene hydrates **17**·1.5H$_2$O and **13**·5H$_2$O, respectively, zinc oxide clusters are generated in situ and encapsulated by two segments **V** (Scheme 13) [88]. The size of the cluster is controlled by the size of

(a)   (b)   (c)

FIGURE 24 Crystal structures of (a) **32**, (b) **33**, and (c) **34**.

the R group. The *n*-propyl derivative **33** hosts the planar $\{(ZnO)_3\}$ arrangement, while the cyclohexyl derivative **34** provides sufficient space for the $\{(ZnO)_6\}$ prism (Fig. 24).

CONCLUSIONS

Cyclotriphosphazenes support a plethora of supramolecular structures, which has led to interesting applications in various fields of chemistry. Rigid trispirocyclic systems form stable clathrate structures that accommodate a variety of guests, ranging from gas molecules to linear polymers, while dendritic systems show liquid-crystal behavior. The phosphazene ring is able to participate in noncovalent bonding such as hydrogen bonding and metal coordination. The donor strength of the ring nitrogen functions can be finely tuned either to engage them actively in supramolecular interactions or to render them innocent bystanders. In particular, amino derivatives $(RNH)_6P_3N_3$ exhibit a rich supramolecular chemistry. They contain both hydrogen-donor sites in the form of exocyclic amino groups and fairly basic nitrogen centers that can operate as hydrogen-acceptors. A common structural feature of these compounds, as well as of their solvate structures, is the lipid-bilayer arrangement, in which the polar network of hydrogen-bonded substrates is sandwiched between the hemispheres of lipophilic R groups. Subtle alterations of the steric demand of R have a great impact on the topology of the supramolecular structure. The smooth switch between hydrogen-bonding modes greatly assists single-crystal growth. In their protonated form these systems host anions via complex networks of hydrogen-bonds. The ability to form well-ordered structures in the presence of very large anions could aid the crystallization of nanoassemblies or macromolecules. Furthermore, cyclotriphosphazenes display a rich coordination chemistry. They can act as nodal ligands in coordination polymers via the trigonal arrangement of ring N centers. Further addition of donor sites to the side chains led to a variety of multisite ligands. Finally, multianionic phosphazenates are able to accommodate large arrays of metal ions in well-defined cavities as well as to template cation–anion arrangements.

REFERENCES

1. Lehn, J.-M. *Supramolecular Chemistry*. Wiley-VCH, Weinheim, Germany, 1995. Steed, J.W., Atwood, J.L. *Supramolecular Chemistry*. Wiley, Chichester, UK, 2000.

2. Allcock, H.R. *Chem. Rev.*, 1972, 72:315. Allen, C.W. *Chem. Rev.*, 1991, 91:119.

3. Chandrasekhar, V., Krishnan, V. *Adv. Inorg. Chem.*, 2002, 53:159. Elias, A., Shreeve, J.M. *Adv. Inorg. Chem.*, 2001, 52:335.

4. Mark, J.E., Allcock, H.R., West, R. *Inorganic Polymers*, 2nd ed., Oxford University Press, New York, 2005, p. 99. Allcock, H.R., Al-Shali, S., Ngo, D.C., Visscher, K.B., Parvez, M. *Dalton Trans.*, 1996, 3549.

5. MacNicol, D.D., McKendrick, J.J., Wilson, D.R. *Chem. Soc. Rev.*, 1978, 7:65.

6. Allcock, H.R. *J. Am. Chem. Soc.*, 1964, 86:2591. Allcock, H.R., Siegel, L.A. *J. Am. Chem. Soc.*, 1964, 86:5140. Allcock, H.R. In Atwood, J.L., Davies, J.E.D. and MacNicol, D.D., eds., *Inclusion Compounds*. Academic Press, New York, 1984, Vol. 1 p. 351.

7. Davis, M.E. *Nature*, 2002, 417:813.

8. Allcock, H.R. *Acc. Chem. Res.*, 1978, 11:81, and references therein.

9. Allcock, H.R., Levin, M.L., Whittle, R.R. *Inorg. Chem.*, 1986, 25:41.

10. Sozzani, P., Comotti, A., Simonutti, R., Meersmann, T., Logan, J.W., Pines, A. *Angew. Chem. Int. Ed.*, 2000, 39:2695. Comotti, A., Bracco, S., Ferretti, L., Mauri, M., Simonutti, R., Sozzani, P. *Chem. Commun.*, 2007, 350.

11. Sozzani, P., Bracco, S., Comotti, A., Ferretti, L., Simonutti, R. *Angew. Chem. Int. Ed.*, 2005, 44:1816.

12. Hertzsch, T., Budde, F., Weber, E., Hulliger, J. *Angew. Chem. Int. Ed.*, 2002, 41:2282.

13. Primrose, A.P., Parvez, M., Allcock, H.R. *Macromolecules*, 1997, 30:670. Allcock, H.R., Primrose, A.P., Sunderland, N.J., Rheingold, A.L., Guzei, I.A., Parvez, M. *Chem. Mater.*, 1999, 11:1243. Comotti, A., Simonutti, R., Catel, G., Sozzani, P. *Chem. Mater.*, 1999, 11:1476. Sozzani, P., Comotti, A., Bracco, S., Simonutti, R. *Chem. Commun.*, 2004, 768.

14. Allcock, H.R., Sunderland, N.J. *Macromolecules*, 2001, 34:3069.

15. Sozzani, P., Comotti, A., Bracco, S., Simonutti, R. *Angew. Chem. Int. Ed.*, 2004, 43:2792.

16. Allcock, H.R., Ferrar, W.T., Levin, M.L. *Macromolecules*, 1982, 15:697. Allcock, H.R., Dudley, G.K., Silverberg, E.N. *Macromolecules*, 1994, 27:1033.

17. Allcock, H.R., Dudley, G.K., Silverberg, E.N. *Macromolecules*, 1994, 27:1039.

18. Allcock, H.R., Stein, M.T. *J. Am. Chem. Soc.*, 1974, 96:49.

19. Kubono, K., Asaka, N., Taga, T., Isoda, S., Kobayashi, T. *J. Mater. Chem.*, 1994, 4:291. Kubono, K., Asaka, N., Isoda, S., Kobayashi, T. *Acta Crystallogr.*, 1993, C49:404.

20. Kubono, K., Asaka, N., Isoda, S., Kobayashi, T. *Acta Cryst, Callogr.* 1994, 50:324.

21. Allcock, H.R., Sunderland, N.J., Primrose, A.P., Rheingold, A.L., Guzei, I.A., Parvez, M. *Chem. Mater.*, 1999, 11:2478.

22. Allcock, H.R., Primrose, A.P., Silverberg, E.N., Visscher, K.B., Rheingold, A.L., Guzei, I.A., Parvez, M. *Chem. Mater.*, 2000, 12:2530.

23. Vicente, V., Fruchier, A., Taillefer, M., Combes-Chamalet, C., Scowen, I.J., Plenat, F., Cristau, H.-J. *New J. Chem.*, 2004, 28:418. Combes-Chamalet, C., Cristau, H.-J., McPartlin, M., Plenat, F., Scowen, I.J., Woodroffe, T.M. *J. Chem. Soc. Perkin Trans.* 1997, 2:15.

24. Allcock, H.R., Kugel, R.L. *Inorg. Chem.*, 1966, 5:1016.

25. Allcock, H.R., Stein, M.T., Bissell, E.C. *J. Am. Chem. Soc.*, 1974, 96:4795.

26. Allcock, H.R., Stein, M.T., Stanko, J.A. *J. Am. Chem. Soc.*, 1971, 93:3173.

27. Ainscough, E.W., Brodie, A.M., Chaplin, A.B., Derwahl, A., Harrison, J.A., Otter, C.A. *Inorg. Chem.*, 2007, 46:2575.

28. Amato, M.E., Caminiti, R., Carriedo, G.A., García-Alonso, F.J., García-Alvarez, J.L., Lombardo, G.M., Pappalardo, G.C. *Chem. Eur. J.*, 2001, 7:1486.

29. Böhmer, V. *Angew. Chem. Int. Ed.*, 1995, 34:713.

30. Dietrich, A., Neumüller, B., Dehnicke, K. *Z. Anorg. Allgem. Chem.*, 2000, 626:2035.

31. Ahmed, F.R., Singh, P., Barnes, W.H. *Acta Crystallogr.*, 1969, B25:316.

32. Wisian-Neilson, P., Johnson, R.S., Zhang, H., Jung, J.-H., Neilson, R.-H., Ji, J., Watson, W.H., Krawiec, M. *Inorg. Chem.*, 2002, 41:4775. Jung, J.-H., Zhang, H., Wisian-Neilson, P. *Inorg. Chem.*, 2002, 41:6720. Jung, J.-H., Potluri, S.K., Zhang, H., Wisian-Neilson, P. *Inorg. Chem.*, 2004, 43:7784. Jung, J.-H., Kmecko, T., Claypool, C.L., Zhang, H., Wisian-Neilson, P. *Macromolecules*, 2005, 38:2122.

33. Cameron, T.S., Labarre, J.-F., Graffeuil, M. *Acta Crystallogr.*, 1982, B38:168. Cameron, T.S., Borecka, B., Kwiatkowski, W. *J. Am. Chem. Soc.*, 1994, 116:1211.

34. Galy, J., Enjalbert, R., Labarre, J.-F. *Acta Crystallogr.*, 1980, B36:392. Herbstein, F.H., Marsh, R.E. *Acta Crystallogr.*, 1982, B38:1051.

35. Cameron, T.S., Labarre, J.-F., Graffeuil, M. *Acta Crystallogr., B*, 1982, 38:2000.

36. Boomishankar, R., Ledger, J., Guilbaud, J.-B., Campbell, N.L., Bacsa, J., Bonar-Law, R., Khimyak, Y.Z., Steiner, A. *Chem. Commun.*, 2007, 5152.

37. Demus, D., Goodby, J.W., Gray, G.W., Spiess, H.-W., Vill, V. eds., *Handbook of Liquid Crystals*. Wiley-VCH, Weinheim, Germany, 1998. Special issue on liquid crystals: *Chem. Soc. Rev.*, 2007, 11:1845–2128. Laschat, S., Baro, A., Steinke, N., Giesselmann, F., Hägele, C., Scalia, G., Judele, R., Kapatsina, E., Sauer, S., Schreivogel, A., Tosoni, M. *Angew. Chem. Int. Ed.*, 2007, 46:4832. Brunsveld, L., Folmer, B.J.B., Meijer, E.W., Sijbesma, R.P. *Chem. Rev.*, 2001, 101:4071.

38. Kato, T., Mizoshita, N., Kishimoto, K. *Angew. Chem. Int. Ed.*, 2006, 45:38.

39. Laschat, S., Baro, A., Steinke, N., Giesselmann, F., Hägele, C., Scalia, G., Judele, R., Kapatsina, E., Sauer, S., Schreivogel, A., Tosoni, M. *Angew. Chem. Int. Ed.*, 2007, 46:4832.

40. Moriya, K., Mizusaki, H., Kato, M., Suzuki, T., Yano, S., Kajiwara, M., Tashiro, K. *Chem. Mater.*, 1997, 9:255. Moriya, K., Suzuki, T., Kawanishi, Y., Masuda, T., Mizusaki, H., Nakagawa, S., Ikematsu, H., Mizuno, K., Yano, S., Kajiwara, M. *Appl. Organomet. Chem.*, 1998, 12:771. Moriya, K., Suzuki, T., Yano, S., Miyajima, S. *J. Phys. Chem. B*, 2001, 105:7920.

41. Barbera, J., Bardajı, M., Jimenez, J., Laguna, A., Martınez, M.P., Oriol, L., Serrano, J.L., Zaragozano, I. *J. Am. Chem. Soc.*, 2005, 127:8994. Barbera, J., Jimenez, J., Laguna, A., Oriol, L., Perez, S., Serrano, J.L. *Chem. Mater.*, 2006, 18:5437.

42. Inoue, K., Itaya, T. *Bull. Chem. Soc. Jpn.*, 2001, 74:1.

43. Itaya, T., Azuma, N., Inoue, K. *Bull. Chem. Soc. Jpn.*, 2002, 75:2275.

44. Bertani, R., Chaux, F., Gleria, M., Metrangolo, P., Milani, R., Pilati, T., Resnati, G., Sansotera, M., Venzo, A. *Inorg. Chim. Acta*, 2007, 360:1191. Bertani, R., Ghedini, E., Gleria, M., Liantonio, R., Marras, G., Metrangolo, P., Meyer, F., Pilatid, T., Resnati, G. *CrystEngComm*, 2005, 7:511.

45. Inoue, K., Itaya, T., Azuma, N. *Supramol. Sci.*, 1998, 5:163.

46. Itaya, T., Inoue, K. *Bull. Chem. Soc. Jpn.*, 2000, 73:2615. Itaya, T., Inoue, K. *Polyhedron*, 2002, 21:1573.

47. Chandrasekhar, V., Thilagar, P., Krishnan, V., Bickley, J.F., Steiner, A. *Cryst. Growth Des.*, 2007, 7:675.

48. Allcock, H.R., Kwon, S. *Macromolecules*, 1989, 22:75.

49. Allcock, H.R., Fitzpatrick, R.J., Salvati, L. *Chem. Mater.*, 1991, 3:1120.

49a. Montoneri, E., Gleria, M., Ricca, G., Pappalardo, G.C. *Makromol. Chem.*, 1989, 190:191.

50. Allcock, H.R., Hofmann, M.A., Wood, R.M. *Macromolecules*, 2001, 34:6915. Lejeune, N., Dez, I., Jaffres, P.-A., Lohier, J.-F., Madec, P.-J., Sopkova-de Oliveira Santos, J. *Eur. J. Inorg. Chem.*, 2008, 138.

51. Omotowa, B.A., Phillips, B.S., Zabinski, J.S., Shreeve, J.M. *Inorg. Chem.*, 2004, 43:5466.

52. Caminade, A.-M., Majoral, J.-P. *Acc. Chem. Res.*, 2004, 37:341. Caminade, A.-M., Majoral, J.-P. *Prog. Polym. Sci.*, 2005, 30:491. Caminade, A.-M., Laurent, R., Majoral, J.-P. *Adv. Drug Deliv. Rev.*, 2005, 57:2130. Sournies, F., Crasnier, F., Graffeuil, M., Faucher, J.-P., Lahana, R., Labarre, M.-C., Labarre, J.-F. *Angew. Chem. Int. Ed.*, 1995, 34:578.

53. Feakins, D., Last, W.A., Shaw, R.A. *J. Chem. Soc.*, 1964, 4464. Feakins, D., Last, W.A., Neemuchwala, N., Shaw, R.A. *J. Chem. Soc.*, 1965, 2804.

54. Heston, A.J., Panzner, M.J., Youngs, W.J., Tessier, C.A. *Inorg. Chem.*, 2005, 44:6518. Heston, A.J., Panzner, M.J., Youngs, W.J., Tessier, C.A. *Phosphorus Sulfur Silicon Relat. Elem.*, 2004, 179:831. Zhang, Y., Tham, F.S., Reed, C.A. *Inorg. Chem.*, 2006, 45:10446. Gonsier, M., Antonijevic, S., Krossing, I. *Chem. Eur. J.*, 2006, 12:1997.

55. Steiner, T. *Angew. Chem. Int. Ed.*, 2002, 41:48–76.

56. Desiraju, G.R. *Angew. Chem. Int. Ed.*, 1995, 34:2311. Desiraju, G.R. *Chem. Commun.*, 1997, 1475. Zaworotko, M.J. *Chem. Commun.*, 2001, 1.

57. Wichmann, K., Antonioli, B., Söhnel, T., Wenzel, M., Gloe, K., Price, J.R., Lindoy, L.F., Blake, A.J., Schröder, M. *Coord. Chem. Rev.*, 2006, 250:2987.

58. Bickley, J.F., Bonar-Law, R., Lawson, G.T., Richards, P.I., Rivals, F., Steiner, A., Zacchini, S. *Dalton Trans.*, 2003, 1235.

59. Golinski, F., Jacobs, H. *Z. Anorg. Allgem. Chem.*, 1994, 620:965.

60. Jacobs, H., Kirchgässner, R. *Z. Anorg. Allgem. Chem.*, 1990, 581:125.

61. Bartlett, S.W., Coles, S.J., Davies, D.B., Hursthouse, M.B., Ibisoglu, H., Kilic, A., Shaw, R.A., Ün, I. *Acta Crystallogr.*, 2006, B 62:321. Fincham, J.K., Parkes, H.G., Shaw, L.S., Shaw, R.A. *J. Chem. Soc. Dalton Trans.*, 1988, 1169. Satish Kumar, N., Kumara Swamy, K.C. *Polyhedron*, 2004, 23:979. Rivals, F., Steiner, A. *Z. Anorg. Allgem. Chem.*, 2003, 629:139.

62. Muralidharan, K., Omotowa, B.A., Twamley, B., Piekarski, C., Shreeve, J.M. *Chem. Commun.*, 2005, 5193. Chandrasekhar, V., Vivekanandan, K., Nagendran, S., Senthil Andavan, G.T., Weathers, N.R., Yarbrough, J.C., Cordes, A.W. *Inorg. Chem.*, 1998, 37:6192. Cameron, T.S., Cordes, R.E., Jackman, F.A. *Acta Crystallogr.*, 1979, B35:980.

63. Lawson, G.T., Richards, P.I., Zacchini, S., Steiner, A., Unpublished results.

64. Boomishankar, R., Nichols, J., Bickley, J.F., Steiner, A. Unpublished results.

65. Benson, M.A., Zacchini, S., Boomishankar, R., Chan, Y., Steiner, A. *Inorg. Chem.*, 2007, 46:7097.

66. Benson, M.A., Steiner, A. *Chem. Commun.*, 2005, 5026.

67. Benson, M.A., Boomishankar, R., Wright, D.S., Steiner, A. *J. Organomet. Chem.*, 2007, 692:2768.

68. Zacchini, S., Steiner, A. Unpublished results.

69. Chandrasekhar, V., Krishnan, V., Thangavelu, G., Andavan, S., Steiner, A., Zacchini, S. *CrystEngComm*, 2003, 5:245.

70. Steel, P.J. *Acc. Chem. Res.*, 2005, 38:243. Brammer, L. *Chem. Soc. Rev.*, 2004, 33:476. Kitagawa, S., Kitaura, R., Noro, S.-I. *Angew. Chem. Int. Ed.*, 2004, 43:2334. Janiak, C. *Dalton Trans.* 2003, 2781. Leininger, S., Olenyuk, B., Stang, P.J. *Chem. Rev.*, 2000, 100:853.

71. Khlobystov, A.N., Blake, A.J., Champness, N.R., Lemenovskii, D.A., Majouga, A.G., Zyk, N.V., Schröder, M. *Coord. Chem. Rev.*, 2001, 222:155.

72. Richards, P.I., Steiner, A. *Inorg. Chem.*, 2004, 43:2810.

73. Richards, P.I., Bickley, J.F., Boomishankar, R., Steiner, A. *Chem. Commun.*, 2008, 1656.

74. Chandrasekhar, V., Nagendran, S. *Chem. Soc. Rev.*, 2001, 30:193. Chandrasekhar, V., Thilagar, P., Pandian, B.M. *Coord. Chem. Rev.*, 2007, 251:1045. Allcock, H.R., Desorcie, J.L., Riding, G.H. *Polyhedron*, 1987, 6:119.

75. Justin Thomas, K.R., Chandrasekhar, V., Pal, P.S., Scott, S.R., Hallford, R., Cordes, A.W. *Inorg. Chem.*, 1993, 32:606. Justin Thomas, K.R., Tharmaraj, P., Chandrasekhar, V., Bryan, C.D., Cordes, A.W. *Inorg. Chem.*, 1994, 33:5382. Justin Thomas, K.R., Chandrasekhar, V., Scott, S.R., Hallford, R., Cordes, A.W. *Dalton. Trans.*, 1993, 2589. Chandrasekaran, A., Krishnamurthy, S.S., Nethaji, M. *Dalton. Trans.*, 1994, 63. Harmjanz, M., Scott, B.L., Burns, C.J. *Chem. Commun.*, 2002, 1386. Chandrasekhar, V., Krishnan, V., Steiner, A., Bickley, J.F. *Inorg. Chem.*, 2004, 43:166.

76. Ainscough, E.W., Brodie, A.M., Depree, C.V. *Dalton Trans.*, 1999, 4123. Ainscough, E.W., Brodie, A.M., Depree, C.V., Otter, C.A. *Polyhedron*, 2006, 25:2341. Ainscough, E.W., Brodie, A.M., Depree, C.V., Moubaraki, B., Murray, K.S., Otter, C.A. *Dalton Trans.*, 2005, 3337. Chandrasekhar, V., Murugesa Pandian, B., Azhakar, R. *Inorg. Chem.*, 2006, 45:3510.

77. Ainscough, E.W., Brodie, A.M., Depree, C.V., Jameson, G.B., Otter, C.A. *Inorg. Chem.*, 2005, 44:7325.

78. Harmjanz, M., Piglosiewicz, I.M., Scott, B.L., Burns, C.J. *Inorg. Chem.*, 2004, 43:642.

79. Ainscough, E.W., Brodie, A.M., Jameson, G.B., Otter, C.A. *Polyhedron*, 2007, 26:460.

80. Brandt, K., Porwolik, I., Olejnik, A., Shaw, R.A., Davies, D.B. *J. Am. Chem. Soc.*, 1996, 118:4496. Brandt, K., Porwolik, I., Siwy, M., Kupka, T., Shaw, R.A., Davies, D.B., Hursthouse, M.B., Sykara, G.D. *J. Am. Chem. Soc.*, 1997, 119:1143. Brandt, K., Porwolik-Czomperlik, I., Siwy, M., Kupka, T., Shaw, R.A., Davies, D.B., Hursthouse, M.B., Sykara, G.D. *J. Am. Chem. Soc.*, 1997, 119:12432. Brandt, K., Siwy, M., Porwolik-Czomperlik, I., Silberring, J. *J. Org. Chem.*, 2001, 66:5701.

Maia, A., Landini, D., Penso, M., Brandt, K., Siwy, M., Schroeder, G., Gierczyk, B. *New J. Chem.*, 2001, 25:1078.

81. Brandt, K., Seliger, P., Grzejdziak, A., Bartczak, T.J., Kruszynski, R., Lach, D., Silberring, J. *Inorg. Chem.*, 2001, 40:3704. Kruszynski, R., Bartczak, T.J., Brandt, K., Lach, D. *Inorg. Chim. Acta*, 2001, 321:185.

82. Sournies, F., Castera, P., El Bakili, A., Faucher, J.P., Graffeuil, M., Labarre, J.-F. *J. Mol. Struct.*, 1990, 221:245. Enjalbert, R., Galy, J., Sournies, F., Labarre, J.-F. *J. Mol. Struct.*, 1990, 221:253. Shaw, R.A., Ture, S. *Phosphorus Sulfur Silicon Relat. Elem.*, 1991, 57:103. Caminade, A.-M., Majoral, J.-P. *Chem. Rev.*, 1994, 94:1183.

83. Allcock, H.R., Napierala, M.E., Olmeijer, D.L., Cameron, C.G., Kuharcik, S.E., Reed, C.S., O'Connor, S.J.M. *Electrochim. Acta*, 1998, 43:1145. Allcock, H.R., O'Connor, S.J.M., Olmeijer, D.L., Napierala, M.E., Cameron, C.G. *Macromolecules*, 1996, 29:7544. Chen-Yang, Y.W., Hwang, J.J., Huang, A.Y. *Macromolecules*, 2000, 33:1237.

84. Richards, P.I., Benson, M.A., Steiner, A. *Chem. Commun.*, 2003, 1392.

85. Steiner, A., Zacchini, S., Richards, P.I. *Coord. Chem. Rev.*, 2002, 227:193.

86. Steiner, A., Wright, D.S. *Angew. Chem. Int. Ed.*, 1996, 35:636. Lawson, G.T., Rivals, F., Tascher, M., Jacob, C., Bickley, J.F., Steiner, A. *Chem. Commun.*, 2000, 341.

87. Rivals, F., Steiner, A. *Chem. Commun.*, 2001, 1426.

88. Boomishankar, R., Richards, P.I., Steiner, A. *Angew. Chem. Int. Ed.*, 2006, 45:4632.

APPENDIX A

DHCP Message Types
Reference: [RFC3315]

Registry:

Value	Description	Reference
0	Reserved	
1	SOLICIT	[RFC3315]
2	ADVERTISE	[RFC3315]
3	REQUEST	[RFC3315]
4	CONFIRM	[RFC3315]
5	RENEW	[RFC3315]
6	REBIND	[RFC3315]
7	REPLY	[RFC3315]
8	RELEASE	[RFC3315]
9	DECLINE	[RFC3315]
10	RECONFIGURE	[RFC3315]
11	INFORMATION-REQUEST	[RFC3315]
12	RELAY-FORW	[RFC3315]
13	RELAY-REPL	[RFC3315]
14	LEASEQUERY	[RFC5007]
15	LEASEQUERY-REPLY	[RFC5007]
16–255	Unassigned	

Polyphosphazenes for Biomedical Applications, Edited by Alexander K. Andrianov
Copyright © 2009 John Wiley & Sons, Inc.

DHCP Option Codes
Reference: [RFC3315]

Registry:

Value	Description	Reference
0	Reserved	
1	OPTION_CLIENTID	[RFC3315]
2	OPTION_SERVERID	[RFC3315]
3	OPTION_IA_NA	[RFC3315]
4	OPTION_IA_TA	[RFC3315]
5	OPTION_IAADDR	[RFC3315]
6	OPTION_ORO	[RFC3315]
7	OPTION_PREFERENCE	[RFC3315]
8	OPTION_ELAPSED_TIME	[RFC3315]
9	OPTION_RELAY_MSG	[RFC3315]
10	Unassigned	
11	OPTION_AUTH	[RFC3315]
12	OPTION_UNICAST	[RFC3315]
13	OPTION_STATUS_CODE	[RFC3315]
14	OPTION_RAPID_COMMIT	[RFC3315]
15	OPTION_USER_CLASS	[RFC3315]
16	OPTION_VENDOR_CLASS	[RFC3315]
17	OPTION_VENDOR_OPTS	[RFC3315]
18	OPTION_INTERFACE_ID	[RFC3315]
19	OPTION_RECONF_MSG	[RFC3315]
20	OPTION_RECONF_ACCEPT	[RFC3315]
21	SIP Servers Domain Name List	[RFC3319]
22	SIP Servers IPv6 Address List	[RFC3319]
23	DNS Recursive Name Server Option	[RFC3646]
24	Domain Search List option	[RFC3646]
25	OPTION_IA_PD	[RFC3633]
26	OPTION_IAPREFIX	[RFC3633]
27	OPTION_NIS_SERVERS	[RFC3898]
28	OPTION_NISP_SERVERS	[RFC3898]
29	OPTION_NIS_DOMAIN_NAME	[RFC3898]
30	OPTION_NISP_DOMAIN_NAME	[RFC3898]
31	OPTION_SNTP_SERVERS	[RFC4075]
32	OPTION_INFORMATION_REFRESH_TIME	[RFC4242]
33	OPTION_BCMCS_SERVER_D	[RFC4280]
34	OPTION_BCMCS_SERVER_A	[RFC4280]
35	Unassigned	
36	OPTION_GEOCONF_CIVIC	[RFC4776]
37	OPTION_REMOTE_ID	[RFC4649]
38	OPTION_SUBSCRIBER_ID	[RFC4580]
39	OPTION_CLIENT_FQDN	[RFC4704]
40	OPTION_PANA_AGENT	[RFC5192]
41	OPTION_NEW_POSIX_TIMEZONE	[RFC4833]

Registry: Value	Description	Reference
42	OPTION_NEW_TZDB_TIMEZONE	[RFC4833]
43	OPTION_ERO	[RFC4994]
44	OPTION_LQ_QUERY	[RFC5007]
45	OPTION_CLIENT_DATA	[RFC5007]
46	OPTION_CLT_TIME	[RFC5007]
47	OPTION_LQ_RELAY_DATA	[RFC5007]
48	OPTION_LQ_CLIENT_LINK	[RFC5007]
49	OPTION_MIP6_HNINF	[RFC-ietf-mip6-hiopt-17.txt]
50	OPTION_MIP6_RELAY	[RFC-ietf-mip6-hiopt-17.txt]
51	OPTION_V6_LOST	[RFC5223]
52	OPTION_CAPWAP_AC_V6	[RFC-ietf-capwap-dhc-ac-option-02.txt]
53–255	Unassigned	

INDEX

Polyphosphazenes for Biomedical Applications, Edited by Alexander K. Andrianov
Copyright © 2009 John Wiley & Sons, Inc.